AF901118

Mat 70-
1/16

MÄNGELEXEMPLAR

Association for Women in Mathematics Series

Volume 3

More information about this series at http://www.springer.com/series/13764

Association for Women in Mathematics Series

Focusing on the groundbreaking work of women in mathematics past, present, and future, Springer's Association for Women in Mathematics Series presents the latest research and proceedings of conferences worldwide organized by the Association for Women in Mathematics (AWM). All works are peer-reviewed to meet the highest standards of scientific literature, while presenting topics at the cutting edge of pure and applied mathematics. Since its inception in 1971, The Association for Women in Mathematics has been a non-profit organization designed to help encourage women and girls to study and pursue active careers in mathematics and the mathematical sciences and to promote equal opportunity and equal treatment of women and girls in the mathematical sciences. Currently, the organization represents more than 3000 members and 200 institutions constituting a broad spectrum of the mathematical community, in the United States and around the world.

Ellen E. Eischen • Ling Long • Rachel Pries
Katherine E. Stange
Editors

Directions in Number Theory

Proceedings of the 2014 WIN3 Workshop

Springer

Editors
Ellen E. Eischen
Department of Mathematics
University of Oregon
Eugene, OR, USA

Ling Long
Department of Mathematics
Louisiana State University
Baton Rouge, LA, USA

Rachel Pries
Department of Mathematics
Colorado State University
Fort Collins, CO, USA

Katherine E. Stange
Department of Mathematics
University of Colorado
Boulder, CO, USA

ISSN 2364-5733 ISSN 2364-5741 (electronic)
Association for Women in Mathematics Series
ISBN 978-3-319-30974-3 ISBN 978-3-319-30976-7 (eBook)
DOI 10.1007/978-3-319-30976-7

Library of Congress Control Number: 2016940911

© Springer International Publishing Switzerland 2016
This work is subject to copyright. All rights are reserved by the Publisher, whether the whole or part of the material is concerned, specifically the rights of translation, reprinting, reuse of illustrations, recitation, broadcasting, reproduction on microfilms or in any other physical way, and transmission or information storage and retrieval, electronic adaptation, computer software, or by similar or dissimilar methodology now known or hereafter developed.
The use of general descriptive names, registered names, trademarks, service marks, etc. in this publication does not imply, even in the absence of a specific statement, that such names are exempt from the relevant protective laws and regulations and therefore free for general use.
The publisher, the authors and the editors are safe to assume that the advice and information in this book are believed to be true and accurate at the date of publication. Neither the publisher nor the authors or the editors give a warranty, express or implied, with respect to the material contained herein or for any errors or omissions that may have been made.

Printed on acid-free paper

This Springer imprint is published by Springer Nature
The registered company is Springer International Publishing AG Switzerland

Preface

This volume is a compilation of research and survey papers in number theory, written by members of the *Women in Numbers* (WIN) network, principally by the collaborative research groups formed at *Women in Numbers 3*, a conference at the Banff International Research Station in Banff, Alberta, on April 21–25, 2014.

The WIN conference series began in 2008, with the aim of strengthening the research careers of female number theorists. The series introduced a novel research-mentorship model: women at all career stages, from graduate students to senior members of the community, joined forces to work in focused research groups on cutting-edge projects designed and led by experienced researchers. This model had tremendous success, branching out not only to *WINE (Women in Numbers Europe)* but also to *Algebraic Combinatorixx*, *WIT (Women in Topology)*, and others. The Association for Women in Mathematics (AWM), funded by the National Science Foundation, is now supporting this research-mentorship model under the umbrella of the *Research Collaboration Conferences for Women* initiative.

The goals for *Women In Numbers 3* were to establish ambitious new collaborations between women in number theory, to train junior participants about topics of current importance, and to continue to build a vibrant community of women in number theory. The majority of the week was devoted to research activities. Before the conference, the participants were organized into nine project groups by research interest and asked to learn background for their project topics. This led to more productive on-site research conversations and the groups were able to share preliminary results on the last day. The workshop also included a lecture series about arithmetic of curves, including elliptic curves, modular curves, and Shimura curves.

Forty-two women attended the WIN3 workshop, which was organized by the last three editors of this volume. This included 15 senior and mid-level faculty, 15 junior faculty and postdocs, and 12 graduate students. This volume is the fourth proceedings to come out of the WIN conference series. It is also the first in the series published by Springer for AWM.

The editors invited WIN3 research groups and members of the larger WIN3 community to submit articles in 2014. After a thorough referee process by external experts, we accepted 10 papers for the volume. One interesting attribute of the collection is the interplay between deep theory and intricate computation. The papers span a wide range of research areas: arithmetic geometry, analytic number theory, algebraic number theory, and applications to coding and cryptography. In this preface, we point out a few connections between the papers.

A major theme of the volume is the study of rational points on varieties via cohomological methods. Three papers on this theme are about rational points over number fields. The paper *Insufficiency of the Brauer-Manin obstruction for rational points on Enriques surfaces* (Balestrieri et al.) is about the failure of the Hasse principle for surfaces. In the paper *Shadow lines in the arithmetic of elliptic curves* (Balakrishnan et al.), the authors use information about analytic ranks and Tate-Shafarevich groups to develop an algorithm for computing anticyclotomic p-adic heights and shadow lines cast by rational points on elliptic curves over imaginary quadratic fields. In the paper *Galois action on the homology of Fermat curves* (Davis et al.), the authors use topology and the étale fundamental group to study obstructions for points on Fermat curves defined over cyclotomic fields.

The paper *Zeta functions of a class of Artin-Schreier curves with many automorphisms over finite fields* (Bouw et al.) is a bridge between several of the disparate topics. It fits in the vein of studying rational points via cohomological methods, because the ℓ-adic cohomology provides information about points on curves defined over finite fields. It connects to the topic of applications to coding theory and cryptography, because the class of Artin-Schreier curves produces large families of supersingular curves useful for error-correcting codes. Similarly, the paper *Hypergeometric series, truncated hypergeometric series, and Gaussian hypergeometric functions* (Deines et al.) draws together several topics. The hypergeometric varieties are higher-dimensional analogues of Legendre curves and the authors obtain information about the number of points on these varieties defined over finite fields. This paper also connects to the more analytic papers in the volume.

There are two other papers with an analytic and geometric focus. The paper *A generalization of S. Zhang's local Gross-Zagier formula for* GL_2 (Maurischat) is about Hecke operators and contains a fundamental lemma for some relative trace formulae. The paper *p-adic q-expansion principles on unitary Shimura varieties* (Caraiani et al.) has results about vanishing theorems for p-adic automorphic forms on unitary groups of arbitrary signature.

The final three papers are about applications of algebraic number theory. The paper *Kneser-Hecke-operators for codes over finite chain rings* (Feaver et al.) is about theta series for lattices for codes over finite fields and an analogue for Hecke operators in this context. In *Ring-LWE cryptography for the number theorist* (Elias et al.), the authors give a survey about attacks on the ring and polynomial learning with errors problems and discuss connections with open problems about algebraic number fields. Finally, the volume ends with a survey about arithmetic statistics in algebraic number theory, *Asymptotics for number fields and class groups* (Wood). This survey is an extended version of Wood's lecture notes for the Arizona Winter School in 2014, on the topic of counting number fields and the distribution of class groups.

Acknowledgments

It was a pleasure to work with BIRS to organize the WIN3 conference and with Springer to prepare this volume. We would like to thank the following sponsoring organizations for their generous financial support of the workshop: Banff International Research Station, Clay Math Institute, Microsoft Research, Pacific Institute for the Mathematical Sciences, and the Number Theory Foundation. We would also like to thank the many referees, whose intelligence and effort helped the authors improve the papers for this volume.
WIN Editorial Committee: December 2015

Ellen Eischen, University of Oregon
Ling Long, Louisiana State University
Rachel Pries, Colorado State University
Katherine E. Stange, University of Colorado Boulder

Workshop Participants and Affiliations at the Time of the Workshop:

Jennifer Balakrishnan, University of Oxford, United Kingdom
Jennifer Berg, University of Texas at Austin, USA
Irene Bouw, Universität Ulm, Germany
Alina Bucur, University of California San Diego, USA
Mirela Ciperiani, University of Texas at Austin, USA
Alina Carmen Cojocaru, University of Illinois at Chicago, USA
Rachel Davis, Purdue University, USA
Alyson Deines, University of Washington, USA
Ellen Eischen, University of North Carolina at Chapel Hill, USA
Yara Elias, McGill University, Canada
Amy Feaver, University of Colorado Boulder, USA
Jessica Fintzen, Harvard University, USA
Jenny Fuselier, High Point University, USA
Bonita Graham, Wesleyan University, USA
Anna Haensch, Max Planck Institute for Mathematics, Germany
Wei Ho, Columbia University, USA
Matilde Lalín, Université de Montrèal, Canada
Jaclyn Lang, University of California at Los Angeles, USA
Kristin Lauter, Microsoft Research, USA
Jingbo Liu, Wesleyan University, USA
Ling Long, Louisiana State University, USA
Beth Malmskog, Colorado College, USA
Michelle Manes, University of Hawai'i at Mānoa, USA
Elena Mantovan, California Institute of Technology, USA

Bahare Mirza, McGill University, Canada
Gabriele Nebe, RWTH Aachen, Germany
Rachel Newton, University of Leiden, Netherlands
Ekin Ozman, University of Texas at Austin, USA
Jennifer Park, McGill University, Canada
Lillian Pierce, Hausdorff Center for Mathematics, Bonn, Germany
Rachel Pries, Colorado State University, USA
Renate Scheidler, University of Calgary, Canada
Padmavathi Srinivasan, Massachusetts Institute of Technology, USA
Katherine E. Stange, University of Colorado Boulder, USA
Vesna Stojanoska, University of Illinois at Urbana-Champaign, USA
Holly Swisher, Oregon State University, USA
Fang-Ting Tu, National Center of Theoretical Sciences in Taiwan, Taiwan
Ila Varma, Princeton University, USA
Christelle Vincent, Stanford University, USA
Bianca Viray, Brown University, USA
Kirsten Wickelgren, Georgia Institute of Technology, USA

Workshop Website

http://www.birs.ca/events/2014/5-day-workshops/14w5009

Eugene, OR, USA — Ellen E. Eischen
Baton Rouge, LA, USA — Ling Long
Fort Collins, CO, USA — Rachel Pries
Boulder, CO, USA — Katherine E. Stange
May 2016

Contents

Contributors

J.S. Balakrishnan Mathematical Institute, University of Oxford, UK

Francesca Balestrieri Mathematical Institute, University of Oxford, UK

Jennifer Berg Department of Mathematics, The University of Texas at Austin, TX, USA

Irene Bouw Institute of Pure Mathematics, Ulm University, Ulm, Germany

Ana Caraiani Department of Mathematics, Princeton University, Princeton, NJ, USA

M. Çiperiani Department of Mathematics, The University of Texas at Austin, TX, USA

Rachel Davis Department of Mathematics, Purdue University, West Lafayette, IN, USA

Alyson Deines Center for Communications Research, San Diego, CA, USA

Yara Elias Department of Mathematics and Statistics, McGill University, Montréal, QC, Canada

Ellen Eischen Department of Mathematics, University of Oregon, Eugene, OR, USA

Amy Feaver Department of Mathematics and Computing Science, The King's University, Edmonton, AB, Canada

Jessica Fintzen Department of Mathematics, Harvard University, Cambridge, MA, USA

Jenny G. Fuselier Department of Mathematics and Computer Science, High Point University, High Point, NC, USA

Anna Haensch Department of Mathematics and Computer Science, Duquesne University, Pittsburgh, PA, USA

Wei Ho Department of Mathematics, University of Michigan, Ann Arbor, MI, USA

J. Lang UCLA Mathematics Department, Los Angeles, CA, USA

Kristin E. Lauter Microsoft Research, Redmond, WA, USA

Jingbo Liu Department of Mathematics and Computer Science, Wesleyan University, Middletown, CT, USA

Ling Long Department of Mathematics, Louisiana State University, Baton Rouge, LA, USA

Beth Malmskog Department of Mathematics and Statistics, Villanova University, Villanova, PA, USA

Michelle Manes Department of Mathematics, University of Hawai'i at Manoa, Honolulu, HI, USA

Elena Mantovan Department of Mathematics, CalTech, Pasadena, CA, USA

Kathrin Maurischat Mathematisches Institut, Universitat Heidelberg, Germany

B. Mirza Department of Mathematics and Statistics, McGill University, Montréal, QC, Canada

Gabriele Nebe Lehrstuhl D für Mathematik, RWTH Aachen University, Aachen, Germany

R. Newton Department of Mathematics and Statistics, The University of Reading, Whiteknights, Reading, UK

Ekin Ozman Department of Mathematics, Faculty of Arts and Science, Bogazici University, Bebek-Istanbul, Turkey

Jennifer Park Department of Mathematics, University of Michigan, Ann Arbor, MI, USA

Rachel Pries Department of Mathematics, Colorado State University, Fort Collins, CO, USA

Renate Scheidler Department of Mathematics and Statistics, University of Calgary, AB, Canada

Padmavathi Srinivasan Department of Mathematics, Massachusetts Institute of Technology, Cambridge, MA, USA

Katherine E. Stange Department of Mathematics, University of Colorado, Boulder, CO, USA

Vesna Stojanoska Department of Mathematics, University of Illinois at Urbana-Champaign, Urbana, IL, USA

Holly Swisher Oregon State University, Corvallis, OR, USA

Fang-Ting Tu National Center for Theoretical Sciences, Hsinchu, Taiwan, R.O.C.

Ila Varma Department of Mathematics, Princeton University, Princeton, NJ, USA

Christelle Vincent Department of Mathematics and Statistics, University of Vermont, 16 Colchester Avenue, Burlington, VT 05401, USA

Bianca Viray Department of Mathematics, University of Washington, Seattle, WA, USA

Kirsten Wickelgren School of Mathematics, Georgia Institute of Technology, Atlanta, GA, USA

Melanie Matchett Wood Department of Mathematics, University of Wisconsin, Madison, WI, USA

American Institute of Mathematics, Palo Alto, CA, USA

Insufficiency of the Brauer–Manin Obstruction for Rational Points on Enriques Surfaces

Francesca Balestrieri, Jennifer Berg, Michelle Manes, Jennifer Park, and Bianca Viray

Abstract In Várilly-Alvarado and Viray (Adv. Math. 226(6):4884–4901, 2011), the authors constructed an Enriques surface X over $\mathbb{Q}$ with an étale-Brauer obstruction to the Hasse principle and no *algebraic* Brauer–Manin obstruction. In this paper, we show that the nontrivial Brauer class of $X_{\overline{\mathbb{Q}}}$ does not descend to $\mathbb{Q}$. Together with the results of Várilly-Alvarado and Viray (Adv. Math. 226(6):4884–4901, 2011), this proves that the Brauer–Manin obstruction is insufficient to explain all failures of the Hasse principle on Enriques surfaces.

The methods of this paper build on the ideas in Creutz and Viray (Math. Ann. 362(3–4):1169–1200, 2015; Manuscripta Math. 147(1–2): 139–167, 2015) and Ingalls et al., (Unramified Brauer classes on cyclic covers of the projective plane, Preprint): we study geometrically unramified Brauer classes on X via pullback of ramified Brauer classes on a rational surface. Notably, we develop techniques which work over fields which are not necessarily separably closed, in particular, over number fields.

F. Balestrieri (✉)
Mathematical Institute, University of Oxford, Oxford, OX2 6HD, UK
e-mail: balestrieri@maths.ox.ac.uk URL http://people.maths.ox.ac.uk/~balestrieri/

J. Berg
Department of Mathematics, The University of Texas at Austin, 2515 Speedway, RLM 8.100, Austin, TX 78712, USA
e-mail: jberg@math.utexas.edu URL http://ma.utexas.edu/users/jberg

M. Manes
Department of Mathematics, University of Hawai'i at Mānoa, 2565 McCarthy Mall Keller 401A, Honolulu, HI 96822, USA
e-mail: mmanes@math.hawaii.edu URL http://math.hawaii.edu/~mmanes

J. Park
Department of Mathematics, University of Michigan, 530 Church Street, Ann Arbor, MI 48109, USA
e-mail: jmypark@umich.edu URL http://math.mcgill.ca/jpark/

B. Viray
Department of Mathematics, University of Washington, Box 354350, Seattle, WA 98195, USA
e-mail: bviray@math.washington.edu URL http://math.washington.edu/~bviray

© Springer International Publishing Switzerland 2016

E.E. Eischen et al. (eds.), *Directions in Number Theory*, Association for Women in Mathematics Series 3, DOI 10.1007/978-3-319-30976-7_1

Keywords Hasse principle • $K3$ surface • Enriques surface • Brauer–Manin obstruction

2010 *Mathematics Subject Classification.* 14F22 (Primary), 14J28 (Secondary), 14G05

1 Introduction

Given a smooth, projective, geometrically integral variety X over a global field k, one may ask whether X has a k-rational point, that is, whether $X(k) \neq \emptyset$. Since k embeds into each of its completions, a necessary condition for $X(k) \neq \emptyset$ is that $X(\mathbb{A}_k) \neq \emptyset$. However, this condition is often not sufficient; varieties X with $X(\mathbb{A}_k) \neq \emptyset$ and $X(k) = \emptyset$ exist, and these are said to fail the Hasse principle.

In 1970, Manin [12] significantly advanced the study of failures of the Hasse principle by use of the Brauer group and class field theory. More precisely, he defined a subset $X(\mathbb{A}_k)^{\mathrm{Br}}$ of $X(\mathbb{A}_k)$, now known as the Brauer–Manin set, with the property that

$$X(k) \subset X(\mathbb{A}_k)^{\mathrm{Br}} \subset X(\mathbb{A}_k).$$

Thus, we may think of an empty Brauer–Manin set as an obstruction to the existence of rational points.

In 1999, Skorobogatov [14] defined a refinement of the Brauer–Manin set, the étale-Brauer set $X(\mathbb{A}_k)^{\text{ét},\mathrm{Br}}$, which still contains $X(k)$. He proved that this new obstruction can be stronger than the Brauer–Manin obstruction, by constructing a bielliptic surface $X/\mathbb{Q}$ such that $X(\mathbb{A}_{\mathbb{Q}})^{\text{ét},\mathrm{Br}} = \emptyset$ and $X(\mathbb{A}_{\mathbb{Q}})^{\mathrm{Br}} \neq \emptyset$.

Bielliptic surfaces have a number of geometric properties in common with Enriques surfaces: both have Kodaira dimension 0 and nontrivial étale covers. This raises the natural question of whether the étale-Brauer obstruction is stronger than the Brauer–Manin obstruction *for Enriques surfaces*. Harari and Skorobogatov took up this question in 2005; they constructed an Enriques surface $X/\mathbb{Q}$ whose étale-Brauer set was strictly smaller than the Brauer–Manin set [9], thereby showing that the Brauer–Manin obstruction is insufficient to explain all failures of weak approximation[1] on Enriques surfaces. Their surface, however, had a $\mathbb{Q}$-rational point, so it did not fail the Hasse principle.

The main result of this paper is the analogue of Harari and Skorobogatov's result for the Hasse principle. Precisely, we prove

Theorem 1.1. *The Brauer–Manin obstruction is insufficient to explain all failures of the Hasse principle on Enriques surfaces.*

[1]A smooth projective variety X satisfies weak approximation if $X(k)$ is dense in $X(\mathbb{A}_k)$ in the adelic topology.

This theorem builds on work by Várilly-Alvarado and Viray. To explain the connection, we must first provide more information about the Brauer group. For any variety X/k, we have the following filtration of the Brauer group:

$$\operatorname{Br}_0 X := \operatorname{im}(\operatorname{Br} k \to \operatorname{Br} X) \subset \operatorname{Br}_1 X := \ker(\operatorname{Br} X \to \operatorname{Br} X_{k^{\mathrm{sep}}}) \subset \operatorname{Br} X = \mathrm{H}^2_{\text{ét}}(X, \mathbb{G}_m).$$

Elements in $\operatorname{Br}_0 X$ are said to be **constant**, elements in $\operatorname{Br}_1 X$ are said to be **algebraic**, and the remaining elements are said to be **transcendental**. The Brauer–Manin set $X(\mathbb{A}_k)^{\operatorname{Br}}$ depends only on the quotient $\operatorname{Br} X/\operatorname{Br}_0 X$ (this follows from the fundamental exact sequence of class field theory, see [15, Sect. 5.2] for more details). As transcendental Brauer elements have historically been difficult to study, one sometimes instead considers the (possibly larger) **algebraic Brauer–Manin set** $X(\mathbb{A}_k)^{\operatorname{Br}_1}$, defined in terms of the subquotient $\operatorname{Br}_1 X/\operatorname{Br}_0 X$.

We now recall the main result of [16].

Theorem ([16, Theorem 1.1]). *There exists an Enriques surface $X/\mathbb{Q}$ such that*

$$X(\mathbb{A}_{\mathbb{Q}})^{\text{ét},\operatorname{Br}} = \emptyset \quad \text{and} \quad X(\mathbb{A}_{\mathbb{Q}})^{\operatorname{Br}_1} \neq \emptyset.$$

The proof of [16, Theorem 1.1] is constructive. Precisely, for any $\mathbf{a} = (a, b, c) \in \mathbb{Z}^3$ with

$$abc(5a+5b+c)(20a+5b+2c)(4a^2+b^2)(c^2-100ab)(c^2+5bc+10ac+25ab) \neq 0, \tag{1}$$

the authors consider $Y_{\mathbf{a}} \subset \mathbb{P}^5$, the smooth degree-8 K3 surface given by

$$\begin{aligned} v_0 v_1 + 5v_2^2 &= w_0^2, \\ (v_0 + v_1)(v_0 + 2v_1) &= w_0^2 - 5w_1^2, \\ av_0^2 + bv_1^2 + cv_2^2 &= w_2^2. \end{aligned}$$

The involution $\sigma\colon \mathbb{P}^5 \to \mathbb{P}^5, (v_0 : v_1 : v_2 : w_0 : w_1 : w_2) \mapsto (-v_0 : -v_1 : -v_2 : w_0 : w_1 : w_2)$ has no fixed points on $Y_{\mathbf{a}}$ so the quotient $X_{\mathbf{a}} := Y_{\mathbf{a}}/\sigma$ is an Enriques surface.

Theorem ([16, Theorem 1.2]). *Let $\mathbf{a} = (a, b, c) \in \mathbb{Z}^3_{>0}$ satisfy the following conditions:*

(1) *for all prime numbers $p \mid (5a + 5b + c)$, 5 is not a square modulo p,*
(2) *for all prime numbers $p \mid (20a + 5b + 2c)$, 10 is not a square modulo p,*
(3) *the quadratic form $av_0^2 + bv_1^2 + cv_2^2 + w_2^2$ is anisotropic over $\mathbb{Q}_3$,*
(4) *the integer $-bc$ is not a square modulo 5,*
(5) *the triplet (a, b, c) is congruent to $(5, 6, 6)$ modulo 7,*
(6) *the triplet (a, b, c) is congruent to $(1, 1, 2)$ modulo 11,*
(7) *$Y_{\mathbf{a}}(\mathbb{A}_{\mathbb{Q}}) \neq \emptyset$,*

(8) *the triplet* (a, b, c) *is* Galois general *(meaning that a certain number field defined in terms of a, b, c is as large as possible).*

Then

$$X_{\mathbf{a}}(\mathbb{A}_{\mathbb{Q}})^{\text{ét,Br}} = \emptyset \quad \textit{and} \quad X_{\mathbf{a}}(\mathbb{A}_{\mathbb{Q}})^{\text{Br}_1} \neq \emptyset.$$

Várilly-Alvarado and Viray deduce [16, Theorem 1.1] from [16, Theorem 1.2] by showing that the triplet $\mathbf{a} = (12, 111, 13)$ satisfies conditions (1)–(8). Henceforth, when we refer to "conditions" by number, we mean the conditions given in the theorem above.

In [16], the authors left open the question of a transcendental obstruction to the Hasse principle for the surfaces $X_{\mathbf{a}}$, due to the "difficulty [...] in finding an explicit representative for [the nontrivial] Brauer class of $[\overline{X}_{\mathbf{a}}]$." Recent work of Creutz and Viray [4, 5], and Ingalls et al. [10] makes this problem more tractable. Building on techniques from [4, 5, 10], we prove

Theorem 1.2. *If* $\mathbf{a} = (a, b, c) \in \mathbb{Z}^3_{>0}$ *satisfies conditions* (5), (6), *and* (8), *then* $\operatorname{Br} X_{\mathbf{a}} = \operatorname{Br}_1 X_{\mathbf{a}}$. *In particular, if* $\mathbf{a}$ *satisfies conditions* (1)–(8), *then*

$$X_{\mathbf{a}}(\mathbb{A}_{\mathbb{Q}})^{\text{ét,Br}} = \emptyset \quad \text{and} \quad X_{\mathbf{a}}(\mathbb{A}_{\mathbb{Q}})^{\text{Br}} \neq \emptyset.$$

1.1 Strategy and Outline

Theorem 1.1 and the second statement of Theorem 1.2 both follow immediately from the first statement of Theorem 1.2 and [16], since the triplet $\mathbf{a} = (12, 111, 13)$ satisfies conditions (1)–(8) [16, Lemma 6.1 and Proof of Theorem 1.1]. Thus, we reduce to proving the first statement of Theorem 1.2.

For any variety X over a field k, the quotient $\operatorname{Br} X / \operatorname{Br}_1 X$ injects into $(\operatorname{Br} X_{k^{\text{sep}}})^{\operatorname{Gal}(k^{\text{sep}}/k)}$. In Skorobogatov's pioneering paper [14], his construction $X/\mathbb{Q}$ had the additional property that $(\operatorname{Br} X_{\overline{\mathbb{Q}}})^{\operatorname{Gal}(\overline{\mathbb{Q}}/\mathbb{Q})} = 0$, so $\operatorname{Br} X = \operatorname{Br}_1 X$. Unfortunately, this strategy cannot be applied to an Enriques surface X, as $\operatorname{Br} X_{k^{\text{sep}}} \cong \mathbb{Z}/2\mathbb{Z}$ [9, p. 3223] and hence the unique nontrivial element is always fixed by the Galois action.

Instead, we will find a Galois extension $K_1/\mathbb{Q}$ and an open set $U' \subset X_{\mathbf{a}}$ such that

(1) the geometrically unramified subgroup $\operatorname{Br}^{\text{g.unr.}} U'_{K_1} \subset \operatorname{Br} U'_{K_1}$ (i.e., the subgroup of elements in $\operatorname{Br} U'_{K_1}$ which are contained in $\operatorname{Br} \overline{X}_{\mathbf{a}} \subset \operatorname{Br} \overline{U'}$ upon base change to $\overline{\mathbb{Q}}$) surjects onto $\operatorname{Br} \overline{X}_{\mathbf{a}}$, and
(2) $(\operatorname{Br}^{\text{g.unr.}} U'_{K_1} / \operatorname{Br} K_1)^{\operatorname{Gal}(K_1/\mathbb{Q})}$ is contained in $\operatorname{Br}_1 U'_{K_1} / \operatorname{Br} K_1$.

The key step is proving (2) *without* necessarily having central simple $\mathbf{k}(U'_{K_1})$-algebra representatives for all of the elements of $\operatorname{Br}^{\text{g.unr.}} U'_{K_1} / \operatorname{Br} K_1$.

Our approach follows the philosophy laid out in [4, 5, 10]: we study geometrically unramified Brauer classes on U'_{K_1} via pullback of ramified Brauer classes on a simpler surface S', of which U' is a double cover. However, in contrast to the work of [4, 5, 10], we carry this out over a field that is not necessarily separably closed. In particular, our methods can be carried out over a number field. As we expect this approach to be of independent interest, we build up some general results in Sect. 2 which can be applied to a double cover of a rational ruled surface, assuming mild conditions on the branch locus.

Remark 1.3. For convenience, we carry out the above strategy on the K3 surface $Y_{\mathbf{a}}$ instead of on the Enriques surface $X_{\mathbf{a}}$. We then descend the results to $X_{\mathbf{a}}$.

Starting in Sect. 3, we restrict our attention to the specific varieties $X_{\mathbf{a}}$ and $Y_{\mathbf{a}}$. After recalling relevant results from [16], we construct double cover maps $\pi\colon Y_{\mathbf{a}} \to S$ and $\tilde{\pi}\colon X_{\mathbf{a}} \to \widetilde{S}$, where S and $\widetilde{S}$ are ruled surfaces, and we study the geometry of these morphisms. These maps allow us to apply the results of [4] to construct, in Sect. 4, an explicit central simple $\mathbf{k}(\overline{X}_{\mathbf{a}})$-algebra representative $\mathcal{A}$ of the nontrivial Brauer class of $\overline{X}_{\mathbf{a}}$. This representative $\mathcal{A}$ will necessarily be defined over a number field K_1, be unramified over an open set U'_{K_1}, and be geometrically unramified. Furthermore, the number field K_1 and the open set U' can be explicitly computed from the representative $\mathcal{A}$.

Section 5 uses the results from Sect. 2 to study the action of $\mathrm{Gal}(\overline{\mathbb{Q}}/K_1)$ on $\mathrm{Br}^{\text{g.unr.}}\, U'_{K_1} / \operatorname{Br} K_1$ and hence prove Theorem 1.2. Namely, by repeated application of the commutative diagram in Theorem 2.2, we demonstrate that no σ-invariant transcendental Brauer class can exist for $Y_{\mathbf{a}}$. Indeed, if such a class existed, the explicit central simple algebra from Sect. 4 would relate it to a function $\tilde{\ell}$ fixed by the Galois action. However a direct computation (given in the Appendix) shows that $\tilde{\ell}$ must be moved by some Galois action, providing the required contradiction.

1.2 General Notation

Throughout, k will be a field with characteristic not equal to 2, with fixed separable closure $\bar{k}$. For any k-scheme X and field extension k'/k, we write $X_{k'}$ for the base change $X \times_{\operatorname{Spec} k} \operatorname{Spec} k'$ and $\overline{X}$ for the base change $X \times_{\operatorname{Spec} k} \operatorname{Spec} \bar{k}$. If X is integral, we write $\mathbf{k}(X)$ for the function field of X. We also denote the absolute Galois group of k by $G_k = \mathrm{Gal}(\bar{k}/k)$. For any k-variety W, we use the term splitting field (of W) to mean the smallest Galois extension of k over which every geometrically irreducible component of W is defined.

The Picard group of X is $\operatorname{Pic} X := \operatorname{Div} X / \operatorname{Princ} X$, where $\operatorname{Div} X$ is the group of Weil divisors on X and $\operatorname{Princ} X$ is the group of principal divisors on X; when X is projective, Pic X is representable by a scheme, called the Picard scheme [7, Cor. 6.6, p. 232–17]. If X is projective, let Pic X denote the subgroup of Pic X that maps to the connected component of the identity in the Picard scheme of X then; the

Néron–Severi group of X is $\operatorname{NS} X := \operatorname{Pic} X/\operatorname{Pic}^0 X$. For a divisor $D \in \operatorname{Div} X$, we write $[D]$ for its equivalence class in $\operatorname{Pic} X$. When X is a curve, the Jacobian of X satisfies $\operatorname{Jac} X = \operatorname{Pic}^0 X$.

For a k-scheme Y, we write $\operatorname{Br} Y$ for the étale cohomology group $\operatorname{Br} Y := \mathrm{H}^2_{\text{ét}}(Y, \mathbb{G}_m)$. If Y is projective, we additionally consider the geometrically unramified subgroup $\operatorname{Br}^{\text{g.unr.}} \mathbf{k}(Y) \subset \operatorname{Br} \mathbf{k}(Y)$ consisting of those Brauer classes which are contained in $\operatorname{Br} \overline{Y}$ upon base change to $\overline{k}$. For an open subscheme $U \subset Y$, we have $\operatorname{Br}^{\text{g.unr.}} U := \operatorname{Br} U \cap \operatorname{Br}^{\text{g.unr.}} \mathbf{k}(Y)$. If A is an étale k-algebra, then we write $\operatorname{Br} A$ for $\operatorname{Br}(\operatorname{Spec} A)$. Given invertible elements a and b in such an A, we define the quaternion algebra $(a, b) := A[i,j]/\langle i^2 = a, j^2 = b, ij = -ji\rangle$. We will identify the algebra (a, b) with its class in $\operatorname{Br} A$.

Now assume that Y is smooth and quasi-projective. Then the following sequence is exact:

$$0 \to \operatorname{Br} Y[2] \to \operatorname{Br} \mathbf{k}(Y)[2] \xrightarrow{\oplus_y \partial_y} \bigoplus_y \mathrm{H}^1(\mathbf{k}(y), \mathbb{Z}/2\mathbb{Z}), \tag{2}$$

where the sum is taken over the set of all codimension-1 points y on Y [8, Theorem 6.1]. As $\operatorname{Br} \mathbf{k}(Y)[2]$ is generated by quaternion algebras, we will only describe the residue map ∂_y on quaternion algebras: for any $a, b \in \mathbf{k}(Y)^\times$, we have

$$\partial_y\left((a, b)\right) = (-1)^{v_y(a)v_y(b)} a^{v_y(b)} b^{-v_y(a)} \in \mathbf{k}(y)^\times/\mathbf{k}(y)^{\times 2} \cong \mathrm{H}^1\left(\mathbf{k}(y), \mathbb{Z}/2\mathbb{Z}\right),$$

where v_y denotes the valuation corresponding to y; as $\mathbf{k}(y)^\times/\mathbf{k}(y)^{\times 2} \cong \mathrm{H}^1(\mathbf{k}(y), Z/2Z)$, we move freely between additive and multiplicative notation when computing residues, depending on the context.

2 Brauer Classes on Double Covers Arising Via Pullback

Let $\pi^0 \colon Y^0 \to S^0$ be a double cover of a smooth, projective, rational, geometrically ruled surface $\varpi \colon S^0 \to \mathbb{P}^1_t$ defined over k and let $B^0 \subset S^0$ denote the branch locus of π^0. (Throughout, $\mathbb{P}^1_t$ is shorthand for $\mathbb{P}^1_{[t_0:t_1]}$, with $t := t_0/t_1$.) We assume that B^0 is reduced, geometrically irreducible, and has at worst ADE singularities. The canonical resolution [1, Theorem 7.2] $\nu \colon Y \to Y^0$ of $\pi^0 \colon Y^0 \to S^0$ has a 2-to-1 k-morphism $\pi \colon Y \to S$ to a smooth rational *generically* ruled surface S; the branch curve $B \subset S$ of π is a smooth proper model of B^0. In summary, we have the following diagram:

$$\begin{array}{ccc} Y & \xrightarrow{\quad \pi \quad} & S \supset B \\ \downarrow{\scriptstyle \nu} & & \downarrow{\scriptstyle \nu_S} \\ Y^0 & \xrightarrow{\quad \pi^0 \quad} & S^0 \supset B^0 \end{array}$$

Since B^0 is geometrically irreducible, $\operatorname{Pic}^0 Y$ is trivial by [4, Corollary 6.3] and so we may conflate $\operatorname{Pic} Y$ and $\operatorname{NS} Y$.

The generic fiber of $\varpi \circ \pi^0$ is a double cover $C \to \mathbb{P}^1_{k(t)} \to \operatorname{Spec} k(t)$. Since $k(t)$ is infinite, by changing coordinates if necessary, we may assume that the double cover is unramified above $\infty \in S^0_{k(t)}$. Then C has a model

$$y^2 = c'h(x),$$

for some $c' \in k(t)$ and $h \in k(t)[x]$ square-free, monic, and with $\deg(h) = 2g(C)+2$, where $g(C)$ denotes the genus of C. Note that $\mathbf{k}(B) = \mathbf{k}(B^0) \cong k(t)[\theta]/(h(\theta))$; we write α for the image of θ in $\mathbf{k}(B)$.

As S^0 is rational and geometrically ruled, $\operatorname{Pic} S^0 \cong \mathbb{Z}^2$ and is generated by a fiber S^0_∞ and a section $\mathfrak{S}$, which we may take to be the closure of $x = \infty$. Since $\nu_S \colon S \to S^0$ is a birational map, $\operatorname{Pic} \overline{S}$ is generated by the strict transforms of $\mathfrak{S}$ and S^0_∞, and the curves $E_1, \dots, E_n$ that are contracted by the map $S \to S^0$. We will often abuse notation and conflate $\mathfrak{S}$ and S^0_∞ with their strict transforms. In any case, by $\mathfrak{S}$, B, E_i, or S^0_∞, we always mean the actual divisors and *not* the divisor classes in the Picard group.

Let

$$\mathcal{E} = \{\mathfrak{S}, S^0_\infty, E_1, \dots, E_n\}$$

denote the aforementioned set of $n+2$ generators and define

$$V := S \setminus \left(B \cup \bigcup_{E \in \mathcal{E}} E\right) \subset S.$$

Possibly after replacing k with a finite extension, we may assume that all elements of $\mathcal{E}$ are defined over k and, in particular, that $\operatorname{Pic} S = \operatorname{Pic} \overline{S}$. Since ν_S is defined over k, we additionally have that S is k-rational and so $\operatorname{Br} S = \operatorname{Br} S^0 = \operatorname{Br} k$.

For any $\ell \in \mathbf{k}(B)^\times$, we define

$$\mathcal{A}_\ell := \operatorname{Cor}_{\mathbf{k}(B)(x)/k(t,x)}\left((\ell, x - \alpha)\right) \in \operatorname{Br} \mathbf{k}(S).$$

We will be particularly concerned with $\mathcal{A}_\ell$ when ℓ is contained in the subgroup

$$\begin{aligned}\mathbf{k}(B)^\times_{\mathcal{E}} &:= \left\{\ell \in \mathbf{k}(B)^\times : \operatorname{div}(\ell) \in \operatorname{im}(\mathbb{Z}^{\mathcal{E}} \to \operatorname{Div}(S) \to \operatorname{Div}(B) \otimes \mathbb{Z}/2\mathbb{Z})\right\}\\ &= \left\{\ell \in \mathbf{k}(B)^\times : \nu_* \operatorname{div}(\ell) \in \langle \mathfrak{S}, S^0_\infty \rangle \subset \operatorname{Div}(B^0) \otimes \mathbb{Z}/2\mathbb{Z}\right\}.\end{aligned}$$

By [4, Proof of Theorem 5.2], this subgroup is exactly the set of functions ℓ such that $\pi^*\mathcal{A}_\ell$ is geometrically unramified. Note that $\mathbf{k}(B)^\times_{\mathcal{E}}$ contains $k^\times \mathbf{k}(B)^{\times 2}$.

Let

$$U := Y \setminus \left(\bigcup_{E \in \mathcal{E}} \pi^{-1}(E)\right) \subset Y.$$

The goal of this section is to prove the following two theorems:

Theorem 2.1. *Let k' be any Galois extension of k. Then we have the following exact sequence of* $\mathrm{Gal}(k'/k)$*-modules:*

$$0 \to \frac{\operatorname{Pic} Y_{k'}}{\pi^* \operatorname{Pic} S + 2 \operatorname{Pic} Y_{k'}} \xrightarrow{j} \frac{\mathbf{k}(B_{k'})^{\times}_{\mathcal{E}}}{k'^{\times}\mathbf{k}(B_{k'})^{\times 2}} \xrightarrow{\beta} \left(\frac{\mathrm{Br}^{\mathrm{g.\,unr.}} U_{k'}}{\operatorname{Br} k'}\right)[2], \tag{3}$$

where j is as in Sect. 2.3 and β is as in Sect. 2.2. Furthermore, if k' is separably closed, then the last map surjects onto $\operatorname{Br} Y[2]$.

Theorem 2.2. *We retain the notation from Theorem 2.1. If* $\operatorname{Br} k' \to \operatorname{Br} \mathbf{k}(S_{k'})$ *is injective and* $\operatorname{Pic} \overline{U}[2] = 0$*, then there is a commutative diagram of* $\mathrm{Gal}(k'/k)$*-modules with exact rows and columns:*

$$\begin{array}{ccccccccc}
 & & \dfrac{\operatorname{Pic} Y_{k'}}{\pi^* \operatorname{Pic} S+2\operatorname{Pic} Y_{k'}} & \xrightarrow{j} & j(\operatorname{Pic} Y_{k'}) & & & & \\
 & & \big\downarrow & & \big\downarrow & & & & \\
0 & \longrightarrow & \left(\dfrac{\operatorname{Pic} \overline{Y}}{\pi^* \operatorname{Pic} S+2\operatorname{Pic} \overline{Y}}\right)^{G_{k'}} & \xrightarrow{j} & \dfrac{\mathbf{k}(B_{k'})_{\mathcal{E}}}{k'^{\times}\mathbf{k}(B_{k'})^{\times 2}} & \xrightarrow{\beta} & \dfrac{\mathrm{Br}^{\mathrm{g.unr.}} U_{k'}}{\mathrm{Br}_1 U_{k'}} & & \\
 & & \big\downarrow{\scriptstyle \beta\circ j} & & \big\downarrow{\scriptstyle \beta} & & \big\| & & \\
0 & \longrightarrow & \dfrac{\mathrm{Br}_1 U_{k'}}{\operatorname{Br} k'} & \longrightarrow & \dfrac{\mathrm{Br}^{\mathrm{g.unr.}} U_{k'}}{\operatorname{Br} k'} & \longrightarrow & \dfrac{\mathrm{Br}^{\mathrm{g.unr.}} U_{k'}}{\mathrm{Br}_1 U_{k'}} & \longrightarrow & 0.
\end{array}$$

The structure of the section is as follows. In Sect. 2.1, we prove some preliminary results about the residues of $\mathcal{A}_\ell$; these are used in Sect. 2.2 to define the map β. Next, in Sect. 2.3, we define j and prove that it is injective. In Sect. 2.4, we characterize the elements of $\operatorname{Br} V$ that pull back to constant algebras under π^*. In Sect. 2.5, we combine the results from the earlier sections to prove Theorem 2.1, and, finally, in Sect. 2.6, we prove Theorem 2.2.

2.1 Residues of $\mathcal{A}_\ell$

In order to define the homomorphism β, we will need to know certain properties about the residues of $\mathcal{A}_\ell$ at various divisors of S^0. We first compute residues associated to horizontal divisors.

Lemma 2.3. *Let $\ell \in \mathbf{k}(B)^\times$, and let F be an irreducible horizontal curve in S^0, i.e., a curve that maps dominantly onto $\mathbb{P}^1_t$.*

(1) *If $F \neq B, \mathfrak{S}$, then $\partial_F(\mathcal{A}_\ell) = 1 \in \mathbf{k}(F)^\times/\mathbf{k}(F)^{\times 2}$.*
(2) $\partial_B(\mathcal{A}_\ell) = [\ell] \in \mathbf{k}(B)^\times/\mathbf{k}(B)^{\times 2}$.

Proof. The arguments in this proof follow those in [5, Proofs of Theorem 1.1 and Prop. 2.3]; as the situation is not identical, we restate the arguments here for the reader's convenience.

Let v be the valuation on $\mathbf{k}(S^0)$ associated to F. By [5, Lemma 2.1], we have

$$\partial_F(\mathcal{A}_\ell) = \prod_{w|v} \mathrm{Norm}_{\mathbf{k}(w)/\mathbf{k}(v)} \left((-1)^{w(\ell)w(x-\alpha)} \ell^{w(x-\alpha)} (x-\alpha)^{-w(\ell)}\right), \tag{4}$$

where w runs over all valuations on $\mathbf{k}(B \times_{\mathbb{P}^1_t} S^0)$ extending v. As F is a horizontal divisor, $v|_{k(t)}$ is trivial and hence $w|_{\mathbf{k}(B)}$ is trivial for all $w|v$. Therefore, (4) simplifies to $\prod_{w|v} \mathrm{Norm}_{\mathbf{k}(w)/\mathbf{k}(v)}(\ell^{w(x-\alpha)})$.

By definition of α, $\mathrm{Norm}_{\mathbf{k}(B)(x)/k(t)(x)}(x-\alpha) = h(x)$. Thus, $w(x-\alpha) = 0$ for all $w|v$ if $v(h(x)) = 0$, or equivalently, if $F \neq B, \mathfrak{S}$. This completes the proof of (1).

Now assume that $F = B$. We know that $h(x)$ factors as $(x-\alpha)h_1(x)$ over $\mathbf{k}(B)(x)$, where $h_1 \in k(t)[x]$ is possibly reducible. Hence, $x-\alpha$ determines a valuation $w_{x-\alpha}$ on $\mathbf{k}(B)(x)$ lying over v; similarly, the other irreducible factors of h_1 also determine valuations lying over v. Notice that since $h(x)$ is separable (as B is reduced), we have that $h_1(\alpha) \neq 0$, and hence that $w(x-\alpha) = 0$ for any valuation w over v corresponding to the irreducible factors of $h_1(x)$. Thus, (4) simplifies to

$$\prod_{w|v} \mathrm{Norm}_{\mathbf{k}(w)/\mathbf{k}(v)}(\ell^{w(x-\alpha)}) = \mathrm{Norm}_{\mathbf{k}(w_{x-\alpha})/\mathbf{k}(v)}(\ell) = \ell,$$

as required. □

Now we compute the residues associated to vertical divisors.

Lemma 2.4. *Let $\ell \in \mathbf{k}(B)^\times_{\mathcal{E}}$, $t_0 \in \mathbb{A}^1_t \subset \mathbb{P}^1_t$ be a closed point, and $F = S^0_{t_0}$. Then,*

$$\partial_F(\mathcal{A}_\ell) \in \mathrm{im}\left(\frac{\mathbf{k}(t_0)^\times}{\mathbf{k}(t_0)^{\times 2}} \to \frac{\mathbf{k}(F)^\times}{\mathbf{k}(F)^{\times 2}}\right).$$

Remark 2.5. If k is separably closed, then $\mathbf{k}(t_0)^{\times 2} = \mathbf{k}(t_0)^\times$ and the result follows from [5, Prop. 3.1].

Proof. It suffices to show that $\partial_F(\mathcal{A}_\ell) \in \mathbf{k}(F)^{\times 2}\mathbf{k}(t_0)^\times$. We repeat [4, Proof of Prop. 3.1] while keeping track of scalars to accommodate the fact that k is not necessarily separably closed.

By [5, Lemma 2.1], we have

$$\partial_F(\mathcal{A}_\ell) = \prod_{\substack{F' \subset S^0 \times_{\mathbb{P}^1_t} B \\ F' \mapsto F \text{ dominantly}}} \mathrm{Norm}_{\mathbf{k}(F')/\mathbf{k}(F)}((-1)^{w'(x-\alpha)w'(\ell)} \ell^{w'(x-\alpha)} (x-\alpha)^{-w'(\ell)}), \tag{5}$$

where F' is an irreducible curve and w' denotes the valuation associated to F'. The surface $S^0 \times_{\mathbb{P}^1_t} B$ is a geometrically ruled surface over B, so the irreducible curves F' are in one-to-one correspondence with points $b' \in B$ mapping to t_0. Furthermore, $\mathbf{k}(F') = \mathbf{k}(b')(x)$ and $\mathbf{k}(F) = \mathbf{k}(t_0)(x)$, so $\mathrm{Norm}_{\mathbf{k}(F')/\mathbf{k}(F)}$ is induced from $\mathrm{Norm}_{\mathbf{k}(b')/\mathbf{k}(t_0)}$. Thus, we may rewrite (5) as

$$\partial_F(\mathcal{A}_\ell) = \prod_{b' \in B, b' \mapsto t_0} \mathrm{Norm}_{\mathbf{k}(b')/\mathbf{k}(t_0)}((-1)^{w'(x-\alpha)w'(\ell)} \ell^{w'(x-\alpha)} (x-\alpha)^{-w'(\ell)}). \tag{6}$$

By [4, Lemma 3.3], there exists an open set $W \subset \mathbb{A}^1$ containing t_0 and constants $d \in \mathbf{k}(t)^\times$, $e \in \mathbf{k}(t)$ such that

$$S^0_W \to \mathbb{P}^1_k \times W, \quad s \mapsto (dx(s) + e, \varpi(s))$$

is an isomorphism. In particular, $dx + e$ is a horizontal function on S^0_W. Consider the following equality:

$$\begin{aligned}\mathrm{Cor}_{\mathbf{k}(B)(x)/\mathbf{k}(S^0)}((dx + e - (d\alpha + e), \ell)) &= \mathcal{A}_\ell + \mathrm{Cor}_{\mathbf{k}(B)(x)/\mathbf{k}(S^0)}((d, \ell)) \\ &= \mathcal{A}_\ell + (d, \mathrm{Norm}(\ell)).\end{aligned}$$

Since $(d, \mathrm{Norm}(\ell)) \in \varpi^* \operatorname{Br} k(t)$, we have

$$\partial_F(\mathcal{A}_\ell) \in \partial_F\left(\mathrm{Cor}_{\mathbf{k}(B)(x)/\mathbf{k}(S^0)}((dx + e - (d\alpha + e), \ell))\right) \mathbf{k}(t_0)^\times.$$

Thus, we may assume that x is a horizontal function, in particular, that x has no zeros or poles along F, and that it restricts to a non-constant function along F. It is then immediate that $w'(x - \alpha) \leq 0$, and that the inequality is strict if and only if $w'(\alpha) < 0$, which in turn happens if and only if b' lies over $B^0_{t_0} \cap \mathfrak{S}$.

We first consider the factor of (6) that corresponds to points that do not lie over $B^0_{t_0} \cap \mathfrak{S}$. If b' does not lie over $B^0_{t_0} \cap \mathfrak{S}$, then (as stated above) $w'(x - \alpha) = 0$, where w' denotes the valuation associated to b'. Therefore, the corresponding factor of (6) simplifies to

$$\prod_{b' \in B \setminus \nu^{-1}(B^0 \cap \mathfrak{S}), b' \mapsto t_0} \mathrm{Norm}_{\mathbf{k}(b')/\mathbf{k}(t_0)} \left((x - \alpha(b'))^{-w'(\ell)}\right).$$

By definition, $\ell \in \mathbf{k}(B)^\times_{\mathcal{E}}$ implies that for all $b'' \in B^0 \setminus (B^0 \cap \mathfrak{S})$, $\sum_{b' \in B, b' \mapsto b''} w'(\ell) \equiv 0 \bmod 2$. Since $\alpha(b')$ depends only on the image of b' in B^0, this shows that the above factor is contained in $\mathbf{k}(F)^{\times 2}$.

Now consider the case that b' lies over $B^0_{t_0} \cap \mathfrak{S}$. We claim that, since $w'(x) = 0$,

$$\mathrm{Norm}_{\mathbf{k}(F')/\mathbf{k}(F)} \left((-1)^{w'(x-\alpha)w'(\ell)} \ell^{w'(x-\alpha)} (x-\alpha)^{-w'(\ell)}\right) \tag{7}$$

reduces to a constant in $\mathbf{k}(F)$. Indeed, if $w'(\ell) = 0$, then we obtain $\ell^{w'(x-\alpha)}$, which reduces (after taking $\mathrm{Norm}_{\mathbf{k}(F')/\mathbf{k}(F)}$) to an element of $\mathbf{k}(t_0)^\times$. If $w'(\ell) \neq 0$, let $\pi_{F'}$ be a uniformizer for F'. Since $w'(x) = 0 > w'(\alpha)$, we have

$$\left(\frac{\ell}{\pi_{F'}^{w'(\ell)}}\right)^{w'(x-\alpha)}\left(\frac{x-\alpha}{\pi_{F'}^{w'(x-\alpha)}}\right)^{-w(\ell)}=\left(\frac{\ell}{\pi_{F'}^{w'(\ell)}}\right)^{w'(x-\alpha)}\left(\frac{-\alpha}{\pi_{F'}^{w'(x-\alpha)}}\right)^{-w(\ell)}\mod \pi_{F'}$$

and so (7) reduces (again, after taking $\mathrm{Norm}_{\mathbf{k}(F')/\mathbf{k}(F)}$) to an element in $\mathbf{k}(t_0)^\times$. Thus, every factor of (6) corresponding to points b' lying over $B^0_{t_0}\cap\mathfrak{S}$ is contained in $\mathbf{k}(t_0)^\times$, and every other factor is an element of $\mathbf{k}(F)^{\times 2}$. This completes the proof. □

2.2 The Morphism β

Proposition 2.6. *Let $\ell\in\mathbf{k}(B)^\times$. There exists an element $\mathcal{A}'=\mathcal{A}'(\ell)\in\operatorname{Br}k(t)$, unique modulo $\operatorname{Br}k$, such that*

$$\mathcal{A}_\ell+\varpi^*\mathcal{A}'\in\operatorname{Br}V.$$

This induces a well-defined homomorphism

$$\beta\colon\frac{\mathbf{k}(B)^\times_{\mathcal{E}}}{k^\times\mathbf{k}(B)^{\times 2}}\to\frac{\operatorname{Br}^{\text{g.unr.}}U}{\operatorname{Br}k}[2],\quad \ell\mapsto\pi^*\left(\mathcal{A}_\ell+\varpi^*\mathcal{A}'\right),$$

which is surjective if k is separably closed.

Proof. Recall that $V=S\setminus\left(B\cup\bigcup_{E\in\mathcal{E}}E\right)\subset S$. Therefore, as a subgroup of $\operatorname{Br}\mathbf{k}(S)=\operatorname{Br}\mathbf{k}(S^0)$, $\operatorname{Br}V$ is equal to $\operatorname{Br}\left(S^0\setminus(\mathfrak{S}\cup S^0_\infty\cup B)\right)$, since the Brauer group of a surface is unchanged under removal of a codimension 2 closed subscheme [8, Theorem 6.1]. Thus, to prove the first statement, it suffices to show that there exists an element $\mathcal{A}'\in\operatorname{Br}k(t)$, unique up to constant algebras, such that $\partial_F(\mathcal{A}_\ell)=\partial_F(\varpi^*\mathcal{A}')$ for all irreducible curves $F\subset S^0$ with $F\neq\mathfrak{S},S^0_\infty,B$.

If F is any horizontal curve, i.e., F maps dominantly to $\mathbb{P}^1_t$, then $\partial_F(\varpi^*\mathcal{A}')=1$ for all $\mathcal{A}'\in\operatorname{Br}k(t)$. If we further assume that $F\neq\mathfrak{S},B$, then Lemma 2.3 gives $\partial_F(\mathcal{A}_\ell)=1$. Thus, for all $\mathcal{A}'\in\operatorname{Br}k(t)$, we have $\partial_F(\mathcal{A}_\ell)=\partial_F(\varpi^*\mathcal{A}')$ for all horizontal curves $F\neq\mathfrak{S},B$.

Now we turn our attention to the vertical curves. Recall Faddeev's exact sequence [6, Corollary 6.4.6]:

$$0\to\operatorname{Br}k\to\operatorname{Br}k(t)\xrightarrow{\oplus\partial_{t_0}}\bigoplus_{t_0\in\mathbb{P}^1_t}\mathrm{H}^1(G_{\mathbf{k}(t_0)},\mathbb{Q}/\mathbb{Z})\xrightarrow{\sum_{t_0}\mathrm{Cor}_{\mathbf{k}(t_0)/k}}\mathrm{H}^1(G_k,\mathbb{Q}/\mathbb{Z})\to 0.\tag{8}$$

Since the residue field at $t_0=\infty$ is equal to k, this sequence implies that for any element $(r_{t_0})\in\bigoplus_{t_0\in\mathbb{A}^1}\mathbf{k}(t_0)^\times/\mathbf{k}(t_0)^{\times 2}$, there exists a Brauer class in $\mathcal{A}'\in\operatorname{Br}k(t)$, unique modulo elements of $\operatorname{Br}k$, such that $\partial_{t_0}(\mathcal{A}')=r_{t_0}$ for all closed points $t_0 \in \mathbb{A}^1$. By Lemma 2.4, for all $t_0 \in \mathbb{A}^1$, we have $\partial_F(\mathcal{A}_\ell) \in \operatorname{im}\left(\mathbf{k}(t_0)^\times/\mathbf{k}(t_0)^{\times 2} \to \mathbf{k}(F)^\times/\mathbf{k}(F)^{\times 2}\right)$, where $F = S^0_{t_0}$. Hence, there exists an $\mathcal{A}' \in \operatorname{Br} k(t)$, unique modulo $\operatorname{Br} k$, such that $\partial_F(\varpi^*\mathcal{A}') = \partial_F(\mathcal{A}_\ell)$ for all $F \neq \mathfrak{S}, B, S^0_\infty$, as desired.

It remains to prove the second statement. The first statement immediately implies the existence of a well-defined homomorphism

$$\frac{\mathbf{k}(B)^\times_{\mathcal{E}}}{\mathbf{k}(B)^{\times 2}} \to \frac{\operatorname{Br} \pi^{-1}(V)}{\operatorname{Br} k}[2], \quad \ell \mapsto \pi^*\left(\mathcal{A}_\ell + \varpi^*\mathcal{A}'\right).$$

In order to complete the proof, we must prove that

(1) $\pi^*\left(\mathcal{A}_d + \varpi^*\mathcal{A}'\right) \in \operatorname{Br} k$ if $d \in k^\times$,
(2) the image lands in $\operatorname{Br}^{\text{g.unr.}} U/\operatorname{Br} k$, and
(3) the image is equal to $\operatorname{Br} Y[2]$ if k is separably closed.

We begin with (1). Let $d \in k^\times$. Then

$$\begin{aligned}\mathcal{A}_d = \operatorname{Cor}_{\mathbf{k}(B)(x)/k(t,x)}(d, x-\alpha) &= (d, \operatorname{Norm}_{\mathbf{k}(B)(x)/k(t,x)}(x-\alpha)) \\ &= (d, h(x)) = (d, c'h(x)) + (d, c').\end{aligned}$$

Since $\sqrt{c'h(x)}$ generates $\mathbf{k}(Y^0)/\mathbf{k}(S^0)$, $\operatorname{div}(c'h(x)) = B + 2Z$ for some divisor Z on S^0. Thus, $(d, c'h(x))$ is unramified away from B; in particular, $(d, c'h(x)) \in \operatorname{Br} V$. Since $\mathcal{A}'$ is the unique element in $\operatorname{Br} k(t)/\operatorname{Br} k$ such that $\mathcal{A}_d + \varpi^*\mathcal{A}'$, then $\mathcal{A}' = (d, c') + \mathcal{B}$ for some $\mathcal{B} \in \operatorname{Br} k$. Hence,

$$\begin{aligned}\pi^*\left(\mathcal{A}_d + \varpi^*\mathcal{A}'\right) &= \pi^*\left((d, c'h(x)) + (d, c') + \varpi^*(d, c') + \varpi^*\mathcal{B}\right), \\ &= \pi^*(d, c'h(x)) + \pi^*\varpi^*\mathcal{B}.\end{aligned}$$

Furthermore, since $c'h(x)$ is a square in $\mathbf{k}(Y^0)$, then $\pi^*\left(\mathcal{A}_d + \varpi^*\mathcal{A}'\right) = \pi^*\varpi^*\mathcal{B} \in \operatorname{Br} k$, as desired.

Now we turn to (2) and (3). Since B is the branch locus of π and π is 2-to-1, any 2-torsion Brauer class in $\operatorname{im}(\pi^*\colon \operatorname{Br}\mathbf{k}(S) \to \operatorname{Br}\mathbf{k}(Y))$ is unramified at $\pi^{-1}(B)_{\text{red}}$. Thus, the image is contained in $\operatorname{Br} U/\operatorname{Br} k$. To prove that it is contained in $\operatorname{Br}^{\text{g.unr.}} U$, we must show that $\pi^*\left(\mathcal{A}_\ell + \varpi^*\mathcal{A}'\right)_{\overline{k}}$ is contained in $\operatorname{Br}\overline{Y}$. By Tsen's theorem, $\pi^*\left(\mathcal{A}_\ell + \varpi^*\mathcal{A}'\right)_{\overline{k}} = (\pi^*\mathcal{A}_\ell)_{\overline{k}}$. This element is contained in $\operatorname{Br}\overline{Y}$ by [4, Theorem I], which yields (2). In fact, [4, Theorem I] shows that $\operatorname{Br}\overline{Y}[2]$ is generated by $\pi^*\mathcal{A}_\ell$ where ℓ runs over the elements in $\mathbf{k}(B_{\overline{k}})_{\mathcal{E}}$, which proves (3). □

2.3 The Morphism j

In this section, we define the map j and prove that it is injective. The map j will be induced by the following homomorphism:

$$\begin{aligned} j'\colon \operatorname{Div}(Y \setminus \pi^{-1}(B)) &\to \mathbf{k}(B)^\times/k^\times \\ D &\mapsto \ell|_B \end{aligned}$$

where $\ell \in \mathbf{k}(S)^\times$ is such that $\operatorname{div}_S(\ell) = \pi_*D - m_1E_1 - \cdots - m_nE_n - d\mathfrak{S} - eS^0_\infty$. (Recall that $E_1, \ldots, E_n$, $\mathfrak{S}$, and S^0_∞ form an integral basis for $\operatorname{Pic} S = \operatorname{Pic} \overline{S}$.)

Lemma 2.7. *The homomorphism j' induces a well-defined injective homomorphism*

$$j : \frac{\operatorname{Pic} Y}{\pi^* \operatorname{Pic} S + 2 \operatorname{Pic} Y} \to \frac{\mathbf{k}(B)^\times_{\mathcal{E}}}{k^\times \mathbf{k}(B)^{\times 2}}.$$

Proof. For any divisor $D \in \operatorname{Div} Y \setminus \pi^{-1}(B)$, the projection formula [11, p.399] yields

$$2\pi_*(D \cap \pi^{-1}(B)_{\mathrm{red}}) = \pi_*(D \cap 2\pi^{-1}(B)_{\mathrm{red}}) = \pi_*(D \cap \pi^*(B)) = (\pi_*D) \cap B.$$

Thus, for any divisor $D \in \operatorname{Div} Y \setminus \pi^{-1}(B)$, we have that $[D \cap \pi^{-1}(B)_{\mathrm{red}}] \in \left(\frac{\operatorname{Pic} B}{\operatorname{im} \operatorname{Pic} S \to \operatorname{Pic} B}\right)[2]$. By the same argument as in proof of [10, Lemma 4.8], this induces a well-defined injective homomorphism

$$\frac{\operatorname{Pic} Y}{\pi^* \operatorname{Pic} S + 2 \operatorname{Pic} Y} \to \left(\frac{\operatorname{Pic} B}{\operatorname{im} \operatorname{Pic} S \to \operatorname{Pic} B}\right)[2], \quad [D] \mapsto [D \cap \pi^{-1}(B)_{\mathrm{red}}]. \tag{9}$$

One can also check that there is a well-defined injective homomorphism

$$\left(\frac{\operatorname{Pic} B}{\operatorname{im} \operatorname{Pic} S \to \operatorname{Pic} B}\right)[2] \to \frac{\mathbf{k}(B)^\times_{\mathcal{E}}}{k^\times \mathbf{k}(B)^{\times 2}} \tag{10}$$

that sends a divisor D which represents a class in $\left(\frac{\operatorname{Pic} B}{\operatorname{im} \operatorname{Pic} S \to \operatorname{Pic} B}\right)[2]$ to a function ℓ such that $\operatorname{div}(\ell) = 2D + \sum_{C \in \operatorname{Pic} S} n_C C \cap B$. Since j is the composition of (9) and (10), this completes the proof that j is well-defined and injective. □

2.4 Brauer Classes on V That Become Constant Under π^*

Proposition 2.8. *If $\mathcal{A} \in \operatorname{Br} V$ is such that $\pi^*\mathcal{A} \in \operatorname{Br} k \subset \operatorname{Br} \mathbf{k}(Y)$, then there exists a divisor $D \in \operatorname{Div} Y$ such that $j([D]) = \partial_B(\mathcal{A})$ in $\mathbf{k}(B)^\times / k^\times \mathbf{k}(B)^{\times 2}$.*

Proof. Recall that $\mathbf{k}(Y_k) = \mathbf{k}(S_k)(\sqrt{c'h(x)})$. Thus, if $\pi^*\mathcal{A} \in \operatorname{Br} k$, then

$$\mathcal{A} = (c'h(x), G) + \mathcal{B} \tag{11}$$

for some $G \in \mathbf{k}(S_k)^\times$ and some $\mathcal{B} \in \operatorname{Br} k$. Since B is the branch locus of π, $v_B(c'h(x))$ must be odd. Therefore, without loss of generality, we may assume that B is not contained in the support of G; write

$$\operatorname{div}(G) = \sum_i n_i C_i + d(\mathfrak{S}) + e(S^0_\infty) + m_1E_1 + \cdots + m_nE_n,$$

where C_i are k-irreducible curves of S distinct from $\mathfrak{S}$, S^0_∞, and $E_1, \ldots, E_n$.

Now we consider the residue of $\mathcal{A}$ at C_i. By (11), the residue of $\mathcal{A}$ at C_i is $[c'h(x)] \in \mathbf{k}(C_i)^\times/\mathbf{k}(C_i)^{\times 2}$. On the other hand, $\mathcal{A} \in \operatorname{Br} V$, so the residue is trivial at C_i. Together, these statements imply that $\pi^{-1}(C_i)$ consists of two irreducible components C_i' and C_i''. As this is true for all C_i, we have that $\operatorname{div}(G) = \pi_*(\sum_i n_i C_i') + m(\mathfrak{S}) + m_0(S_\infty^0) + m_1 E_1 + \cdots + m_n E_n$, and so $j'(\sum n_i C_i') = G|_B$ modulo $k^\times$. Since the residue of $\mathcal{A}$ at B is equal to $G|_B$, this completes the proof. □

2.5 Proof of Theorem 2.1

We note that much of this proof is very similar to proofs in [10, Lemmas 4.4 and 4.8].

We will first prove the sequence is exact, and then show that the maps are compatible with the Galois action. Since all assumed properties of k are preserved under field extension, we may, for the moment, assume that $k = k'$. Then Lemma 2.7 yields an injective homomorphism

$$j\colon \frac{\operatorname{Pic} Y_{k'}}{\pi^* \operatorname{Pic} S + 2 \operatorname{Pic} Y_{k'}} \longrightarrow \frac{\mathbf{k}(B_{k'})_{\mathcal{E}}^\times}{k'^\times \mathbf{k}(B_{k'})^{\times 2}},$$

and Proposition 2.6 yields a homomorphism

$$\beta\colon \frac{\mathbf{k}(B_{k'})_{\mathcal{E}}^\times}{k'^\times \mathbf{k}(B_{k'})^{\times 2}} \longrightarrow \left(\frac{\operatorname{Br}^{\text{g. unr.}} U_{k'}}{\operatorname{Br} k'}\right)[2],$$

which is surjective if k' is separably closed. We now show that $\operatorname{im}(j) = \ker(\beta)$.

Let $\ell \in \mathbf{k}(B_{k'})_{\mathcal{E}}^\times$ be such that $\beta(\ell) \in \operatorname{Br} k'$. Recall that β factors through $\operatorname{Br} V/\operatorname{Br} k$ by the map

$$\ell \mapsto \underbrace{\mathcal{A} := \mathcal{A}_\ell + \varpi^* \mathcal{A}'}_{\in \operatorname{Br} V/\operatorname{Br} k} \mapsto \pi^* \mathcal{A},$$

where $\mathcal{A}' \in \operatorname{Br} k'(t)$ is as in Proposition 2.6. By assumption, $\pi^* \mathcal{A} \in \operatorname{Br} k$, thus, by Proposition 2.8, there is some $D \in \operatorname{Div} Y_k$ such that $j([D]) = \partial_B(\mathcal{A}) = \partial_B(\mathcal{A}_\ell)\partial_B(\varpi^* \mathcal{A}') \bmod k^\times$. However, $\partial_B(\varpi^* \mathcal{A}') = 1$ since B is a horizontal divisor, and $\partial_B(\mathcal{A}_\ell) = [\ell]$ by Lemma 2.3. Hence, $\ell \in \operatorname{im}(j)$, and so $\operatorname{im}(j) \supset \ker(\beta)$.

For the opposite inclusion, it suffices to prove that $\beta(j([D])) \in \operatorname{Br} k'$ for any prime divisor $D \in \operatorname{Div}(Y_{k'} \setminus \pi^{-1}(B))$. Let $\ell = j'(D)$; recall that ℓ is the restriction to B of a function $\ell_S \in \mathbf{k}(S_{k'})$ such that $\operatorname{div}(\ell_S) = \pi_* D - m_1 E_1 - \cdots - m_n E_n - d\mathfrak{S} - e S_\infty^0$. As above, let $\mathcal{A} := \mathcal{A}_\ell + \varpi^* \mathcal{A}'$. We claim that

$$\mathcal{A} = (c'h(x), \ell_S) + \mathcal{B} \quad \in \operatorname{Br} \mathbf{k}(S_{k'}) = \operatorname{Br} \mathbf{k}(S_{k'}^0)$$

for some $\mathcal{B} \in \operatorname{Br} k'$. Since $c'h(x) \in \mathbf{k}(Y_{k'})^{\times 2}$, this equality implies that $\pi^*(\mathcal{A}) = \pi^* \mathcal{B} \in \operatorname{Br} k'$. To prove the claim, we will compare residues of $\mathcal{A}$ and $(c'h(x), \ell_S)$ on S^0. Repeated application of Faddeev's exact sequence [6, Corollary 6.4.6] shows that $\operatorname{Br}\mathbb{A}^n_{k'}$ is trivial; in particular, the Brauer group of $S^0 \setminus (S^0_\infty \cup \mathfrak{S})$, which is isomorphic to $\mathbb{A}^2$, consists only of constant algebras. Hence, if $\mathcal{A} - (c'h(x), \ell_S)$ is unramified everywhere on $S^0 \setminus (\mathfrak{S} \cup S^0_\infty)$, then it must be constant. By Lemma 2.3(2) and the assumption that $\mathcal{A} \in \operatorname{Br} V$, it suffices to show that $\partial_B((c'h(x), \ell_S)) = [\ell]$ and that $(c'h(x), \ell_S)$ is unramified along all other curves irreducible curves contained in V.

Let R be a prime divisor of S^0 different from $S^0_\infty, \mathfrak{S}$, and B. Since B is the branch locus of π, we may assume that $v_R(c'h(x)) = 0$. Hence, $\partial_R((c'h(x), \ell_S)) = (c'h(x))^{v_R(\ell_S)}$. Now, we know that

$$\operatorname{div}_S(\ell_S) = \pi_* D - m_1 E_1 - \cdots - m_n E_n - d\mathfrak{S} - eS^0_\infty.$$

Thus, if $R \neq \pi(D)$, then $v_R(\ell_S) = 0$ and hence $\partial_R((c'h(x), \ell_S))$ is trivial. It remains to consider the case $R = \pi(D)$. Note that $\pi_*(D)$ is equal to $\pi(D)$ if π maps D isomorphically to its image, and is equal to $2\pi(D)$ otherwise. In the latter case, we must have that $v_{\pi(D)}(\ell_S)$ is even, meaning that $\partial_{\pi(D)}((c'h(x), \ell_S))$ is trivial (up to squares). On the other hand, if $\pi_*(D) = \pi(D)$, then $c'h(x)$ must be a square in $\mathbf{k}(\pi(D))$, and hence $\partial_{\pi(D)}((c'h(x), \ell_S))$ is again trivial (up to squares).

Finally, $\partial_B((c'h(x), \ell_S)) = [\ell_S|_B] = [\ell]$. Indeed, since we know that

$$\operatorname{div}(\ell_S) = \pi_* D - m_1 E_1 - \cdots - m_n E_n - d\mathfrak{S} - eS^0_\infty$$

for some $D \in \operatorname{Div}(Y \setminus \pi^{-1}(B))$, we see that B is not in the support of $\operatorname{div}(\ell_S)$. Hence $v_B(\ell_S) = 0$. Moreover, since $v_B(c'h(x))$ is odd, the usual residue formula allows us to deduce that $\partial_B((c'h(x), \ell_S)) = [\ell]$ modulo squares. Hence, $\mathcal{A} = (c'h(x), \ell_S) + \mathcal{B}$ and the sequence is exact, as desired.

Now we consider the Galois action. That j respects the Galois action is clear from the definition. To see that β is a homomorphism of Galois modules, we note that every geometrically irreducible curve outside of V is irreducible over k. Thus, the residue maps ∂_F for F outside of V are defined over k, which shows that β is a homomorphism of Galois modules. This completes the proof. □

2.6 *Proof of Theorem 2.2*

Applying Theorem 2.1 to $k' = \bar{k}$ and $k = k'$ and taking the subgroups of Galois invariant elements gives an exact sequence

$$0 \longrightarrow \left(\frac{\operatorname{Pic}\overline{Y}}{\pi^*\operatorname{Pic} S + 2\operatorname{Pic}\overline{Y}}\right)^{G_{k'}} \xrightarrow{j_{\bar{k}}} \left(\frac{\mathbf{k}(\overline{B})^\times_{\mathcal{E}}}{\mathbf{k}(\overline{B})^{\times 2}}\right)^{G_{k'}} \xrightarrow{\beta_{\bar{k}}} (\operatorname{Br}\overline{Y})^{G_{k'}}. \tag{12}$$

Recall that $j_{\bar{k}}$ factors through $\mathbf{k}(\overline{S})/\mathbf{k}(\overline{S})^{\times 2}$, so the middle term may be replaced with $\left(\mathbf{k}(\overline{B})_{\mathcal{E}}/\mathbf{k}(\overline{B})^{\times 2}\right)^{G_{k'}} \cap \operatorname{im}(\mathbf{k}(\overline{S})/\mathbf{k}(\overline{S})^{\times 2})^{G_{k'}}$.

To determine $(\mathbf{k}(\overline{S})/\mathbf{k}(\overline{S})^{\times 2})^{G_{k'}}$, we consider the exact sequence

$$0 \longrightarrow \overline{k}^{\times} \longrightarrow \mathbf{k}(\overline{S})^{\times} \longrightarrow \mathbf{k}(\overline{S})^{\times}/\overline{k}^{\times} \longrightarrow 0.$$

After taking the cohomological long exact sequence, applying inflation-restriction and Hilbert's Theorem 90 (twice), and applying the assumption that $\operatorname{Br} k' \longrightarrow \operatorname{Br} \mathbf{k}(S_{k'})$ is injective, we obtain

$$\mathrm{H}^0\left(G_{k'}, \mathbf{k}(\overline{S})^{\times}/\overline{k}^{\times}\right) = \mathbf{k}(S_{k'})^{\times}/k'^{\times}, \quad \text{and} \quad \mathrm{H}^1\left(G_{k'}, \mathbf{k}(\overline{S})^{\times}/\overline{k}^{\times}\right) = 0.$$

Then the cohomological long exact sequence associated to

$$0 \longrightarrow \mathbf{k}(\overline{S})^{\times}/\overline{k}^{\times} \xrightarrow{\times 2} \mathbf{k}(\overline{S})^{\times}/\overline{k}^{\times} \longrightarrow \mathbf{k}(\overline{S})^{\times}/\mathbf{k}(\overline{S})^{\times 2} \longrightarrow 0$$

yields $\left(\mathbf{k}(\overline{S})^{\times}/\mathbf{k}(\overline{S})^{\times 2}\right)^{G_{k'}} = \mathbf{k}(S_{k'})/(k'^{\times}\mathbf{k}(S_{k'})^{\times 2})$. Thus, we may replace (12) with

$$0 \longrightarrow \left(\frac{\operatorname{Pic} \overline{Y}}{\pi^* \operatorname{Pic} S + 2 \operatorname{Pic} \overline{Y}}\right)^{G_{k'}} \xrightarrow{j_{\overline{k}}} \frac{\mathbf{k}(B_{k'})_{\mathcal{E}}}{k'^{\times}\mathbf{k}(B_{k'})^{\times 2}} \xrightarrow{\beta_{\overline{k}}} (\operatorname{Br} \overline{Y})^{G_{k'}}.$$

Note that by Proposition 2.6 $\beta_{\overline{k}}|_{\mathbf{k}(B_{k'})_{\mathcal{E}}}$ factors through $\operatorname{Br}^{\text{g.unr.}} U_{k'} / \operatorname{Br} k'$. Hence, we obtain the following commutative diagram:

$$\begin{array}{ccccccccc}
 & & \frac{\operatorname{Pic} Y_{k'}}{\pi^* \operatorname{Pic} S + 2 \operatorname{Pic} Y_{k'}} & \xrightarrow{j} & j(\operatorname{Pic} Y_{k'}) & & & & \\
 & & \downarrow & & \downarrow & & & & \\
0 & \longrightarrow & \left(\frac{\operatorname{Pic} \overline{Y}}{\pi^* \operatorname{Pic} S + 2 \operatorname{Pic} \overline{Y}}\right)^{G_{k'}} & \xrightarrow{j} & \frac{\mathbf{k}(B_{k'})_{\mathcal{E}}}{k'^{\times}\mathbf{k}(B_{k'})^{\times 2}} & \xrightarrow{\beta} & \frac{\operatorname{Br}^{\text{g.unr.}} U_{k'}}{\operatorname{Br}_1 U_{k'}} & & \\
 & & \downarrow{\scriptstyle \beta\circ j} & & \downarrow{\scriptstyle \beta} & & \| & & \\
0 & \longrightarrow & \frac{\operatorname{Br}_1 U_{k'}}{\operatorname{Br} k'} & \longrightarrow & \frac{\operatorname{Br}^{\text{g.unr.}} U_{k'}}{\operatorname{Br} k'} & \longrightarrow & \frac{\operatorname{Br}^{\text{g.unr.}} U_{k'}}{\operatorname{Br}_1 U_{k'}} & \longrightarrow & 0.
\end{array}$$

To complete the proof of the theorem, it remains to prove that the rows and columns are exact and that the leftmost bottom vertical arrow is surjective. The exactness of the middle row follows from the above discussion and the exactness of the bottom row follows from the definitions. The middle column is exact by Theorem 2.1; Theorem 2.1 and Proposition 2.8 together imply that the leftmost column is exact.

Consider the map induced by $\beta \circ j$

$$\frac{\left(\operatorname{Pic} \overline{Y}/(\pi^* \operatorname{Pic} S + 2 \operatorname{Pic} \overline{Y})\right)^{G_{k'}}}{\operatorname{Pic} Y_{k'}/(\pi^* \operatorname{Pic} S + 2 \operatorname{Pic} Y_{k'})} \hookrightarrow \frac{\operatorname{Br}_1 U_{k'}}{\operatorname{Br} k'}[2];$$

we would like to show that it is surjective. Since $U = Y \setminus \bigcup_{E\in\mathcal{E}} \pi^{-1}(E)$ and the elements of $\mathcal{E}$ form an integral basis for $\operatorname{Pic} S = \operatorname{Pic} \overline{S}$, we have that $\operatorname{Pic} \overline{U} \cong \operatorname{Pic} \overline{Y}/\pi^* \operatorname{Pic} S$ and $\operatorname{Pic} U_{k'} \cong \operatorname{Pic} Y_{k'}/\pi^* \operatorname{Pic} S$. In particular,

$$\frac{\operatorname{Pic} \overline{Y}}{\pi^* \operatorname{Pic} S + 2 \operatorname{Pic} \overline{Y}} \xrightarrow{\sim} \frac{\operatorname{Pic} \overline{U}}{2 \operatorname{Pic} \overline{U}} \quad \text{and} \quad \frac{\operatorname{Pic} Y_{k'}}{\pi^* \operatorname{Pic} S + 2 \operatorname{Pic} Y_{k'}} \xrightarrow{\sim} \frac{\operatorname{Pic} U_{k'}}{2 \operatorname{Pic} U_{k'}}.$$

Furthermore, since $\operatorname{Pic} \overline{U}[2] = 0$, the cohomological long exact sequence associated to the multiplication by 2 map yields the isomorphism

$$\frac{(\operatorname{Pic} \overline{U}/2 \operatorname{Pic} \overline{U})^{G_{k'}}}{(\operatorname{Pic} \overline{U})^{G_{k'}}/(2 \operatorname{Pic} \overline{U})^{G_{k'}}} \cong \mathrm{H}^1(G_{k'}, \operatorname{Pic} \overline{U})[2].$$

In addition, since $\mathrm{H}^0(\overline{U}, \mathbb{G}_m) = \overline{k}^\times$, the long exact sequence of low degree terms from the Hochschild–Serre spectral sequence [13, Theorem 2.20] gives the isomorphism

$$\frac{\operatorname{Br}_1 U_{k'}}{\operatorname{Br} k'} \xrightarrow{\sim} \mathrm{H}^1(G_{k'}, \operatorname{Pic} \overline{U}).$$

Since all groups in question are finite, a cardinality argument completes the proof of surjectivity, and hence the proof of the theorem. □

3 Geometry of $X_{\mathbf{a}}$ and $Y_{\mathbf{a}}$

3.1 Review of [16, Sect. 4]

The K3 surface $Y_{\mathbf{a}}$ is defined as the base locus of a net of quadrics. As explained in [2, Example IX.4.5] and [16, Sect. 4.1], each isolated singular point in the degeneracy locus of the net gives rise to two genus 1 fibrations on $\overline{Y}_{\mathbf{a}}$. As the degeneracy locus of the net has 14 singular points, we obtain 28 classes of curves in $\operatorname{Pic} \overline{Y_{\mathbf{a}}}$, which we denote $F_1, G_1, \ldots, F_{14}, G_{14}$; for all i, j we have the relation $F_i + G_i = F_j + G_j$.

Nine of these singular points define fibrations which descend to $\overline{X}_{\mathbf{a}}$. On $\overline{X}_{\mathbf{a}}$, these fibrations have exactly two nonreduced fibers. For a fixed point P_i in the degeneracy locus, we let $C_i, \widetilde{C}_i, D_i$, and $\widetilde{D}_i$ denote the reduced subschemes of the nonreduced fibers of the corresponding fibrations. After possibly relabeling, we may assume that we have the following relations in $\operatorname{Pic} \overline{X}_{\mathbf{a}}$:

$$C_i + D_i = \widetilde{C}_j + \widetilde{D}_j, \quad 2(C_i - \widetilde{C}_i) = 2(D_i - \widetilde{D}_i) = 0,$$
$$f^* C_i = f^* \widetilde{C}_i = F_i, \quad f^* D_i = f^* \widetilde{D}_i = G_i.$$

Defining equations for curves representing each of these classes is given in Table A.4. The intersection numbers of these curves are as follows:

$$F_i^2 = G_i^2 = 0,\ \ F_i \cdot G_i = 4,\ \ F_i \cdot G_j = F_i \cdot F_j = G_i \cdot G_j = 2 \text{ for all } i \neq j,$$
$$C_i^2 = D_i^2 = 0,\ \ C_i \cdot D_i = 2,\ \ C_i \cdot D_j = C_i \cdot C_j = D_i \cdot D_j = 1 \text{ for all } i \neq j.$$

Proposition 3.1 ([16, Corollary 4.3]). *Let* $\mathbf{a} \in \mathbb{Z}^3_{>0}$ *satisfy conditions* (5), (6), *and* (8). *Then* $\operatorname{Pic} \overline{Y}_{\mathbf{a}} \cong \mathbb{Z}^{15}$ *and is generated by* $G_1, F_1, \ldots, F_{14}$ *and*

$$Z_1 := \frac{1}{2}\left(F_1 + F_2 + F_3 + F_{10} + F_{12}\right),$$

$$Z_2 := \frac{1}{2}\left(F_1 + G_1 + F_4 + F_5 + F_6 + F_{10} + F_{11}\right),$$

$$Z_3 := \frac{1}{2}\left(F_1 + F_4 + F_7 + F_{13} + F_{14}\right),$$

$$Z_4 := \frac{1}{2}\left(F_1 + G_1 + F_7 + F_8 + F_9 + F_{11} + F_{12}\right).$$

As in [16], we let $K/\mathbb{Q}$ denote the splitting field of the curves F_i, G_i. We will be concerned with two particular subfields $K_0 \subset K_1 \subset K$; we give generators for these fields in Appendix and describe their defining properties in Sects. 3.2 and 3.3.

3.2 Double Cover Morphisms

In order to apply the results of Sect. 2, we must realize $Y_{\mathbf{a}}$ as a double cover of a rational ruled surface. We will be able to do so over the Galois extension $K_0 := \mathbb{Q}(i, \sqrt{2}, \sqrt{5}, \sqrt{-2+2\sqrt{2}})$; throughout this section, we work over K_0.

Consider the morphism

$$\phi\colon Y_{\mathbf{a}} \to S^0 := \mathbb{P}^1_x \times \mathbb{P}^1_t,$$

which sends a point $[v_0 : v_1 : v_2 : w_0 : w_1 : w_2]$ to

$$\left(\left[w_0 - \sqrt{5}w_1 : v_0 + 2v_1\right], \left[\sqrt{-2+2\sqrt{2}}w_0 - \sqrt{5}w_1, v_0 + \sqrt{2}v_1 + \sqrt{5}(1-\sqrt{2})v_2\right]\right). \tag{13}$$

For any $P \in Y_{\mathbf{a}}$, we have $\phi(\sigma(P)) = [-1](\phi(P))$, where $[-1]$ denotes the automorphism of S^0 that sends (x, t) to $(-x, -t)$. Thus, we have an induced morphism

$$\tilde{\phi}\colon X_{\mathbf{a}} \to \left(S^0\right)/[-1],$$

obtained by quotienting $Y_{\mathbf{a}}$ by σ and S^0 by $[-1]$.

The morphism ϕ factors through a double cover morphism $\pi\colon Y_{\mathbf{a}} \to S$, where $S := Y_{\mathbf{a}}/(w_2 \mapsto -w_2)$. The quotient S is a smooth del Pezzo surface of degree 4 given by

$$\{v_0v_1 + 5v_2^2 - w_0^2 = v_0^2 + 3v_0v_1 + 2v_1^2 - w_0^2 + 5w_1^2 = 0\} \subset \mathbb{P}^4_{(v_0:v_1:v_2:w_0:w_1)}.$$

Using (13), one can check that ϕ induces a birational map $\nu_S: S \to S^0$ which contracts four (-1)-curves; we denote these curves by E_1, E_2, E_3, and E_4. (Defining equations for the curves are given in Table A.3.)

The preimages of the E_i under π are irreducible (-2)-curves in $Y_{\mathbf{a}}$. Thus, we may also factor ϕ by first blowing-down these four (-2)-curves to obtain a (singular) surface $Y_{\mathbf{a}}^0$ and then quotienting by an involution to obtain a double cover morphism. Hence, we have the following commutative diagram:

$$\begin{array}{ccc} Y_{\mathbf{a}} & \xrightarrow{\pi} & S \\ \downarrow \nu & \searrow^{\phi} & \downarrow \nu_S \\ Y_{\mathbf{a}}^0 & \xrightarrow{\pi^0} & S^0 \end{array}$$

where the vertical maps are birational and the horizontal maps are 2-to-1. Note that over $k = K_0$, these varieties and morphisms satisfy all assumptions from Sect. 2. In particular, all morphisms are defined over K_0 and $\operatorname{Pic} S_{K_0} = \operatorname{Pic} \overline{S}$.

Since σ induces an involution on $Y_{\mathbf{a}}^0$, S, and S^0 which is compatible with the morphisms, the above commutative diagram descends to the following commutative diagram involving the Enriques surface:

$$\begin{array}{ccc} X_{\mathbf{a}} & \xrightarrow{\tilde{\pi}} & \tilde{S} \\ \downarrow \tilde{\nu} & \searrow^{\tilde{\phi}} & \downarrow \tilde{\nu}_S \\ X_{\mathbf{a}}^0 & \xrightarrow{\tilde{\pi}^0} & (S^0)/[-1] \end{array}$$

3.3 Branch Loci of the Double Covers

Let $B, B^0, \widetilde{B}, \widetilde{B}^0$ denote the branch loci of $\pi, \pi^0, \tilde{\pi}$, and $\tilde{\pi}^0$, respectively.

Proposition 3.2.

(1) $B = V(w_2) \subset \mathbb{P}^4$ *is a smooth genus* 5 *curve,*
(2) $\widetilde{B}$ *is a smooth genus* 3 *curve,*
(3) B^0 *is an arithmetic genus* 9 *curve with four nodal singularities, and*
(4) $\widetilde{B}^0$ *is an arithmetic genus* 5 *curve with two nodal singularities.*

Furthermore, the projection morphism $\widetilde{B} \to V(av_0^2 + bv_1^2 + cv_2^2) \subset \mathbb{P}^2$ *is a double cover map, so* $\widetilde{B}$ *is geometrically hyperelliptic.*

Proof. From the definition of π, one can immediately see that B is the image of $V(w_2) \cap Y_{\mathbf{a}}$, so B is given by an intersection of three quadrics in $\mathbb{P}^4$.

Since condition (1) holds, B is smooth by the Jacobian criterion. Therefore, B is a smooth genus 5 curve.

As $B/\sigma = \widetilde{B}$ and $\sigma|_B$ has no fixed points, the Riemann–Hurwitz formula implies that $\widetilde{B}$ is a smooth genus 3 curve. Furthermore, the curve B has a 4-to-1 projection map to the plane conic $\{av_0^2 + bv_1^2 + cv_2^2 = 0\} \subset \mathbb{P}^2$. This induces a double cover map from $\widetilde{B}$ to the same conic, so $\widetilde{B}$ is (geometrically) hyperelliptic.

Using the equations given in Table A.3 and computer algebra software, one can check that each (-1)-curve E_i intersects B transversely in two distinct points. Thus, the curve B^0 has four nodal singularities, and so has arithmetic genus 9. As $B^0 \to \widetilde{B}^0$ is an étale double cover, B^0 has two nodal singularities and the Riemann–Hurwitz formula implies that $\widetilde{B}^0$ has arithmetic genus 5. □

Since $\widetilde{B}$ is geometrically hyperelliptic, we may identify $\mathrm{Jac}(\widetilde{B}_{\overline{\mathbb{Q}}})[2]$ with subsets of the Weierstrass points of $\widetilde{B}$ of even order, modulo identifying complementary subsets. Under this identification, the group operation is given by the symmetric difference, i.e., $A + B = (A \cup B) \setminus (A \cap B)$. One can easily see that the ramification locus of the projection morphism $B \to V(av_0^2 + bv_1^2 + cv_2^2) \subset \mathbb{P}^2$ is given by $V(w_0) \cup V(w_1) \subset B$. Since this projection morphism factors as

$$B \longrightarrow \widetilde{B} \longrightarrow V(av_0^2 + bv_1^2 + cv_2^2) \subset \mathbb{P}^2,$$

where the first morphism is étale, the Weierstrass points of $\widetilde{B}$ are $f(V(w_0)) \cup f(V(w_1))$. There are four points in $f(V(w_0)) \subset \widetilde{B}$, which we denote by P_1, P_2, P_3, and P_4, and four points in $f(V(w_1)) \subset \widetilde{B}$, which we denote by Q_1, Q_2, Q_3, and Q_4. We let

$$K_1/K_0 := \text{the splitting field of the Weierstrass points.}$$

Proposition 3.3.

(1) *The kernel of* $f^*\colon \mathrm{Jac}(\widetilde{B}_{\overline{\mathbb{Q}}})[2] \to \mathrm{Jac}(B_{\overline{\mathbb{Q}}})[2]$ *is isomorphic to* $\mathbb{Z}/2\mathbb{Z}$, *and the unique nontrivial element is* $\{P_1, P_2, P_3, P_4\}$.

(2) *The image of* f^* *fits in the following* non-split *exact sequence of* G_{K_0}*-modules:*

$$0 \to \mathrm{Ind}_{K_0(\theta_0)}^{K_0}(\mathbb{Z}/2\mathbb{Z}) \times \mathrm{Ind}_{K_0(\sqrt{ab})}^{K_0}(\mathbb{Z}/2\mathbb{Z}) \to \mathrm{im} f^* \to \mathbb{Z}/2\mathbb{Z} \to 0.$$

Proof. **(1)** Let P_i' and P_i'' be the inverse image of P_i under the map $B \to \widetilde{B}$. The vanishing locus $V(w_0)$ on B consists of P_i', P_i'' for $i = 1, \dots, 4$. For any distinct pair of numbers $i_0, i_1 \in \{1, 2, 3, 4\}$, there exists a linear form $L = L(v_0, v_1, v_2)$ such that $V(L) \subset \mathbb{P}^2$ contains the images of P_{i_0} and P_{i_1}. Since the points P_i are Weierstrass points of $\widetilde{B}$ and the map $B \to \widetilde{B}$ is étale, this implies that

$$\mathrm{div}_B(w_0/L) = \sum_{i=1}^{4}(P_i + P_i') - 2(P_{i_0}' + P_{i_0}'' + P_{i_1}' + P_{i_1}'').$$

Thus $f^*\{P_1, P_2, P_3, P_4\}$ is principal on B.

Now let D be a divisor such that $[D] \in \mathrm{Jac}\,\widetilde{B}[2]$ is a nontrivial element of $\ker f^*$. Then $f^{-1}(D) = \mathrm{div}_B(g_0)$, for some function $g_0 \in \mathbf{k}(B)$. On the other hand, $2D = \operatorname{div}_{\widetilde{B}}(g_1)$ for some function $g_1 \in \mathbf{k}(\widetilde{B})$. Thus, $g_1 = g_0^2$ in $\mathbf{k}(B)^\times/k^\times$. Since $\mathbf{k}(B)$ is a quadratic extension of $\mathbf{k}(\widetilde{B})$ generated by w_0/L for $L = L(v_0, v_1, v_2)$ a linear form, this implies that $g_0 = h \cdot (w_0/L)$ in $\mathbf{k}(B)$ for some $h \in \mathbf{k}(\widetilde{B})^{\times 2}$. (Note that $w_0/L \notin \mathbf{k}(\widetilde{B})$.) Hence $D = \{P_1, P_2, P_3, P_4\}$ in $\operatorname{Jac}\widetilde{B}$.

(2) By (1) and the description of $\operatorname{Jac}\widetilde{B}[2]$ in terms of Weierstrass points, we see that the image of f^* is isomorphic to $(\mathbb{Z}/2\mathbb{Z})^5$ (as a group) and is generated by

$$\{P_1, P_3\},\ \{P_1, P_4\},\ \{Q_1, Q_3\},\ \{Q_1, Q_4\},\ \text{and}\ \{P_1, Q_1\}.$$

Using Table A.5, one can check that (as G_{K_0}-modules) we have

$$\langle \{P_1, P_3\},\ \{P_1, P_4\} \rangle \cong \operatorname{Ind}_{K_0(\sqrt{ab})}^{K_0}(\mathbb{Z}/2\mathbb{Z}), \quad \text{and}$$
$$\langle \{Q_1, Q_3\},\ \{Q_1, Q_4\} \rangle \cong \operatorname{Ind}_{K_0(\theta_0)}^{K_0}(\mathbb{Z}/2\mathbb{Z}),$$

thus proving the existence of the exact sequence. Since $\operatorname{Gal}(K_1/K_0)$ acts transitively on $\{P_1, P_2, P_3, P_4\}$, no element of the form $\{P_i, Q_j\}$ is fixed by G_{K_0} and so the exact sequence does not split. □

3.4 *The Quotient Group* $\operatorname{Pic}\overline{Y}_{\mathbf{a}}/\left(\pi^* \operatorname{Pic} S + 2 \operatorname{Pic}\overline{Y}_{\mathbf{a}}\right)$

The following lemma about the structure of $\operatorname{Pic}\overline{Y}_{\mathbf{a}}/\left(\pi^* \operatorname{Pic} S + 2 \operatorname{Pic}\overline{Y}_{\mathbf{a}}\right)$ as a Galois module will be useful in later sections:

Lemma 3.4. *We have an isomorphism of* $\operatorname{Gal}(K_1/K_0)$*-modules*

$$\left(\frac{\operatorname{Pic}\overline{Y}_{\mathbf{a}}}{\pi^* \operatorname{Pic} S + 2 \operatorname{Pic}\overline{Y}_{\mathbf{a}}}\right)^{G_{K_1}} \xrightarrow{\sim} \mathbb{Z}/2\mathbb{Z} \times \operatorname{Ind}_{K_0(\theta_0)}^{K_0}(\mathbb{Z}/2\mathbb{Z}) \times \operatorname{Ind}_{K_0(\sqrt{ab})}^{K_0}(\mathbb{Z}/2\mathbb{Z}).$$

Proof. By computing intersection numbers, we find that $\pi^* \operatorname{Pic} S$ is generated by

$$G_1,\ F_1 - F_2,\ Z_1 - F_2 - F_{10} - F_{12},\ Z_1 - F_1 - F_2,\ -Z_1 + F_1 + F_{12},\ \text{and}\ -Z_1 + F_1 + F_{10}.$$

Therefore, $\operatorname{Pic}\overline{Y}_{\mathbf{a}}/\left(\pi^* \operatorname{Pic} S + 2 \operatorname{Pic}\overline{Y}_{\mathbf{a}}\right)$ is a 9-dimensional $\mathbb{F}_2$ vector space with basis

$$F_5,\ F_6,\ F_8,\ F_9,\ F_{11},\ F_{13},\ Z_2,\ Z_3,\ \text{and}\ Z_4.$$

In this quotient we have the relations $G_i = F_i$, $F_1 = F_2 = F_{10} = F_{12} = Z_1 = G_1 = F_3 = 0$,

$$F_4 = F_5+F_6+F_{11}, \quad F_7 = F_8+F_9+F_{11}, \quad \text{and} \quad F_{14} = F_5+F_6+F_8+F_9+F_{13}.$$

Using this basis and Table A.5, we compute that $(\operatorname{Pic} \overline{Y}_{\mathbf{a}}/\left(\pi^* \operatorname{Pic} S + 2 \operatorname{Pic} \overline{Y}_{\mathbf{a}}\right))^{G_{K_1}}$ is a 5-dimensional $\mathbb{F}_2$-vector space with basis F_5, F_6, F_8, F_9, and F_{11}. The isomorphism then follows after noting that in the quotient F_{11} is fixed by $\operatorname{Gal}(K_1/K_0)$, F_5 and F_6 are interchanged by $\operatorname{Gal}(K_0(\sqrt{ab})/K_0)$ and fixed by all other elements, and similarly for F_8 and F_9 and $\operatorname{Gal}(K_0(\theta_0)/K_0)$. □

4 A Representative of the Nontrivial Brauer Class on $\overline{X}_{\mathbf{a}}$

As mentioned in the introduction, $\operatorname{Br} \overline{X}_{\mathbf{a}} \cong \mathbb{Z}/2\mathbb{Z}$. Therefore, there is at most one nontrivial class in $\operatorname{Br} X_{\mathbf{a}}/\operatorname{Br}_1 X_{\mathbf{a}}$. To determine the existence of such a class, we must first obtain an explicit representative for the unique Brauer class in $\operatorname{Br} \overline{X}_{\mathbf{a}}$. Using Beauville's criterion [3, Corollary 5.7], Várilly-Alvarado and the last author showed that if $\mathbf{a}$ satisfies conditions (5), (6) and (8), then

$$\ker\left(f^*\colon \operatorname{Br} \overline{X}_{\mathbf{a}} \to \operatorname{Br} \overline{Y}_{\mathbf{a}}\right) = 0. \quad [16, \text{Proof of Prop. 5.2}]$$

Let D be a divisor on $\widetilde{B}_{\overline{\mathbb{Q}}}$ such that $[D] \in \operatorname{Jac}(\widetilde{B})[2]$ and such that $[D]$ corresponds to a subset of the Weierstrass points containing an *odd* number of the points $\{P_i\}_{i=1,\dots,4}$. Let $\tilde{\ell} \in \mathbf{k}(\widetilde{B}_{\overline{\mathbb{Q}}})$ be such that $\operatorname{div}(\tilde{\ell}) = 2D$.

Proposition 4.1. *Assume that* $\mathbf{a}$ *satisfies conditions* (5), (6), *and* (8). *Then the Brauer class*

$$\pi^* \mathcal{A}_{\tilde{\ell}} = \pi^* \operatorname{Cor}_{\mathbf{k}(\overline{B})(x)/\overline{\mathbb{Q}}(t,x)}\left((\tilde{\ell}, x-\alpha)_{\overline{\mathbb{Q}}}\right)$$

defines an element of $\operatorname{Br} \overline{Y}_{\mathbf{a}}$ *and generates the order* 2 *subgroup* $f^* \operatorname{Br} \overline{X}_{\mathbf{a}}$.

Proof. By [4, Theorem 7.2], the subgroup $f^* \operatorname{Br} X_{\mathbf{a}}[2] \subset \operatorname{Br} \overline{Y}_{\mathbf{a}}$ is generated by $\mathcal{A}_{\tilde{\ell}'}$, where $\tilde{\ell}' \in \mathbf{k}(\widetilde{B}_{\overline{\mathbb{Q}}})$ is such that $\nu_*(\operatorname{div}(\tilde{\ell}')) \in 2\operatorname{Div}(\widetilde{B}^0)$. Furthermore, by Theorem 2.1 applied to $k' = \overline{k}$ we have the following exact sequence:

$$0 \longrightarrow \frac{\operatorname{Pic} \overline{Y}_{\mathbf{a}}}{\pi^* \operatorname{Pic} S + 2 \operatorname{Pic} \overline{Y}} \xrightarrow{\ j\ } \frac{\mathbf{k}(B_{\overline{\mathbb{Q}}})^{\times}_{\mathcal{E}}}{\mathbf{k}(B_{\overline{\mathbb{Q}}})^{\times 2}} \longrightarrow \operatorname{Br} \overline{Y}_{\mathbf{a}}[2] \longrightarrow 0.$$

Thus, to prove the proposition, it remains to prove that $[\tilde{\ell}]$ is not in the image of j.

For convenience, we set

$$\tilde{G} := \{\ell \in \mathbf{k}(\widetilde{B}_{\overline{\mathbb{Q}}})^{\times} : \nu_* \operatorname{div}(\ell) \in 2\operatorname{Div}(\widetilde{B}^0_{\overline{\mathbb{Q}}})\}.$$

We have the following commutative diagram with exact rows [4, Proposition 4.5], where $\operatorname{Sing}(B_{\overline{\mathbb{Q}}})$ and $\operatorname{Sing}(\tilde{B}_{\overline{\mathbb{Q}}})$ denote the set of singular points of $B_{\overline{\mathbb{Q}}}$ and $\widetilde{B}_{\overline{\mathbb{Q}}}$, respectively, and the last vertical map is the diagonal embedding on the first factor, i.e., $(m,n) \mapsto (m,m,n,n,0,0)$.

$$\begin{array}{ccccccccc} 0 \longrightarrow & \operatorname{Jac} \widetilde{B}_{\overline{\mathbb{Q}}}[2] & \longrightarrow & \dfrac{\tilde{G}}{\mathbf{k}(\widetilde{B}_{\overline{\mathbb{Q}}})^{\times 2}} & \longrightarrow & (\mathbb{Z}/2\mathbb{Z})^{\operatorname{Sing}(\widetilde{B}_{\overline{\mathbb{Q}}})} & & \longrightarrow 0 \\ & \downarrow f^* & & \downarrow f^* & & \downarrow & & \\ 0 \longrightarrow & \operatorname{Jac} B_{\overline{\mathbb{Q}}}[2] & \longrightarrow & \dfrac{\mathbf{k}(B_{\overline{\mathbb{Q}}})_{\mathcal{E}}}{\mathbf{k}(B_{\overline{\mathbb{Q}}})^{\times 2}} & \longrightarrow & (\mathbb{Z}/2\mathbb{Z})^{\operatorname{Sing}(B_{\overline{\mathbb{Q}}})} \times (\mathbb{Z}/2\mathbb{Z})^2 & & \longrightarrow 0. \end{array}$$

In the rest of this section, we will view elements of $\operatorname{Jac} \widetilde{B}_{\overline{\mathbb{Q}}}[2]$ and $\operatorname{Jac} B_{\overline{\mathbb{Q}}}[2]$ as elements of $\tilde{G}/\mathbf{k}(\widetilde{B}_{\overline{\mathbb{Q}}})^{\times 2}$ and $\mathbf{k}(B_{\overline{\mathbb{Q}}})_{\mathcal{E}}/\mathbf{k}(B_{\overline{\mathbb{Q}}})^{\times 2}$, respectively.

Lemma 4.2. *Assume that* $\mathbf{a}$ *satisfies conditions* (5), (6), *and* (8). *Then the group* $\operatorname{im} j \cap f^*(\operatorname{Jac} \widetilde{B}_{\overline{\mathbb{Q}}}[2])$ *is an index* 2 $\operatorname{Gal}(\overline{\mathbb{Q}}/K_0)$*-invariant subgroup of* $f^*(\operatorname{Jac} \widetilde{B}_{\overline{\mathbb{Q}}}[2])$.

Proof. Since $\operatorname{Br} \overline{X}_{\mathbf{a}} \cong \mathbb{Z}/2\mathbb{Z}$, the subgroup $\operatorname{im} j \cap f^*(\tilde{G})$ has index 2 in $f^*(\tilde{G})$. Therefore, the subgroup $\operatorname{im} j \cap f^*(\operatorname{Jac} \widetilde{B}_{\overline{\mathbb{Q}}}[2])$ has index at most 2 in $f^*(\operatorname{Jac} \widetilde{B}_{\overline{\mathbb{Q}}}[2])$.

Recall that K_1 is splitting field of the Weierstrass points over K_0. Thus $\operatorname{Gal}(\overline{\mathbb{Q}}/K_1)$ fixes $f^*(\operatorname{Jac} \widetilde{B}_{\overline{\mathbb{Q}}})$ and hence $\operatorname{im} j \cap f^*(\operatorname{Jac} \widetilde{B}_{\overline{\mathbb{Q}}}[2]) = (\operatorname{im} j)^{G_{K_1}} \cap f^*(\operatorname{Jac} \widetilde{B}_{\overline{\mathbb{Q}}}[2])$. Since j and f^* are G_{K_0}-equivariant homomorphisms, the intersection is G_{K_0}-invariant and the elements in the intersection must have compatible G_{K_0} action. Then Proposition 3.3 and Lemma 3.4 together imply that the intersection is a submodule of $\operatorname{Ind}_{K_0(\theta_0)}^{K_0}(\mathbb{Z}/2\mathbb{Z}) \times \operatorname{Ind}_{K_0(\sqrt{ab})}^{K_0}(\mathbb{Z}/2\mathbb{Z})$ and hence a proper subgroup of $f^*(\operatorname{Jac} \widetilde{B}[2])$ with index equal to 2. □

Now we resume the proof of Proposition 4.1. Assume that $\tilde{\ell}$ is contained in the image of j and hence in $\operatorname{im} j \cap f^*(\operatorname{Jac} \widetilde{B}[2])$. From Tables A.1 and A.5, we see that $\operatorname{Gal}(\overline{\mathbb{Q}}/K_0)$ acts transitively on the Weierstrass points of $\widetilde{B}$. Therefore, the subgroup of $\mathbf{k}(B_{\overline{\mathbb{Q}}})_{\mathcal{E}}$ generated by $\tilde{\ell}$ and all of its $\operatorname{Gal}(\overline{\mathbb{Q}}/K_0)$ conjugates contains all of $f^*(\operatorname{Jac} \widetilde{B}_{\overline{\mathbb{Q}}}[2])$. Since $\operatorname{im} j \cap f^*(\operatorname{Jac} \widetilde{B}[2])$ is $\operatorname{Gal}(\overline{\mathbb{Q}}/K_0)$-invariant, this implies that

$$\operatorname{im} j \cap f^*(\operatorname{Jac} \widetilde{B}_{\overline{\mathbb{Q}}}[2]) = f^*(\operatorname{Jac} \widetilde{B}_{\overline{\mathbb{Q}}}[2])$$

which contradicts Lemma 4.2. □

5 Proof of Theorem 1.2

Assume that $\mathbf{a} \in \mathbb{Z}^3_{>0}$ satisfies conditions (5), (6), and (8). The last statement of the theorem follows immediately from the first statement together with [16, Theorem 1.2]. Thus our goal is to prove that $\operatorname{Br} X_{\mathbf{a}} = \operatorname{Br}_1 X_{\mathbf{a}}$. Since $\operatorname{Br} X_{\mathbf{a}} \to \operatorname{Br} \overline{X}_{\mathbf{a}}$ factors through $(\operatorname{Br} X_{\mathbf{a},K_1})^{\operatorname{Gal}(K_1/K_0)}$, it suffices to prove that $(\operatorname{Br} X_{\mathbf{a},K_1})^{\operatorname{Gal}(K_1/K_0)} = (\operatorname{Br}_1 X_{\mathbf{a},K_1})^{\operatorname{Gal}(K_1/K_0)}$.

Recall from Sect. 4 that $f^* \operatorname{Br} \overline{X}_{\mathbf{a}}$ is a nontrivial subgroup of $\operatorname{Br} \overline{Y}_{\mathbf{a}}$. So if $(\operatorname{Br} X_{\mathbf{a},K_1})^{\operatorname{Gal}(K_1/K_0)}$ is strictly larger than $(\operatorname{Br}_1 X_{\mathbf{a},K_1})^{\operatorname{Gal}(K_1/K_0)}$, then there exists an element $\mathcal{B} \in (\operatorname{Br} Y_{\mathbf{a},K_1})^{\operatorname{Gal}(K_1/K_0)}$ such that $\mathcal{B}_{\overline{\mathbb{Q}}}$ is the unique nontrivial element in $f^* \operatorname{Br} \overline{X}_{\mathbf{a}}$. Let $\tilde{\ell} \in \mathbf{k}(\widetilde{B_{\overline{\mathbb{Q}}}})$ be such that $\operatorname{div}(\tilde{\ell}) = 2D$ where $[D] \in \operatorname{Jac}(\widetilde{B})[2]$ corresponds to a subset of the Weierstrass points containing an odd number of the points $\{P_i\}_{i=1}^4$. Note that we may choose $\tilde{\ell}$ so that it is contained in $\mathbf{k}(\widetilde{B}_{K_1})^\times$. In what follows, we will view $\tilde{\ell}$ as a function on B under the natural inclusion $\mathbf{k}(\widetilde{B}) \subset \mathbf{k}(B)$. We remark that, under this inclusion, $\tilde{\ell} \in \mathbf{k}(B_{K_1})^\times_{\mathcal{E}}$.

By Proposition 4.1, $f^* \operatorname{Br} \overline{X}_{\mathbf{a}}$ is generated by $(\pi^* \mathcal{A}_{\tilde{\ell}})_{\overline{\mathbb{Q}}}$, so $\mathcal{B}_{\overline{\mathbb{Q}}} = (\pi^* \mathcal{A}_{\tilde{\ell}})_{\overline{\mathbb{Q}}}$. Let $\mathcal{A}' := \mathcal{A}'(\tilde{\ell})$ be as in Proposition 2.6. Then an application of Tsen's theorem shows $(\pi^*(\varpi^* \mathcal{A}'))_{\overline{\mathbb{Q}}} = 0$. Hence, $\beta(\tilde{\ell})_{\overline{\mathbb{Q}}} = (\pi^* \mathcal{A}_{\tilde{\ell}})_{\overline{\mathbb{Q}}} = \mathcal{B}_{\overline{\mathbb{Q}}}$. Also, since $\beta(\tilde{\ell}) \in \operatorname{Br} U_{K_1}$, we must have that

$$\mathcal{B} - \beta(\tilde{\ell}) \in \operatorname{Br}_1 U_{K_1}.$$

By Theorem 2.2, $\operatorname{Br}_1 U_{K_1} / \operatorname{Br} K_1$ is contained in $\beta(\mathbf{k}(B_{K_1})^\times_{\mathcal{E}})$, meaning that $\mathcal{B} - \beta(\tilde{\ell}) = \beta(\ell')$ for some $\ell' \in \mathbf{k}(B_{K_1})^\times_{\mathcal{E}}$. Since both ℓ' and $\tilde{\ell}$ are in $\mathbf{k}(B_{K_1})^\times_{\mathcal{E}}$, we then have $\mathcal{B} = \beta(\tilde{\ell}\ell') =: \beta(\ell_{\mathcal{B}})$ for some $\ell_{\mathcal{B}} \in \mathbf{k}(B_{K_1})_{\mathcal{E}}$. Furthermore, since $\mathcal{B}$ is $\operatorname{Gal}(K_1/K_0)$-invariant and is equal to $\beta(\tilde{\ell})$ modulo $\operatorname{Br}_1 U_{K_1}$, Theorem 2.2 implies that

(a) the class of $\ell_{\mathcal{B}}$ in $\mathbf{k}(B_{K_1})/j(\operatorname{Pic} Y_{\mathbf{a},K_1}) K_1^\times \mathbf{k}(B_{K_1})^{\times 2}$ must be $\operatorname{Gal}(K_1/K_0)$-invariant, and
(b) $\ell_{\mathcal{B}} \tilde{\ell}^{-1} \in j\left((\operatorname{Pic} \overline{Y}_{\mathbf{a}}/(\pi^* \operatorname{Pic} S + 2 \operatorname{Pic} \overline{Y}_{\mathbf{a}}))^{G_{K_1}}\right) K_1^\times \mathbf{k}(B_{K_1})^{\times 2}$.

An inspection of Table A.5 reveals that $\operatorname{Pic} Y_{\mathbf{a},K_1} = \operatorname{Pic} S_{K_1}$, so $j(\operatorname{Pic} Y_{\mathbf{a},K_1}) \subset K_1^\times \mathbf{k}(B_{K_1})^{\times 2}$. In addition, Lemma 3.4 shows that every element of $j\left((\operatorname{Pic} \overline{Y}_{\mathbf{a}}/(\pi^* \operatorname{Pic} S + 2 \operatorname{Pic} \overline{Y}_{\mathbf{a}}))^{G_{K_1}}\right)$ is $\operatorname{Gal}(K_0(\theta_0, \sqrt{ab})/K_0)$-invariant. Thus, conditions (*a*) and (*b*) imply that

$$\tilde{\ell} \in \left(\frac{\mathbf{k}(B_{K_1})^\times}{K_1^\times \mathbf{k}(B_{K_1})^{\times 2}}\right)^{\operatorname{Gal}(K_0(\theta_0, \sqrt{ab})/K_0)}.$$

Given our assumption on $\tilde{\ell}$, this results in a contradiction, as demonstrated in Table A.5. □

Appendix: Fields, Defining Equations, and Galois Actions

The splitting field K of the genus 1 curves $C_i, \widetilde{C}_i, \widetilde{D}_i, F_i, G_i$ is generated by

$$i, \sqrt{2}, \sqrt{5}, \sqrt{a}, \sqrt{c}, \eta_0 := \sqrt{c^2 - 100ab}, \gamma_0 := \sqrt{-c^2 - 5bc - 10ac - 25ab},$$

$$\sqrt[4]{ab}, \sqrt{-2 + 2\sqrt{2}}, \sqrt{-c - 10\sqrt{ab}},$$

$$\theta_0 := \sqrt{4a^2+b^2},\ \xi_0 := \sqrt{a+b+c/5},\ \xi_0' := \sqrt{a+b/4+c/10},$$
$$\theta_1^+ := \sqrt{20a^2-10ab-2bc+(10a+2c)\theta_0},\ \theta_2^+ := \sqrt{-5a-5/2b-5/2\theta_0},$$
$$\xi_1^+ := \sqrt{20a+10b+3c+20\xi_0\xi_0'},\ \xi_2^+ := \sqrt{4a+2b+2/5c+4\xi_0\xi_0'}.$$

Define

$$\eta_1^+ := \frac{c-\eta_0+10\sqrt{ab}}{10\sqrt{a}\sqrt{-c-10\sqrt{ab}}}\quad \gamma_1^+ := (\theta_1^+)^{-1}\left(10a^2-5ab-bc+2a\gamma_0+(c+5a)\theta_0\right)$$
$$\eta_1^- := \frac{c+\eta_0+10\sqrt{ab}}{10\sqrt{a}\sqrt{-c-10\sqrt{ab}}}\quad \gamma_1^- := (\theta_1^+)^{-1}\left(10a^2-5ab-bc-2a\gamma_0+(c+5a)\theta_0\right).$$

The following subfields of K are of particular interest:

$$\mathbb{Q} \subset K_0 := \mathbb{Q}\left(i, \sqrt{2}, \sqrt{5}, \sqrt{-2+2\sqrt{2}}\right) \subset K_1 := K_0\left(\theta_0, \sqrt{ab}, \eta_1^+, \gamma_1^+\right) \subset K.$$

The field extensions $K/\mathbb{Q}$ and $K_1/\mathbb{Q}$ are Galois, as the following relations show:

$$\sqrt{-2-2\sqrt{2}} = \frac{2i}{\sqrt{-2+2\sqrt{2}}},$$
$$\sqrt{-c+10\sqrt{ab}} = \frac{\eta_0}{\sqrt{-c-10\sqrt{ab}}},$$
$$\theta_1^- := \sqrt{20a^2-10ab-2bc+(10a+2c)\theta_0} = \frac{4a\gamma_0}{\theta_1^+},$$
$$\theta_2^- := \sqrt{-5a-5/2b+5/2\theta_0} = \frac{5\sqrt{ab}}{\theta_2^+},$$
$$\xi_1^- := \sqrt{20a+10b+3c-20\xi_0\xi_0'} = \frac{\eta_0}{\xi_1^+},$$
$$\xi_2^- := \sqrt{4a+2b+2/5c-4\xi_0\xi_0'} = \frac{2\gamma_0}{5\xi_2^+},$$
$$(\gamma_1^+)^2 = 10a^2-5ab-bc+2a\gamma_0,\quad (\eta_1^+)^2 = \frac{-c+\eta_0}{50a},$$
$$(\gamma_1^-)^2 = 10a^2-5ab-bc-2a\gamma_0,\quad (\eta_1^-)^2 = \frac{-c-\eta_0}{50a},$$
$$\gamma_1^+\gamma_1^- = (5a+c)\theta_0,\quad \eta_1^+\eta_1^- = \frac{-1}{5a}\sqrt{ab}.$$

Tables A.1 and A.4 show that these fields have the properties claimed in Sect. 3.

Table A.1 Defining equations for the Weierstrass points on $\widetilde{B}$

$\widetilde{B}$	Defining equations	$\widetilde{B}$	Defining equations
P_1	$10av_0 - (c - \eta_0)v_1,\ v_2 - \eta_1^+ v_1$	Q_1	$(c+5a)v_0 + (c - \gamma_0)v_1,\ (c+5a)v_2 - \gamma_1^+ v_1$
P_2	$10av_0 - (c - \eta_0)v_1,\ v_2 + \eta_1^+ v_1$	Q_2	$(c+5a)v_0 + (c - \gamma_0)v_1,\ (c+5a)v_2 + \gamma_1^+ v_1$
P_3	$10av_0 - (c + \eta_0)v_1,\ v_2 - \eta_1^- v_1$	Q_3	$(c+5a)v_0 + (c + \gamma_0)v_1,\ (c+5a)v_2 - \gamma_1^- v_1$
P_4	$10av_0 - (c + \eta_0)v_1,\ v_2 + \eta_1^- v_1$	Q_4	$(c+5a)v_0 + (c + \gamma_0)v_1,\ (c+5a)v_2 + \gamma_1^- v_1$

Table A.2 Defining equation for the exceptional curves on S

$\nu_S(E_1)$	$\left[1-\sqrt{2}-i\sqrt{-2+2\sqrt{2}} : 1\right],\quad \left[(i-1+i\sqrt{2})\sqrt{-2+2\sqrt{2}} : 2\sqrt{2}\right]$
$\nu_S(E_2)$	$\left[-1+\sqrt{2}+i\sqrt{-2+2\sqrt{2}} : 1\right],\quad \left[(1-i-i\sqrt{2})\sqrt{-2+2\sqrt{2}} : 2\sqrt{2}\right]$
$\nu_S(E_3)$	$\left[1-\sqrt{2}+i\sqrt{-2+2\sqrt{2}} : 1\right],\quad \left[-(i+1+i\sqrt{2})\sqrt{-2+2\sqrt{2}} : 2\sqrt{2}\right]$
$\nu_S(E_4)$	$\left[-1+\sqrt{2}-i\sqrt{-2+2\sqrt{2}} : 1\right],\quad \left[(i+1+i\sqrt{2})\sqrt{-2+2\sqrt{2}} : 2\sqrt{2}\right]$

Table A.3 Pullbacks of exceptional curves in terms of F_i, G_i

$2\pi^*E_1$	$F_1 + 2G_1 - F_2 + F_3 - F_{10} - F_{12}$
$2\pi^*E_2$	$-F_1 - F_2 + F_3 + F_{10} + F_{12}$
$2\pi^*E_3$	$F_1 + 2G_1 - F_2 - F_3 - F_{10} + F_{12}$
$2\pi^*E_4$	$F_1 + 2G_1 - F_2 - F_3 + F_{10} - F_{12}$

Remark 1. Tables A.1 and A.4 list defining equations of a subvariety of the $Y_{\mathbf{a}}$. The image of this subvariety gives the corresponding object in $X_{\mathbf{a}}$ (Table A.5).

Acknowledgements This project began at the Women in Numbers 3 conference at the Banff International Research Station. We thank BIRS for providing excellent working conditions and the organizers of WIN3, Ling Long, Rachel Pries, and Katherine Stange, for their support. We also thank Anthony Várilly-Alvarado for allowing us to reproduce some of the tables from [16] in this paper for the convenience of the reader. F.B. supported by EPSRC scholarship EP/L505031/1. M.M. partially supported by NSF grant DMS-1102858. J.P. partially supported by NSERC PDF grant. B.V. partially supported by NSF grant DMS-1002933.

Table A.4 Defining equations of a curve representing a divisor class

$\overline{Y}_\mathbf{a}$	$\overline{X}_\mathbf{a}$	Defining equations	$\overline{Y}_\mathbf{a}$	$\overline{X}_\mathbf{a}$	Defining equations
F_1	C_1	$w_0 - \sqrt{5}w_1,\ v_0 + 2v_1$	F_7	C_7	$\sqrt{c}w_1 - w_2,\ (5a+c)v_0 + (c+\gamma_0)v_1$
	$\widetilde{C}_1$	$w_0 + \sqrt{5}w_1,\ v_0 + v_1$		$\widetilde{C}_7$	$\sqrt{c}w_1 + w_2,\ (5a+c)v_0 + (c-\gamma_0)v_1$
G_1	D_1	$w_0 - \sqrt{5}w_1,\ v_0 + v_1$	G_7	D_7	$\sqrt{c}w_1 - w_2,\ (5a+c)v_0 + (c-\gamma_0)v_1$
	$\widetilde{D}_1$	$w_0 + \sqrt{5}w_1,\ v_0 + 2v_1$		$\widetilde{D}_7$	$\sqrt{c}w_1 + w_2,\ (5a+c)v_0 + (c+\gamma_0)v_1$
F_2	C_2	$\sqrt{-2+2\sqrt{2}}w_0 - \sqrt{5}w_1,$ $v_0 + \sqrt{2}v_1 + \sqrt{5}(1-\sqrt{2})v_2$	F_8	C_8	$\theta_2^+ w_1 - w_2,$ $v_0 + (2a+b+\theta_0)/(b+\theta_0)v_1 - \theta_1^+/(2a)v_2$
	$\widetilde{C}_2$	$\sqrt{-2+2\sqrt{2}}w_0 + \sqrt{5}w_1,$ $v_0 + \sqrt{2}v_1 - \sqrt{5}(1-\sqrt{2})v_2$		$\widetilde{C}_8$	$\theta_2^+ w_1 + w_2,$ $v_0 + (2a+b+\theta_0)/(b+\theta_0)v_1 + \theta_1^+/(2a)v_2$
G_2	D_2	$\sqrt{-2+2\sqrt{2}}w_0 - \sqrt{5}w_1,$ $v_0 + \sqrt{2}v_1 - \sqrt{5}(1-\sqrt{2})v_2$	G_8	D_8	$\theta_2^+ w_1 - w_2,$ $v_0 + (2a+b+\theta_0)/(b+\theta_0)v_1 + \theta_1^+/(2a)v_2$
	$\widetilde{D}_2$	$\sqrt{-2+2\sqrt{2}}w_0 + \sqrt{5}w_1,$ $v_0 + \sqrt{2}v_1 + \sqrt{5}(1-\sqrt{2})v_2$		$\widetilde{D}_8$	$\theta_2^+ w_1 + w_2,$ $v_0 + (2a+b+\theta_0)/(b+\theta_0)v_1 - \theta_1^+/(2a)v_2$
F_3	C_3	$\sqrt{-2-2\sqrt{2}}w_0 - \sqrt{5}w_1,$ $v_0 - \sqrt{2}v_1 + \sqrt{5}(1+\sqrt{2})v_2$	F_9	C_9	$\theta_2^- w_1 - w_2,$ $v_0 + (2a+b-\theta_0)/(b-\theta_0)v_1 - \theta_1^-/(2a)v_2$
	$\widetilde{C}_3$	$\sqrt{-2-2\sqrt{2}}w_0 + \sqrt{5}w_1,$ $v_0 - \sqrt{2}v_1 - \sqrt{5}(1+\sqrt{2})v_2$		$\widetilde{C}_9$	$\theta_2^- w_1 + w_2,$ $v_0 + (2a+b-\theta_0)/(b-\theta_0)v_1 + \theta_1^-/(2a)v_2$
G_3	D_3	$\sqrt{-2-2\sqrt{2}}w_0 - \sqrt{5}w_1,$ $v_0 - \sqrt{2}v_1 - \sqrt{5}(1+\sqrt{2})v_2$	G_9	D_9	$\theta_2^- w_1 - w_2,$ $v_0 + (2a+b-\theta_0)/(b-\theta_0)v_1 + \theta_1^-/(2a)v_2$
	$\widetilde{D}_3$	$\sqrt{-2-2\sqrt{2}}w_0 + \sqrt{5}w_1,$ $v_0 - \sqrt{2}v_1 + \sqrt{5}(1+\sqrt{2})v_2$		$\widetilde{D}_9$	$\theta_2^- w_1 + w_2,$ $v_0 + (2a+b-\theta_0)/(b-\theta_0)v_1 - \theta_1^-/(2a)v_2$

(continued)

Table A.4 (continued)

$\overline{Y}_{\mathbf{a}}$	$\overline{X}_{\mathbf{a}}$	Defining equations	$\overline{Y}_{\mathbf{a}}$	$\overline{X}_{\mathbf{a}}$	Defining equations
F_4	C_4	$\sqrt{c}w_0 - \sqrt{5}w_2$, $10av_0 - (c + \sqrt{c^2 - 100ab})v_1$	F_{10}		$w_0 - \sqrt{5}v_2, v_0$
	$\widetilde{C}_4$	$\sqrt{c}w_0 + \sqrt{5}w_2$, $10av_0 - (c - \sqrt{c^2 - 100ab})v_1$	G_{10}		$w_0 - \sqrt{5}v_2, v_1$
G_4	D_4	$\sqrt{c}w_0 - \sqrt{5}w_2$, $10av_0 - (c - \sqrt{c^2 - 100ab})v_1$	F_{11}		$w_2 - \sqrt{c}v_2, \sqrt{a}v_0 + i\sqrt{b}v_1$
	$\widetilde{D}_4$	$\sqrt{c}w_0 + \sqrt{5}w_2$, $10av_0 - (c + \sqrt{c^2 - 100ab})v_1$	G_{11}		$w_2 - \sqrt{c}v_2, \sqrt{a}v_0 - i\sqrt{b}v_1$
F_5	C_5	$i\sqrt{2}\sqrt[4]{ab}w_0 - w_2$, $\sqrt{a}v_0 + \sqrt{b}v_1 + \sqrt{-c - 10\sqrt{ab}}v_2$	F_{12}		$w_1 - v_2, v_0 + (1 - i)v_1$
	$\widetilde{C}_5$	$i\sqrt{2}\sqrt[4]{ab}w_0 + w_2$, $\sqrt{a}v_0 + \sqrt{b}v_1 - \sqrt{-c - 10\sqrt{ab}}v_2$	G_{12}		$w_1 - v_2, v_0 + (1 + i)v_1$
G_5	D_5	$i\sqrt{2}\sqrt[4]{ab}w_0 - w_2$, $\sqrt{a}v_0 + \sqrt{b}v_1 - \sqrt{-c - 10\sqrt{ab}}v_2$	F_{13}		$\xi_2^+ w_0 - \xi_1^+ w_1$, $(\xi_0 + 2\xi_0')v_0 + (2\xi_0 + 2\xi_0')v_1 - w_2$
	$\widetilde{D}_5$	$i\sqrt{2}\sqrt[4]{ab}w_0 + w_2$, $\sqrt{a}v_0 + \sqrt{b}v_1 + \sqrt{-c - 10\sqrt{ab}}v_2$	G_{13}		$\xi_2^+ w_0 - \xi_1^+ w_1$, $(\xi_0 + 2\xi_0')v_0 + (2\xi_0 + 2\xi_0')v_1 + w_2$
F_6	C_6	$\sqrt{2}\sqrt[4]{ab}w_0 + w_2$, $\sqrt{a}v_0 - \sqrt{b}v_1 + \sqrt{-c + 10\sqrt{ab}}v_2$	F_{14}		$\xi_2^- w_0 - \xi_1^- w_1$, $(\xi_0 - 2\xi_0')v_0 + (2\xi_0 - 2\xi_0')v_1 - w_2$
	$\widetilde{C}_6$	$\sqrt{2}\sqrt[4]{ab}w_0 - w_2$, $\sqrt{a}v_0 - \sqrt{b}v_1 - \sqrt{-c + 10\sqrt{ab}}v_2$	G_{14}		$\xi_2^- w_0 - \xi_1^- w_1$, $(\xi_0 - 2\xi_0')v_0 + (2\xi_0 - 2\xi_0')v_1 + w_2$
G_6	D_6	$\sqrt{2}\sqrt[4]{ab}w_0 + w_2$, $\sqrt{a}v_0 - \sqrt{b}v_1 - \sqrt{-c + 10\sqrt{ab}}v_2$			
	$\widetilde{D}_6$	$\sqrt{2}\sqrt[4]{ab}w_0 - w_2$, $\sqrt{a}v_0 - \sqrt{b}v_1 + \sqrt{-c + 10\sqrt{ab}}v_2$			

Table A.5 The Galois action on the fibers of the genus 1 fibrations and the Weierstrass points

Action on splitting field	Action on Pic $\overline{X}_{\mathbf{a}}$	Action on Pic $\overline{Y}_{\mathbf{a}}$	Action on Weierstrass points	Action on splitting field	Action on Pic $\overline{X}_{\mathbf{a}}$	Action on Pic $\overline{Y}_{\mathbf{a}}$	Action on Weierstrass points
$\sqrt{5} \mapsto -\sqrt{5}$	$C_1 \mapsto D_1$ $D_1 \leftrightarrow \widetilde{C}_1$ $C_2 \mapsto \widetilde{C}_2$ $C_3 \mapsto \widetilde{C}_3$	$F_1 \leftrightarrow G_1$		$i \mapsto -i$	$C_3 \mapsto \widetilde{D}_3$ $C_5 \mapsto \widetilde{D}_5$	$F_3 \mapsto G_3$ $F_5 \mapsto G_5$ $F_{11} \mapsto G_{11}$ $F_{12} \mapsto G_{12}$	
	$C_4 \mapsto \widetilde{D}_4$	$F_4 \mapsto G_4$ $F_{10} \mapsto G_{10}$		$\sqrt{-2+2\sqrt{2}}$ $\mapsto -\sqrt{-2+2\sqrt{2}}$	$C_2 \mapsto \widetilde{D}_2$ $C_3 \mapsto \widetilde{D}_3$	$F_2 \mapsto G_2$ $F_3 \mapsto G_3$	
$\sqrt{2} \mapsto -\sqrt{2}$ $\sqrt{-2+2\sqrt{2}}$ $\mapsto \sqrt{-2-2\sqrt{2}}$	$C_2 \leftrightarrow C_3$ $C_5 \mapsto \widetilde{D}_5$ $C_6 \mapsto \widetilde{D}_6$	$F_2 \leftrightarrow F_3$ $F_5 \mapsto G_5$ $F_6 \mapsto G_6$		$\sqrt{c} \mapsto -\sqrt{c}$	$C_4 \mapsto \widetilde{D}_4$ $C_7 \mapsto \widetilde{D}_7$	$F_4 \mapsto G_4$ $F_7 \mapsto G_7$ $F_{11} \mapsto G_{11}$	
$\gamma_0 \mapsto -\gamma_0$	$C_7 \mapsto D_7$ $C_9 \mapsto D_9$	$F_7 \mapsto G_7$ $F_9 \mapsto G_9$ $F_{14} \mapsto G_{14}$	$Q_1 \leftrightarrow Q_3$ $Q_2 \leftrightarrow Q_4$	$\eta_0 \mapsto -\eta_0$	$C_4 \rightarrow D_4$ $C_6 \rightarrow D_6$	$F_4 \mapsto G_4$ $F_6 \mapsto G_6$ $F_{14} \mapsto G_{14}$	$P_1 \leftrightarrow P_3$ $P_2 \leftrightarrow P_4$
$\sqrt[4]{ab} \mapsto i\sqrt[4]{ab}$ $\sqrt{-c-10\sqrt{ab}}$ $\mapsto \sqrt{-c+10\sqrt{ab}}$	$C_5 \mapsto C_6$ $C_6 \mapsto \widetilde{D}_5$ $C_9 \mapsto \widetilde{D}_9$	$F_5 \mapsto G_6$ $F_6 \mapsto G_5$ $F_9 \mapsto G_9$ $F_{11} \mapsto G_{11}$	$P_3 \leftrightarrow P_4$	$\sqrt{a} \mapsto -\sqrt{a}$	$C_5 \mapsto D_5$ $C_6 \mapsto D_6$	$F_5 \mapsto G_5$ $F_6 \mapsto G_6$	$P_1 \leftrightarrow P_2$ $P_3 \leftrightarrow P_4$
				$\sqrt{-c-10\sqrt{ab}}$ $\mapsto -\sqrt{-c-10\sqrt{ab}}$	$C_5 \mapsto D_5$ $C_6 \mapsto D_6$	$F_5 \mapsto G_5$ $F_6 \mapsto G_6$	
$\xi_0 \mapsto -\xi_0$ $\xi_i^+ \mapsto \xi_i^-$		$F_{13} \mapsto G_{14}$ $F_{14} \mapsto G_{13}$		$\theta_0 \mapsto -\theta_0$ $\theta_i^+ \mapsto \theta_i^-$	$C_8 \leftrightarrow C_9$	$F_8 \leftrightarrow F_9$	$Q_3 \leftrightarrow Q_4$

(continued)

Table A.5 (continued)

Action on splitting field	Action on Pic $\overline{X}_{\mathbf{a}}$	Action on Pic $\overline{Y}_{\mathbf{a}}$	Action on Weierstrass points	Action on splitting field	Action on Pic $\overline{X}_{\mathbf{a}}$	Action on Pic $\overline{Y}_{\mathbf{a}}$	Action on Weierstrass points
$\xi_0' \mapsto -\xi_0'$ $\xi_i^+ \mapsto \xi_i^-$		$F_{13} \leftrightarrow F_{14}$		$\theta_1^+ \mapsto -\theta_1^+$	$C_8 \mapsto D_8$ $C_9 \mapsto D_9$	$F_8 \mapsto G_8$ $F_9 \mapsto G_9$	$Q_1 \leftrightarrow Q_2$ $Q_3 \leftrightarrow Q_4$
$\xi_1^+ \mapsto -\xi_1^+$		$F_{13} \mapsto G_{13}$ $F_{14} \mapsto G_{14}$		$\theta_2^+ \mapsto -\theta_2^+$	$C_8 \mapsto \widetilde{D}_8$ $C_9 \mapsto \widetilde{D}_9$	$F_8 \mapsto G_8$ $F_9 \mapsto G_9$	
$\xi_2^+ \mapsto -\xi_2^+$		$F_{13} \mapsto G_{13}$ $F_{14} \mapsto G_{14}$					

The action on the splitting field is described by the action on the generators listed at the beginning of the Appendix. If a generator of K is not listed, then we assume that it is fixed. We use the same convention for the curve classes and the Weierstrass points.

References

1. Barth, W., Hulek, K., Peters, C., Van de Ven, A.: Compact complex surfaces, Ergebnisse der Mathematik und ihrer Grenzgebiete (3) [Results in Mathematics and Related Areas (3)], vol. 4, Springer, Berlin (1984)
2. Beauville, A.: Complex algebraic surfaces. London Mathematical Society Student Texts, vol. 34, 2nd ed. Cambridge University Press, Cambridge, 1996; Translated from the 1978 French original by R. Barlow, with assistance from N.I. Shepherd-Barron and M. Reid
3. Beauville, A.: On the Brauer group of Enriques surfaces. Math. Res. Lett. **16**(6), 927–934
4. Creutz, B., Viray, B.: On Brauer groups of double covers of ruled surfaces. Math. Ann. **362**(3–4), 1169–1200 (2015). doi:10.1007/s00208-014-1153-0. MR3368096
5. Creutz, B., Viray, B.: Two torsion in the Brauer group of a hyperelliptic curve. Manuscripta Math. **147**(1–2), 139–167 (2015). doi:10.1007/s00229-014-0721-7. MR3336942
6. Gille, P., Szamuely, T.: Central simple algebras and Galois cohomology, Cambridge Studies in Advanced Mathematics, vol. 101, Cambridge University Press, Cambridge (2006)
7. Grothendieck, A.: Fondements de la géométrie algébrique. [Extraits du Séminaire Bourbaki, 1957–1962.]. Secrétariat mathématique, Paris (1962) (French). MR0146040 (26 #3566)
8. Grothendieck, A.: Le groupe de Brauer. III. Exemples et compléments. Dix Exposés sur la Cohomologie des Schémas, pp. 88–188. North-Holland/Masson, Amsterdam/Paris (1968) (French)
9. Harari, D., Skorobogatov, A.: Non-abelian descent and the arithmetic of Enriques surfaces. Int. Math. Res. Not. **52**, 3203–3228 (2005)
10. Ingalls, C., Obus, A., Ozman, E., Viray, B.: Unramified Brauer classes on cyclic covers of the projective plane. Preprint. arXiv:1310.8005
11. Liu, Q.: Algebraic geometry and arithmetic curves, Oxford Graduate Texts in Mathematics, vol. 6, Oxford University Press, Oxford (2002); Translated from the French by Reinie Erné; Oxford Science Publications.
12. Manin, Y.I., Le groupe de Brauer-Grothendieck en géométrie diophantienne. Actes du Congrés International des Mathématiciens (Nice, 1970), Tome 1, pp. 401–411 (1971)
13. Milne, J.S.: Étale cohomology. Princeton Mathematical Series, vol. 33, Princeton University Press, Princeton, NJ (1980). MR559531 (81j:14002)
14. Skorobogatov, A.N.: Beyond the Manin obstruction. Invent. Math. **135**(2), 399–424 (1999)
15. Skorobogatov, A.N.: Torsors and rational points. Cambridge Tracts in Mathematics, vol. 144, Cambridge University Press, Cambridge (2001). MR1845760 (2002d:14032)
16. Várilly-Alvarado, A., Viray, B.: Failure of the Hasse principle for Enriques surfaces. Adv. Math. **226** (2011)(6), 4884–4901. doi:10.1016/j.aim.2010.12.020

Shadow Lines in the Arithmetic of Elliptic Curves

J.S. Balakrishnan, M. Çiperiani, J. Lang, B. Mirza, and R. Newton

Abstract Let $E/\mathbb{Q}$ be an elliptic curve and p a rational prime of good ordinary reduction. For every imaginary quadratic field $K/\mathbb{Q}$ satisfying the Heegner hypothesis for E we have a corresponding line in $E(K) \otimes \mathbb{Q}_p$, known as a shadow line. When $E/\mathbb{Q}$ has analytic rank 2 and E/K has analytic rank 3, shadow lines are expected to lie in $E(\mathbb{Q}) \otimes \mathbb{Q}_p$. If, in addition, p splits in $K/\mathbb{Q}$, then shadow lines can be determined using the anticyclotomic p-adic height pairing. We develop an algorithm to compute anticyclotomic p-adic heights which we then use to provide an algorithm to compute shadow lines. We conclude by illustrating these algorithms in a collection of examples.

Keywords Elliptic curve • Universal norm • Anticyclotomic p-adic height • Shadow line

2010 *Mathematics Subject Classification.* 11G05, 11G50, 11Y40

J.S. Balakrishnan
Mathematical Institute, University of Oxford, Woodstock Road, Oxford OX2 6GG, UK
e-mail: balakrishnan@maths.ox.ac.uk

M. Çiperiani (✉)
Department of Mathematics, The University of Texas at Austin, 1 University Station, C1200
Austin, TX 78712, USA
e-mail: mirela@math.utexas.edu

J. Lang
UCLA Mathematics Department, Box 951555, Los Angeles, CA 90095-1555, USA
e-mail: jaclynlang@math.ucla.edu

B. Mirza
Department of Mathematics and Statistics, McGill University, 805 Sherbrooke Street West,
Montreal, QC, Canada H3A 0B9
e-mail: mirza@math.mcgill.ca

R. Newton
Department of Mathematics and Statistics, The University of Reading, Whiteknights, PO Box
220, Reading RG6 6AX, UK
e-mail: r.d.newton@reading.ac.uk

© Springer International Publishing Switzerland 2016

E.E. Eischen et al. (eds.), *Directions in Number Theory*, Association
for Women in Mathematics Series 3, DOI 10.1007/978-3-319-30976-7_2

1 Introduction

Fix an elliptic curve $E/\mathbb{Q}$ of analytic rank 2 and an odd prime p of good ordinary reduction. Assume that the p-primary part of the Tate–Shafarevich group of $E/\mathbb{Q}$ is finite. Let K be an imaginary quadratic field such that the analytic rank of E/K is 3 and the Heegner hypothesis holds for E, i.e., all primes dividing the conductor of $E/\mathbb{Q}$ split in K. We are interested in computing the subspace of $E(K) \otimes \mathbb{Q}_p$ generated by the anticyclotomic universal norms. To define this space, let K_∞ be the anticyclotomic $\mathbb{Z}_p$-extension of K and K_n denote the subfield of K_∞ whose Galois group over K is isomorphic to $\mathbb{Z}/p^n\mathbb{Z}$. The module of *universal norms* is defined by

$$\mathcal{U} = \bigcap_{n \geq 0} N_{K_n/K}(E(K_n) \otimes \mathbb{Z}_p),$$

where $N_{K_n/K}$ is the norm map induced by the map $E(K_n) \to E(K)$ given by $P \mapsto \sum_{\sigma \in \mathrm{Gal}(K_n/K)} P^\sigma$.

Consider

$$L_K := \mathcal{U} \otimes \mathbb{Q}_p \subseteq E(K) \otimes \mathbb{Q}_p.$$

By work of Cornut [6] and Vatsal [18], our assumptions on the analytic ranks of $E/\mathbb{Q}$ and E/K together with the assumed finiteness of the p-primary part of the Tate–Shafarevich group of $E/\mathbb{Q}$ imply that $\dim L_K \geq 1$. Bertolini [2] showed that $\dim L_K = 1$ under certain conditions on the prime p. Wiles and Çiperiani [4, 5] have shown that Bertolini's result is valid whenever $\mathrm{Gal}(\mathbb{Q}(E_p)/\mathbb{Q})$ is not solvable; here E_p denotes the full p-torsion of E and $\mathbb{Q}(E_p)$ is its field of definition. The 1-dimensional $\mathbb{Q}_p$-vector space L_K is known as the *shadow line* associated to the triple (E, K, p).

Complex conjugation acts on $E(K) \otimes \mathbb{Q}_p$, and we consider its two eigenspaces $E(K)^+ \otimes \mathbb{Q}_p$ and $E(K)^- \otimes \mathbb{Q}_p$. Observe that $E(K)^+ \otimes \mathbb{Q}_p = E(\mathbb{Q}) \otimes \mathbb{Q}_p$. By work of Skinner–Urban [15], Nekovář [14], Gross–Zagier [8], and Kolyvagin [10] we know that

$$\dim E(K)^+ \otimes \mathbb{Q}_p \geq 2 \quad \text{and} \quad \dim E(K)^- \otimes \mathbb{Q}_p = 1.$$

Then by the Sign Conjecture [11] we expect that

$$L_K \subseteq E(\mathbb{Q}) \otimes \mathbb{Q}_p.$$

Our main motivating question is the following:

Question (Mazur and Rubin). *As K varies, we presumably get different shadow lines L_K – what are these lines, and how are they distributed in $E(\mathbb{Q}) \otimes \mathbb{Q}_p$?*

In order to gather data about this question one can add the assumption that p splits in $K/\mathbb{Q}$ and then make use of the *anticyclotomic p-adic height pairing* on $E(K) \otimes \mathbb{Q}_p$. It is known that $\mathcal{U}$ is contained in the kernel of this pairing [12]. In fact, in our situation we expect that $\mathcal{U}$ equals the kernel of the anticyclotomic p-adic height pairing. Indeed we have $\dim E(K)^- \otimes \mathbb{Q}_p = 1$ and the weak Birch and Swinnerton–Dyer Conjecture for $E/\mathbb{Q}$ predicts that $\dim E(\mathbb{Q}) \otimes \mathbb{Q}_p = 2$, from which the statement about $\mathcal{U}$ follows by the properties of the anticyclotomic p-adic height pairing and its expected non-triviality. (This is discussed in Sect. 4 in further detail.) Thus computing the anticyclotomic p-adic height pairing allows us to determine the shadow line L_K.

Let $\Gamma(K)$ be the Galois group of the maximal $\mathbb{Z}_p$-power extension of K, and let $I(K) = \Gamma(K) \otimes_{\mathbb{Z}_p} \mathbb{Q}_p$. Identifying $\Gamma(K)$ with an appropriate quotient of the idele class group of K, Mazur et al. [13, §2.6] gave an explicit description of the universal p-adic height pairing

$$(\, ,) : E(K) \times E(K) \to I(K).$$

One obtains various $\mathbb{Q}_p$-valued height pairings on E by composing this universal pairing with $\mathbb{Q}_p$-linear maps $I(K) \to \mathbb{Q}_p$. The kernel of such a (non-zero) $\mathbb{Q}_p$-linear map corresponds to a $\mathbb{Z}_p$-extension of K.

In particular, the anticyclotomic $\mathbb{Z}_p$-extension of K corresponds to a $\mathbb{Q}_p$-linear map $\rho : I(K) \to \mathbb{Q}_p$ such that $\rho \circ c = -\rho$, where c denotes complex conjugation. The resulting anticyclotomic p-adic height pairing is denoted by $(\, ,)_\rho$. One key step of our work is an explicit description of the map ρ, see Sect. 2. As in [13], for $P \in E(K)$ we define the anticyclotomic p-adic height of P to be $h_\rho(P) = -\frac{1}{2}(P, P)_\rho$. Mazur et al. [13, §2.9] provide the following formula[1] for the anticyclotomic p-adic height of a point $P \in E(K)$:

$$h_\rho(P) = \rho_\pi(\sigma_\pi(P)) - \rho_\pi(\sigma_\pi(P^c)) + \sum_{w \nmid p\infty} \rho_w(d_w(P)),$$

where π is one of the prime divisors of p in K and the remaining notation is defined in Sect. 3. An algorithm for computing σ_π was given in [13]. Using our explicit description of ρ, in Sect. 3 we find a computationally feasible way of determining the contribution of finite primes w which do not divide p. This enables us to compute anticyclotomic p-adic height pairings.

We then proceed with a general discussion of shadow lines and their identification in $E(\mathbb{Q}) \otimes \mathbb{Q}_p$, see Sect. 4. In Sect. 5 we present the algorithms that we use to compute anticyclotomic p-adic heights and shadow lines. We conclude by displaying in Sect. 6 two examples of the computation of shadow lines L_K on the elliptic curve "389.a1" with the prime $p = 5$ and listing the results of several additional shadow line computations.

[1]The formula appearing in [13, §2.9] contains a sign error which is corrected here.

2 Anticyclotomic Character

Let K be an imaginary quadratic field with ring of integers $\mathcal{O}_K$ in which p splits as $p\mathcal{O}_K = \pi\pi^c$, where c denotes complex conjugation on K. Let $\mathbb{A}^\times$ be the group of ideles of K. We also use c to denote the involution of $\mathbb{A}^\times$ induced by complex conjugation on K. For any finite place v of K, denote by K_v the completion of K at v, $\mathcal{O}_v$ the ring of integers of K_v, and μ_v the group of roots of unity in $\mathcal{O}_v$. Let $\Gamma(K)$ be the Galois group of the maximal $\mathbb{Z}_p$-power extension of K. As in [13], we consider the idele class $\mathbb{Q}_p$-vector space $I(K) = \Gamma(K) \otimes_{\mathbb{Z}_p} \mathbb{Q}_p$. By class field theory $\Gamma(K)$ is a quotient of $J' := \mathbb{A}^\times / \overline{K^\times \mathbb{C}^\times \prod_{w\nmid p} \mathcal{O}_w^\times}$ by its finite torsion subgroup T, see the proof of Theorem 13.4 in [19]. The bar in the definition of J' denotes closure in the idelic topology, and the subgroup T is the kernel of the Nth power map on J' where N is the order of the finite group

$$\mathbb{A}^\times / \overline{K^\times \mathbb{C}^\times \prod_{w\nmid p} \mathcal{O}_w^\times (1+\pi\mathcal{O}_\pi)(1+\pi^c\mathcal{O}_{\pi^c})}.$$

Thus we have

$$I(K) = J'/T \otimes_{\mathbb{Z}_p} \mathbb{Q}_p. \tag{1}$$

We shall use this idelic description of $\Gamma(K)$ in what follows.

Definition 2.1 (Anticyclotomic p-adic Idele Class Character). An *anticyclotomic p-adic* idele class character is a continuous homomorphism

$$\rho : \mathbb{A}^\times / K^\times \to \mathbb{Z}_p$$

such that $\rho \circ c = -\rho$.

Lemma 2.2. *Every p-adic idele class character*

$$\rho : \mathbb{A}^\times / K^\times \to \mathbb{Z}_p$$

factors via the natural projection

$$\mathbb{A}^\times / K^\times \twoheadrightarrow \mathbb{A}^\times / \Big(K^\times \mathbb{C}^\times \prod_{w\nmid p} \mathcal{O}_w^\times \prod_{v\mid p} \mu_v \Big).$$

Proof. This is an immediate consequence of the fact that $\mathbb{Z}_p$ is a torsion-free pro-p group. □

The aim of this section is to define a non-trivial anticyclotomic p-adic idele class character. By the identification (1), such a character will give rise to a $\mathbb{Q}_p$-linear map $I(K) \to \mathbb{Q}_p$ which cuts out the anticyclotomic $\mathbb{Z}_p$-extension of K.

2.1 The Class Number One Case

We now explicitly construct an anticyclotomic p-adic idele class character ρ in the case when the class number of K is 1.

Recall our assumption that p splits in $K/\mathbb{Q}$ as $p\mathcal{O}_K = \pi\pi^c$ and let

$$U_\pi = 1 + \pi\mathcal{O}_\pi \quad \text{and} \quad U_{\pi^c} = 1 + \pi^c\mathcal{O}_{\pi^c}.$$

Define a continuous homomorphism

$$\varphi : \mathbb{A}^\times \to U_\pi \times U_{\pi^c}$$

as follows. Let $(x_v)_v \in \mathbb{A}^\times$. Under our assumption that K has class number 1, we can find $\alpha \in K^\times$ such that

$$\alpha x_v \subset \mathcal{O}_v^\times \qquad \text{for all finite } v.$$

Indeed, the ideal $\mathfrak{a}_v$ corresponding to the place v is principal, say generated by $\varpi_v \in \mathcal{O}_K$. Then take $\alpha = \prod_v \varpi_v^{-\mathrm{ord}_v(x_v)}$, where the product is taken over all finite places v of K. We define

$$\varphi((x_v)_v) = ((\alpha x_\pi)^{p-1}, (\alpha x_{\pi^c})^{p-1}). \tag{2}$$

Note that since p is split in K we have $\mathcal{O}_\pi^\times \cong \mathbb{Z}_p^\times \cong \mu_{p-1} \times U_\pi$, and similarly for π^c. To see that φ is independent of the choice of α, we note that any other choice $\alpha' \in K^\times$ differs from α by an element of $\mathcal{O}_K^\times$. Since K is an imaginary quadratic field, $\mathcal{O}_K^\times$ consists entirely of roots of unity. In particular, under the embedding $K \hookrightarrow K_\pi$ we see that $\mathcal{O}_K^\times \hookrightarrow \mu_{p-1}$. Thus, any ambiguity about α is killed when we raise α to the $(p-1)$-power. Therefore, φ is well-defined. The continuity of φ is easily verified.

Proposition 2.3. *Suppose that K has class number* 1. *Then the map φ defined in* (2) *induces an isomorphism of topological groups*

$$\mathbb{A}^\times / \Big(K^\times\mathbb{C}^\times \prod_{w\nmid p} \mathcal{O}_w^\times \prod_{v\mid p} \mu_v\Big) \to U_\pi \times U_{\pi^c}.$$

Proof. For $v \in \{\pi, \pi^c\}$, the p-adic logarithm gives an isomorphism $U_v \cong 1 + p\mathbb{Z}_p \to p\mathbb{Z}_p$. Hence, raising to the power $(p-1)$ is an automorphism on U_v for $v \in \{\pi, \pi^c\}$ and consequently φ is surjective. It is easy to see that $K^\times\mathbb{C}^\times \prod_{w\nmid p} \mathcal{O}_w^\times \subset \ker\varphi$. Since $\mu_v \cong \mathbb{F}_p^\times$ for $v \in \{\pi, \pi^c\}$, we have $\prod_{v\mid p} \mu_v \subset \ker\varphi$. We claim that $\ker\varphi = K^\times\mathbb{C}^\times \prod_{w\nmid p} \mathcal{O}_w^\times \prod_{v\mid p} \mu_v$. Let $(x_v)_v \in \ker\varphi$ and let $\alpha \in K^\times$ be such that $\alpha x_v \in \mathcal{O}_v^\times$ for all finite v. It suffices to show that $(\alpha x_v)_v \in \mathbb{C}^\times \prod_{w\nmid p} \mathcal{O}_w^\times \prod_{v\mid p} \mu_v$. This is clear: since $(x_v)_v \in \ker\varphi$, we have $\alpha x_v \in \mu_v$ for $v \in \{\pi, \pi^c\}$.

Finally, since φ is a continuous open map, it follows that φ induces the desired homeomorphism. □

By Lemma 2.2 we have reduced the problem of constructing an anticyclotomic p-adic idele class character to the problem of constructing a character

$$\chi : U_\pi \times U_{\pi^c} \to \mathbb{Z}_p \tag{3}$$

satisfying $\chi \circ c = -\chi$. Note that this last condition implies that $\chi(x, y) = \chi(x/y^c, 1)$. Explicitly:

$$\begin{aligned} \chi(x, y) &= -\chi \circ c(x, y) = -\chi(y^c, x^c) = -\chi(y^c, 1) - \chi(1, x^c) \\ &= -\chi(y^c, 1) + \chi(x, 1) = \chi(x/y^c, 1). \end{aligned} \tag{4}$$

In other words, χ factors via the surjection

$$\begin{aligned} f_\pi : U_\pi \times U_{\pi^c} &\twoheadrightarrow U_\pi \\ (x, y) &\mapsto x/y^c. \end{aligned}$$

Therefore, it is enough to define a character $U_\pi \to \mathbb{Z}_p$. Fixing an isomorphism of valued fields $\psi : K_\pi \to \mathbb{Q}_p$ gives an identification $U_\pi \cong 1 + p\mathbb{Z}_p$. Now, up to scaling, there is only one choice of character, namely $\log_p : 1 + p\mathbb{Z}_p \to p\mathbb{Z}_p$. We write $\log_p$ for the unique group homomorphism $\log_p : \mathbb{Q}_p^\times \to (\mathbb{Q}_p, +)$ with $\log_p(p) = 0$ extending $\log_p : 1 + p\mathbb{Z}_p \to p\mathbb{Z}_p$. The extension to $\mathbb{Z}_p^\times$ of the map $\log_p$ is explicitly given by

$$\log_p(u) = \frac{1}{p-1} \log_p(u^{p-1}).$$

We choose the normalization $\rho = \frac{1}{p(p-1)} \log_p \circ \psi \circ f_\pi \circ \varphi$. We summarize our construction of the anticyclotomic p-adic idele class character ρ in the following proposition:

Proposition 2.4. *Suppose that K has class number 1. Fix a choice of isomorphism $\psi : K_\pi \to \mathbb{Q}_p$. Consider the map $\rho : \mathbb{A}^\times/K^\times \to \mathbb{Z}_p$ such that*

$$\rho((x_v)_v) = \frac{1}{p} \log_p \circ \psi \left(\frac{\alpha x_\pi}{\alpha^c x_{\pi^c}^c} \right)$$

where $\alpha \in K^\times$ is such that $\alpha x_v \in \mathcal{O}_v^\times$ for all finite v. Then ρ is the unique (up to scaling) non-trivial anticyclotomic p-adic idele class character.

Proof. Let $\alpha \in K^\times$ be such that $\alpha x_v \in \mathcal{O}_v^\times$ for all finite v. By our earlier discussion and the definition of the extension of $\log_p$ to $\mathbb{Z}_p^\times$, we have

$$\rho((x_v)_v) = \frac{1}{p(p-1)} \log_p \circ \psi \left(\frac{(\alpha x_\pi)^{p-1}}{(\alpha^c x_{\pi^c}^c)^{p-1}} \right)$$
$$= \frac{1}{p} \log_p \circ \psi \left(\frac{\alpha x_\pi}{\alpha^c x_{\pi^c}^c} \right).$$

□

2.2 *The General Case*

There is a simple generalization of the construction of ρ to the case when the class number of K may be greater than one. Let h be the class number of K. We can no longer define the homomorphism φ of (2) on the whole of $\mathbb{A}^\times$ because $\mathcal{O}_K$ is no longer assumed to be a principal ideal domain. However, we can define

$$\varphi_h : (\mathbb{A}^\times)^h \to U_\pi \times U_{\pi^c}$$

in a similar way, as follows. Let $\mathfrak{a}_v$ be the ideal of K corresponding to the place v. Then $\mathfrak{a}_v^h$ is principal, say generated by $\varpi_v \in \mathcal{O}_K$. For $(x_v)_v \in \mathbb{A}^\times$ we set $\alpha(v) = {\varpi_v}^{-\mathrm{ord}_v(x_v)}$. Then $\alpha(v)x_v^h \in \mathcal{O}_v^\times$ and $\alpha(v) \in \mathcal{O}_w^\times$ for all $w \neq v$. Note that $\alpha(v) = 1$ for all but finitely many v. Set $\alpha = \prod_v \alpha(v)$ and observe that $\alpha x_v^h \in \mathcal{O}_v^\times$ for all v. Then we define φ_h by

$$\varphi_h((x_v)_v^h) = ((\alpha x_\pi^h)^{p-1}, (\alpha x_{\pi^c}^h)^{p-1}). \tag{5}$$

Fix an isomorphism $\psi : K_\pi \to \mathbb{Q}_p$. As before, we can now use the p-adic logarithm to define an anticyclotomic character $\rho : (\mathbb{A}^\times)^h \to \mathbb{Z}_p$ by setting

$$\rho = \frac{1}{p(p-1)} \log_p \circ \psi \circ f_\pi \circ \varphi_h.$$

We extend the definition of ρ to the whole of $\mathbb{A}^\times$ by setting $\rho((x_v)_v) = \frac{1}{h}\rho((x_v)_v^h)$.

As in Proposition 2.4, we now summarize our construction of the anticyclotomic p-adic idele class character in this more general setting.

Proposition 2.5. *Let h be the class number of K, and fix a choice of isomorphism $\psi : K_\pi \to \mathbb{Q}_p$. Consider the map $\rho : \mathbb{A}^\times / K^\times \to \frac{1}{h}\mathbb{Z}_p$ such that*

$$\rho((x_v)_v) = \frac{1}{hp} \log_p \circ \psi \left(\frac{\alpha x_\pi^h}{\alpha^c x_{\pi^c}^{ch}} \right)$$

where $\alpha \in K^\times$ is such that $\alpha x_v^h \in \mathcal{O}_v^\times$ for all finite v. Then ρ is the unique (up to scaling) non-trivial anticyclotomic p-adic idele class character.

Remark 2.6. Note that $\rho : \mathbb{A}^\times/K^\times \to \frac{1}{h}\mathbb{Z}_p$, so if $p \mid h$, then ρ is not strictly an anticyclotomic idele class character in the sense of Definition 2.1. However, the choice of scaling of ρ is of no great importance since our purpose is to use ρ to define an anticyclotomic height pairing on $E(K)$ and compute the kernel of this pairing.

Remark 2.7. The ideal $\prod_v \mathfrak{a}_v^{-h\,\mathrm{ord}_v(x_v)}$ is principal and a generator of this ideal is the element $\alpha \in K$ that we use when evaluating the character ρ defined in Proposition 2.5.

3 Anticyclotomic p-adic Height Pairing

We wish to compute the anticyclotomic p-adic height h_ρ using our explicit description of the anticyclotomic idele class character ρ given in Proposition 2.5. For any finite prime w of K, the natural inclusion $K_w^\times \hookrightarrow \mathbb{A}^\times$ induces a map $\iota_w : K_w^\times \to I(K)$, and we write $\rho_w = \rho \circ \iota_w$. For every finite place w of K and every non-zero point $P \in E(K)$ we can find $d_w(P) \in \mathcal{O}_w$ and $a_w(P), b_w(P) \in \mathcal{O}_w$, each relatively prime to $d_w(P)$, such that

$$(\iota_w(x(P)), \iota_w(y(P))) = \left(\frac{a_w(P)}{d_w(P)^2}, \frac{b_w(P)}{d_w(P)^3}\right). \tag{6}$$

We refer to $d_w(P)$ as a *local denominator* of P at w. The existence of $d_w(P)$ follows from the Weierstrass equation for E and the fact that $\mathcal{O}_w$ is a principal ideal domain. Finally, we let σ_π denote the π-adic σ-function of E.

Given a non-torsion point $P \in E(K)$ such that

- P reduces to 0 modulo primes dividing p, and
- P reduces to the connected component of all special fibers of the Neron model of E,

we can compute its anticyclotomic p-adic height using the following formula[2] [13, §2.9]:

$$h_\rho(P) = \rho_\pi(\sigma_\pi(P)) - \rho_{\bar{\pi}}(\sigma_{\bar{\pi}}(P^c)) + \sum_{w \nmid p\infty} \rho_w(d_w(P)). \tag{7}$$

In the following lemmas, we make some observations which simplify the computation of $h_\rho(P)$.

Lemma 3.1. *Let w be a finite prime such that $w \nmid p$. Let $x_w \in K_w^\times$. Then $\rho_w(x_w)$ only depends on* $\mathrm{ord}_w(x_w)$. *In particular, if $x_w \in \mathcal{O}_w^\times$, then $\rho_w(x_w) = 0$.*

[2]The formula appearing in [13, §2.9] contains a sign error which is corrected here.

Proof. This follows immediately from Lemma 2.2. Alternatively, note that the auxiliary element α used in the definition of ρ only depends on the valuation of x_w. □

Lemma 3.2. *Let w be a finite prime of K. Then $\rho_{w^c} = -\rho_w \circ c$. In particular, if $w = w^c$, then $\rho_w = 0$.*

Proof. This is an immediate consequence of the relations $\rho \circ c = -\rho$ and $c \circ \iota_{\lambda^c} = \iota_\lambda \circ c$. □

Lemma 3.2 allows us to write the formula (7) for the anticyclotomic p-adic height as follows:

$$h_\rho(P) = \rho_\pi\left(\frac{\sigma_\pi(P)}{\sigma_\pi(P^c)}\right) + \sum_{\substack{\ell=\lambda\lambda^c \\ \ell\neq p}} \rho_\lambda\left(\frac{d_\lambda(P)}{d_{\lambda^c}(P)^c}\right). \tag{8}$$

Remark 3.3. In order to implement an algorithm for calculating the anticyclotomic p-adic height h_ρ, we must determine a finite set of primes which includes all the split primes $\ell = \lambda\lambda^c \nmid p$ for which $\rho_\lambda\left(\frac{d_\lambda(P)}{d_{\lambda^c}(P)^c}\right) \neq 0$. Let k_λ be the residue field of K at λ and set $\mathcal{D}(P) = \prod_{\lambda\nmid p\infty}(\#k_\lambda)^{\mathrm{ord}_\lambda(d_\lambda(P))}$. It turns out that $\mathcal{D}(P)$ can be computed easily from the leading coefficient of the minimal polynomial of the x-coordinate of P [1, Proposition 4.2]. Observe that $\rho_\lambda\left(\frac{d_\lambda(P)}{d_{\lambda^c}(P)^c}\right) \neq 0$ implies that $\mathrm{ord}_\lambda(d_\lambda(P)) \neq 0$ or $\mathrm{ord}_{\lambda^c}(d_{\lambda^c}(P)) \neq 0$. Hence, the only primes $\ell \neq p$ which contribute to the sum in (8) are those that are split in $K/\mathbb{Q}$ and divide $\mathcal{D}(P)$. However, in the examples that we have attempted, factoring $\mathcal{D}(P)$ is difficult due to its size.

We now package together the contribution to the anticyclotomic p-adic height coming from primes not dividing p. Consider the ideal $x(P)\mathcal{O}_K$ and denote by $\delta(P) \subset \mathcal{O}_K$ its denominator ideal. Observe that by (6) we know that all prime factors of $\delta(P)$ appear with even powers. Fix $\mathbf{d}_h(P) \in \mathcal{O}_K$ as follows:

$$\mathbf{d}_h(P)\mathcal{O}_K = \prod_{\mathfrak{q}} \mathfrak{q}^{h\,\mathrm{ord}_{\mathfrak{q}}(\delta(P))/2} \tag{9}$$

where h is the class number of K, and the product is over all prime ideals $\mathfrak{q}$ in $\mathcal{O}_K$.

Proposition 3.4. *Let $P \in E(K)$ be a non-torsion point which reduces to 0 modulo primes dividing p, and to the connected component of all special fibers of the Neron model of E. Then the anticyclotomic p-adic height of P is*

$$h_\rho(P) = \frac{1}{p}\log_p\left(\psi\left(\frac{\sigma_\pi(P)}{\sigma_\pi(P^c)}\right)\right) + \frac{1}{hp}\log_p\left(\psi\left(\frac{\mathbf{d}_h(P)^c}{\mathbf{d}_h(P)}\right)\right),$$

where $\psi : K_\pi \to \mathbb{Q}_p$ is the fixed automorphism.

Proof. By (7) we have

$$h_\rho(P) = \rho_\pi\left(\frac{\sigma_\pi(P)}{\sigma_\pi(P^c)}\right) + \sum_{w\nmid p\infty} \rho_w(d_w(P)). \tag{10}$$

Let $P = (x, y) \in E(K)$. Since P reduces to the identity modulo π and π^c, we have

$$\operatorname{ord}_\pi(x) = -2e_\pi, \operatorname{ord}_\pi(y) = -3e_\pi,$$
$$\operatorname{ord}_{\pi^c}(x) = -2e_{\pi^c}, \operatorname{ord}_{\pi^c}(y) = -3e_{\pi^c},$$

for positive integers e_π and e_{π^c}. Since the p-adic σ function has the form $\sigma(t) = t + \cdots \in t\mathbb{Z}_p[[t]]$, we see that

$$\operatorname{ord}_\pi(\sigma_\pi(P)) = \operatorname{ord}_\pi\left(\sigma_\pi\left(\frac{-x}{y}\right)\right) = \operatorname{ord}_\pi\left(\frac{-x}{y}\right) = e_\pi$$

and similarly

$$\operatorname{ord}_\pi(\sigma_\pi(P^c)) = \operatorname{ord}_\pi\left(\frac{-x^c}{y^c}\right) = \operatorname{ord}_{\pi^c}\left(\frac{-x}{y}\right) = e_{\pi^c}.$$

Thus,

$$\operatorname{ord}_\pi\left(\frac{\sigma_\pi(P)}{\sigma_\pi(P^c)}\right) = e_\pi - e_{\pi^c}. \tag{11}$$

Let $\alpha \in K^\times$ generate the principal ideal π^h. By (11) and the definition of the anticyclotomic p-adic idele class character, we have

$$\begin{aligned}\rho_\pi\left(\frac{\sigma_\pi(P)}{\sigma_\pi(P^c)}\right) &= \frac{1}{hp}\log_p \circ\psi\left(\frac{\alpha^{e_{\pi^c}-e_\pi}\sigma_\pi(P)^h}{(\alpha^c)^{e_{\pi^c}-e_\pi}\sigma_\pi(P^c)^h}\right)\\ &= \frac{1}{p}\log_p\left(\psi\left(\frac{\sigma_\pi(P)}{\sigma_\pi(P^c)}\right)\right) + \frac{1}{hp}\log_p\left(\psi\left(\frac{\alpha}{\alpha^c}\right)^{e_{\pi^c}-e_\pi}\right).\end{aligned}$$

Now it remains to show that

$$\sum_{w\nmid p\infty} \rho_w(d_w(P)) = \frac{1}{hp}\log_p\left(\psi\left(\frac{\mathbf{d}_h(P)^c}{\mathbf{d}_h(P)}\right)\right) - \frac{1}{hp}\log_p\left(\psi\left(\frac{\alpha}{\alpha^c}\right)^{e_{\pi^c}-e_\pi}\right). \tag{12}$$

By the definition of ρ, we have

$$\sum_{w\nmid p\infty} \rho_w(d_w(P)) = \frac{1}{h}\sum_{w\nmid p\infty} \rho_w(d_w(P)^h). \tag{13}$$

Since $\mathrm{ord}_w(d_w(P)^h) = \mathrm{ord}_w(\mathbf{d}_h(P))$, Lemma 3.1 gives $\rho_w(d_w(P)^h) = \rho_w(\mathbf{d}_h(P))$ for every $w \nmid p\infty$. Substituting this into (13) gives

$$\begin{aligned}\sum_{w\nmid p\infty} \rho_w(d_w(P)) &= \frac{1}{h}\sum_{w\nmid p\infty} \rho_w(\mathbf{d}_h(P))\\ &= \frac{1}{h}\sum_{w\nmid p\infty} \rho\circ\iota_w(\mathbf{d}_h(P))\\ &= \frac{1}{h}\rho\Big(\prod_{w\nmid p\infty}\iota_w(\mathbf{d}_h(P))\Big).\end{aligned}$$

Now $\prod_{w\nmid p\infty}\iota_w(\mathbf{d}_h(P))$ is the idele with entry $\mathbf{d}_h(P)$ at every place $w \nmid p\infty$ and entry 1 at all other places. Define $\beta \in \mathcal{O}_K$ by $\mathbf{d}_h(P) = \alpha^{e_\pi}(\alpha^c)^{e_{\pi^c}}\beta$. Thus, by Proposition 2.5 and Remark 2.7, we get

$$\begin{aligned}\frac{1}{h}\rho\Big(\prod_{w\nmid p\infty}\iota_w(\mathbf{d}_h(P))\Big) &= \frac{1}{hp}\log_p\left(\psi\left(\frac{\beta^c}{\beta}\right)\right)\\ &= \frac{1}{hp}\log_p\left(\psi\left(\frac{\mathbf{d}_h(P)^c}{\mathbf{d}_h(P)}\right)\right) - \frac{1}{hp}\log_p\left(\psi\left(\frac{\alpha}{\alpha^c}\right)^{e_{\pi^c}-e_\pi}\right)\end{aligned}$$

as required. This concludes the proof. □

In [13], the authors describe the "universal" p-adic height pairing $(P,Q) \in I(K)$ of two points $P, Q \in E(K)$. Composition of the universal height pairing with any $\mathbb{Q}_p$-linear map $\rho : I(K) \to \mathbb{Q}_p$ gives rise to a canonical symmetric bilinear pairing

$$(\ ,\ )_\rho : E(K) \times E(K) \to \mathbb{Q}_p$$

called the *ρ-height pairing*. The ρ-height of a point $P \in E(K)$ is defined to be $-\frac{1}{2}(P,P)_\rho$.

Henceforth, we fix ρ to be the anticyclotomic p-adic idele class character defined in Sect. 2. The corresponding ρ-height pairing is referred to as the *anticyclotomic p-adic height pairing*, and it is denoted as follows:

$$\langle\ ,\ \rangle = (\ ,\ )_\rho : E(K) \times E(K) \to \mathbb{Q}_p$$

Observe that

$$\langle P, Q\rangle = h_\rho(P) + h_\rho(Q) - h_\rho(P+Q).$$

Let $E(K)^+$ and $E(K)^-$ denote the $+1$-eigenspace and the -1-eigenspace, respectively, for the action of complex conjugation on $E(K)$. Since σ_π is an odd function, using (8) we see that the anticyclotomic height satisfies

$$h_\rho(P) = 0 \quad \text{for all } P \in E(K)^+ \cup E(K)^-.$$

Therefore, the anticyclotomic p-adic height pairing satisfies

$$\langle E(K)^{+}, E(K)^{+}\rangle = \langle E(K)^{-}, E(K)^{-}\rangle = 0. \tag{14}$$

Consequently, if $P \in E(K)^{+}$ and $Q \in E(K)^{-}$, then

$$\begin{aligned} \langle P, Q\rangle &= h_\rho(P) + h_\rho(Q) - h_\rho(P+Q) \\ &= -\frac{1}{2}\langle P, P\rangle - \frac{1}{2}\langle Q, Q\rangle - h_\rho(P+Q) \\ &= -h_\rho(P+Q). \end{aligned} \tag{15}$$

4 The Shadow Line

Let E be an elliptic curve defined over $\mathbb{Q}$ and p an odd prime of good ordinary reduction. Fix an imaginary quadratic extension $K/\mathbb{Q}$ satisfying the Heegner hypothesis for $E/\mathbb{Q}$ (i.e., all primes dividing the conductor of $E/\mathbb{Q}$ split in K). Consider the anticyclotomic $\mathbb{Z}_p$-extension K_∞ of K. Let K_n denote the subfield of K_∞ whose Galois group over K is isomorphic to $\mathbb{Z}/p^n\mathbb{Z}$. The module of *universal norms* for this $\mathbb{Z}_p$-extension is defined as follows:

$$\mathcal{U} := \bigcap_{n\geq 0} N_{K_n/K}(E(K_n)\otimes \mathbb{Z}_p) \subseteq E(K)\otimes \mathbb{Z}_p,$$

where $N_{K_n/K}$ is the norm map induced by the map $E(K_n) \to E(K)$ given by $P \mapsto \sum_{\sigma\in \mathrm{Gal}(K_n/K)} P^\sigma$.

By work of Cornut [6] and Vatsal [18] we know that for n large enough, we have a non-torsion Heegner point in $E(K_n)$. Since p is a prime of good ordinary reduction, the trace down to K_{n-1} of the Heegner points defined over K_n is related to Heegner points defined over K_{n-1}, see [1, §2] for further details. Due to this relation among Heegner points defined over the different layers of K_∞, if the p-primary part of the Tate–Shafarevich group of E/K is finite, then these points give rise to non-trivial universal norms. Hence, if the p-primary part of the Tate–Shafarevich group of E/K is finite, then $\mathcal{U}$ is non-trivial whenever the Heegner hypothesis holds. By Bertolini [2], Ciperiani and Wiles [5], and Ciperiani [4] we know that if $\mathrm{Gal}(\mathbb{Q}(E_p)/\mathbb{Q})$ is not solvable, then $\mathcal{U} \simeq \mathbb{Z}_p$.

Consider

$$L_K := \mathcal{U}\otimes \mathbb{Q}_p.$$

If the p-primary part of the Tate–Shafarevich group of E/K is finite, then L_K is a line in the vector space $E(K)\otimes \mathbb{Q}_p$ known as the *shadow line* associated to the triple

(E, K, p). The space $E(K) \otimes \mathbb{Q}_p$ splits as the direct sum of two eigenspaces under the action of complex conjugation

$$E(K) \otimes \mathbb{Q}_p = E(K)^+ \otimes \mathbb{Q}_p \oplus E(K)^- \otimes \mathbb{Q}_p.$$

Observe that

$$E(K)^+ \otimes \mathbb{Q}_p = E(\mathbb{Q}) \otimes \mathbb{Q}_p \quad \text{and} \quad E(K)^- \otimes \mathbb{Q}_p \simeq E^K(\mathbb{Q}) \otimes \mathbb{Q}_p,$$

where E^K denotes the quadratic twist of E with respect to K. Since the module $\mathcal{U}$ is fixed by complex conjugation, the shadow line L_K lies in one of the eigenspaces:

$$L_K \subseteq E(\mathbb{Q}) \otimes \mathbb{Q}_p \quad \text{or} \quad L_K \subseteq E(K)^- \otimes \mathbb{Q}_p.$$

The assumption of the Heegner hypothesis forces the analytic rank of E/K to be odd, and hence the dimension of $E(K) \otimes \mathbb{Q}_p$ is odd by the Parity Conjecture [14] and our assumption of the finiteness of the p-primary part of the Tate–Shafarevich group of E/K. Hence, $\dim E(K)^- \otimes \mathbb{Q}_p \neq \dim E(\mathbb{Q}) \otimes \mathbb{Q}_p$. The Sign Conjecture states that L_K is expected to lie in the eigenspace of higher dimension [11].

Our main motivating question is the following:

Question 4.1 (Mazur and Rubin). *Consider an elliptic curve $E/\mathbb{Q}$ of positive even analytic rank r, an imaginary quadratic field K such that E/K has analytic rank $r+1$, and a prime p of good ordinary reduction such that the p-primary part of the Tate–Shafarevich group of $E/\mathbb{Q}$ is finite. By the Sign Conjecture, we expect L_K to lie in $E(\mathbb{Q}) \otimes \mathbb{Q}_p$. As K varies, we presumably get different shadow lines L_K. What are these lines and how are they distributed in $E(\mathbb{Q}) \otimes \mathbb{Q}_p$?*

Note that in the statement of the above question we make use of the following results:

1. Since $E/\mathbb{Q}$ has positive even analytic rank we know that $\dim E(\mathbb{Q}) \otimes \mathbb{Q}_p \geq 2$ by work of Skinner–Urban [15, Theorem 2] and work of Nekovar [14] on the Parity Conjecture.
2. Since our assumptions on the analytic ranks of $E/\mathbb{Q}$ and E/K imply that the analytic rank of $E^K/\mathbb{Q}$ is 1, by work of Gross–Zagier [8] and Kolyvagin [10] we know that
   (a) $\dim E(K)^- \otimes \mathbb{Q}_p = 1$;
   (b) the p-primary part of the Tate–Shafarevich group of $E^K/\mathbb{Q}$ is finite, and hence the finiteness of the p-primary part of the Tate–Shafarevich group of E/K follows from the finiteness of the p-primary part of the Tate–Shafarevich group of $E/\mathbb{Q}$.

Thus by (2b) we know that $L_K \subseteq E(K) \otimes \mathbb{Q}_p$, while (1) and (2a) are the input to the Sign Conjecture.

It is natural to start the study of Question 4.1 by considering elliptic curves $E/\mathbb{Q}$ of analytic rank 2. In this case, assuming that

$$\mathrm{rank}_{\mathbb{Z}} E(\mathbb{Q}) = 2, \tag{16}$$

we identify L_K in $E(\mathbb{Q}) \otimes \mathbb{Q}_p$ by making use of the anticyclotomic p-adic height pairing, viewing it as a pairing on $E(K) \otimes \mathbb{Z}_p$. This method forces us to restrict our attention to quadratic fields K where p splits. It is known that $\mathcal{U}$ is contained in the kernel of the anticyclotomic p-adic height pairing [12, Proposition 4.5.2]. In fact, in our situation, the properties of this pairing and (16) together with the fact that $\dim E(K)^- \otimes \mathbb{Q}_p = 1$ imply that either $\mathcal{U}$ is the kernel of the pairing or the pairing is trivial. Thus computing the anticyclotomic p-adic height pairing allows us to verify the Sign Conjecture and determine the shadow line L_K.

In order to describe the lines L_K for multiple quadratic fields K, we fix two independent generators P_1, P_2 of $E(\mathbb{Q}) \otimes \mathbb{Q}_p$ (with E given by its reduced minimal model) and compute the slope of $L_K \otimes \mathbb{Q}_p$ in the corresponding coordinate system. For each quadratic field K we compute a non-torsion point $R \in E(K)^-$ (on the reduced minimal model of E). The kernel of the anticyclotomic p-adic height pairing on $E(K) \otimes \mathbb{Z}_p$ is generated by $aP_1 + bP_2$ for $a, b \in \mathbb{Z}_p$ such that $\langle aP_1 + bP_2, R \rangle = 0$. Then by (15) the shadow line $L_K \otimes \mathbb{Q}_p$ in $E(\mathbb{Q}) \otimes \mathbb{Q}_p$ is generated by $h_p(P_2 + R)P_1 - h_p(P_1 + R)P_2$ and its slope with respect to the coordinate system induced by $\{P_1, P_2\}$ equals

$$-h_p(P_1 + R)/h_p(P_2 + R).$$

5 Algorithms

Let $E/\mathbb{Q}$ be an elliptic curve of analytic rank 2; see [3, Chap. 4] for an algorithm that can provably verify the non-triviality of the second derivative of the L-function. Our aim is to compute shadow lines on the elliptic curve E. In order to do this using the method described in Sect. 4 we need to

- verify that $\mathrm{rank}_{\mathbb{Z}} E(\mathbb{Q}) = 2$, and
- compute two $\mathbb{Z}$-independent points $P_1, P_2 \in E(\mathbb{Q})$.

By work of Kato [9, Theorem 17.4], computing the ℓ-adic analytic rank of $E/\mathbb{Q}$ for any prime ℓ of good ordinary reduction gives an upper bound on $\mathrm{rank}_{\mathbb{Z}} E(\mathbb{Q})$ (see [16, Proposition 10.1]). Using the techniques in [16, §3], which have been implemented in `Sage`, one can compute an upper bound on the ℓ-adic analytic rank using an approximation of the ℓ-adic L-series, thereby obtaining an upper bound on $\mathrm{rank}_{\mathbb{Z}} E(\mathbb{Q})$. Since the analytic rank of $E/\mathbb{Q}$ is 2, barring the failure of standard conjectures we find that $\mathrm{rank}_{\mathbb{Z}} E(\mathbb{Q}) \leq 2$. Then using work of Cremona [7, Sect. 3.5] implemented in `Sage`, we search for points of bounded height, increasing the height until we find two $\mathbb{Z}$-independent points $P_1, P_2 \in E(\mathbb{Q})$. We have thus computed a basis of $E(\mathbb{Q}) \otimes \mathbb{Q}_p$.

We will now proceed to describe the algorithms that allow us to compute shadow lines on the elliptic curve $E/\mathbb{Q}$.

Algorithm 5.1. *Generator of* $E(K)^- \otimes \mathbb{Q}_p$.

Input:

- *an elliptic curve* $E/\mathbb{Q}$ *(given by its reduced minimal model) of analytic rank* 2;
- *an odd prime* p *of good ordinary reduction;*
- *an imaginary quadratic field* K *such that*
  - *the analytic rank of* E/K *equals* 3, *and*
  - *all rational primes dividing the conductor of* $E/\mathbb{Q}$ *split in* K.

Output: *A generator of* $E(K)^- \otimes \mathbb{Q}_p$ *(given as a point on the reduced minimal model of* $E/\mathbb{Q}$*).*

(1) *Let* $d \in \mathbb{Z}$ *such that* $K = \mathbb{Q}(\sqrt{d})$. *Compute a short model of* E^K, *of the form* $y^2 = x^3 + ad^2x + bd^3$.
(2) *Our assumption on the analytic ranks of* $E/\mathbb{Q}$ *and* E/K *implies that the analytic rank of* $E^K/\mathbb{Q}$ *is* 1. *Compute a non-torsion point*[3] *of* $E^K(\mathbb{Q})$ *and denote it* (x_0, y_0). *Then* $(\frac{x_0}{d}, \frac{y_0\sqrt{d}}{d^2})$ *is an element of* $E(K)$ *on the model* $y^2 = x^3 + ax + b$.
(3) *Output the image of* $(\frac{x_0}{d}, \frac{y_0\sqrt{d}}{d^2})$ *on the reduced minimal model of* E.

Algorithm 5.2. *Computing the anticyclotomic* p*-adic height associated to* (E, K, p).

Input:

- *elliptic curve* $E/\mathbb{Q}$ *(given by its reduced minimal model);*
- *an odd prime* p *of good ordinary reduction;*
- *an imaginary quadratic field* K *such that* p *splits in* $K/\mathbb{Q}$;
- *a non-torsion point* $P \in E(K)$.

Output: *The anticyclotomic* p*-adic height of* P.

(1) *Let* $p\mathcal{O}_K = \pi\pi^c$. *Fix an identification* $\psi : K_\pi \simeq \mathbb{Q}_p$. *In particular,* $v_p(\psi(\pi)) = 1$.
(2) *Let* $m_0 = \mathrm{lcm}\{c_\ell\}$, *where* ℓ *runs through the primes of bad reduction for* $E/\mathbb{Q}$ *and* c_ℓ *is the Tamagawa number at* ℓ. *Compute*[4] $R = m_0P$.

---

[3]Note that by Gross and Zagier [8] and Kolyvagin [10] the analytic rank of $E^K/\mathbb{Q}$ being 1 implies that the algebraic rank of $E^K/\mathbb{Q}$ is 1 and the Tate–Shafarevich group of $E^K/\mathbb{Q}$ is finite. Furthermore, in this case, computing a non-torsion point in $E^K(\mathbb{Q})$ can be done by choosing an auxiliary imaginary quadratic field F satisfying the Heegner hypothesis for $E^K/\mathbb{Q}$ such that the analytic rank of E^K/F is 1 and computing the corresponding basic Heegner point in $E^K(F)$.

[4]Note that Steps 2 and 3 are needed to ensure that the point whose anticyclotomic p-adic height we will compute using formula (7) satisfies the required conditions.

(3) *Determine the smallest positive integer n such that nR and nR^c reduce to $0 \in E(\mathbb{F}_p)$ modulo π. Note that n is a divisor of $\#E(\mathbb{F}_p)$. Compute $T = nR$.*

(4) *Compute $\mathbf{d}_h(R) \in \mathcal{O}_K$ defined in (9) as a generator of the ideal*

$$\prod_{\mathfrak{q}} \mathfrak{q}^{h \operatorname{ord}_{\mathfrak{q}}(\delta(R))/2}$$

where h is the class number of K, the product is over all prime ideals $\mathfrak{q}$ of $\mathcal{O}_K$, and $\delta(R)$ is the denominator ideal of $x(R)\mathcal{O}_K$.

(5) *Let f_n denote the nth division polynomial associated to E. Compute $\mathbf{d}_h(T) = \mathbf{d}_h(nR) = f_n(R)^h \mathbf{d}_h(R)^{n^2}$. Note that by Step (2) and Proposition 1 of Wuthrich [20] we see that $f_n(R)^h \mathbf{d}_h(R)^{n^2} \in \mathcal{O}_K$ since $\mathbf{d}_h(T)$ is an element of K that is integral at every finite prime.*

(6) *Compute $\sigma_\pi(t) := \sigma_p(t)$ as a formal power series in $t\mathbb{Z}_p[[t]]$ with sufficient precision. This equality holds since our elliptic curve E is defined over $\mathbb{Q}$.*

(7) *We use Proposition 3.4 to determine the anticyclotomic p-adic height of T: compute*

$$\begin{aligned} h_\rho(T) &= \frac{1}{p}\log_p\left(\psi\left(\frac{\sigma_\pi(T)}{\sigma_\pi(T^c)}\right)\right) + \frac{1}{hp}\log_p\left(\psi\left(\frac{\mathbf{d}_h(T)^c}{\mathbf{d}_h(T)}\right)\right) \\ &= \frac{1}{p}\log_p\left(\psi\left(\frac{\sigma_p\left(\frac{-x(T)}{y(T)}\right)}{\sigma_p\left(\frac{-x(T)^c}{y(T)^c}\right)}\right)\right) + \frac{1}{hp}\log_p\left(\psi\left(\frac{\mathbf{d}_h(T)^c}{\mathbf{d}_h(T)}\right)\right) \\ &= \frac{1}{p}\log_p\left(\frac{\sigma_p\left(\psi\left(\frac{-x(T)}{y(T)}\right)\right)}{\sigma_p\left(\psi\left(\frac{-x(T)^c}{y(T)^c}\right)\right)}\right) + \frac{1}{hp}\log_p\left(\psi\left(\frac{\mathbf{d}_h(T)^c}{\mathbf{d}_h(T)}\right)\right). \end{aligned}$$

(8) *Output the anticyclotomic p-adic height of P: compute*[5]

$$h_\rho(P) = \frac{1}{n^2 m_0^2} h_\rho(T).$$

Algorithm 5.3. *Shadow line attached to (E, K, p).*

Input:

- *an elliptic curve $E/\mathbb{Q}$ (given by its reduced minimal model) of analytic rank* 2 *such that* $\operatorname{rank}_{\mathbb{Z}} E(\mathbb{Q}) = 2$;
- *an odd prime p of good ordinary reduction such that the p-primary part of the Tate–Shafarevich group of $E/\mathbb{Q}$ is finite;*
- *two $\mathbb{Z}$-independent points $P_1, P_2 \in E(\mathbb{Q})$;*

[5]As a consistency check we compute the height of nP and verify that $h_\rho(nP) = \frac{1}{n^2} h_\rho(P)$ for positive integers $n \leq 5$.

- *an imaginary quadratic field K such that*
  - *the analytic rank of E/K equals 3, and*
  - *p and all rational primes dividing the conductor of $E/\mathbb{Q}$ split in K.*

Output: *The slope of the shadow line $L_K \subseteq E(\mathbb{Q}) \otimes \mathbb{Q}_p$ with respect to the coordinate system induced by $\{P_1, P_2\}$.*

(1) *Use Algorithm 5.1 to compute a non-torsion point $S \in E(K)^-$. We then have generators P_1, P_2, S of $E(K) \otimes \mathbb{Q}_p$ such that $P_1, P_2 \in E(\mathbb{Q})$ and $S \in E(K)^-$ (given as points on the reduced minimal model of $E/\mathbb{Q}$).*
(2) *Compute $P_1 + S$ and $P_2 + S$.*
(3) *Use Algorithm 5.2 to compute[6] the anticyclotomic p-adic heights: $h_\rho(P_1 + S)$ and $h_\rho(P_2 + S)$. Finding that at least one of these heights is non-trivial implies that the shadow line associated to (E, K, p) lies in $E(\mathbb{Q}) \otimes \mathbb{Q}_p$, i.e., the Sign Conjecture holds for (E, K, p).*
(4) *The point $h_\rho(P_2 + S)P_1 - h_\rho(P_1 + S)P_2$ is a generator of the shadow line associated to (E, K, p). Output the slope of the shadow line $L_K \subseteq E(\mathbb{Q}) \otimes \mathbb{Q}_p$ with respect to the coordinate system induced by $\{P_1, P_2\}$: compute*

$$-h_\rho(P_1 + S)/h_\rho(P_2 + S) \in \mathbb{Q}_p.$$

6 Examples

Let E be the elliptic curve "389.a1" [17, Elliptic Curve 389.a1] given by the model

$$y^2 + y = x^3 + x^2 - 2x.$$

We know that the analytic rank of $E/\mathbb{Q}$ equals 2 [3, §6.1] and $\text{rank}_{\mathbb{Z}} E(\mathbb{Q}) = 2$, see [7]. In addition, 5 and 7 are good ordinary primes for E. We find two $\mathbb{Z}$-independent points

$$P_1 = (-1, 1), P_2 = (0, 0) \in E(\mathbb{Q}).$$

We will now use the algorithms described in Sect. 5 to compute the slopes of two shadow lines on $E(\mathbb{Q}) \otimes \mathbb{Q}_5$ with respect to the coordinate system induced by $\{P_1, P_2\}$.

6.1 *Shadow Line Attached to* ("389.a1", $\mathbb{Q}(\sqrt{-11})$, 5)

The imaginary quadratic field $K = \mathbb{Q}(\sqrt{-11})$ satisfies the Heegner hypothesis for E and the quadratic twist E^K has analytic rank 1. Moreover, the prime 5 splits in K.

[6] We compute the height of $P_1 + P_2 + S$ as a consistency check.

We use Algorithm 5.1 to find a non-torsion point $S = (\frac{1}{4}, \frac{1}{8}\sqrt{-11} - \frac{1}{2}) \in E(K)^-$. We now proceed to compute the anticyclotomic p-adic heights of $P_1 + S$ and $P_2 + S$ which are needed to determine the slope of the shadow line associated to the triple $(\text{“}389.a1\text{”}, \mathbb{Q}(\sqrt{-11}), 5)$. We begin by computing

$$A_1 := P_1 + S = \left(-\frac{6}{25}\sqrt{-11} + \frac{27}{25}, -\frac{62}{125}\sqrt{-11} + \frac{29}{125}\right),$$

$$A_2 := P_2 + S = (-2\sqrt{-11}, -4\sqrt{-11} - 12).$$

We carry out the steps of Algorithm 5.2 to compute $h_p(A_1)$:

(1) Let $5\mathcal{O}_K = \pi\pi^c$, where $\pi = (\frac{1}{2}\sqrt{-11} + \frac{3}{2})$ and $\pi^c = (-\frac{1}{2}\sqrt{-11} + \frac{3}{2})$. This allows us to fix an identification

$$\psi : K_\pi \to \mathbb{Q}_5$$

that sends

$$\frac{1}{2}\sqrt{-11} + \frac{3}{2} \mapsto 2 \cdot 5 + 5^2 + 3 \cdot 5^3 + 4 \cdot 5^4 + 4 \cdot 5^5 + 3 \cdot 5^7 + 5^8 + 5^9 + O(5^{10}).$$

(2) Since the Tamagawa number at 389 is trivial, i.e., $c_{389} = 1$, we have $m_0 = 1$. Thus $R = A_1$.

(3) We find that $n = 9$ is the smallest multiple of R and R^c such that both points reduce to 0 in $E(\mathcal{O}_K/\pi)$. Set $T = 9R$.

(4) Note that the class number of K is $h = 1$. We find $\mathbf{d}_h(R) = \frac{1}{2}\sqrt{-11} - \frac{3}{2}$.

(5) Let f_9 denote the 9th division polynomial associated to E. We compute

$$\begin{aligned}\mathbf{d}_h(T) &= \mathbf{d}_h(9R)\\ &= f_9(R)\mathbf{d}_h(R)^{9^2}\\ &= 24227041862247516754088925710922259344570\sqrt{-11}\\ &\quad - 147355399895912034115896942557395263175125.\end{aligned}$$

(6) We compute

$$\begin{aligned}\sigma_\pi(t) :&= \sigma_5(t)\\ &= t + \left(4 + 5 + 3 \cdot 5^2 + 5^3 + 2 \cdot 5^4 + 3 \cdot 5^5 + 2 \cdot 5^6 + O(5^8)\right) t^3\\ &\quad + \left(3 + 2 \cdot 5 + 2 \cdot 5^2 + 2 \cdot 5^3 + 2 \cdot 5^4 + 2 \cdot 5^5 + 2 \cdot 5^6 + O(5^7)\right) t^4\\ &\quad + \left(1 + 5 + 5^2 + 5^3 + 3 \cdot 5^4 + 3 \cdot 5^5 + O(5^6)\right) t^5\\ &\quad + \left(4 + 2 \cdot 5 + 2 \cdot 5^2 + 2 \cdot 5^3 + 3 \cdot 5^4 + O(5^5)\right) t^6\\ &\quad + \left(4 + 3 \cdot 5 + 4 \cdot 5^2 + O(5^4)\right) t^7 + \left(3 + 3 \cdot 5^2 + O(5^3)\right) t^8\\ &\quad + \left(3 \cdot 5 + O(5^2)\right) t^9 + (2 + O(5))\, t^{10} + O(t^{11}).\end{aligned}$$

(7) We use Proposition 3.4 to determine the anticyclotomic p-adic height of T: we compute

$$h_\rho(T) = \frac{1}{p}\log_p\left(\psi\left(\frac{\sigma_\pi(T)}{\sigma_\pi(T^c)}\right)\right) + \frac{1}{hp}\log_p\left(\psi\left(\frac{\mathbf{d}_h(T)^c}{\mathbf{d}_h(T)}\right)\right)$$

$$= \frac{1}{p}\log_p\left(\frac{\sigma_p\left(\psi\left(\frac{-x(T)}{y(T)}\right)\right)}{\sigma_p\left(\psi\left(\frac{-x(T)^c}{y(T)^c}\right)\right)}\right) + \frac{1}{hp}\log_p\left(\psi\left(\frac{\mathbf{d}_h(T)^c}{\mathbf{d}_h(T)}\right)\right)$$

$$= 3 + 5 + 5^2 + 4\cdot 5^4 + 3\cdot 5^5 + 4\cdot 5^7 + 3\cdot 5^8 + 5^9 + O(5^{10}).$$

(8) We output the anticyclotomic p-adic height of A_1:

$$h_\rho(A_1) = \frac{1}{9^2}h_\rho(T)$$

$$= 3 + 3\cdot 5 + 3\cdot 5^2 + 2\cdot 5^4 + 4\cdot 5^5 + 4\cdot 5^6 + 3\cdot 5^8 + O(5^{10}).$$

Repeating Steps (1)–(8) for A_2 yields

$$h_\rho(A_2) = 3 + 2\cdot 5 + 4\cdot 5^2 + 2\cdot 5^5 + 5^6 + 4\cdot 5^7 + 4\cdot 5^9 + O(5^{10}).$$

As a consistency check, we also compute

$$h_\rho(P_1 + P_2 + S) = 1 + 5 + 3\cdot 5^2 + 5^3 + 2\cdot 5^4 + 5^5 + 5^6 + 4\cdot 5^8 + 4\cdot 5^9 + O(5^{10}).$$

Observe that, numerically, we have

$$h_\rho(P_1 + P_2 + S) = h_\rho(P_1 + S) + h_\rho(P_2 + S).$$

The slope of the shadow line $L_K \subseteq E(\mathbb{Q}) \otimes \mathbb{Q}_p$ with respect to the coordinate system induced by $\{P_1, P_2\}$ is thus

$$-\frac{h_\rho(P_1 + S)}{h_\rho(P_2 + S)} = 4 + 2\cdot 5 + 5^2 + 3\cdot 5^3 + 5^4 + 5^6 + 5^7 + O(5^{10}).$$

6.2 *Shadow Line Attached to* ("389.*a*1", $\mathbb{Q}(\sqrt{-24})$, 5)

Consider the imaginary quadratic field $K = \mathbb{Q}(\sqrt{-24})$. Note that K satisfies the Heegner hypothesis for E, the twist E^K has analytic rank 1, and the prime 5 splits in K.

Using Algorithm 5.1 we find a non-torsion point

$$S = \left(\frac{1}{2}, \frac{1}{8}\sqrt{-24} - \frac{1}{2}\right) \in E(K)^{-}.$$

We then compute

$$P_1 + S = \left(-\frac{1}{6}\sqrt{-24} + \frac{1}{3}, -\frac{5}{18}\sqrt{-24} - 1\right)$$

$$P_2 + S = \left(-\frac{1}{2}\sqrt{-24} - 2, -6\right).$$

Many of the steps taken to compute $h_\rho(P_1 + S)$ and $h_\rho(P_2 + S)$ are quite similar to those in Sect. 6.1. One notable difference is that in this example the class number h of K is equal to 2. We find that

$$h_\rho(P_1 + S) = 4 + 2 \cdot 5 + 3 \cdot 5^4 + 2 \cdot 5^5 + 4 \cdot 5^6 + 2 \cdot 5^7 + 5^8 + 2 \cdot 5^9 + O(5^{10}),$$

$$h_\rho(P_2 + S) = 1 + 5 + 5^3 + 5^5 + 2 \cdot 5^6 + 4 \cdot 5^7 + 2 \cdot 5^8 + 3 \cdot 5^9 + O(5^{10}).$$

In addition, we compute $h_\rho(P_1 + P_2 + S)$ and verify that

$$\begin{aligned} h_\rho(P_1 + P_2 + S) &= 4 \cdot 5 + 5^3 + 3 \cdot 5^4 + 3 \cdot 5^5 + 5^6 + 2 \cdot 5^7 + 4 \cdot 5^8 + O(5^{10}) \\ &= h_\rho(P_1 + S) + h_\rho(P_2 + S). \end{aligned}$$

This gives that the slope of the shadow line $L_K \subseteq E(\mathbb{Q}) \otimes \mathbb{Q}_p$ with respect to the coordinate system induced by $\{P_1, P_2\}$ is

$$-\frac{h_\rho(P_1 + S)}{h_\rho(P_2 + S)} = 1 + 5 + 3 \cdot 5^2 + 3 \cdot 5^5 + 3 \cdot 5^6 + 3 \cdot 5^7 + 2 \cdot 5^8 + 5^9 + O(5^{10}).$$

6.3 *Summary of Results of Additional Computations of Shadow Lines*

The algorithms developed in Sect. 5 enable us to compute shadow lines in many examples which is what is needed to initiate a study of Question 4.1. We will now list some results of additional computations of slopes of shadow lines on the elliptic curve "389.a1". In Tables 1 and 2 we fix the prime $p = 5, 7$, respectively, and vary the quadratic field.

Table 1 Slopes of shadow lines for $\left(\text{“}389.a1\text{”}, K, 5\right)$

K	Slope
$\mathbb{Q}(\sqrt{-11})$	$4 + 2 \cdot 5 + 5^2 + 3 \cdot 5^3 + 5^4 + 5^6 + 5^7 + O(5^{10})$
$\mathbb{Q}(\sqrt{-19})$	$1 + 4 \cdot 5 + 2 \cdot 5^2 + 2 \cdot 5^3 + 2 \cdot 5^5 + 5^6 + 4 \cdot 5^7 + 3 \cdot 5^8 + 4 \cdot 5^9 + O(5^{10})$
$\mathbb{Q}(\sqrt{-24})$	$1 + 5 + 3 \cdot 5^2 + 3 \cdot 5^5 + 3 \cdot 5^6 + 3 \cdot 5^7 + 2 \cdot 5^8 + 5^9 + O(5^{10})$
$\mathbb{Q}(\sqrt{-59})$	$4 + 5 + 4 \cdot 5^2 + 5^3 + 2 \cdot 5^4 + 2 \cdot 5^5 + 3 \cdot 5^7 + 4 \cdot 5^8 + 2 \cdot 5^9 + O(5^{10})$
$\mathbb{Q}(\sqrt{-79})$	$2 + 5 + 2 \cdot 5^2 + 2 \cdot 5^3 + 4 \cdot 5^4 + 4 \cdot 5^5 + 3 \cdot 5^6 + 3 \cdot 5^7 + 3 \cdot 5^8 + 2 \cdot 5^9 + O(5^{10})$
$\mathbb{Q}(\sqrt{-91})$	$4 + 3 \cdot 5 + 5^2 + 5^4 + 2 \cdot 5^5 + 4 \cdot 5^6 + 5^7 + 2 \cdot 5^9 + O(5^{10})$
$\mathbb{Q}(\sqrt{-111})$	$5^{-2} + 4 \cdot 5^{-1} + 4 + 4 \cdot 5 + 2 \cdot 5^2 + 2 \cdot 5^3 + 4 \cdot 5^4 + 2 \cdot 5^5 + 3 \cdot 5^6 + 5^7 + 2 \cdot 5^8 + 5^9 + O(5^{10})$
$\mathbb{Q}(\sqrt{-119})$	$4 \cdot 5^{-1} + 2 + 2 \cdot 5 + 2 \cdot 5^2 + 4 \cdot 5^3 + 4 \cdot 5^4 + 2 \cdot 5^5 + 5^6 + 4 \cdot 5^7 + 4 \cdot 5^8 + 4 \cdot 5^9 + O(5^{10})$
$\mathbb{Q}(\sqrt{-159})$	$2 \cdot 5 + 4 \cdot 5^4 + 4 \cdot 5^5 + 5^6 + 5^7 + 4 \cdot 5^8 + 5^9 + O(5^{10})$
$\mathbb{Q}(\sqrt{-164})$	$3 + 2 \cdot 5 + 4 \cdot 5^2 + 5^3 + 4 \cdot 5^4 + 3 \cdot 5^5 + 3 \cdot 5^6 + 3 \cdot 5^8 + 4 \cdot 5^9 + O(5^{10})$

Table 2 Slopes of shadow lines for $(\text{“}389.a1\text{”}, K, 7)$

K	Slope
$\mathbb{Q}(\sqrt{-19})$	$3+2\cdot 7+2\cdot 7^2+3\cdot 7^3+7^4+7^5+4\cdot 7^7+6\cdot 7^9+O(7^{10})$
$\mathbb{Q}(\sqrt{-20})$	$1+5\cdot 7+6\cdot 7^2+6\cdot 7^3+2\cdot 7^4+3\cdot 7^5+3\cdot 7^6+3\cdot 7^7+O(7^{10})$
$\mathbb{Q}(\sqrt{-24})$	$1+3\cdot 7+3\cdot 7^2+2\cdot 7^3+6\cdot 7^4+2\cdot 7^5+2\cdot 7^6+6\cdot 7^7+2\cdot 7^8+O(7^{10})$
$\mathbb{Q}(\sqrt{-52})$	$1+5\cdot 7+7^2+3\cdot 7^3+3\cdot 7^4+2\cdot 7^5+5\cdot 7^6+3\cdot 7^9+O(7^{10})$
$\mathbb{Q}(\sqrt{-55})$	$1+7+6\cdot 7^2+3\cdot 7^3+5\cdot 7^4+3\cdot 7^5+7^7+4\cdot 7^9+O(7^{10})$
$\mathbb{Q}(\sqrt{-59})$	$2+7+3\cdot 7^2+3\cdot 7^3+5\cdot 7^4+5\cdot 7^5+2\cdot 7^6+4\cdot 7^7+7^8+6\cdot 7^9+O(7^{10})$
$\mathbb{Q}(\sqrt{-68})$	$4+4\cdot 7+2\cdot 7^3+5\cdot 7^4+5\cdot 7^5+7^6+7^7+5\cdot 7^8+5\cdot 7^9+O(7^{10})$
$\mathbb{Q}(\sqrt{-87})$	$3\cdot 7+4\cdot 7^2+7^3+2\cdot 7^4+2\cdot 7^5+7^6+5\cdot 7^7+7^9+O(7^{10})$
$\mathbb{Q}(\sqrt{-111})$	$7^{-2}+2\cdot 7^{-1}+5+2\cdot 7+7^2+2\cdot 7^3+6\cdot 7^4+5\cdot 7^5+7^6+2\cdot 7^7+2\cdot 7^9+O(7^{10})$
$\mathbb{Q}(\sqrt{-143})$	$5+5\cdot 7+2\cdot 7^3+4\cdot 7^4+3\cdot 7^5+3\cdot 7^6+2\cdot 7^7+2\cdot 7^9+O(7^{10})$

Acknowledgements The authors are grateful to the organizers of the conference "WIN3: Women in Numbers 3" for facilitating this collaboration and acknowledge the hospitality and support provided by the Banff International Research Station. During the preparation of this manuscript: the second author was partially supported by NSA grant H98230-12-1-0208 and NSF grant DMS-1352598; the third author was partially supported by NSF grant DGE-1144087.

References

1. Balakrishnan, J.S., Çiperiani, M., Stein, W.: p-adic heights of Heegner points and Λ-adic regulators. Math. Comp. **84**(292), 923–954 (2015)
2. Bertolini, M.: Selmer groups and Heegner points in anticyclotomic $\mathbb{Z}_p$-extensions. Compos. Math. **99**(2), 153–182 (1995)
3. Bradshaw, R.W.: Provable computation of motivic L-functions. Ph.D. Thesis, University of Washington (2010)
4. Çiperiani, M.: Tate-Shafarevich groups in anticyclotomic $\mathbb{Z}_p$-extensions at supersingular primes. Compos. Math. **145**, 293–308 (2009)
5. Çiperiani, M., Wiles, A.: Solvable points on genus one curves. Duke Math. J. **142**, 381–464 (2008)
6. Cornut, C.: Mazur's conjecture on higher Heegner points. Invent. Math. **148**, 495–523 (2002)
7. Cremona, J.E.: Algorithms for Modular Elliptic Curves, 2nd edn. Cambridge University Press, Cambridge (1997)
8. Gross, B., Zagier, D.: Heegner points and derivatives of L-series. Invent. Math. **84**(2), 225–320 (1986)
9. Kato, K.: p-adic Hodge theory and values of zeta functions of modular forms, Cohomologies p-adiques et application arithmétiques. III. In: Astérisque, vol. 295. Société Mathématique de France, Paris (2004)
10. Kolyvagin, V.A.: Euler systems. In: The Grothendieck Festschrift, vol. II. Progress in Mathematics, vol. 87, pp. 435–483. Birkhauser, Boston, MA (1990)
11. Mazur, B., Rubin, K.: Studying the growth of Mordell-Weil. Doc. Math. Extra Volume, 585–607 (2003)
12. Mazur, B., Tate, J.: Canonical height pairings via biextensions. In: Arithmetic and Geometry, vol. I, pp. 195–237. Progress in Mathematics, vol. 35. Birkhäuser, Boston, MA (1983)
13. Mazur, B., Stein, W., Tate, J.: Computation of p-adic heights and log convergence. Doc. Math. Extra Volume, 577–614 (2006)
14. Nekovář, J.: On the parity of ranks of Selmer groups. II. C. R. Acad. Sci. Paris Sèr. I Math. **332**(2), 99–104 (2001)
15. Skinner, C., Urban, C.: The Iwasawa main conjectures for GL_2. Invent. Math. **195**(1), 1–277 (2014)
16. Stein, W., Wuthrich, C.: Algorithms for the arithmetic of elliptic curves using Iwasawa theory. Math. Comp. **82**(283), 1757–1792 (2013)
17. The LMFDB Collaboration, The L-functions and Modular Forms Database (2014). http://www.lmfdb.org (Online; accessed 14 October 2014)
18. Vatsal, V.: Special values of anticyclotomic L-functions. Duke Math. J. **116**, 219–261 (2003)
19. Washington, L.C.: Introduction to Cyclotomic Fields, 2nd edn. Graduate Texts in Mathematics, vol. 83. Springer, New York (1997)
20. Wuthrich, C.: On p-adic heights in families of elliptic curves. J. Lond. Math. Soc. (2) **70**(1), 23–40 (2004)

Galois Action on the Homology of Fermat Curves

Rachel Davis, Rachel Pries, Vesna Stojanoska, and Kirsten Wickelgren

Abstract In Anderson (Duke Math J 54(2):501 – 561, 1987), the author determines the homology of the degree n Fermat curve as a Galois module for the action of the absolute Galois group $G_{\mathbb{Q}(\zeta_n)}$. In particular, when n is an odd prime p, he shows that the action of $G_{\mathbb{Q}(\zeta_p)}$ on a more powerful relative homology group factors through the Galois group of the splitting field of the polynomial $1-(1-x^p)^p$. If p satisfies Vandiver's conjecture, we give a proof that the Galois group G of this splitting field over $\mathbb{Q}(\zeta_p)$ is an elementary abelian p-group of rank $(p+1)/2$. Using an explicit basis for G, we completely compute the relative homology, the homology, and the homology of an open subset of the degree 3 Fermat curve as Galois modules. We then compute several Galois cohomology groups which arise in connection with obstructions to rational points.

Keywords Fermat curve • Cyclotomic field • Homology • Cohomology • Galois module • Étale fundamental group

MSC2010: 11D41, 11R18, 11R34, 14F35, 14G25, 55S35

R. Davis
Department of Mathematics, Purdue University, West Lafayette, IN, USA
e-mail: davis705@math.purdue.edu

R. Pries (✉)
Department of Mathematics, Colorado State University, Fort Collins, CO, USA
e-mail: pries@math.colostate.edu

V. Stojanoska
Department of Mathematics, University of Illinois at Urbana-Champaign, Urbana, IL, USA,
e-mail: vesna@illinois.edu

K. Wickelgren
School of Mathematics, Georgia Institute of Technology, Atlanta, GA, USA
e-mail: kwickelgren3@math.gatech.edu

© Springer International Publishing Switzerland 2016

E.E. Eischen et al. (eds.), *Directions in Number Theory*, Association for Women in Mathematics Series 3, DOI 10.1007/978-3-319-30976-7_3

1 Introduction

The Galois actions on the étale homology, cohomology, and homotopy groups of varieties carry information about rational points. We revisit results of Anderson [2] on a relative homology group of the Fermat curve of prime exponent to make his results amenable to computations of groups such as $H^1(G_S, \pi_1^{ab})$ and $H^2(G_S, \pi_1^{ab} \wedge \pi_1^{ab})$ where G_S denotes a Galois group of a maximal extension of a number field with restricted ramification and π_1^{ab} denotes the abelianized geometric fundamental group of the Fermat curve, or of an open subset. These groups arise in obstructions of Ellenberg to rational points [8] as well as in McCallum's application of the method of Coleman and Chabauty to Fermat curves [13].

Let k be a number field. The Fermat curve of exponent n is the smooth projective curve $X \subset \mathbb{P}^2_k$ of genus $g = (n-1)(n-2)/2$ given by the equation

$$x^n + y^n = z^n.$$

The affine open $U \subset X$ given by $z \neq 0$ has affine equation $x^n + y^n = 1$. The closed subscheme $Y \subset X$ defined by $xy = 0$ consists of $2n$ points. Let $H_1(U, Y; \mathbb{Z}/n)$ denote the étale homology group of the pair $(U \otimes \overline{k}, Y \otimes \overline{k})$, which is a continuous module over the absolute Galois group G_k of k. The $\mu_n \times \mu_n$ action on X given by

$$(\zeta^i, \zeta^j) \cdot [x, y, z] = [\zeta^i x, \zeta^j y, z], \qquad (\zeta^i, \zeta^j) \in \mu_n \times \mu_n$$

determines an action on U and Y. These actions give $H_1(U, Y; \mathbb{Z}/n)$ the structure of a $(\mathbb{Z}/n)[\mu_n \times \mu_n]$ module. As a $(\mathbb{Z}/n)[\mu_n \times \mu_n]$ module, $H_1(U, Y; \mathbb{Z}/n)$ is free of rank one [2, Theorem 6], with generator denoted by β. It follows that the Galois action of $\sigma \in G_k$ is determined by $\sigma\beta = B_\sigma \beta$ for some $B_\sigma \in (\mathbb{Z}/n)[\mu_n \times \mu_n]$.

Anderson shows that B_σ is determined by an analogue of the classical gamma function $\Gamma_\sigma \in \mathbb{Z}/n^{\text{sh}}[\mu_n]$, where $\mathbb{Z}/n^{\text{sh}}$ denotes the strict Henselization of $\mathbb{Z}/n$. In particular, there is a formula [2, Theorems 9 and 7] recalled in (2) as the equation $d'^{sh}(\Gamma_\sigma) = B_\sigma$ with d'^{sh} defined in (1) and immediately below. The canonical derivation $d : \mathbb{Z}/n^{\text{sh}}[\mu_n] \to \Omega\mathbb{Z}/n^{\text{sh}}[\mu_n]$ from the ring $\mathbb{Z}/n^{\text{sh}}[\mu_n]$ to its module of Kähler differentials allows one to take the logarithmic derivative $\operatorname{dlog} \Gamma_\sigma$ of Γ_σ, which is convenient to view as an element of a particular quotient of $\Omega\mathbb{Z}/n^{\text{sh}}[\mu_n]$. See Sect. 2. For n prime, $\operatorname{dlog} \Gamma_\sigma$ determines B_σ uniquely [2, 10.5.2,10.5.3]. The function $\sigma \mapsto \operatorname{dlog} \Gamma_\sigma$ is in turn determined by a relative homology group of the punctured affine line $H_1(\mathbb{A}^1 - V(\sum_{i=0}^{n-1} x^i), \{0, 1\}; \mathbb{Z}/n)$ [2, Theorem 10]. Putting this together, Anderson shows that, for $n = p$ a prime, the $G_{\mathbb{Q}(\zeta_p)}$ action on $H_1(U, Y; \mathbb{Z}/p)$ factors through $\text{Gal}(L/\mathbb{Q}(\zeta_p))$ where L is the splitting field of $1 - (1 - x^p)^p$. Ihara [11] and Coleman [6] obtain similar results from different viewpoints.

Let K denote the cyclotomic field $K = \mathbb{Q}(\zeta_n)$, where ζ_n denotes a primitive nth root of unity, and let G_K be its absolute Galois group. When the exponent is clear, let ζ denote ζ_n or ζ_p for a prime p. Let κ denote the classical Kummer map; for $\theta \in K^*$, let $\kappa(\theta) : G_K \to \mu_n$ be defined by

$$\kappa(\theta)(\sigma) = \frac{\sigma\sqrt[n]{\theta}}{\sqrt[n]{\theta}}.$$

In Proposition 2, we determine dlog Γ_σ in terms of the classical Kummer map for all $n \geq 3$, modulo indeterminacy which does not affect B_σ, with the answer being dlog $\Gamma_\sigma = \sum_{i=1}^{n-1} \kappa(1-\zeta^{-i})(\sigma)\zeta^i \operatorname{dlog} \zeta$.

Recall that Vandiver's conjecture for a prime p is that p does not divide h^+, where h^+ is the order of the class group of $\mathbb{Q}(\zeta_p + \zeta_p^{-1})$. It has been verified for all p less than 163 million. For $n = p$ a prime satisfying Vandiver's conjecture, we give a proof that Gal(L/K) is isomorphic to $(\mathbb{Z}/p)^r$ with $r = (p+1)/2$ in Proposition 1. This is false for p not satisfying Vandiver's conjecture as seen in Remark 4. There are a couple of natural choices for such an isomorphism. In Corollary 1, we show that the following map gives an isomorphism:

$$\Phi = \kappa(\zeta) \times \prod_{i=1}^{\frac{p-1}{2}} \kappa(1-\zeta^{-i}) : \mathrm{Gal}(L/K) \to (\mu_p)^{\frac{p+1}{2}}.$$

For $p = 3$, we use the formula for dlog Γ_σ to compute B_σ explicitly in Lemma 6. It is possible to extend this calculation to compute B_σ for all primes p and we will make this computation available in a forthcoming paper. (As seen in Remark 6, the element dlog Γ_σ and [2, 10.5.2] do not determine B_σ when n is not prime so the calculation of B_σ when n is not prime will require further input.) Combining the above, we obtain:

Theorem 1. *Let $p = 3$ and $K = \mathbb{Q}(\zeta_p)$. The G_K-action on $H_1(U, Y; \mathbb{Z}/p)$ factors through $G_K \to \mathrm{Gal}(L/K)$, where L denotes the splitting field of $1-(1-x^p)^p$ (or equivalently of $x^6 - 3x^3 + 3$). Write $H_1(U, Y; \mathbb{Z}/p) \cong \mathbb{Z}_p[\zeta_0, \zeta_1]/\langle \zeta_0^p - 1, \zeta_1^p - 1\rangle$ and $\mathrm{Gal}(L/K) \cong \mathbb{Z}/p \times \mathbb{Z}/p$. Then $(c_0, c_1) \in \mathbb{Z}/p \times \mathbb{Z}/p$ acts on $\mathbb{Z}_p[\zeta_0, \zeta_1]/\langle \zeta_0^p - 1, \zeta_1^p - 1\rangle$ by multiplication by $B_\sigma = \sum_{i,j=0}^{p-1} b_{i,j}\zeta_0^i\zeta_1^j$ where*

$$b_{0,0} = 1 + c_0 - c_0^2$$

$$b_{0,1} = c_1 - c_0^2$$

$$b_{1,1} = -c_1 - c_0^2.$$

and where the rest of the coefficients $b_{i,j}$ are determined by $b_{i,j} = b_{j,i}$, and the fact that $b_{0,0} + b_{0,1} + b_{0,2} = 1$, $b_{1,0} + b_{1,1} + b_{1,2} = 0$, and $b_{2,0} + b_{2,1} + b_{2,2} = 0$.

We have an analogous calculation of $H_1(U, Y; \mathbb{Z}/p)$ for all primes p satisfying Vandiver's conjecture, which we will make available shortly.

Given the Galois action on $H_1(U, Y; \mathbb{Z}/n)$, we compute the Galois actions on $H_1(U; \mathbb{Z}/n)$ and $H_1(X; \mathbb{Z}/n)$ for all $n \geq 3$ in Sect. 6.

These computations can be used to study rational points on varieties in the following way. Let Z be a scheme over k, and for simplicity assume that Z has a rational point b. (This assumption is unnecessary, but it is satisfied in the situations encountered in this paper and it simplifies the exposition.) Choose a geometric point of Z with image b and let $\pi = \pi_1(Z_{\bar{k}}, b)$ denote the geometric étale fundamental group of Z based at the chosen geometric point. The generalized *Kummer map* associated with Z and b is the map $\kappa : Z(k) \to H^1(G_k, \pi)$ defined by

$$\kappa(x) = [\sigma \mapsto \gamma^{-1}\sigma\gamma]$$

where γ is an étale path from b to a geometric point above x. Before returning to the potential application to rational points, we remark that the map κ is functorial and the computation of dlog Γ_σ in Proposition 2 is obtained by applying κ to the K-map $\mathbb{A}^1 - V(\sum_{i=0}^{n-1} x^i) \to \mathbb{G}_m^{n-1}$.

From κ, we also obtain a map $\kappa^{\mathrm{ab,p}} : Z(k) \to H^1(G_k, \pi^{\mathrm{ab}} \otimes \mathbb{Z}_p)$ defined to be the composition of κ with the map $H^1(G_k, \pi) \to H^1(G_k, \pi^{\mathrm{ab}} \otimes \mathbb{Z}_p)$ induced by the quotient map $\pi \to \pi^{\mathrm{ab}} \otimes \mathbb{Z}_p$, where $\mathbb{Z}_p$ denotes the p-adic integers. For Z a curve or abelian variety over a number field, $\kappa^{\mathrm{ab,p}}$ is well-known to be injective. Let S denote a set of places of k including the infinite places, all the primes of bad reduction of Z and a place above p. Let $G_S = \pi_1(\mathcal{O}_k[1/S])$ denote the Galois group of the maximal extension of k ramified only over S. Assume that Z is proper to simplify exposition. Then $\kappa^{\mathrm{ab,p}}$ factors through a map $\kappa^{\mathrm{ab,p}} : Z(k) \to H^1(G_S, \pi^{\mathrm{ab}} \otimes \mathbb{Z}_p)$. Let $\pi = [\pi]_1 \supseteq [\pi]_2 \supseteq \ldots$ denote the lower central series of the profinite group π, where $[\pi]_m$ is the closure of the subgroup $[[\pi]_{m-1}, \pi]$ generated by commutators of elements of π with elements of $[\pi]_{m-1}$. Using work of Schmidt and Wingberg [17], Ellenberg [8] defines a series of obstructions to a point of the Jacobian of a curve Z lying in the image of the Abel–Jacobi map associated with b. The first of these obstructions is defined using a map

$$\delta_2 : H^1(G_S, \pi^{\mathrm{ab}} \otimes \mathbb{Z}_p) \to H^2(G_S, ([\pi]_2/[\pi]_3) \otimes \mathbb{Z}_p)$$

such that $\operatorname{Ker}\delta_2 \supset Z(k)$. Zarkhin defines a similar map [21]. The group $([\pi]_2/[\pi]_3) \otimes \mathbb{Z}_p$ fits into a short exact sequence

$$0 \to \mathbb{Z}_p(1) \to (\pi^{\mathrm{ab}} \wedge \pi^{\mathrm{ab}}) \otimes \mathbb{Z}_p \to ([\pi]_2/[\pi]_3) \otimes \mathbb{Z}_p \to 0.$$

There are mod p versions of δ_2 and the generalized Kummer maps. A more detailed account of Ellenberg's obstructions is in [20].

Thus computations of $H^1(G_S, \pi^{\mathrm{ab}} \otimes \mathbb{Z}/p)$ and $H^2(G_S, (\pi^{\mathrm{ab}} \wedge \pi^{\mathrm{ab}}) \otimes \mathbb{Z}/p)$ give information about rational points. Groups closely related to $H^1(G_S, \pi^{\mathrm{ab}} \otimes \mathbb{Z}/p)$ also appear in [5, 13].

The final section of this paper includes calculations of $H^1(\mathrm{Gal}(L/K), M)$ and $H^2(\mathrm{Gal}(L/K), M)$ for M each of $H_1(U, Y; \mathbb{Z}/n)$, $H_1(U; \mathbb{Z}/n)$, $H_1(X; \mathbb{Z}/n)$, and

$$H_1(U, \mathbb{Z}/n) \wedge H_1(U, \mathbb{Z}/n).$$

These can be inserted into the Hochschild–Serre spectral sequence

$$H^i(\mathrm{Gal}(L/K), H^j(G_{S,L}, M)) \Rightarrow H^{i+j}(G_{S,K}, M),$$

where $G_{S,L}$ denotes the Galois group of the maximal extension of L only ramified at places above S, and $G_{S,K} = G_S$. Since $H_1(U, Y; \mathbb{Z}/n)$, $H_1(U; \mathbb{Z}/n)$, and $H_1(X; \mathbb{Z}/n)$ are $\pi^{\mathrm{ab}} \otimes \mathbb{Z}/n$ for $Z = U/Y$, $Z = U$, and $Z = X$, respectively, these are groups mentioned above, and appear in Ellenberg's obstructions. This is the subject of on-going work.

1.1 Notation

Let $n \geq 3$ be an integer; often n will be a prime p. Let ζ be a fixed primitive nth root of unity and $K = \mathbb{Q}(\zeta)$. For brevity, let $A = \mathbb{Z}/n$, and let A^{sh} denote the strict Henselization of A. If $n = p$ is prime, then the field A is a Henselian local ring and its strict Henselization is the separable closure $A^{sh} \simeq \bar{\mathbb{F}}_p$.

If k is any number field, G_k denotes the absolute Galois group of k.

Definition 1. Given a primitive nth root $\sqrt[n]{\theta}$ of $\theta \in k$ and $\sigma \in G_k$, then $\kappa(\theta)\sigma$ is the element of A such that

$$\sigma \sqrt[n]{\theta} = \zeta^{\kappa(\theta)\sigma} \sqrt[n]{\theta}.$$

Remark 1. The map $\kappa : k^* \to H^1(G_k, \mathbb{Z}/n(1))$ defined by letting $\kappa(\theta)$ be represented by the twisted homomorphism $\sigma \mapsto \kappa(\theta)\sigma$ is the generalized Kummer map of $\mathbb{G}_{m,k}$ with base point $1 \in \mathbb{G}_{m,k}(k)$. Here $\mathbb{Z}/n(1)$ is the Galois module with underlying group $\mathbb{Z}/n$ and Galois action given by the cyclotomic character. See, for example, [20, 12.2.1, Example 1].

For $\theta \in K^*$ and $n = p$, the map $\kappa(\theta) : G_K \to \mathbb{Z}/p$ is a homomorphism and is independent of the choice of pth root of θ because $\mu_p \subset K$.

2 Anderson's Results, Revisited

In this section, we recall results from [2] that are relevant for this paper. Recall that $K = \mathbb{Q}(\zeta)$, that $U \subset \mathbb{A}^2_K$ denotes the affine Fermat curve over K with equation $x^n + y^n = 1$, and that $Y \subset U$ is the divisor defined by $xy = 0$. The path $\beta : [0, 1] \to U(\mathbb{C})$ given by $t \mapsto (\sqrt[n]{t}, \sqrt[n]{1-t})$, where $\sqrt[n]{-}$ denotes the real nth root, determines a singular 1-simplex in the homology of U relative to Y whose class we denote by the same name.

For $m \in \mathbb{N}$, let Λ_m denote the group ring over A of the finite group $\mu_n(\mathbb{C})^{\times(m+1)}$. Then Λ_m has a natural G_K-action. For $0 \leq i \leq m$, let ζ_i denote a primitive nth root of unity in the ith copy of $\mu_n(\mathbb{C})$. Then

$$\Lambda_m = A[\zeta_0, \ldots, \zeta_m]/(\zeta_0^n - 1, \ldots, \zeta_m^n - 1).$$

There is an action of Λ_1 on U given by $\zeta_0^i \times \zeta_1^j : (x, y) \mapsto (\zeta_0^i x, \zeta_1^j y)$. This action stabilizes Y. Thus the relative homology group $H_1(U, Y; A)$ is a Λ_1-module. Note that $H_1(U, Y; A)$ has rank n^2 over A.

Anderson describes the G_K-action on $H_1(U, Y; A)$. First, [2, Theorem 6] states that $H_1(U, Y; A)$ is a free rank one module over Λ_1 generated by the class β.

Specifically, $\sigma \in G_K$ acts A-linearly, and

$$\sigma \cdot (\zeta_0^i \zeta_1^j \beta) = (\sigma \cdot \zeta_0^i)(\sigma \cdot \zeta_1^j) B_\sigma \beta,$$

where B_σ is a unit in Λ_1 defined by

$$\sigma \cdot \beta = B_\sigma \beta.$$

Thus to describe the G_K-action on $H_1(U, Y; A)$, it is necessary and sufficient to describe the action on the element β.

Anderson also proves that the action of the absolute Galois group $G_{\mathbb{Q}}$ on $H_1(U, Y; A)$ factors through a finite quotient. This result is a consequence of the analysis in the rest of the section. In particular, if n is a prime p, then $\sigma \in G_K$ acts trivially on $H_1(U, Y; A)$ if and only if σ fixes the splitting field L of the polynomial $f_p = 1-(1-x^p)^p$ [2, Sect. 10.5]. In Sect. 3, we prove that $\mathrm{Gal}(L/K)$ is an elementary abelian p-group of rank at most $(p + 1)/2$.

Anderson highlights the following application of this result. By [2, Lemma, page 558], there is a connection between the action of $\sigma \in G_{\mathbb{Q}}$ on $H_1(U, Y; A)$ and the action of σ on the fields of definition of points of a generalized Jacobian of X.

Theorem 2 ([2], Theorem 0). *Let S be the generalized Jacobian of X with conductor ∞. Let b denote the $\mathbb{Q}$-rational point of S corresponding to the difference of the points* $(0, 1)$ *and* $(1, 0)$*. The number field generated by the coordinates of the nth roots of b in $S(\overline{\mathbb{Q}})$ contains the splitting field L of the polynomial* $1 - (1 - x^n)^n$*, with equality if n is prime.*

Information on fields generated by points of the Jacobian of quotients of Fermat curves is also contained in [5, 7, 10, 18].

In the remainder of this section, we describe Anderson's method for determining B_σ. Let $b_{i,j}$ denote the coefficients of B_σ, so that

$$B_\sigma = \sum_{0 \le i,j < n} b_{i,j} \zeta_0^i \zeta_1^j.$$

It will often be convenient to arrange the coefficients of B_σ in an $n \times n$ matrix.

Let $w : \Lambda_1 \to \Lambda_1$ be the map induced by swapping the two copies of $\mu_n(\mathbb{C})$, i.e., by swapping ζ_0 and ζ_1. Then w preserves the units in Λ_1. Let $(\Lambda_1^\times)^w$ denote the symmetric units, i.e., the units fixed by w. If $a_{i,j} \in A$, then an element

$$\sum_{0 \le i,j < n} a_{i,j} \zeta_0^i \zeta_1^j \in \Lambda_1^\times$$

is in $(\Lambda_1^\times)^w$ precisely when $a_{i,j} = a_{j,i}$ for all i, j.

Fact 3 ([2, Theorem 7]). *If $\sigma \in G_{\mathbb{Q}}$, then $B_\sigma \in (\Lambda_1^\times)^w$. In other words, the coefficients of B_σ are symmetric; $b_{i,j} = b_{j,i}$ for any $0 \le i, j < n$.*

Next, consider the map $d'' : (\Lambda_1^\times)^w \to \Lambda_2^\times$ given by

$$\sum a_{i,j}\zeta_0^i\zeta_1^j \mapsto \frac{\left(\sum a_{i,j}\zeta_0^j\zeta_1^i\zeta_2^j\right)\left(\sum a_{i,j}\zeta_0^i\zeta_2^j\right)}{\left(\sum a_{i,j}\zeta_0^i\zeta_1^j\zeta_2^j\right)\left(\sum a_{i,j}\zeta_1^i\zeta_2^j\right)}.$$

By [2, Theorem 7], B_σ is in the kernel of d''. In particular, there is an equality in $\Lambda_2^\times$, given by

$$\left(\sum b_{i,j}\zeta_0^j\zeta_1^i\zeta_2^j\right)\left(\sum b_{i,j}\zeta_0^i\zeta_2^j\right) = \left(\sum b_{i,j}\zeta_0^i\zeta_1^j\zeta_2^j\right)\left(\sum b_{i,j}\zeta_1^i\zeta_2^j\right).$$

This gives, via the map $\Lambda_2^\times \to \Lambda_1^\times$ sending $\zeta_2 \mapsto 1$, the equality

$$\left(\sum b_{i,j}\zeta_0^j\zeta_1^i\right)\left(\sum b_{i,j}\zeta_0^i\right) = \left(\sum b_{i,j}\zeta_0^i\zeta_1^j\right)\left(\sum b_{i,j}\zeta_1^i\right).$$

By Fact 3, the first terms on each side cancel giving

$$\sum b_{i,j}\zeta_0^i = \sum b_{i,j}\zeta_1^j.$$

This is only possible if the following is true.

Fact 4 ([2, 10.5.4]). *If $1 \le i \le n$, then $\sum_{0 \le j < n} b_{i,j} = 0$.*

In other words, the entries of each column of the matrix B_σ sum up to zero, for all but the zeroth column. By Fact 3, the entries of each row of the matrix B_σ also sum up to zero, for all but the zeroth row.

Furthermore, consider the map $d' : \Lambda_0^\times \to (\Lambda_1^\times)^w$ given by

$$\sum a_i\zeta_0^i \mapsto \frac{\left(\sum a_i\zeta_0^i\right)\left(\sum a_i\zeta_1^i\right)}{\left(\sum a_i\zeta_0^i\zeta_1^i\right)}, \tag{1}$$

as well as its extension $d'^{sh} : \bar{\Lambda}_0^\times \to (\bar{\Lambda}_1^\times)^w$, where $\bar{\Lambda}_i = \Lambda_i \otimes_A A^{sh}$. The kernel $\mathrm{Ker}(d'^{sh})$ is determined in [2, Proposition 8.3.1]; when $n = p$ is prime, it is the cyclic subgroup of order p multiplicatively generated by ζ_0.

Fact 5 ([2, Theorem 9]). *Let $n \ge 3$ and let $\mathrm{Ker}(d'^{sh})$ denote the kernel of d'^{sh}. In $\bar{\Lambda}_0^\times/\mathrm{Ker}(d'^{sh})$, there exists a unique element Γ_σ which maps to B_σ under d'^{sh}.*

In the sequel, the notation Γ_σ will also be used to denote an element of $\bar{\Lambda}_0^\times$ representing this coset in $\bar{\Lambda}_0^\times/\mathrm{Ker}(d'^{sh})$.

Fact 6 ([2, 9.6 and 10.5.2]). *The difference $B_\sigma - 1$ lies in the augmentation ideal* $(1-\epsilon_0)(1-\epsilon_1)\Lambda_1$.

Consider the element Γ_σ such that

$$d'^{sh}(\Gamma_\sigma) = B_\sigma. \tag{2}$$

By Fact 5, in order to determine B_σ, it suffices to find the preimage Γ_σ. To accomplish this, Anderson looks at the logarithmic derivative homomorphisms from the groups of units $\Lambda_k^\times$ to the Kähler differentials $\Omega(\Lambda_k)$. This has the geometric meaning of comparing with "the circular motive," where the Galois action is more transparent.

There is a commutative square

$$\begin{array}{ccc} \bar{\Lambda}_0^\times & \xrightarrow{d'} & (\bar{\Lambda}_1^\times)^w \\ \downarrow{\scriptstyle \mathrm{dlog}} & & \downarrow{\scriptstyle \mathrm{dlog}} \\ \Omega(\bar{\Lambda}_0) & \longrightarrow & \Omega(\bar{\Lambda}_1)^w, \end{array}$$

where the bottom horizontal map is defined analogously to d'. Note that for each m, the $\bar{\Lambda}_m$-module $\Omega(\bar{\Lambda}_m)$ is free on generators $\{\mathrm{dlog}\,\zeta_i\}_{0\le i\le m}$.

Here is some notation needed to describe $\mathrm{dlog}\,\Gamma_\sigma$. Let $\tilde{V} = \mathbb{A}^1 - \mu_n$ and let $V = \tilde{V} \cup \{1\}$. Let λ_0 be a small counterclockwise loop around 1. Choose the isomorphism

$$H_1(\tilde{V};A) = A[\mu_n]\lambda_0 \simeq \Omega A[\Lambda_0],$$

where $\lambda_0 \mapsto \frac{d\epsilon_0}{\epsilon_0}$.

Consider the exact sequence from [2, §9]

$$0 \to A\lambda_0 \to H_1(\tilde{V};A) \to H_1(V;A) \to 0,$$

or

$$0 \to A\frac{d\epsilon_0}{\epsilon_0} \to \Omega A[\Lambda_0] \to H_1(V;A) \to 0, \tag{3}$$

which identifies $H_1(V;A)$ as a quotient of $\Omega(\Lambda_0)$.

Let Z denote the subscheme of V defined by the vanishing of $x_0(1-x_0)$, i.e., the points 0 and 1 in V. Let $\psi \in H_1(V,Z;A)$ denote the homology class represented by the cycle given by the interval $[0,1]$. Let $(\sigma-1)\psi$ denote the cycle given by concatenating the path $\sigma\psi$ and the path ψ traveled in reverse. Since G_K fixes the endpoints of ψ, the cycle $(\sigma-1)\psi$ represents a class in $H_1(V;A) = H_1(V,\emptyset;A)$.

Let Ψ_σ denote the coset in $\Omega(\Lambda_0)/A\,\mathrm{dlog}\,\zeta_0$ which corresponds to the homology class of $(\sigma-1)\psi$ under (3). The following theorem computes $\mathrm{dlog}\,\Gamma_\sigma$ to be Ψ_σ.

Theorem 7 ([2, Theorem 10]). $\text{dlog}\,\Gamma_\sigma \in \Omega(\bar{\Lambda}_0)$ *represents the* A^{sh} dlog ζ_0*-coset* Ψ_σ.

For this paper, the importance of Theorem 7 lies in the geometric description of Ψ_σ. This description shows that $\sigma \mapsto \Psi_\sigma$ is the image of a rational point under a generalized Kummer map of the sort which arises in the section conjecture. We use this observation to compute Ψ_σ in Sect. 4. By Theorem 7, we have therefore also computed dlog Γ_σ.

To give a complete description of $H_1(U, Y; A)$ as a Galois module, it thus suffices to achieve the following goal, which we complete in Sect. 5 for the case $n = 3$ and in future work for n an odd prime.

Goal: reconstruct B_σ from Ψ_σ.

3 Galois Group of the Splitting Field of $1 - (1 - x^p)^p$ over K

Let $n = p$ be an odd prime and let ζ be a primitive pth root of unity. The choice of ζ fixes an identification $\mathbb{Z}/p \to \mu_p$ by sending i to ζ^i. Let $K = \mathbb{Q}(\zeta)$.

Let L be the splitting field of the polynomial $f_p(x) = 1 - (1 - x^p)^p$. In Proposition 1, we determine the structure of the Galois group $G = \text{Gal}(L/K)$ for primes p satisfying Vandiver's conjecture. The techniques in this section are well-known to experts but we could not find an off-the-shelf reference for this result. Before starting the proof, we describe some motivation for it in the next remark.

Remark 2.

1. As seen in Theorem 2, L is the field of definition of the pth roots of a point b in a certain generalized Jacobian. By [2, Sect. 10.5], an automorphism $\sigma \in G_K$ acts trivially on $H_1(U, Y; A)$ if and only if $\sigma \in G_L$. In view of this result, to determine the action of G_K on $H_1(U, Y; A)$, it remains to determine the action of the finite Galois group $\text{Gal}(L/K)$.
2. We would like to thank the referee for pointing out related work in [10]. Recall that the Jacobian of the Fermat curve X of exponent p is isogenous to $\mathbb{J} = \prod_{a=1}^{p-2} J_a$ where J_a is the Jacobian of the curve $y^p = x^a(1-x)$. Consider the field extension $L_{\mathbb{J}}$ of $\mathbb{Q}$ generated by the points of order p on $\mathbb{J}$. In [10, Theorem 4], Greenberg proves that $L_{\mathbb{J}}$ is the field $K(\{\sqrt[p]{\eta} \mid \eta \in C^+)$ generated over K by the pth roots of the real cyclotomic units. (Note that Lemma 2 below implies that $L_{\mathbb{J}} \subset L$.) He remarks that $\text{Gal}(L_{\mathbb{J}}/K) \simeq (\mathbb{Z}/p)^t$ with $t \leq (p-3)/2$ and that $t = (p-3)/2$ when p satisfies Vandiver's conjecture.
3. We would like to thank Sharifi for pointing out similar work in [1, Sect. 2.8], where the authors determine the Galois group of the Galois closure of $\sqrt[p]{1 - \sqrt[p]{1-\zeta}}$ over K. That extension is non-abelian over $\mathbb{Q}(\mu_{p^2})$, in contrast with the extension in this paper which is abelian even over K.

3.1 The Splitting Field of $1-(1-x^p)^p$

The prime p is totally ramified in K with $p = \langle 1-\zeta\rangle^{p-1}$ [19, Lemma 1.4]. Thus there is a unique place $\nu = \langle 1-\zeta\rangle$ above p in K. Also, $p = \prod_{i=1}^{p-1}(1-\zeta^i)$ and $(1-\zeta^i)/(1-\zeta)$ is a unit of $\mathcal{O}_K$ by [19, Lemma 1.3]. Thus $\nu = \langle 1-\zeta^i\rangle$ for all $i = 1,2,\ldots,p-1$. Since $1 = v_\nu(p) = (p-1)v_\nu(1-\zeta)$, it follows that $v_\nu(1-\zeta^i) = 1/(p-1)$ for $1 \le i \le p-1$.

Let L' be the maximal elementary abelian p-group extension of K unramified except over $\nu = \langle 1-\zeta\rangle$.

Lemma 1.

1. $L = K(\sqrt[p]{1-\zeta^i}, 1 \le i \le p-1)$.
2. $L \subset L'$ *and* $\mathrm{Gal}(L/K)$ *is an elementary abelian p-group.*

Proof.

1. Let $z = 1-x^p$ where x is a root of $1-(1-x^p)^p$. The equality $z^p = 1$ implies that $K \subset L$. The p^2-p non-zero roots of $f_p(x)$ are the pth roots of $1-\zeta^i$ for $1 \le i \le p-1$. Thus $L = K(\sqrt[p]{1-\zeta^i}, 1 \le i \le p-1)$.
2. The field L is the compositum of the fields $K(\sqrt[p]{1-\zeta^i})$. For each i, the extension $K(\sqrt[p]{1-\zeta^i})/K$ is a Galois degree p extension ramified only above $1-\zeta^i$ and ∞. This proves both statements.

Lemma 2. *The field L is the same as the fields L_2 and L_3 where*

$$L_2 = K(\sqrt[p]{1-\zeta^i}, 1 \le i \le \frac{p-1}{2}, \sqrt[p]{p});$$

$$L_3 = K(\sqrt[p]{1-\zeta^i}, 1 \le i \le \frac{p-1}{2}, \sqrt[p]{\zeta}).$$

Proof. The idea of the proof is to show $L \subseteq L_3 \subseteq L_2 \subseteq L$.

$L \subseteq L_3$: For $\frac{p-1}{2} < i \le p-1$, write $j = -i$. Then

$$\sqrt[p]{1-\zeta^j} = \sqrt[p]{1-\zeta^{-i}} = \sqrt[p]{\zeta^{-i}-1} \cdot \sqrt[p]{-1}.$$

Since p is odd, $\sqrt[p]{-1} \in K$. So

$$\sqrt[p]{1-\zeta^j} = \sqrt[p]{\zeta^{-i}(1-\zeta^i)} \cdot \sqrt[p]{-1} = \sqrt[p]{1-\zeta^i} \cdot (\sqrt[p]{\zeta})^{-i} \cdot \sqrt[p]{-1} \in L_3.$$

$L_3 \subseteq L_2$:
Let ζ_{p^2} denote a pth root of ζ. It suffices to show that $\zeta_{p^2} \in L_2$. Write $p = bc$ with

$$b = \prod_{i=1}^{\frac{p-1}{2}}(1-\zeta^i),\ c = \prod_{i=\frac{p+1}{2}}^{p-1}(1-\zeta^i).$$

Note that $(1-\zeta^i)/(1-\zeta^{-i}) = -\zeta^i$. Thus, $\frac{b}{c} = (-1)^{\frac{p-1}{2}}\zeta^{\frac{(p-1)(p+1)}{8}}$ and

$$b^2 = \frac{b}{c}\cdot bc = (-1)^{\frac{p-1}{2}}\zeta^{\frac{(p-1)(p+1)}{8}}\cdot p.$$

Then

$$\zeta^{\frac{(p-1)(p+1)}{8}} = (-1)^{\frac{p-1}{2}}p^{-1}\prod_{i=1}^{\frac{p-1}{2}}(1-\zeta^i)^2.$$

Let $J = (p-1)^2(p+1)/16$ and note that $p \nmid J$. Raising both sides of the previous equation to the power $\frac{p-1}{2}\frac{1}{p}$ shows that

$$\zeta_{p^2}^J = \zeta'(\sqrt[p]{-1})^{\frac{(p-1)^2}{4}}(\sqrt[p]{p})^{\frac{1-p}{2}}\prod_{i=1}^{\frac{p-1}{2}}\left(\sqrt[p]{(1-\zeta^i)^2}\right)^{\frac{p-1}{2}},$$

for some pth root of unity ζ'. Thus $\zeta_{p^2}^J \in L_2$ and $\zeta_{p^2} \in L_2$.

$L_2 \subseteq L$: This follows from the equality $\sqrt[p]{p} = \prod_{i=1}^{p-1}\sqrt[p]{1-\zeta^i}$.

3.2 Background on Units in Cyclotomic Fields

Let $K = \mathbb{Q}(\zeta)$ and let $K^+ = \mathbb{Q}(\zeta+\zeta^{-1})$. Let $E = \mathcal{O}_K^*$ (resp., $E^+ = \mathcal{O}_{K^+}^*$) denote the group of units in $\mathcal{O}_K$ (resp., $\mathcal{O}_{K^+}$). Let V denote the subgroup of K^* generated by $\{\pm\zeta, 1-\zeta^i : i = 1, 2, \ldots, p-1\}$. Let W be the group of roots of unity in K.

Consider the cyclotomic units $C = V \cap \mathcal{O}^*$ of K and the cyclotomic units $C^+ = C \cap (\mathcal{O}^+)^*$ of K^+ [19, page 143]. By [19, Lemma 8.1], C is generated by ζ and C^+; and C^+ is generated by -1 and the units

$$\epsilon_a = \zeta^{(1-a)/2}(1-\zeta^a)/(1-\zeta),$$

for $1 < a < p/2$. By [19, Theorem 4.12], the index of WE^+ in E is 1 or 2. Let h^+ denote the order of the class group of K^+.

Theorem 8 ([19, Theorem 8.2]). *The index of the cyclotomic units C^+ in E^+ is the class number h^+ of K^+. Thus if Vandiver's conjecture is true for the prime p, then E/E^p is generated by C.*

Remark 3. Vandiver's conjecture (first conjectured by Kummer in 1849) states that p does not divide the class number h^+. It has been verified for all p less than 163 million [4]. It is also true for all regular primes.

3.3 The Galois Group of $1-(1-x^p)^p$

Proposition 1. *If Vandiver's conjecture is true for the prime p, then the Galois group of L/K is an elementary abelian p-group of rank $(p+1)/2$.*

Proof. By Lemma 1, $\mathrm{Gal}(L/K)$ is an elementary abelian p-group. Let r be the integer such that $\mathrm{Gal}(L/K) \simeq (\mathbb{Z}/p)^r$. The field L is obtained by adjoining pth roots of elements in some subgroup $B \subset K^*/(K^*)^p$, and by Kummer theory $B \simeq (\mathbb{Z}/p)^r$. By Lemma 2, B is generated by ζ and $1-\zeta^i$ for $1 \leq i \leq (p-1)/2$. Thus $r \leq (p+1)/2$. Thus it suffices to show that $r \geq (p+1)/2$.

Note that B is generated by ζ and $1-\zeta^i$ for $1 \leq i \leq (p-1)/2$. Thus B is also generated by ζ, $1-\zeta$, and ϵ_a for $1 < a < p/2$. Consider the subgroup B' of $K^*/(K^*)^p$ generated by ζ and ϵ_a for $1 < a < p/2$. Let r' be the rank of B' over $\mathbb{Z}/p$. Since ζ and ϵ_a are units, and $1-\zeta$ has positive valuation at the prime above p, it suffices to show that $r' \geq (p-1)/2$.

Since -1 is a pth power, B' is also the subgroup generated by the cyclotomic units C. By hypothesis, p satisfies Vandiver's conjecture and so Theorem 8 implies that $B' \simeq E/E^p$. By Dirichlet's unit theorem, $E \simeq \mathbb{Z}^{\frac{p-1}{2}-1} \times \mu_p$. Thus $r' = \frac{p-1}{2} - 1 + 1 = (p-1)/2$.

We now describe an explicit set of generators for $\mathrm{Gal}(L/K)$. Given a primitive pth root $\sqrt[p]{\theta}$ of $\theta \in K$ and $\sigma \in G_K$, recall from Definition 1 that $\kappa(\theta)\sigma$ is the element of $\mathbb{Z}/p$ such that

$$\sigma \sqrt[p]{\theta} = \zeta^{\kappa(\theta)\sigma} \sqrt[p]{\theta}.$$

Corollary 1. *Let p be an odd prime such that $p \nmid h^+$. Then the following map is an isomorphism:*

$$\Phi = \kappa(\zeta) \times \prod_{i=1}^{\frac{p-1}{2}} \kappa(1-\zeta^{-i}) : \mathrm{Gal}(L/K) \to (\mathbb{Z}/p)^{\frac{p+1}{2}}.$$

Proof. By Lemma 2, $L = K(\sqrt[p]{\zeta}, \sqrt[p]{1-\zeta^{-i}} \,:\, i = 1, 2, \ldots, \frac{p-1}{2})$. Let $G \subseteq K^*/(K^*)^p$ denote the subgroup generated by $S = \{\zeta, 1-\zeta^{-i} : i = 1, 2, \ldots, \frac{p-1}{2}\}$. By Kummer theory, it suffices to show that S is a $\mathbb{Z}/p$-basis for the $\mathbb{Z}/p$-vector space $\mathrm{Gal}(L/K)$, which follows from Proposition 1.

Remark 4. If $p \mid h^+$, then p divides $[E^+ : C^+]$ by [19, Theorem 8.2]. Since E^+ does not contain the pth roots of unity, E^+ has no p-torsion, and it follows that there is an element c of C^+ which is a pth power of an element in E^+, but not a pth power of any element of C^+. Since -1 is a pth power and C^+ is generated by -1 and $\{\epsilon_a : 1 < i < p/2\}$, c may be taken to be $c = \prod_{a=2}^{(p-1)/2} \epsilon_a^{e_a}$ with $0 \leq e_a \leq p-1$. Since B' in the previous proof is generated by C, it follows that B' is generated by $\{\zeta, \epsilon_a : 1 < i < p/2\}$. Since c maps to 0 in E/E^p, this implies that the rank r' of B' is less than the cardinality of $\{\zeta, \epsilon_a : 1 < i < p/2\}$. Thus $r = r' + 1 < (p+1)/2$. Thus if Vandiver's conjecture is not true for the prime p, then the rank of the elementary abelian p-group $\mathrm{Gal}(L/K)$ is strictly less than $(p+1)/2$.

4 Comparison with an $(n-1)$-Torus

Recall the notation from Sect. 2 that $\tilde{V} = \mathbb{A}^1 - \mu_n$, $V = \tilde{V} \cup \{1\}$, and Z consists of the points 0 and 1 in V. Recall that $\psi \in H_1(V, Z; A)$ denotes the homology class represented by the path from 0 to 1 along the real axis, and that Ψ_σ is defined to be the element of $\Omega(\Lambda_0)/A \operatorname{dlog} \zeta_0$ determined by $(\sigma - 1)\psi$ and the exact sequence

$$0 \to A \operatorname{dlog} \zeta_0 \to \Omega(\Lambda_0) \to H_1(V; A) \to 0,$$

where the quotient map $\Omega(\Lambda_0) \to H_1(V; A)$ is the map of Λ_0 modules mapping $\zeta_0 \operatorname{dlog} \zeta_0$ to a small counterclockwise loop around ζ.

Note that there is a map from V to a torus which induces a Galois equivariant isomorphism on $H_1(-; A)$. For example, this map could be the Abel–Jacobi map to the generalized Jacobian. Furthermore, over K, this torus splits, and it is easy to write down a map to a split torus inducing an isomorphism on $H_1(-; A)$. Namely, the map

$$f : V_K \to (\mathbb{G}_{m,K})^{\times n-1},$$

given by $z \mapsto (z - \zeta, z - \zeta^2, \ldots, z - \zeta^{n-1})$ induces a Galois equivariant isomorphism on $H_1(-; A)$.

In this section, we use the isomorphism $H_1(f; A)$ to compute Ψ_σ in terms of the classical Kummer map, relying on the facts that $H_1((\mathbb{G}_{m,K})^{\times(n-1)}; A) \cong A^{n-1}$ and that the map κ for $\mathbb{G}_m$ can be identified with the classical Kummer map. We will furthermore see in Sect. 4.2 that this computation is compatible with Sect. 3.

4.1 Computation of Ψ_σ

Fix the isomorphism $I : \Omega(\Lambda_0)/A \operatorname{dlog} \zeta_0 \to A^{n-1}$ given by

$$\sum_{i=1}^{n-1} a_i \zeta_0^i \operatorname{dlog} \zeta_0 \mapsto (a_1, a_2, \ldots, a_{n-1}).$$

This isomorphism I can also be obtained by composing the isomorphism described above $\Omega(\Lambda_0)/A \operatorname{dlog} \zeta_0 \cong H_1(V; A)$ with $H_1(f; A)$ and an obvious isomorphism $H_1((\mathbb{G}_{m,K})^{\times(n-1)}; A) \cong A^{n-1}$.

Proposition 2. *With notation as above,*

$$\Psi_\sigma = (\kappa(1-\zeta^{-1})(\sigma), \dots, \kappa(1-\zeta^{-(n-1)})(\sigma)).$$

Proof. Consider the maps $\kappa^{ab}_{V,b} : V(K) \to H^1(G_K, H_1(V))$, defined so that $\kappa^{ab}_{V,b}(x)$ is represented by the cocycle

$$\sigma \mapsto \gamma^{-1}\sigma\gamma$$

where γ is a path from b to x, and composition of paths is written from right to left, so $\gamma^{-1}\sigma\gamma$ is a loop based at b. As in [20, p. 8], the dependency on the choice of base point b in V is

$$\kappa_{b'}(x) = \kappa_b(x) - \kappa_b(b'). \tag{4}$$

By definition, Ψ_σ is the element of $H_1(V; A)$ determined by $(\sigma - 1)\psi$. Note that $(\sigma - 1)\psi = \kappa^{\mathrm{ab}}_{V,0}(1)(\sigma)$.

Since κ is functorial, one sees that $H_1(f)(\sigma - 1)\psi = \kappa^{\mathrm{ab}}_{T,f(0)}(f(1))(\sigma)$, where T is the torus $T = (\mathbb{G}_{m,K})^{\times(n-1)}$ and f is the map $V_K \to T$ defined above.

Since the geometric fundamental group respects products over algebraically closed fields of characteristic 0 [16, XIII Proposition 4.6], the map $\kappa_T = \kappa^{\mathrm{ab}}_T$ for T decomposes as the product of the maps κ for $\mathbb{G}_{m,K}$ which are each given by $\kappa_{\mathbb{G}_{m,K},1}(\theta)(\sigma) = \kappa(\theta)\sigma$ as in Definition 1 and Remark 1. Thus $\kappa^{\mathrm{ab}}_{T,f(0)}(f(1))(\sigma)$ is identified with $\prod_{i=1}^{n-1} \kappa_{\mathbb{G}_{m,K},-\zeta^i}(1-\zeta^i)(\sigma)$ when, via the projection maps, $\pi_1(T_{\overline{k}}, 1)$ is identified with $\prod_{i=1}^{n-1} \pi_1(\mathbb{G}_{m,\overline{k}}, 1)$.

Applying (4) with $b = 1$, using the fact that κ from Definition 1 is a homomorphism, yields that $\prod_{i=1}^{n} \kappa_{\mathbb{G}_{m,K},-\zeta^i}(1-\zeta^i)(\sigma) = \prod_{i=1}^{n} \kappa(\frac{1-\zeta^i}{-\zeta^i})(\sigma)$. The proposition follows from the above, since $(1-\zeta^i)/(-\zeta^i) = 1 - \zeta^{-i}$.

Combining with Theorem 7 (c.f. [2, Theorem 10]), we obtain:

Corollary 2. *Modulo a term of the form* $\alpha\,\mathrm{dlog}\,\zeta$, *with* $\alpha \in A^{sh}$,

$$\mathrm{dlog}(\Gamma_\sigma) = \sum_{i=1}^{n-1} c_i \zeta^i \,\mathrm{dlog}\,\zeta, \text{ with } c_i = \kappa(1-\zeta^{-i})(\sigma).$$

4.2 Compatibility with Sect. 3

Remark 5. In computing $\kappa(1-\zeta^{-i})(\sigma)$ for $k = K$, one can restrict to the image $\overline{\sigma} \in \mathrm{Gal}(L/K)$.

Corollary 3. *Suppose $n = p$ is a prime satisfying Vandiver's conjecture. With respect to the isomorphism $\Phi : \mathrm{Gal}(L/K) \to A^{\frac{p+1}{2}}$ from Corollary 1 and the isomorphism $I : \Omega(\Lambda_0)/A \operatorname{dlog} \zeta_0 \to A^{p-1}$ from Sect. 4.1, the map*

$$\mathrm{Gal}(L/K) \to \Omega(\Lambda_0)/A \operatorname{dlog} \zeta_0, \ \sigma \mapsto \Psi_\sigma$$

is the explicit A-linear map

$$(c_0, c_1, \ldots, c_{\frac{p-1}{2}}) \mapsto (c_1, c_2, \ldots, c_{\frac{p-1}{2}}, c_{\frac{p-1}{2}} + \tfrac{p-1}{2}c_0, \ldots, c_2 + 2c_0, c_1 + c_0).$$

Proof. By Proposition 2, Ψ_σ is computed

$$\Psi_\sigma = (\kappa(1-\zeta^{-1})(\sigma), \ldots, \kappa(1-\zeta^{-(p-1)})(\sigma))$$

with respect to the isomorphism I. For $i = 1, 2, \ldots, \frac{p-1}{2}$, then $\kappa(1-\zeta^{-i})$ is identified with the projection onto c_i, the $(i+1)$st coordinate of $(\mathbb{Z}/p)^{\frac{p+1}{2}} \cong \mathrm{Gal}(L/K)$ via the isomorphism Φ. Recall that $(1-\zeta^i)/(1-\zeta^{-i}) = -\zeta^i$ and -1 is a pth power since p is odd. Thus

$$\kappa(1-\zeta^i) - \kappa(1-\zeta^{-i}) = i\kappa(\zeta) = ic_0.$$

Rearranging terms yields that $\kappa(1-\zeta^{-i}) = \kappa(1-\zeta^i) - ic_0$. Applying this equation when $i = \frac{p-1}{2}+1, \ldots, p-1$ shows that $\kappa(1-\zeta^{-i}) = \kappa(1-\zeta^{-(p-i)}) - ic_0 = c_{p-i} - ic_0$. This implies that $\kappa(1-\zeta^{-i})$ is the projection onto the $(p-i+1)$st coordinate c_{p-i} plus $p-i$ times the projection onto the first coordinate c_0.

4.3 Coordinate Sum of Ψ_σ

We include the following result for its own interest; it is not needed in the computation of $H_1(U, Y; A)$, $H_1(U; A)$, or $H_1(X; A)$ as Galois modules, and it is not needed in the computations of Sect. 7. For $\sigma \in \mathrm{Gal}(L/K)$, write $\Psi_\sigma = (c_1, \ldots, c_{p-1})$ as in Corollary 3.

Lemma 3. *If $\sigma \in G_M$, with $M = \mathbb{Q}(\sqrt[p]{p})$, then $\sum_{i=1}^{p-1} c_i \equiv 0 \bmod p$. More generally, if $M_1 = \mathbb{Q}(\zeta_p, \sqrt[p]{p})$ and if $\tau \in \mathrm{Gal}(M_1, \mathbb{Q}(\zeta_p))$ is such that $\tau(\sqrt[p]{p}) = \zeta_p^j \sqrt[p]{p}$, then $\sum_{i=1}^{p-1} c_i \equiv j \bmod p$.*

Proof. Write $\theta_i = 1 - \zeta_p^{-i}$ and note that $\prod_{i=1}^{p-1} \theta_i = p$. Thus $\prod_{i=1}^{p-1} \sqrt[p]{\theta_i} = \sqrt[p]{p} \in M$ is fixed by $\sigma \in G_M$. So $\prod_{i=1}^{p-1} \sigma(\sqrt[p]{\theta_i}) = \sqrt[p]{p}$. By definition, $\sigma(\sqrt[p]{\theta_i}) = \zeta_p^{\kappa_p(\theta_i)\sigma} \sqrt[p]{\theta_i}$. By Proposition 2, $c_i = \kappa_p(\theta_i)\sigma$. Thus,

$$\sqrt[p]{p} = \prod_{i=1}^{p-1} \zeta_p^{c_i} \sqrt[p]{\theta_i} = \zeta_p^{\sum_{i=1}^{p-1} c_i} \sqrt[p]{p}.$$

It follows that $\sum_{i=1}^{p-1} c_i \equiv 0 \bmod p$.

Similarly,

$$\zeta_p^j \sqrt[p]{p} = \tau(\sqrt[p]{p}) = \prod_{i=1}^{p-1} \zeta_p^{c_{i,\tau}} \sqrt[p]{\theta_i} = \zeta_p^{\sum_{i=1}^{p-1} c_{i,\tau}} \sqrt[p]{p},$$

so $\sum_{i=1}^{p-1} c_{i,\tau} \equiv j \bmod p$.

5 Explicit Computation of B_σ

5.1 Determining B_σ from Ψ_σ

Recall from Fact 5 that Γ_σ is an element of $\bar{\Lambda}_0^\times$, unique modulo the kernel of d'^{sh} : $\bar{\Lambda}_0^\times \to \Lambda_1^\times$, such that

$$d'^{sh}(\Gamma_\sigma) = B_\sigma.$$

Corollary 2 determines the coefficients of the logarithmic derivative dlog Γ_σ; they are the ones appearing in Ψ_σ, and explicitly described in Proposition 2.

When n is prime, the kernel of dlog is easy to manage and thus Ψ_σ determines the action of G_K on $H_1(U, Y; A)$ as seen in the next result. This result is implicit in [2, 10.5].

Proposition 3. *Let $n = p$ be a prime. Then Ψ_σ uniquely determines B_σ.*

The following lemmas will be useful for the proof of Proposition 3.

Lemma 4. *The kernel of* dlog : $\bar{\Lambda}_0^\times \to \Omega(\bar{\Lambda}_0)$ *consists of elements* $x = \sum_{0\le i<n} a_i \zeta_0^i$ *such that* $ia_i = 0 \in A$ *for all* $0 \le i < n$. *In particular, when n is prime, the kernel of* dlog *consists of the constant (in ζ_0) invertible polynomials* $(A^{sh})^\times \subset \bar{\Lambda}_0^\times$.

Remark 6. On the contrary, when n is not prime, this kernel can be significantly larger. For example, when $n = 6$, it contains elements such as $3\zeta_0^2 + 2\zeta_0^3$.

The following characterization of Γ_σ will be used to pinpoint the exact element in a coset that Γ_σ represents.

Lemma 5. *Write* $\Gamma_\sigma = \sum_{0\le i<n} d_i \zeta_0^i$, *with* $d_i \in A^{sh}$, *for an element in* $\bar{\Lambda}_0^\times$ *which is a* d'^{sh}*-preimage of* B_σ. *Then* $d_\Sigma := \sum_{0\le i<n} d_i = 1$.

Proof. By Fact 6, $B_\sigma - 1$ is in the augmentation ideal $(1 - \epsilon_0)(1 - \epsilon_1)\Lambda_1$. Since

$$B_\sigma - 1 = \frac{\left(\sum d_i \zeta_0^i\right)\left(\sum d_i \zeta_1^i\right)}{\left(\sum d_i \zeta_0^i \zeta_1^i\right)} - 1,$$

it lies in the augmentation ideal if and only if the difference

$$\left(\sum d_i\zeta_0^i\right)\left(\sum d_i\zeta_1^i\right)-\left(\sum d_i\zeta_0^i\zeta_1^i\right)$$

does. But the augmentation of the latter is precisely $(d_\Sigma^2 - d_\Sigma) = d_\Sigma(d_\Sigma - 1)$. As Γ_σ is invertible, d_Σ must also be invertible, hence $d_\Sigma = 1$.

We are now ready to prove Proposition 3.

Proof. Consider $\Psi_\sigma = \sum_{0\le i<n} c_i\zeta_0^i \operatorname{dlog}\zeta_0$, with $c_i \in A^{sh}$. By Fact 5, [2, Theorem 9], B_σ is uniquely determined by Γ_σ in an explicit way, as $B_\sigma = d'^{sh}(\Gamma_\sigma)$. Hence it suffices to show that Γ_σ is determined by Ψ_σ in a way unique modulo the kernel of d'^{sh}.

Corollary 2 gives that

$$\operatorname{dlog}\Gamma_\sigma = \alpha \operatorname{dlog}\zeta_0 + \sum_{0\le i<n} c_i\zeta_0^i \operatorname{dlog}\zeta_0,$$

for some $\alpha \in A^{sh}$. Note that the kernel $\operatorname{Ker}(d'^{sh})$ (cf. Fact 5) of $d'^{sh} : \bar{\Lambda}_0^\times \to \bar{\Lambda}_1^\times$ maps under dlog to the kernel $\operatorname{Ker}(d_\Omega'^{sh})$ of the map

$$d_\Omega'^{sh} : \Omega(\bar{\Lambda}_0) \to \Omega(\bar{\Lambda}_1),$$

which is given by $\operatorname{dlog}(d'^{sh})$, i.e.,

$$d_\Omega'^{sh}\left(\sum a_i\zeta_0^i \operatorname{dlog}\zeta_0\right) = \sum a_i\zeta_0^i(1-\zeta_1^i)\operatorname{dlog}\zeta_0 + \sum a_i\zeta_1^i(1-\zeta_0^i)\operatorname{dlog}\zeta_1.$$

By Anderson [2, 8.5.1], $\operatorname{Ker}(d_\Omega'^{sh})$ is precisely $A^{sh}\operatorname{dlog}\zeta_0$. When n is prime, dlog : $\operatorname{Ker}(d'^{sh}) \to \operatorname{Ker}(d_\Omega'^{sh})$ is an isomorphism by Anderson [2, 8.3.1], which determines $\operatorname{Ker}(d'^{sh})$. Hence the ambiguity that α introduces is irrelevant for the computation of B_σ.

The remaining obstruction to reconstructing Γ_σ, and therefore B_σ, is the kernel of dlog : $\bar{\Lambda}_0^\times \to \Omega(\bar{\Lambda}_0)$.

By Lemma 4, when n is prime, the kernel of dlog is $A^{sh} \simeq \bar{\mathbb{F}}_p^\times \subset \Lambda_0^\times$. Suppose a lies in this kernel; this means that $\operatorname{dlog}(a\Gamma_\sigma) = \operatorname{dlog}(\Gamma_\sigma)$. On the other hand, $d'^{sh}(a\Gamma_\sigma) = ad'^{sh}(\Gamma_\sigma) = aB_\sigma$, thus a could introduce an ambiguity.

Nonetheless, this ambiguity can be eliminated using Lemma 5, which asserts that the sum of the coefficients of Γ_σ is fixed and equals one. Hence the sum of the coefficients of $a\Gamma_\sigma$, for $a \in \bar{\mathbb{F}}_p^\times$, must be a. By Lemma 5, this implies that $a\Gamma_\sigma$ is not a preimage of B_σ unless $a = 1$.

In conclusion, when n is prime, dlog Γ_σ uniquely determines Γ_σ and therefore B_σ.

In theory, by Proposition 3, the coefficients c_i of Ψ_σ studied in Sect. 4 uniquely and explicitly determine the coefficients of B_σ, and thus the action of G_K on $H_1(U, Y; A)$. We carry out this computation explicitly when $n = 3$ in the following subsection.

5.2 The Case $n = 3$

Consider the smallest example, i.e., that of $n = 3$. Write

$$\Psi_\sigma = (c_1\zeta_0 + c_2\zeta_0^2)\,\mathrm{dlog}\,\zeta_0,$$

for $c_1, c_2 \in \bar{\mathbb{F}}_p \simeq A^{sh}$. Write

$$\Gamma_\sigma = d_0 + d_1\zeta_0 + d_2\zeta_0^2,$$

with $d_i \in \bar{\mathbb{F}}_p$ such that $d_0 + d_1 + d_2 = 1$. To determine the d_i's in terms of the c_i's, it is easier to work with the nilpotent variable $y = \zeta_0 - 1$ instead of ζ_0 and use the basis $dy = \zeta_0\,\mathrm{dlog}\,\zeta_0$ of $\Omega(\Lambda_0)$.

Indeed, $\Gamma_\sigma = 1 + (d_1 - d_2)y + d_2y^2$, and

$$\Psi_\sigma = (c_1 + c_2 + c_2y)dy.$$

By Fact 7, dlog Γ_σ agrees with Ψ_σ modulo terms in $\bar{\mathbb{F}}_3\,\mathrm{dlog}\,\zeta_0 = \bar{\mathbb{F}}_3(y+1)^2dy$. Therefore, for some $\alpha \in \bar{\mathbb{F}}_3$, one sees that

$$\mathrm{dlog}\,\Gamma_\sigma = \Psi_\sigma + \alpha(y+1)^2dy,$$

which yields the equalities

$$\begin{aligned} d_2 - d_1 &= c + \alpha \\ -d_2 &= (c+\alpha)^2 + c_2 - \alpha \\ 0 &= d_2(c+\alpha) + (c+\alpha)(c_2 - \alpha) + \alpha, \end{aligned}$$

where $c = c_1 + c_2$. In particular, α must be a solution of the polynomial equation

$$\alpha^3 - \alpha + c^3 = 0.$$

For an arbitrary choice of solution α, the coefficients of Γ_σ are

$$\begin{aligned} d_1 &= c_1 - \alpha - (c+\alpha)^2, \\ d_2 &= -c_2 + \alpha - (c+\alpha)^2. \end{aligned}$$

Note that the inverse of Γ_σ expressed in the original ζ_0-basis is

$$\Gamma_\sigma^{-1} = (1 + d_1 + d_2 + (d_2 - d_1)^2) + ((d_2 - d_1)^2 - d_1)\zeta_0 + ((d_2 - d_1)^2 - d_2)\zeta_0^2.$$

In terms of the c's and α, this becomes

$$\Gamma_\sigma^{-1} = (1 + c_1 - c_2 - (c+\alpha)^2) + (c_2 + c - (c+\alpha)^2)\zeta_0 + (c_2 - \alpha - (c+\alpha)^2)\zeta_0^2.$$

Now $B_\sigma = d'^{sh}(\Gamma_\sigma)$ can be computed.

Lemma 6. *Suppose* $\Psi_\sigma = (c_1\zeta_0 + c_2\zeta_0^2)\,\mathrm{dlog}\,\zeta_0$, *and let* $b_{i,j}$ *be the coefficient of* $\zeta_0^i\zeta_1^j$ *in* B_σ. *Then*

$$\begin{aligned} b_{0,0} &= 1 + c_2 - c_1 - (c_2 - c_1)^2 \\ b_{0,1} &= c_1 - (c_2 - c_1)^2 \\ b_{1,1} &= -c_1 - (c_2 - c_1)^2. \end{aligned} \tag{5}$$

The rest of the coefficients are determined by symmetry $b_{i,j} = b_{j,i}$ *and the fact that* $b_{0,0} + b_{0,1} + b_{0,2} = 1$, $b_{1,0} + b_{1,1} + b_{1,2} = 0$, *and* $b_{2,0} + b_{2,1} + b_{2,2} = 0$.

Remark 7. From the proof of Corollary 3, if $i = 1, \dots, \frac{p-1}{2}$, then $\kappa(1 - \zeta^{-i}) = c_{p-i} - ic_0$. By Proposition 2, $c_2 = \kappa(1 - \zeta^{-2})$. Rearranging terms gives $c_0 = c_2 - c_1$, and it follows that Lemma 6 completes the proof of Theorem 1.

6 Homology of the Affine and Projective Fermat Curve

In this section, we determine the Galois module structure of the homology of the projective Fermat curve X and its affine open $U = X - Y$ with coefficients in $A = \mathbb{Z}/n$ for all $n \geq 3$.

6.1 Homology of the Affine Curve

We first determine the Galois module structure of $H_1(U)$, where $H_i(U)$ abbreviates $H_i(U; A)$, and more generally, all homology groups will be taken with coefficients in A.

The closed subset $Y \subset U$ given by $xy = 0$ consists of the $2n$ points

$$R_i = [\zeta^i : 0 : 1],\ Q_i = [0 : \zeta^i : 1].$$

Thus, $H_0(Y) \simeq \Lambda_0 \oplus \Lambda_0$ is generated by $\zeta_0 \oplus 0$ and $0 \oplus \zeta_1$. The first copy indexes the points R_i and the second copy indexes the points Q_i. The homomorphism $H_0(Y) \to H_0(U) \simeq A$ sends both $\zeta_0 \oplus 0$ and $0 \oplus \zeta_1$ to 1.

Note that $H_0(Y)$ is a Λ_1-module via $\zeta_0 \mapsto \zeta_0 \oplus 1$ and $\zeta_1 \mapsto 1 \oplus \zeta_1$. The boundary map $\delta : H_1(U, Y) \to H_0(Y)$ is a Λ_1-module map given by

$$\beta \mapsto 1 \oplus 0 - 0 \oplus 1. \tag{6}$$

Lemma 7. *There is an exact sequence of Galois modules*

$$0 \to H_1(U) \to H_1(U, Y) \xrightarrow{\delta} H_0(Y) \to H_0(U) \to 0. \tag{7}$$

The first Betti number of U *is* $(n-1)^2$.

Proof. This follows from the long exact sequence for relative homology, using the facts that $H_1(Y) = 0$ and $H_0(U, Y) = 0$. The Betti number is the A-rank of $H_1(U)$; note that $H_1(U, Y)$, $H_0(Y)$, and $H_0(U)$ are all free A-modules, hence

$$\operatorname{rank}(H_1(U)) = \operatorname{rank}(H_1(U, Y)) - \operatorname{rank}(H_0(Y)) + \operatorname{rank}(H_0(U)).$$

So, the rank of $H_1(U)$ is $n^2 - 2n + 1 = (n-1)^2$.

An element $W \in \Lambda_1$ will be written as $W = \sum_{0 \le i,j \le n-1} a_{ij} \zeta_0^i \zeta_1^j$.

Proposition 4. *Let $W = \sum_{0 \le i,j \le n-1} a_{ij} \zeta_0^i \zeta_1^j$ be an element of Λ_1, and consider the corresponding element $W\beta$ of $H_1(U, Y)$. Then $W\beta$ restricts to $H_1(U)$ if and only if for each $0 \le j \le n-1$, $\sum_{i=0}^{n-1} a_{ij} = 0$, and for each $0 \le i \le n-1$, $\sum_{j=0}^{n-1} a_{ij} = 0$.*

Proof. By Lemma 7, $W\beta \in H_1(U)$ if and only if $W\beta \in \ker(\delta)$. Note that, by (6),

$$\delta(\zeta_0 \beta) = (\zeta_0 \oplus 1)(1 \oplus 0 - 0 \oplus 1) = \zeta_0 \oplus 0 - 0 \oplus 1,$$

and similarly,

$$\delta(\zeta_1 \beta) = (1 \oplus \zeta_1)(1 \oplus 0 - 0 \oplus 1) = 1 \oplus 0 - 0 \oplus \zeta_1.$$

Thus

$$\begin{aligned} \delta(W\beta) &= \sum_{0 \le i,j \le n-1} a_{ij} \delta(\zeta_0^i \zeta_1^j \beta) \\ &= \sum a_{ij} (\zeta_0^i \oplus 1)(1 \oplus \zeta_1^j)(1 \oplus 0 - 0 \oplus 1) \\ &= \sum a_{ij} (\zeta_0^i \oplus 0 - 0 \oplus \zeta_1^j). \end{aligned}$$

So $W\beta \in \ker(\delta)$ if and only if the rows and columns of W sum to zero.

6.2 *Homology of the Projective Curve*

We next determine the Galois module structure of $H_1(X)$, which has rank $2g = n^2 - 3n + 2$.

Proposition 5.

1. *There is an exact sequence of Galois modules and A-modules:*

$$0 \to H_2(X) \to H_2(X, U) \xrightarrow{D} H_1(U) \to H_1(X) \to 0.$$

2. *The image of D is* $\operatorname{Stab}(\zeta_0 \zeta_1)$ *where* $\operatorname{Stab}(\zeta_0 \zeta_1)$ *consists of $W\beta \in H_1(U)$ which are invariant under $\zeta_0 \zeta_1$, i.e., for which $a_{i+1,j+1} = a_{ij}$, where indices are taken modulo n when necessary.*
3. *As a Galois module and A-module, $H_1(X) = H_1(U)/\operatorname{Stab}(\zeta_0 \zeta_1)$.*

Proof.

(1) The long exact sequence in homology of the pair (X, U) implies that the sequence

$$\cdots \to H_2(U) \to H_2(X) \to H_2(X,U) \xrightarrow{D} H_1(U) \to H_1(X) \to H_1(X,U) \to \cdots$$

is exact. Since U is affine, $H_2(U) = 0$. It thus suffices to show $H_1(X, U) = 0$. This follows from the fact that $H_1(X, U)$ is isomorphic as an abelian group to the singular homology of $(X(\mathbb{C}), U(\mathbb{C}))$, where $X(\mathbb{C})$ and $U(\mathbb{C})$ are given the analytic topology.

Here is an alternative proof that $H_1(X, U) = 0$ which does not use the analytic topology and which will be useful for part (2). It follows from [2, §4 Theorem 1] that Anderson's étale homology with coefficients in A [2, §2] is naturally isomorphic to the homology of the étale homotopy type in the sense of Friedlander [9]. It follows from Voevodsky's purity theorem [15, Theorem 2.23] and the factorization of the étale homotopy type through $\mathbb{A}^1$ algebraic topology [12] that there is a natural isomorphism $H_i(X, U) \cong \tilde{H}_i(\vee_{(X-U)(\overline{K})}\mathbb{P}^1_{\overline{K}})$, where $\tilde{H}_i$ denotes reduced homology. (For this, it is necessary to observe that the proof of [15, Theorem 2.23] goes through with the étale topology replacing the Nisnevich topology.) Thus

$$\tilde{H}_i(\mathbb{P}^1_{\overline{K}}; A) = \begin{cases} A(1) & \text{if } i = 2 \\ 0 & \text{otherwise,} \end{cases}$$

where $A(1)$ denotes the module $A = \mathbb{Z}/n$ with action given by the cyclotomic character. Over K, $A(1) = A$. It follows that

$$H_1(X, U) = \tilde{H}_1(\vee_{(X-U)(\overline{K})}\mathbb{P}^1_{\overline{K}}; A) = \oplus_{(X-U)(\overline{K})}\tilde{H}_i(\mathbb{P}^1_{\overline{K}}; A) = 0.$$

As a third alternative, one can see that $H_1(X, U) = 0$ using [14, VI Theorem 5.1] and universal coefficients argument to change information about cohomology to information about homology.

(2) As above,

$$H_2(X, U) \cong \tilde{H}_1(\vee_{(X-U)(\overline{K})}\mathbb{P}^1_{\overline{K}}) = \oplus_{(X-U)(\overline{K})}\tilde{H}_i(\mathbb{P}^1_{\overline{K}}) = \oplus_{(X-U)(\overline{K})}A(1).$$

For $\eta \in (X - U)(\overline{K})$, let η also represent the corresponding basis element of $\oplus_{(X-U)(\overline{K})}A(1)$. Then $D(\eta)$ is represented by a small loop around η.

Note that the coordinates of η are $[\epsilon : -\epsilon : 0]$ for some nth root of unity ϵ. In particular, η is fixed by $\zeta_0\zeta_1$. The loop $D(\eta)$ is therefore also fixed by $\zeta_0\zeta_1$ because $\zeta_0\zeta_1$ preserves orientation.

Consider the subset $\mathrm{Stab}(\zeta_0\zeta_1)$ of elements of $H_1(U)$ fixed by $\zeta_0\zeta_1$. Then $\mathrm{Stab}(\zeta_0\zeta_1)$ contains the image of D. In fact, $\mathrm{Stab}(\zeta_0\zeta_1) = \mathrm{Image}(D)$. To see this, it suffices to show that both $\mathrm{Stab}(\zeta_0\zeta_1)$ and $\mathrm{Image}(D)$ are isomorphic to A^{n-1}.

By (1), one sees that $\mathrm{Image}(D)$ is isomorphic to the quotient of $H_2(X, U)$ by the image of $H_2(X) \to H_2(X, U)$. Since X is a smooth proper curve, $H_2(X) \cong A(1)$ and $H_2(X, U) \cong \oplus_{(X-U)(\overline{K})}A(1)$. The map $H_2(X) \to H_2(X, U)$ can be described as the map that sends the basis element of $A(1)$ to the diagonal element $\oplus_{(X-U)(\overline{K})}1$. It follows that $\mathrm{Image}(D) \simeq A^{n-1}$ as claimed.

Now ζ_0 and ζ_1 act on $H_1(U, Y)$ via multiplication. Note that these actions have the effect of shifting the columns or rows of W and thus stabilize $H_1(U)$. The stabilizer of $\zeta_0\zeta_1$ is isomorphic to A^{n-1} because an element of the stabilizer is uniquely determined by an arbitrary choice of $a_{01}, a_{02}, \ldots, a_{0n}$.

(3) is immediate from (1) and (2).

7 Computing Galois Cohomology When $p = 3$

In this section, we explicitly compute several cohomology groups when $p = 3$. Let $e = \zeta_0$ and $f = \zeta_1$.

7.1 Computation of B_σ

Let σ and τ denote the generators of $G = \mathrm{Gal}(L/K) \simeq (\mathbb{Z}/3)^2$ such that

$$\sigma : \sqrt[3]{\zeta} \mapsto \zeta\sqrt[3]{\zeta}, \qquad \tau : \sqrt[3]{\zeta} \mapsto \sqrt[3]{\zeta}$$
$$\sqrt[3]{1-\zeta^{-1}} \mapsto \sqrt[3]{1-\zeta^{-1}} \qquad \sqrt[3]{1-\zeta^{-1}} \mapsto \zeta\sqrt[3]{1-\zeta^{-1}}.$$

The equality $-\zeta(1-\zeta^{-1}) = 1-\zeta$ shows that $(c_1)_\sigma = 0$, $(c_2)_\sigma = 1$ and $(c_1)_\tau = 1$, $(c_2)_\tau = 1$. By Lemma 6, this implies that

$$\begin{aligned} B_\sigma &= 1-(e+f)+(e^2-ef+f^2)-(e^2f+ef^2) \\ &= 1-(e+f)(1-e)(1-f), \qquad (8) \\ B_\tau &= 1+(e+f)-(e^2+ef+f^2)+e^2f^2. \end{aligned}$$

7.1.1 The Kernel and Image of B

Let $G = \langle\sigma, \tau\rangle$. Consider the map

$$B : \mathbb{F}_3[G] \to \Lambda_1,\ B(\sigma) = B_\sigma.$$

When $p = 3$, the domain and range of B both have dimension 9. Of course, B is not surjective since its image is contained in the 6-dimensional subspace of symmetric elements.

Lemma 8. *When $p = 3$, the image of B has dimension 4 and the kernel of B has dimension 5. In particular,* $\mathrm{Im}(B)$ *consists of symmetric elements whose 2nd and 3rd rows sum to 0, i.e., elements of the form*

$$a_{00}+a_{01}(e+f)+a_{02}(e^2+f^2)+a_{11}ef-(a_{01}+a_{11})(e^2f+ef^2)+(a_{01}+a_{11}-a_{02})e^2f^2;$$

and $\mathrm{Ker}(B)$ *is determined by the relations:*

$$\begin{aligned} B_{\tau^2} + B_\tau + B_1 &= 0, \\ B_{\sigma^2\tau} - B_{\sigma^2} - B_\tau + B_1 &= 0, \\ B_{\sigma\tau} - B_\sigma - B_\tau + B_1 &= 0, \\ B_{\sigma^2\tau^2} - B_{\sigma^2} - B_{\tau^2} + B_1 &= 0, \\ B_{\sigma\tau^2} - B_\sigma - B_{\tau^2} + B_1 &= 0. \end{aligned}$$

Proof. Magma computation using (8).

7.2 *Cohomology of* Gal(L/K)

Since σ has order 3, by Brown [3, Example 1.1.4, Exercise 1.2.2], the projective resolution of $\mathbb{Z}$ as a $\mathbb{Z}[\langle\sigma\rangle]$-module is

$$\mathbb{Z}[G] \overset{1-\sigma}{\leftarrow} \mathbb{Z}[G] \overset{1+\sigma+\sigma^2}{\leftarrow} \mathbb{Z}[G] \overset{1-\sigma}{\leftarrow} \cdots$$

where $G = \mathrm{Gal}(L/K) = \langle \sigma, \tau \mid \sigma^3 = \tau^3 = [\sigma, \tau] = 1\rangle \simeq (\mathbb{Z}/3)^2$. By Brown [3, Proposition V.1.1], the total complex associated with the following double complex is a projective resolution of $\mathbb{Z}$ as a $\mathbb{Z}[G]$-module:

$$\begin{array}{ccccccc}
\mathbb{Z}[G] & \xleftarrow{1-\sigma} & \mathbb{Z}[G] & \xleftarrow{1+\sigma+\sigma^2} & \mathbb{Z}[G] & \xleftarrow{1-\sigma} & \\
\downarrow{\scriptstyle 1+\tau+\tau^2} & & \downarrow{\scriptstyle 1+\tau+\tau^2} & & \downarrow{\scriptstyle 1+\tau+\tau^2} & & \\
\mathbb{Z}[G] & \xleftarrow{(1-\sigma)} & \mathbb{Z}[G] & \xleftarrow{-(1+\sigma+\sigma^2)} & \mathbb{Z}[G] & \xleftarrow{1-\sigma} & \\
\downarrow{\scriptstyle 1-\tau} & & \downarrow{\scriptstyle 1-\tau} & & \downarrow{\scriptstyle 1-\tau} & & \\
\mathbb{Z}[G] & \xleftarrow{1-\sigma} & \mathbb{Z}[G] & \xleftarrow{1+\sigma+\sigma^2} & \mathbb{Z}[G] & \xleftarrow{1-\sigma} &
\end{array}$$

Therefore, to compute $H^1(G, M)$, one can compute the cohomology of the complex

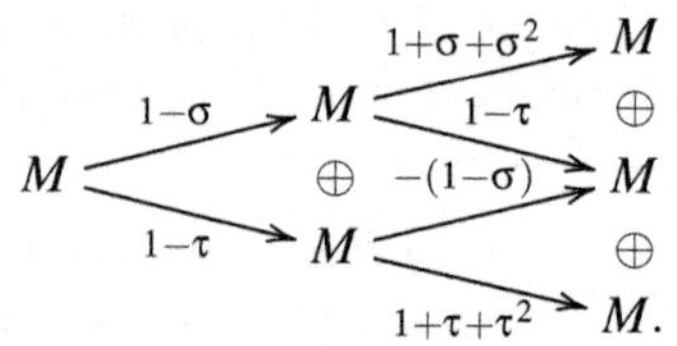

Given $h \in \mathbb{F}_3[G]$, let $\mathrm{Ann}_M(h) = \{m \in M \mid hm = 0\}$. Let $M = \Lambda_1$.

Lemma 9. *Let $M = \Lambda_1$ with $e = \epsilon_0$ and $f = \epsilon_1$.*

1. $\mathrm{Ann}_M(1 + \tau + \tau^2) = M$.
2. $\mathrm{Ann}_M(1 + \sigma + \sigma^2) = (1 - e, 1 - f)$ *consists of all* $m = \sum m_{ij} e^i f^j$ *such that* $\sum m_{ij} = 0$.
3. $\mathrm{Ann}_M(1 - \sigma) = (1 + e + e^2, 1 + f + f^2)$.
4. $\mathrm{Ann}_M(1 - \tau) = (e - f, 1 + f + f^2)$.

Proof.

1. Every $m \in M$ is in the annihilator of $1 + \tau + \tau^2$ because $1 + B_\tau + B_{\tau^2}$ equals

$$(e + f) - (e^2 + ef + f^2) + e^2f^2 + (e + f)^2 + (e^2 + ef + f^2)^2 + ef$$
$$+ (e + f)(e^2 + ef + f^2) - (e + f)e^2f^2 + (e^2 + ef + f^2)e^2f^2,$$

which is zero.

2. Note that $B_\sigma = 1 - (e + f)(1 - e)(1 - f)$, which gives that

$$1 + B_\sigma + B_{\sigma^2} = (e^2 - ef + f^2)(1 + e + e^2)(1 + f + f^2)$$
$$= (1 + e + e^2)(1 + f + f^2).$$

Note that $(1 + B_\sigma + B_{\sigma^2})e^i f^j = (1 + B_\sigma + B_{\sigma^2})$, so for $m = \sum_{i,j} m_{i,j} e^i f^j$,

$$(1 + B_\sigma + B_{\sigma^2})m = (1 + B_\sigma + B_{\sigma^2})(\sum_{i,j} m_{i,j}).$$

Thus $\mathrm{Ann}_M(1 + B_\sigma + B_{\sigma^2})$ consists of $m \in M$ whose entries sum to 0.

3. Note that $(1 - \sigma)m = 0$ if and only if $\sigma m = m$ which in turn simplifies to $(e + f)(1 - e)(1 - f)m = 0$. Thus $\mathrm{Ann}_M(1 - \sigma)$ is generated by the annihilators of $1 - e$ and $1 - f$ and $e + f$, which are $1 + e + e^2$ and $1 + f + f^2$ and 0.
4. A Magma calculation using that $1 - B_\tau = -(e + f) + (e^2 + f^2) + ef - e^2f^2$.

7.3 Preliminary Calculations

Consider the maps $X : M \to M^2$, $Y : M^2 \to M^3$, and $Z : M^3 \to M^4$. The goal is to compute $\ker(Y)/\operatorname{im}(X)$ and $\ker(Z)/\operatorname{im}(Y)$.

After choosing a basis for M, the maps X, Y, and Z can be written in matrix form. The basis of M chosen here is

$$1, f, f^2, e, ef, ef^2, e^2, e^2f, e^2f^2.$$

By Lemma 9, all of the entries of the matrix V for the map $\mathrm{Nm}(\tau) : M \to M$ are 0. All of the entries of the matrix U for the map $\mathrm{Nm}(\sigma) : M \to M$ are 1 since $\mathrm{Nm}(\sigma)$ acts on each element of M by summing its coefficients.

Let S be the matrix for the map $1 - B_\sigma : M \to M$. Let T be the matrix for the map $1 - B_\tau : M \to M$. Here are the matrices S and T:

$$S = \begin{bmatrix} 0\,1\,2\,1\,1\,1\,2\,1\,0 \\ 2\,0\,1\,1\,1\,1\,0\,2\,1 \\ 1\,2\,0\,1\,1\,1\,1\,0\,2 \\ 2\,1\,0\,0\,1\,2\,1\,1\,1 \\ 0\,2\,1\,2\,0\,1\,1\,1\,1 \\ 1\,0\,2\,1\,2\,0\,1\,1\,1 \\ 1\,1\,1\,2\,1\,0\,0\,1\,2 \\ 1\,1\,1\,0\,2\,1\,2\,0\,1 \\ 1\,1\,1\,1\,0\,2\,1\,2\,0 \end{bmatrix};$$

and

$$T = \begin{bmatrix} 0\,2\,1\,2\,1\,0\,1\,0\,2 \\ 1\,0\,2\,0\,2\,1\,2\,1\,0 \\ 2\,1\,0\,1\,0\,2\,0\,2\,1 \\ 1\,0\,2\,0\,2\,1\,2\,1\,0 \\ 2\,1\,0\,1\,0\,2\,0\,2\,1 \\ 0\,2\,1\,2\,1\,0\,1\,0\,2 \\ 2\,1\,0\,1\,0\,2\,0\,2\,1 \\ 0\,2\,1\,2\,1\,0\,1\,0\,2 \\ 1\,0\,2\,0\,2\,1\,2\,1\,0 \end{bmatrix}.$$

The block matrices for X, Y, and Z are given as follows:

$$X = \begin{bmatrix} S\ T \end{bmatrix};$$

$$Y = \begin{bmatrix} U & T & 0 \\ 0 & -S & V = 0 \end{bmatrix};$$

$$Z = \begin{bmatrix} S & T & 0 & 0 \\ 0 & -U & V = 0 & 0 \\ 0 & 0 & S & T \end{bmatrix}.$$

7.4 *Calculation of* $H^1(\mathrm{Gal}(L/K), M)$

The plan is to compute the cohomology of the complex:

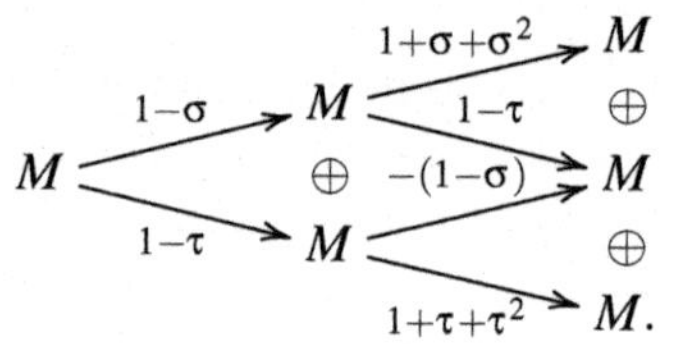

Lemma 10. *The kernel of* $Y : M^2 \to M^3$ *has dimension* 13 *and a basis is*

$$
\begin{aligned}
&(f - e^2f^2) \oplus 0,\\
&(e - e^2f^2) \oplus 0,\\
&(1 - e^2f^2) \oplus (ef - ef^2 - e^2f + e^2f^2),\\
&(f^2 - e^2f^2) \oplus (-ef + ef^2 + e^2f - e^2f^2),\\
&(ef - e^2f^2) \oplus (-ef + ef^2 + e^2f - e^2f^2),\\
&(ef^2 - e^2f^2) \oplus (ef - ef^2 - e^2f + e^2f^2),\\
&(e^2 - e^2f^2) \oplus (-ef + ef^2 + e^2f - e^2f^2),\\
&(e^2f - e^2f^2) \oplus (ef - ef^2 - e^2f + e^2f^2),\\
&0 \oplus (1 - ef - ef^2 - e^2f - e^2f^2),\\
&0 \oplus (f + ef + e^2f),\\
&0 \oplus (f^2 + ef^2 + e^2f^2),\\
&0 \oplus (e + ef + ef^2),\\
&0 \oplus (e^2 + e^2f + e^2f^2).
\end{aligned}
$$

Proof. Magma calculation.

Lemma 11. *The image of* $X : M \to M^2$ *has dimension* 4 *and a basis is*

$$
\begin{aligned}
&(1 - f^2 - e^2 + e^2f^2) \oplus (1 - f^2 - ef + ef^2 - e^2 + e^2f),\\
&(f - f^2 - e^2 + e^2f) \oplus (1 - f + f^2 - e - ef^2 - e^2 + e^2f^2),\\
&(e - ef^2 - e^2 + e^2f^2) \oplus (f - f^2 + e - ef - e^2 + e^2f^2),\\
&(ef - ef^2 - e^2f + e^2f^2) \oplus (-1 + f + e - ef^2 - e^2f + e^2f^2).
\end{aligned}
$$

Proof. Magma calculation.

Note that the image of X is contained in the kernel of Y.

Proposition 6. *The dimension of* $H^1(\mathrm{Gal}(L/K), M)$ *is 9 and a basis is*

$$
\begin{aligned}
&(f^2 - e^2) \oplus 0,\\
&(ef^2 - fe^2) \oplus 0,\\
&(e^2 + e^2 f + e^2 f^2) \oplus 0,\\
&(e^2 f - e^2 f^2) \oplus (ef - ef^2 - e^2 f + e^2 f^2),\\
&0 \oplus (1 - ef - ef^2 - e^2 f - e^2 f^2),\\
&0 \oplus (f + ef + e^2 f),\\
&0 \oplus (f^2 + ef^2 + e^2 f^2),\\
&0 \oplus (e + ef + ef^2),\\
&0 \oplus (e^2 + e^2 f + e^2 f^2).
\end{aligned}
$$

Proof. The quotient $H^1(\mathrm{Gal}(L/K), M) = \ker(Y)/\mathrm{im}(X)$ can be computed using the complement function in Magma.

7.5 Calculation of $H^2(\mathrm{Gal}(L/K), M)$

In this section, we compute the kernel of $Z : M^3 \to M^4$ modulo the image of $Y : M^2 \to M^3$.

Proposition 7. *The dimension of* $H^2(\mathrm{Gal}(L/K), M)$ *is 13 and a basis is*

$$
\begin{array}{rcl}
(f + ef + e^2 f) \oplus & 0 & \oplus\, 0,\\
(f^2 + ef^2 + e^2 f^2) \oplus & 0 & \oplus\, 0,\\
(e + ef + ef^2) \oplus & 0 & \oplus\, 0,\\
(e^2 + e^2 f + e^2 f^2) \oplus & 0 & \oplus\, 0,\\
0 \oplus & (f^2 - e^2 f^2) & \oplus\, 0,\\
0 \oplus & (ef^2 - e^2 f^2) & \oplus\, 0,\\
0 \oplus & (e^2 - e^2 f^2) & \oplus\, 0,\\
0 \oplus & (e^2 f - e^2 f^2) & \oplus\, 0,\\
0 \oplus & 0 & \oplus\, (1 - ef - ef^2 - e^2 f - e^2 f^2),\\
0 \oplus & 0 & \oplus\, (f + ef + e^2 f),\\
0 \oplus & 0 & \oplus\, (f^2 + ef^2 + e^2 f^2),\\
0 \oplus & 0 & \oplus\, (e + ef + ef^2),\\
0 \oplus & 0 & \oplus\, (e^2 + e^2 f + e^2 f^2).
\end{array}
$$

7.6 First and Second Cohomology with Coefficients in $H_1(U)$

Recall that $V = H_1(U)$ has dimension $(p-1)^2$. If $p = 3$, then V is a 4-dimensional subspace of M and a basis for V is

$$v_1 = \begin{bmatrix} 1 & 0 & -1 \\ 0 & 0 & 0 \\ -1 & 0 & 1 \end{bmatrix};$$

$$v_2 = \begin{bmatrix} 0 & 0 & 0 \\ 1 & 0 & -1 \\ -1 & 0 & 1 \end{bmatrix};$$

$$v_3 = \begin{bmatrix} 0 & 1 & -1 \\ 0 & 0 & 0 \\ 0 & -1 & 1 \end{bmatrix};$$

$$v_4 = \begin{bmatrix} 0 & 0 & 0 \\ 0 & 1 & -1 \\ 0 & -1 & 1 \end{bmatrix}.$$

Let X_1, Y_1, and Z_1 be the restriction of X, Y, and Z, respectively. Similarly, let $S_1 = 1 - B_\sigma$, $T_1 = 1 - B_\tau$, $U_1 = \mathrm{Nm}(\sigma)$, and $V_1 = \mathrm{Nm}(\tau)$ be the restrictions of S, T, U, and V, respectively. Then T_1, U_1, and V_1 are each the 4×4 zero matrix and

$$S_1 = \begin{bmatrix} -1 & -1 & -1 & -1 \\ 1 & 1 & 1 & 1 \\ 1 & 1 & 1 & 1 \\ -1 & -1 & -1 & -1 \end{bmatrix}.$$

Then

$$X_1 = \begin{bmatrix} S_1 & T_1 = 0 \end{bmatrix};$$

$$Y_1 = \begin{bmatrix} U_1 = 0 & T_1 = 0 & 0 \\ 0 & -S_1 & V_1 = 0 \end{bmatrix};$$

$$Z_1 = \begin{bmatrix} S_1 & T_1 = 0 & 0 & 0 \\ 0 & -U_1 = 0 & V_1 = 0 & 0 \\ 0 & 0 & S_1 & T_1 = 0 \end{bmatrix}.$$

Proposition 8. *The dimension of $H^1(G, H_1(U))$ is 6 and a basis is*

$$v_2 \oplus 0, v_3 \oplus 0, v_4 \oplus 0, 0 \oplus (v_1 - v_4), 0 \oplus (v_2 - v_4), 0 \oplus (v_3 - v_4).$$

The dimension of $H^2(G, H_1(U))$ is 9 and a basis is

$$\begin{aligned}(v_1 - v_4) \oplus 0 \oplus 0,\\ (v_2 + v_4) \oplus 0 \oplus 0,\\ (v_3 + v_4) \oplus 0 \oplus 0,\\ 0 \oplus v_2 \oplus 0,\\ 0 \oplus v_3 \oplus 0,\\ 0 \oplus v_4 \oplus 0,\\ 0 \oplus 0 \oplus (v_1 - v_4),\\ 0 \oplus 0 \oplus (v_2 + v_4),\\ 0 \oplus 0 \oplus (v_3 + v_4).\end{aligned}$$

7.7 First and Second Cohomology with Coefficients in $H_1(U) \wedge H_1(U)$

The vector space $W = H_1(U) \wedge H_1(U)$ has dimension $\binom{(p-1)^2}{2} = 6$.

Proposition 9. *Let $W = H_1(U) \wedge H_1(U)$. Then $H^1(G, W) = W^2$ (with dimension 12) and $H^2(G, W) = W^3$ (with dimension 18).*

Proof. The map $S^\wedge : V \wedge V \to V \wedge V$ induced by $S = (1 - B_\sigma)$ is the exterior square of S_1. One computes that $S^\wedge$ is the 6×6 zero matrix. Similarly the matrices for $T^\wedge$, $U^\wedge$ and $V^\wedge$ are zero. Then $H^1(G, W) = W^2$ since $\mathrm{Im}(X^\wedge) = 0$ and $\mathrm{Ker}(Y) = W^2$. Also $H^2(G, W) = W^3$ since $\mathrm{Im}(Y^\wedge) = 0$ and $\mathrm{Ker}(Z) = W^3$.

Acknowledgements We would like to thank BIRS for hosting the WIN3 conference where we began this project. Some of this work was done while the third and fourth authors were in residence at MSRI during the spring 2014 Algebraic topology semester, supported by NSF grant 0932078 000. The second author was supported by NSF grant DMS-1101712. The third author was supported by NSF grant DMS-1307390. The fourth author was supported by an American Institute of Mathematics 5 year fellowship and NSF grant DMS-1406380. We would also like to thank Sharifi and the referee for helpful remarks.

References

1. Anderson, G., Ihara, Y.: Pro-l branched coverings of $\mathbf{P}^1$ and higher circular l-units. Ann. Math. (2) **128**(2), 271–293 (1988). doi:10.2307/1971443. http://dx.doi.org/10.2307/1971443
2. Anderson, G.W.: Torsion points on Fermat Jacobians, roots of circular units and relative singular homology. Duke Math. J. **54**(2), 501–561 (1987). doi:10.1215/S0012-7094-87-05422-6. http://dx.doi.org/10.1215/S0012-7094-87-05422-6
3. Brown, K.S.: Cohomology of Groups. Graduate Texts in Mathematics, vol. 87. Springer, New York (1982)

4. Buhler, J.P., Harvey, D.: Irregular primes to 163 million. Math. Comput. **80**(276), 2435–2444 (2011). doi:10.1090/S0025-5718-2011-02461-0. http://dx.doi.org/10.1090/S0025-5718-2011-02461-0
5. Cassou-Nogués, P., Gillibert, J., Jehanne, A.: Galois module structure and Jacobians of Fermat curves. Bull. Lond. Math. Soc. 46(6), 117–1182 (2014). doi:10.1112/blms/bdu071
6. Coleman, R.F.: Anderson-Ihara theory: Gauss sums and circular units. In: Algebraic Number Theory. Advanced Studies in Pure Mathematics, vol. 17, pp. 55–72. Academic Press, Boston, MA (1989)
7. Coleman, R.F., Tamagawa, A., Tzermias, P.: The cuspidal torsion packet on the Fermat curve. J. Reine Angew. Math. **496**, 73–81 (1998). doi:10.1515/crll.1998.033. http://dx.doi.org/10.1515/crll.1998.033
8. Ellenberg, J.: 2-nilpotent quotients of fundamental groups of curves. Preprint (2000)
9. Friedlander, E.M.: Étale homotopy of simplicial schemes. In: Annals of Mathematics Studies, vol. 104. Princeton University Press, Princeton, NJ; University of Tokyo Press, Tokyo (1982)
10. Greenberg, R.: On the Jacobian variety of some algebraic curves. Compos. Math. **42**(3), 345–359 (1980/81). http://www.numdam.org/item?id=CM_1980__42_3_345_0
11. Ihara, Y.: Profinite braid groups, Galois representations and complex multiplications. Ann. Math. (2) **123**(1), 43–106 (1986). doi:10.2307/1971352. http://dx.doi.org/10.2307/1971352
12. Isaksen, D.C.: Etale realization on the $\mathbb{A}^1$-homotopy theory of schemes. Adv. Math. **184**(1), 37–63 (2004). doi:10.1016/S0001-8708(03)00094-X. http://dx.doi.org.prx.library.gatech.edu/10.1016/S0001-8708(03)00094-X
13. McCallum, W.G.: On the method of Coleman and Chabauty. Math. Ann. **299**(3), 565–596 (1994). doi:10.1007/BF01459799. http://dx.doi.org.prx.library.gatech.edu/10.1007/BF01459799
14. Milne, J.S.: Étale Cohomology. Princeton Mathematical Series, vol. 33. Princeton University Press, Princeton, NJ (1980)
15. Morel, F., Voevodsky, V.: $\mathbf{A}^1$-homotopy theory of schemes. Inst. Hautes Études Sci. Publ. Math. **90**, 45–143 (2001)/(1999). http://www.numdam.org/item?id=PMIHES_1999__90__45_0
16. Revêtements étales et groupe fondamental (SGA 1). Documents Mathématiques (Paris) [Mathematical Documents (Paris)], 3. Société Mathématique de France, Paris (2003). Séminaire de géométrie algébrique du Bois Marie 1960–61. [Algebraic Geometry Seminar of Bois Marie 1960–1961], Directed by A. Grothendieck, With two papers by M. Raynaud, Updated and annotated reprint of the 1971 original [Lecture Notes in Mathematics, vol. 224, Springer, Berlin MR0354651 (50 #7129)]
17. Schmidt, A., Wingberg, K.: On the fundamental group of a smooth arithmetic surface. Math. Nachr. **159**, 19–36 (1992). doi:10.1002/mana.19921590103. http://dx.doi.org.prx.library.gatech.edu/10.1002/mana.19921590103
18. Tzermias, P.: Almost rational torsion points and the cuspidal torsion packet on Fermat quotient curves. Math. Res. Lett. **14**(1), 99–105 (2007). doi:10.4310/MRL.2007.v14.n1.a8. http://dx.doi.org/10.4310/MRL.2007.v14.n1.a8
19. Washington, L.C.: Introduction to Cyclotomic Fields, 2nd edn. Graduate Texts in Mathematics, vol. 83. Springer, New York (1997). doi:10.1007/978-1-4612-1934-7. http://dx.doi.org/10.1007/978-1-4612-1934-7
20. Wickelgren, K.: 3-nilpotent obstructions to π_1 sections for $\mathbb{P}^1_{\mathbb{Q}} - \{0, 1, \infty\}$. In: Stix, J. (ed.) The Arithmetic of Fundamental Groups (PIA 2010). Contributions in Mathematical and Computational Sciences, vol. 2, pp. 281–329. Springer, Berlin (2012)
21. Zarhin, J.G.: Noncommutative cohomology and Mumford groups. Mat. Zametki **15**, 415–419 (1974)

Zeta Functions of a Class of Artin–Schreier Curves with Many Automorphisms

Irene Bouw, Wei Ho, Beth Malmskog, Renate Scheidler, Padmavathi Srinivasan, and Christelle Vincent

Abstract This paper describes a class of Artin–Schreier curves, generalizing results of Van der Geer and Van der Vlugt to odd characteristic. The automorphism group of these curves contains a large extraspecial group as a subgroup. Precise knowledge of this subgroup makes it possible to compute the zeta function of the curves in this class over the field of definition of all automorphisms in the subgroup.

Keywords Zeta function • Artin-Schreier curves • Extraspecial groups • Maximal curves

2010 *Mathematics Subject Classification.* Primary: 14G10; Secondary: 11G20, 14H37

I. Bouw (✉)
Institute of Pure Mathematics, Ulm University, D-89069 Ulm, Germany
e-mail: irene.bouw@uni-ulm.de

W. Ho
Department of Mathematics, University of Michigan, 530 Church Street, Ann Arbor, MI 48109, USA
e-mail: weiho@umich.edu

B. Malmskog
Department of Mathematics and Statistics, Villanova University, 800 Lancaster Avenue, Villanova, PA 19085, USA
e-mail: beth.malmskog@villanova.edu

R. Scheidler
Department of Mathematics and Statistics, University of Calgary, 2500 University Drive NW, Calgary, AB, Canada T2N 1N4
e-mail: rscheidl@ucalgary.ca

P. Srinivasan
Department of Mathematics, Massachusetts Institute of Technology, 77 Massachusetts Avenue, Cambridge, MA 02139, USA
e-mail: padma_sk@math.mit.edu

C. Vincent
Department of Mathematics and Statistics, University of Vermont, 16 Colchester Avenue, Burlington, VT 05401, USA
e-mail: christelle.vincent@uvm.edu

© Springer International Publishing Switzerland 2016

E.E. Eischen et al. (eds.), *Directions in Number Theory*, Association for Women in Mathematics Series 3, DOI 10.1007/978-3-319-30976-7_4

1 Introduction

In [24], Van der Geer and Van der Vlugt introduced a class of Artin–Schreier curves over a finite field with a highly rich structure. For example, these curves have a very large automorphism group that contains a large extraspecial p-group as a subgroup. Results of Lehr–Matignon [15] show that the automorphism groups of these curves are "maximal" in a precise sense. (Lehr–Matignon call this a *big action*.) A further remarkable property is that all these curves are supersingular. This yields an easy way of producing large families of supersingular curves.

In [24], the authors explore these curves and their Jacobians over fields of characteristic 2. In this case, there is an intriguing connection between the curves in this class and the weight enumerator of Reed–Müller codes, which was their original motivation for investigating this family of curves. In Sect. 13 of [24], they sketched extensions of some of their results to odd characteristic, but few details are given. The present paper extends the main results and strategy of [24] to the corresponding class of curves in odd characteristic, providing full details and proofs.

The main difference between the two cases is that the aforementioned extraspecial group of automorphisms has exponent p in the case of odd characteristic p, whereas the exponent is 4 in characteristic 2. As a result, some of the arguments in the odd characteristic case are more involved than those of [24]. Moreover, we have streamlined the reasoning of [24] and combined it with ideas from [15] to describe the automorphism group of the curves under investigation.

Arguably the most important object associated to an algebraic curve is its zeta function since it encodes a large amount of information about the curve, including point counts. Our main result is Theorem 8.4 which computes the zeta function of the members of the family of curves under consideration over a sufficiently large field. This not only generalizes the corresponding result in [24] for characteristic 2, but we also note that the authors of [24] do not offer an odd-characteristic analogue in their paper.

The most prominent member of the family of curves considered in this paper is the Hermite curve H_p (Example 9.5), which is well known to be a maximal curve over fields of square cardinality. We discuss other members of the family that are maximal in Sect. 9. More examples along the same lines have also been found by Çakçak and Özbudak in [3].

We now describe the contents of this paper in more detail. Let p be an odd prime and $R(X) \in \overline{\mathbb{F}}_p[X]$ be an additive polynomial of degree p^h, i.e., for indeterminates X and Y we have $R(X+Y) = R(X) + R(Y)$. We denote by C_R the smooth projective curve given by the Artin–Schreier equation

$$Y^p - Y = XR(X).$$

The key to the structure of the curve C_R is the bilinear form $\mathrm{Tr}(XR(Y) + YR(X))$, introduced in Sect. 2, whose kernel W is characterized in Proposition 2.1, Part 2. We obtain an expression for the number of points of C_R over a finite field in terms of W. Over a sufficiently large field $\mathbb{F}_q$ of square cardinality, we conclude that the curve C_R is either maximal or minimal, i.e., either the upper or lower Hasse–Weil bound is attained (Theorem 2.5 and Part 2 of Remark 8.2). To determine which of these cases applies, we use the automorphisms of C_R.

In Sects. 3 and 4, we show that W also determines a large p-subgroup P of the group of automorphisms (Theorem 4.3). With few exceptions, P is the Sylow p-subgroup of $\mathrm{Aut}(C_R)$ (Theorem 4.4). It is an extraspecial group of exponent p and order p^{2h+1}, where $\deg(R) = p^h$ (Theorem 5.3).

In general, the size of the automorphism group restricts the possibilities for the number of rational points of a curve. In our situation, there is a concrete relationship, since both the automorphisms and the rational points of C_R may be described in terms of the space W. We establish a point-counting result that applies to the smallest field $\mathbb{F}_q$ over which all automorphisms in P are defined.

The determination of the zeta function of C_R over $\mathbb{F}_q$ (Theorem 8.4) relies on a decomposition result for the Jacobian $J(C_R)$ of C_R (Proposition 6.3) that is an application of a result of Kani–Rosen [12]. More precisely, we show that $J(C_R)$ is isogenous over $\mathbb{F}_q$ to the product of Jacobians of quotients of C_R by suitable subgroups of P over $\mathbb{F}_q$ (Proposition 6.3). These quotient curves are twists of the curve C_{R_0} with $R_0(X) = X$ (Theorem 7.4) for which we may determine the zeta function by explicit point counting. Putting everything together yields a precise expression for the zeta function of C_R.

Our results also yield explicit examples of maximal curves (Sect. 9). The main technical difficulty here is determining the field $\mathbb{F}_q$ over which all automorphisms in P are defined.

1.1 Notation

Let p denote an odd prime, $\mathbb{F}_p$ be the finite field of order p, and $k = \overline{\mathbb{F}}_p$ be the algebraic closure of $\mathbb{F}_p$. All curves under consideration are assumed to be smooth, projective, and absolutely irreducible. Consider the curve C_R defined by the affine equation

$$Y^p - Y = XR(X), \tag{1}$$

where

$$R(X) = \sum_{i=0}^{h} a_i X^{p^i} \in \mathbb{F}_{p^r}[X]$$

is a fixed additive polynomial of degree p^h with $h \geq 0$ and whose coefficient field is denoted $\mathbb{F}_{p^r}$. Note that R is additive, i.e., $R(X+Y) = R(X) + R(Y)$ in $\mathbb{F}_{p^r}[X]$. Thus, C_R is defined over $\mathbb{F}_{p^r}$ and has genus

$$g(C_R) = \frac{p^h(p-1)}{2}.$$

Of interest will be the polynomial $E(X)$ derived from $R(X)$ via

$$E(X) = (R(X))^{p^h} + \sum_{i=0}^{h} (a_i X)^{p^{h-i}} \in \mathbb{F}_{p^r}[X] \tag{2}$$

with zero locus

$$W = \{c \in k : E(c) = 0\}. \tag{3}$$

Note that the formal derivative of $E(X)$ with respect to X is the constant nonzero polynomial a_h, so $E(X)$ is a separable additive polynomial of degree p^{2h} with coefficients in $\mathbb{F}_{p^r}$. It follows that W is an $\mathbb{F}_p$-vector space of dimension $2h$. When $h = 0$, i.e., $R(X) = a_0 X$, we have $W = \{0\}$.

We denote by $\mathbb{F}_q$ the splitting field of $E(X)$, so $W \subset \mathbb{F}_q$. In Sect. 4 of this paper we will define and investigate a subgroup P of the group of automorphisms of C_R, and the automorphisms contained in P will be defined over this field $\mathbb{F}_q$.

For convenience, we summarize the most frequently used notation in Table 1.

2 The Kernel of the Bilinear Form Associated to C_R

Let $\mathbb{F}_{p^s}$ be any extension of $\mathbb{F}_{p^r}$. For each s a multiple of r, we associate to the curve C_R the s-ary quadratic form

$$x \mapsto \mathrm{Tr}_{\mathbb{F}_{p^s}/\mathbb{F}_p}(xR(x))$$

on $\mathbb{F}_{p^s}$, where $\mathrm{Tr}_{\mathbb{F}_{p^s}/\mathbb{F}_p}\colon \mathbb{F}_{p^s} \to \mathbb{F}_p$ is the trace from the s-dimensional vector space $\mathbb{F}_{p^s}$ down to $\mathbb{F}_p$. The associated symmetric bilinear form on $\mathbb{F}_{p^s} \times \mathbb{F}_{p^s}$ is

$$(x, y) \mapsto \frac{1}{2}\,\mathrm{Tr}_{\mathbb{F}_{p^s}/\mathbb{F}_p}(xR(y) + yR(x)), \tag{4}$$

with kernel

$$W(\mathbb{F}_{p^s}) = \{c \in \mathbb{F}_{p^s} : \mathrm{Tr}_{\mathbb{F}_{p^s}/\mathbb{F}_p}(cR(y) + yR(c)) = 0 \text{ for all } y \in \mathbb{F}_{p^s}\}. \tag{5}$$

Note that $W(\mathbb{F}_{p^s})$ is a vector space over $\mathbb{F}_p$. The following characterizations and properties of $W(\mathbb{F}_{p^s})$ will turn out to be useful:

Proposition 2.1. *Let $c \in \mathbb{F}_{p^s}$. Then the following hold:*

1. *If $c \in W(\mathbb{F}_{p^s})$, then $\mathrm{Tr}_{\mathbb{F}_{p^s}/\mathbb{F}_p}(cR(c)) = 0$.*
2. *We have $c \in W(\mathbb{F}_{p^s})$ if and only if there exists a polynomial $B(X) \in \mathbb{F}_{p^s}[X]$ with*

$$B(X)^p - B(X) = cR(X) + R(c)X. \tag{6}$$

Table 1 Frequently used notation

Symbol	Meaning and place of definition
p	An odd prime
$\mathbb{F}_{p^r}$	Field of definition of $R(X)$ and of C_R (Sect. 1.1)
$\mathbb{F}_{p^s}$	An arbitrary extension of $\mathbb{F}_{p^r}$ (Sect. 2)
$\mathbb{F}_q$	$\mathbb{F}_q \supseteq \mathbb{F}_{p^r}$ splitting field of $E(X)$ (Sect. 1.1)
$k = \overline{\mathbb{F}}_p$	Algebraic closure of $\mathbb{F}_p$ (Sect. 1.1)
C_R	The curve $C_R : Y^p - Y = XR(X)$ over $\mathbb{F}_{p^r}$ (Eq. 1)
$\overline{C}_A$	Quotient curve C_R/A (Theorem 7.4)
$R(X)$	$R(X) = \sum_{i=0}^{h} a_i X^{p^i} \in \mathbb{F}_{p^r}[X]$ an additive polynomial (Eq. 1)
$E(X)$	$E(X) = (R(X))^{p^h} + \sum_{i=0}^{h}(a_i X)^{p^{h-i}} \in \mathbb{F}_{p^r}[X]$ (Eq. 2)
b, c	Elements in k with $b^p - b = cR(c)$ (Remark 3.3)
$B_c(X) = B(X)$	Polynomial s.t. $B(X)^p - B(X) = cR(X) + R(c)X$ (Eqs. 6 and 11)
$W(\mathbb{F}_{p^s})$	$W(\mathbb{F}_{p^s}) = \{c \in \mathbb{F}_{p^s} : \mathrm{Tr}_{\mathbb{F}_{p^s}/\mathbb{F}_p}(cR(y) + y(R(c)) = 0$ for all $y \in \mathbb{F}_{p^s}\}$ (Eq. 5)
W	$W = W(\mathbb{F}_q)$, space of zeros of $E(X)$ (Eq. 3)
$S(f)$	$S(f) = \{(a, c, d) \in k^* \times k \times \mathbb{F}_p^*$: there is $g \in k[X]$ s. t. $f(aX + c) - df(X) = g(X)^p - g(X)\}$ (Eq. 10)
$\mathrm{Aut}^0(C_R)$	Group of automorphisms of C_R that fix ∞ (Sect. 4)
$\sigma_{a,b,c,d}$	Automorphism in $\mathrm{Aut}^0(C_R)$ (Eq. 15)
$\sigma_{b,c}$	$\sigma_{b,c} = \sigma_{1,b,c,1}$ (Sect. 5)
ρ	Artin–Schreier automorphism, $\rho = \sigma_{1,1,0,1}$ (following Eq. 15)
P	Sylow p-subgroup of $\mathrm{Aut}^0(C_R)$ (Theorem 4.3)
H	$H = \mathrm{Aut}^0(C_R)/P$ (Theorem 4.3)
$Z(G)$	Center of a group G
$E(p^3)$	Extraspecial group of order p^3 and exponent p (Corollary 5.4)
$\mathscr{A}$	A maximal abelian subgroup of P (Proposition 5.5)
J_R	$J_R = \mathrm{Jac}(C_R)$, the Jacobian variety of C_R
$J \sim_{\mathbb{F}} J'$	The ab. var. J and J' are isogenous over the field $\mathbb{F}$ (Sect. 6).
$L_{C,\mathbb{F}}(T)$	Numerator of the zeta function of the curve C over the field $\mathbb{F}$ (Sect. 8)

Moreover, there is a unique solution $B_c(X) \in X\mathbb{F}_{p^s}[X]$ to Eq. (6), and

a. The polynomial $B_c(X)$ is additive.
b. Every solution $B(X)$ of (6) is of the form $B(X) = B_c(X) + \beta$ for some $\beta \in \mathbb{F}_p$.
c. If $c_1, c_2 \in W(\mathbb{F}_{p^s})$, then $B_{c_1+c_2}(X) = B_{c_1}(X) + B_{c_2}(X)$.

3. *We have $c \in W(\mathbb{F}_{p^s})$ if and only if $E(c) = 0$, where $E(X)$ is the polynomial of (2) with zero locus W as defined in (3). In other words, $W(\mathbb{F}_{p^s}) = W \cap \mathbb{F}_{p^s}$.*

Proof.

1. Let $c \in W(\mathbb{F}_{p^s})$. Then substituting $y = c$ into (5) yields $\mathrm{Tr}_{\mathbb{F}_{p^s}/\mathbb{F}_p}(2cR(c)) = 0$. Since $\mathrm{Tr}_{\mathbb{F}_{p^s}/\mathbb{F}_p}(X)$ is $\mathbb{F}_p$-linear and p is odd, this forces $\mathrm{Tr}_{\mathbb{F}_{p^s}/\mathbb{F}_p}(cR(c)) = 0$.

2. The proof of Part 2 is analogous to that of Proposition 3.2 of [24]. Assume that $c \in W(\mathbb{F}_{p^s})$. We show the existence of a solution B of (6), and show that statements 2a–2c hold.

   We first recursively define numbers b_i using the following formulas:

$$b_0 = -ca_0 - R(c), \tag{7}$$

$$b_i = -ca_i + b_{i-1}^p \qquad \text{for } 1 \le i \le h-1, \tag{8}$$

   and set $B_c(X) = \sum_{i=0}^{h-1} b_i X^{p^i}$. Then $B_c(X) \in X\mathbb{F}_{p^s}[X]$, $B_c(X)$ is additive, and $B_{c_1+c_2}(X) = B_{c_1}(X) + B_{c_2}(X)$ for all $c_1, c_2 \in W(\mathbb{F}_{p^s})$. Furthermore, a simple calculation reveals that

$$B_c^p(X) - B_c(X) = cR(X) + R(c)X + \epsilon X^{p^h}$$

   with $\epsilon = b_{h-1}^p - ca_h \in \mathbb{F}_{p^s}$. Note that $\mathrm{Tr}_{\mathbb{F}_{p^s}/\mathbb{F}_p}(B_c(y)^p - B_c(y)) = 0$ for all $y \in \mathbb{F}_{p^s}$ by the additive version of Hilbert's Theorem 90.

   If $c \in W(\mathbb{F}_{p^s})$, then $\mathrm{Tr}_{\mathbb{F}_{p^s}/\mathbb{F}_p}(cR(y) + yR(c)) = 0$ for all $y \in \mathbb{F}_{p^s}$, therefore $\mathrm{Tr}_{\mathbb{F}_{p^s}/\mathbb{F}_p}(\epsilon y^{p^h}) = 0$, which forces $\epsilon = 0$. Hence $B_c(X)$ satisfies (6), and

$$b_{h-1}^p = ca_h. \tag{9}$$

   Moreover, if $B(X)$ is any solution to (6), then $(B(X) - B_c(X))^p = B(X) - B_c(X)$, so $B(X) - B_c(X) \in \mathbb{F}_p$.

   Conversely, if (6) has a solution $B(X) \in \mathbb{F}_{p^s}[X]$, then

$$0 = \mathrm{Tr}_{\mathbb{F}_{p^s}/\mathbb{F}_p}(B(y)^p - B(y)) = \mathrm{Tr}_{\mathbb{F}_{p^s}/\mathbb{F}_p}(cR(y) + R(c)y)$$

   for all $y \in \mathbb{F}_{p^s}$, so $c \in W(\mathbb{F}_{p^s})$.
3. This result is stated for p odd in Proposition 13.1 and proved for $p = 2$ in Proposition 3.1 of [24]. It is also addressed in Remark 4.15 of the preprint [14] (the explicit statement is not included in [15], but can readily be deduced from the results therein).

□

Remark 2.2. The characteristic-2 analogue of Part 2 of Proposition 2.1 can be found in Sect. 3 of [24]. We also note that Part 1 of Proposition 2.1 does not hold in characteristic $p = 2$ in general (see Sect. 5 of [24]).

Part 3 of Proposition 2.1 immediately establishes the following corollary:

Corollary 2.3. *$W(\mathbb{F}_{p^s}) \subseteq W$, with equality for any extension $\mathbb{F}_{p^s}$ of the splitting field $\mathbb{F}_q$ of E.*

We conclude this section with a connection between the $\mathbb{F}_p$-dimension of the space $V_s = \mathbb{F}_{p^s}/W(\mathbb{F}_{p^s})$ and the number of $\mathbb{F}_{p^s}$-rational points on the curve C_R.

This is obtained by projecting the bilinear form (4) onto V_s. We write $\overline{x} = x + W(\mathbb{F}_{p^s})$ for the elements in V_s. Proposition 2.6 below is one of the key ingredients in the determination of the zeta function of C_R over $\mathbb{F}_q$ (Theorem 8.4).

Proposition 2.4. *Define a map Q_s on $V_s \times V_s$ via*

$$Q_s(\overline{x}, \overline{y}) = \frac{1}{2}\,\mathrm{Tr}_{\mathbb{F}_{p^s}/\mathbb{F}_p}(xR(y) + yR(x)).$$

Then Q_s is a non-degenerate bilinear form on $V_s \times V_s$.

Proof. We begin by showing that Q_s is well-defined. Let $x_1, x_2 \in \mathbb{F}_{p^s}$. Then

$$\begin{aligned}
\overline{x}_1 = \overline{x}_2 &\iff x_1 - x_2 \in W(\mathbb{F}_{p^s}) \\
&\iff \mathrm{Tr}_{\mathbb{F}_{p^s}/\mathbb{F}_p}((x_1 - x_2)R(y) + yR(x_1 - x_2)) = 0 \text{ for all } y \in \mathbb{F}_{p^s} \\
&\iff \mathrm{Tr}_{\mathbb{F}_{p^s}/\mathbb{F}_p}(x_1R(y) + yR(x_1)) = \mathrm{Tr}_{\mathbb{F}_{p^s}/\mathbb{F}_p}(x_2R(y) + yR(x_2)) \text{ for all } y \in \mathbb{F}_{p^s} \\
&\iff Q_s(\overline{x}_1, \overline{y}) = Q_s(\overline{x}_2, \overline{y}) \text{ for all } \overline{y} \in V_s.
\end{aligned}$$

Similarly, one obtains that $\overline{y}_1 = \overline{y}_2$ if and only if $Q_s(\overline{x}, \overline{y}_1) = Q_s(\overline{x}, \overline{y}_2)$ for all $\overline{x} \in V_s$. So if $(\overline{x}_1, \overline{y}_1) = (\overline{x}_2, \overline{y}_2)$, then $Q_s(\overline{x}_1, \overline{y}_1) = Q_s(\overline{x}_1, \overline{y}_2) = Q_s(\overline{x}_2, \overline{y}_2)$.

It is obvious that Q_s is bilinear. To establish non-degeneracy, let $\overline{x} \in V_s$ with $Q_s(\overline{x}, \overline{y}) = 0$ for all $\overline{y} \in V_s$. Then $\mathrm{Tr}_{\mathbb{F}_{p^s}/\mathbb{F}_p}(xR(y) + yR(x)) = 0$ for all $y \in \mathbb{F}_{p^s}$, so $x \in W(\mathbb{F}_{p^s})$, and hence $\overline{x} = \overline{0}$. □

It follows that the quadratic form $\overline{x} \mapsto Q_s(\overline{x}, \overline{x})$ on V_s is non-degenerate. Therefore, its zero locus

$$\{\overline{x} \in V_s : \mathrm{Tr}_{\mathbb{F}_{p^s}/\mathbb{F}_p}(xR(x)) = 0\}$$

defines a smooth quadric over $\mathbb{F}_p$.

In [11], Joly provides a formula for the cardinality of the zero locus of a non-degenerate quadratic form, which we reproduce here for the convenience of the reader. The case of n odd is treated in Chap. 6, Sect. 3, Proposition 2.7, and the case of n even is Proposition 2.9 of Chap. 6, Sect. 3. Note that in [11], the result is proved for forms over an arbitrary finite field, but we restrict to $\mathbb{F}_p$ here which is sufficient for our purpose.

Theorem 2.5 (Joly [11]). *Let $a_1X_1^2 + \cdots + a_nX_n^2$ be a non-degenerate quadric in n variables with coefficients in $\mathbb{F}_p$, and N be the cardinality of its zero locus. Then*

$$N = \begin{cases} p^{n-1} & \text{if } n \text{ is odd,} \\ p^{n-1} + (p^{n/2} - p^{n/2-1}) & \text{if } n \text{ is even and } (-1)^{n/2}a_1 \cdots a_n \in (\mathbb{F}_p^*)^2, \\ p^{n-1} - (p^{n/2} - p^{n/2-1}) & \text{if } n \text{ is even and } (-1)^{n/2}a_1 \cdots a_n \notin (\mathbb{F}_p^*)^2. \end{cases}$$

Applying this result to the quadric $x \mapsto \mathrm{Tr}_{\mathbb{F}_{p^s}/\mathbb{F}_p}(xR(x))$ on the space $\mathbb{F}_{p^s}/W(\mathbb{F}_{p^s})$, we obtain the following point count for the curve C_R. This result is already presented in [24], but we include it here to provide a proof.

Proposition 2.6 (Proposition 13.4 of [24]). *Let* $w_s = \dim_{\mathbb{F}_p}(W(\mathbb{F}_{p^s}))$ *and* $n_s = s - w_s$. *Then the number of* $\mathbb{F}_{p^s}$*-rational points on* C_R *is*

$$\#C_R(\mathbb{F}_{p^s}) = \begin{cases} p^s + 1 & \text{for } n_s \text{ odd,} \\ p^s + 1 \pm (p-1)\sqrt{p^{s+w_s}} & \text{for } n_s \text{ even,} \end{cases}$$

with the sign depending on the coefficients of the quadratic form Q_s.

Proof. We have $V_s = \mathbb{F}_{p^s}/W(\mathbb{F}_{p^s}) \simeq \mathbb{F}_p^{n_s}$, where $n_s = s - w_s$. Therefore, for $\bar{x} \in V_s$, we may write $\bar{x} = (x_1, \ldots, x_{n_s})$, with each $x_i \in \mathbb{F}_p$. In this way, $Q_s(\bar{x}, \bar{x})$ on the space V_s is a non-degenerate quadric in n_s variables with coefficients in $\mathbb{F}_p$. Furthermore, it is diagonalizable by Chap. 8, Theorem 3.1 of [5] since p is odd, and therefore can be written in the form $\sum_{i=1}^{n_s} a_i X_i^2$ with $a_i \in \mathbb{F}_p$ for $1 \le i \le n_s$. As a consequence we may apply Theorem 2.5 to obtain the cardinality of the set

$$\{\bar{x} \in V_s \simeq \mathbb{F}_p^{n_s} : Q_s(\bar{x}, \bar{x}) = 0\} = \{\bar{x} \in V_s : \mathrm{Tr}_{\mathbb{F}_{p^s}/\mathbb{F}_p}(xR(x)) = 0\}.$$

Each $\bar{x} \in V_s$ with $Q_s(\bar{x}, \bar{x}) = 0$ gives rise to p^{w_s} distinct values $x \in \mathbb{F}_{p^s}$ such that $\mathrm{Tr}_{\mathbb{F}_{p^s}/\mathbb{F}_p}(xR(x)) = 0$. For each of these $x \in \mathbb{F}_{p^s}$, we have p solutions y to the equation $y^p - y = xR(x)$. In addition to these points, C_R has one point at infinity which is defined over any extension of $\mathbb{F}_{p^r}$. Hence $\#C_R(\mathbb{F}_{p^s}) = p^{w_s+1}N + 1$ with N given as in Theorem 2.5 (with $n = n_s$). □

Note that a more general version of Proposition 2.6 can be found in Theorem 4.1 of [4].

3 Connection to Automorphisms of C_R

In this section, we generalize the results of Proposition 2.1 to lay the groundwork for our investigation of the k-automorphisms of C_R that stabilize ∞, the unique point at infinity on C_R. We follow Sect. 3 of [15], but our notation is slightly different. Similar results may also be found in [7].

We define for any polynomial $f(X) \in k[X]$ the set

$$S(f(X)) = \{(a, c, d) \in k^* \times k \times \mathbb{F}_p^* : \text{there exists } g(X) \in Xk[X] \text{ such that} \\ f(aX + c) - df(X) = g(X)^p - g(X)\}. \quad (10)$$

In our situation we take $f(X) = XR(X)$, where $R(X)$ is an additive polynomial of degree p^h. It is easy to verify that if $(a, c, d) \in S(XR(X))$ then the map $(x, y) \mapsto (ax + c, dy + g(x))$ is an automorphism of C_R that fixes ∞. In fact, in Lemma 4.1 we will see that every automorphism of C_R that fixes ∞ is of this form. The elements in $S(XR(X))$, along with the polynomial $g(X)$, can be characterized explicitly as follows:

Proposition 3.1. *If* $h = 0$*, then* $S(XR(X)) = \{(a, 0, a^2) : a^2 \in \mathbb{F}_p^*\}$.

Proof. If $h = 0$, then $R(X) = a_0X$, so

$$(aX + c)R(aX + c) - dXR(X) = a_0\left((a^2 - d)X^2 + 2acX + c^2\right).$$

This polynomial is of the form $g(X)^p - g(X)$ if and only if $g(X)^p - g(X) = 0$, or equivalently, $a^2 = d$, $c = 0$ and $g(X) \in \mathbb{F}_p$. □

Proposition 3.2.

1. *Assume that* $h \geq 1$ *and let* $a \in k^*$, $c \in k$ *and* $d \in \mathbb{F}_p^*$*. Then* $(a, c, d) \in S(XR(X))$ *if and only if there exists* $B(X) \in Xk[X]$ *such that*

$$cR(X) + R(c)X = B(X)^p - B(X), \tag{11}$$

   and

$$aR(aX) = dR(X). \tag{12}$$

2. *If the equivalent conditions of Part 1 are fulfilled, then* c *and* $B(X)$ *satisfy the following conditions:*

   a. *We have that* $c \in W$.
   b. *The polynomial* $B(X) = B_c(X)$ *only depends on* c *and is uniquely determined by (11) and the condition that* $B_c(X) \in Xk[X]$*. It is an additive polynomial with coefficients in* $\mathbb{F}_{p^r}(c) \subseteq \mathbb{F}_q$.
   c. *The polynomial* $B_c(X)$ *is identically zero if and only if* $c = 0$*, and has degree* p^{h-1} *otherwise.*

3. *For a triple* $(a, c, d) \in S(XR(X))$*, all polynomials* $g(X)$ *as given in (10) are of the form*

$$g(X) = B_c(aX) + \frac{B_c(c)}{2} + i,$$

   as i *ranges over* $\mathbb{F}_p$*. In particular, each of these polynomials* $g(X)$ *has coefficients in* $\mathbb{F}_q(a)$.

Proof.

1. Let $(a, c, d) \in k^* \times k \times \mathbb{F}_p^*$. Suppose first that there exists $B(X) \in Xk[X]$ satisfying (11), and that a and d satisfy (12). Then for any $b \in k$ such that $b^p - b = cR(c)$, we have

$$\begin{aligned}(aX + c)R(aX + c) - dXR(X) &= X(aR(aX) - dR(X)) + cR(aX) + aXR(c) + cR(c)\\ &= B(aX)^p - B(aX) + b^p - b,\end{aligned}$$

   and so we may take $g(X) = B(aX) + b$ to show that $(a, c, d) \in S(XR(X))$.

Conversely, suppose that $(a, c, d) \in S(XR(X))$. Then there exists a polynomial $g(X) \in k[X]$ such that

$$X(aR(aX) - dR(X)) + cR(aX) + aR(c)X + cR(c) = g(X)^p - g(X).$$

Writing $g(X) = b + \widetilde{B}(X)$ with $\widetilde{B}(X) \in Xk[X]$, we see that this is equivalent to the existence of a polynomial $\widetilde{B}(X) \in Xk[X]$ such that

$$\widetilde{B}(X)^p - \widetilde{B}(X) = XF(X) + G(X) \tag{13}$$

where $F(X) = aR(aX) - dR(X)$ and $G(X) = cR(aX) + aR(c)X$ are both additive polynomials. We note for future reference during the proof of Part 3 that this also implies $b^p - b = cR(c)$.

Note that (12) holds if and only if $F(X) = 0$, in which case $B(X) = \widetilde{B}(X/a) \in Xk[X]$ satisfies (11). Thus, it suffices to show that $(a, c, d) \in S(XR(X))$ implies $F(X) = 0$ to complete the proof of Part 1.

To this end, we note that all the monomials in $XF(X)$ and $G(X)$ are of the form X^{p^i+1} and X^{p^i} for $0 \leq i \leq h$. If $\widetilde{B}(X) = 0$, then this immediately forces $F(X) = G(X) = 0$, so assume that $\widetilde{B}(X) \neq 0$.

Comparing degrees in (13) shows that $\deg(\widetilde{B}) \leq p^{h-1}$. Put

$$\widetilde{B}(X) = \sum_{j=1}^{p^{h-1}} \tilde{b}_j X^j, \quad \tilde{b}_j \in k \text{ for } 1 \leq j \leq p^{h-1},$$

and consider the polynomial $\widetilde{B}(X)^p - \widetilde{B}(X)$. In this polynomial, the coefficient of X^j for $1 \leq j \leq p^h$ is

$$\begin{cases} -\tilde{b}_j & \text{when } p \nmid j, \\ \tilde{b}^p_{j/p} - \tilde{b}_j & \text{when } p \mid j \text{ and } j \leq p^{h-1}, \\ \tilde{b}^p_{p^{h-1}} & \text{when } j = p^h. \end{cases}$$

All coefficients of X^j for $j \neq p^i, p^i + 1$ must vanish. We conclude that the coefficients $\tilde{b}_j$ of $\widetilde{B}(X)$ are zero for all $j \neq p^i, p^i + 1$, so we may write $\widetilde{B}(X) = XU(X) + V(X)$ where $U(X), V(X) \in k[X]$ are additive polynomials. Then (13) yields

$$X^p U(X)^p + V(X)^p - XU(X) - V(X) = XF(X) + G(X).$$

Except for the monomials in $X^p U(X)^p$, this polynomial identity only contains monomials of the form X^{p^i} and X^{p^i+1}; the monomials in $X^p U(X)^p$ all take the form $X^{p+p^{i+1}}$. This forces $U(X) = 0$. Thus, $XF(X) = V(X)^p - V(X) - G(X)$ is additive, which is only possible if $F(X) = 0$.

2. The proof of Part 2 is now straightforward. We remark that Eq. (11) is identical to Eq. (6). Therefore 2a follows from Part 2 of Proposition 2.1, and $B(X)$ is identical to the polynomial $B_c(X)$ defined in that proposition since $B(X) \in Xk[X]$. Thus, $B(X)$ only depends on c and is unique, and we write $B_c(X)$ for this polynomial from now on. The additivity of $B_c(X)$ was already established in the proof of Part 1, since $B_c(X) = \widetilde{B}(X/a)$, and $\tilde{B}(X) = V(X)$ is additive; note that it also follows from Part 2a of Proposition 2.1. Moreover, the coefficients of B_c satisfy (7)–(9) and thus belong to $\mathbb{F}_{p^r}(c)$. Part 1 and Corollary 2.3 imply that $\mathbb{F}_{p^r}(c) \subseteq \mathbb{F}_q$. This proves 2b.
   If $c = 0$, then $B_c(X) = 0$. If $c \neq 0$, the polynomial $B_c(X)$ is obviously nonzero and (9) shows that $B_c(X)$ has degree p^{h-1}. This proves 2c.
3. Writing $g(X) = b + \widetilde{B}(X)$ with $\widetilde{B}(X) \in Xk[X]$ as in the proof of Part 1, we have already seen that $B_c(X) = \widetilde{B}(X/a)$, and b is any solution to the equation $b^p - b = cR(c)$. Any two such solutions differ by addition of an element in $\mathbb{F}_p$. Furthermore, since $2 \in \mathbb{F}_p^*$, it follows from (11) that $b = B_c(c)/2$ satisfies $b^p - b = cR(c)$, and the first statement of Part 3 follows. The second statement of Part 3 follows from Part 2b.

□

Remark 3.3. We repeat here a remark made in the proof since we will use this throughout the paper. For a triple $(a, c, d) \in S(XR(X))$, all polynomials $g(X)$ as given in (10) can be written as

$$g(X) = B_c(aX) + b,$$

where $B_c(aX) \in \mathbb{F}_q(a)[x]$, and $b \in k$ is a solution of the equation

$$b^p - b = cR(c). \tag{14}$$

Part 3 of Proposition 3.2 implies that every solution b of this equation is of the form $b = B_c(c)/2 + i$ with $i \in \mathbb{F}_p$.

4 Automorphism Group of C_R

In this section we apply the results of the previous section to study the group $\mathrm{Aut}(C_R)$ of k-automorphisms of the curve C_R, and more particularly the subgroup $\mathrm{Aut}^0(C_R)$ of automorphisms of C_R that fix the unique point at infinity, i.e., the unique point of C_R which does not belong to the affine curve defined by (1). The main result is Theorem 4.3, which describes $\mathrm{Aut}^0(C_R)$.

Recall from Sect. 3 that to a triple $(a, c, d) \in S(XR(X))$ we associate the k-automorphism

$$\begin{aligned} \sigma_{a,b,c,d} \colon C_R &\to C_R \\ (x, y) &\mapsto (ax + c, dy + b + B_c(ax)) \end{aligned} \tag{15}$$

of C_R. Here b is a solution of the equation $b^p - b = cR(c)$ (see Remark 3.3) and B_c is as in Proposition 3.2. Note that $\sigma_{a,b,c,d}$ fixes the point ∞. In the rest of the paper, we denote by

$$\rho(x, y) = \sigma_{1,1,0,1}(x, y) = (x, y + 1)$$

the Artin–Schreier automorphism of the curve C_R.

The following lemma summarizes some properties of the automorphisms $\sigma_{a,b,c,d}$:

Lemma 4.1. *With the above notation and assumptions, we have*

1. *Every element of the stabilizer* $\mathrm{Aut}^0(C_R)$ *of the point* ∞ *is of the form* $\sigma_{a,b,c,d}$ *as in (15).*
2. *The automorphisms* $\sigma_{1,b,c,1}$ *with* $(b, c) \neq (0, 0)$ *have order* p. *For* $(a, d) \neq (1, 1)$ *the order of* $\sigma_{a,b,c,d}$ *is not a* p*-power.*

Proof. The lemma follows from Corollaries 3.4 and 3.5 in [15]. We recall the proof.

1. Part 1 follows from Proposition 3.3 of [15] in the case that $g(C_R) \geq 2$. (Since p is odd in our set-up and the genus of C_R is $p^h(p-1)/2$, this only excludes the case that $h = 0$ and $p = 3$. This case is treated in the proof of Corollary 3.4 of [15].) Namely, let $\varphi \in \mathrm{Aut}^0(C_R)$ be an automorphism of C_R fixing ∞. Then the proof of Proposition 3.3 of [15] shows that there exists an isomorphism $\tilde{\varphi}: \mathbb{P}^1 \to \mathbb{P}^1$ together with a commutative diagram

$$\begin{array}{ccc} C_R & \xrightarrow{\varphi} & C_R \\ \downarrow & & \downarrow \\ \mathbb{P}^1 & \xrightarrow{\tilde{\varphi}} & \mathbb{P}^1, \end{array}$$

where the vertical maps are $(x, y) \mapsto x$.

The morphism $\tilde{\varphi}$ fixes $\infty \in \mathbb{P}^1$, hence it is an affine linear transformation and we may write it as $\tilde{\varphi}(x) = ax + c$ with $a \in k^*$ and $c \in k$. The commutative diagram above implies that $\varphi(x, y) = (ax + c, dy + g(x))$ for some $g(X) \in k(X)$ and $d \in k^*$. The assumption that φ fixes the point ∞ implies that $g(X) \in k[X]$ is a polynomial. The statement that $\varphi = \sigma_{a,b,c,d}$ follows since φ is assumed to be an automorphism of C_R.
2. To prove Part 2 we first remark that if $\sigma_{a,b,c,d}$ has p-power order, then $a = d = 1$, since 1 is the only pth root of unity in k. We show that every nontrivial automorphism $\sigma_{1,b,c,1}$ has order p.

   We compute that

$$\sigma_{1,b,c,1}^p(x, y) = (x + pc, y + pb + B_c(x) + B_c(x + c) + \cdots + B_c(x + (p-1)c)).$$

Recall from Proposition 3.2 that B_c is an additive polynomial; in particular, its constant term vanishes. Hence

$$B_c(X) + B_c(X + c) + \cdots + B_c(X + (p-1)c) = \sum_{i=0}^{p-1} B_c(ic) = \sum_{i=0}^{p-1} iB_c(c) = 0.$$

This implies that $\sigma_{1,b,c,1}^p = 1$.

□

Remark 4.2. Part 2 of Lemma 4.1 does not hold for $p = 2$. In [24], Theorem 4.1 it is shown that $\mathrm{Aut}^0(C_R)$ always contains automorphisms of order 4 for $h \geq 1$ and $p = 2$. See also [15], Sect. 7.2 for a concrete example. In Remark 7.3 we give a few more details on the differences between the cases $p = 2$ and p odd.

The following result is Theorem 13.3 of [24], and describes the group $\mathrm{Aut}^0(C_R)$. The structure of the Sylow p-subgroup P of $\mathrm{Aut}^0(C_R)$ will be described in more detail in Sect. 5 below. Again, we include this result here to provide a proof.

Theorem 4.3 (Theorem 13.3 of [24]).

1. *The group* $\mathrm{Aut}^0(C_R)$ *has a unique Sylow p-subgroup, which we denote by P. It is the subgroup consisting of all automorphisms* $\sigma_{1,b,c,1}$ *and has cardinality* p^{2h+1}.
2. *The automorphisms* $\sigma_{a,0,0,d}$ *form a cyclic subgroup* $H \subset \mathrm{Aut}^0(C_R)$ *of order*

$$\frac{e(p-1)}{2} \gcd_{\substack{i \geq 0 \\ a_i \neq 0}}(p^i + 1),$$

where $e = 2$ *if all of the indices i such that* $a_i \neq 0$ *have the same parity, and* $e = 1$ *otherwise.*

3. *The group* $\mathrm{Aut}^0(C_R)$ *is the semi-direct product of the normal subgroup P and the subgroup H.*

Proof.

1. To prove Part 1, one easily checks that $\{\sigma_{1,b,c,1} : \sigma_{1,b,c,1} \in \mathrm{Aut}^0(C_R)\}$ is a subgroup of $\mathrm{Aut}^0(C_R)$. (This is similar to the proof of Lemma 5.2 below.) The statements on the order of $\sigma_{a,b,c,d}$ in Part 2 of Lemma 4.1 imply that P is the unique Sylow p-subgroup of $\mathrm{Aut}^0(C_R)$, which implies that P is a normal subgroup.

   Parts 2a and 3 of Proposition 3.2 imply that the cardinality of P is equal to $p\#W$. The last statement of Part 1 therefore follows from Part 3 of Proposition 2.1, since E is a separable polynomial of degree p^{2h}.
2. To prove Part 2, we consider all elements $(a, 0, d) \in S(XR(X))$. Part 2c of Proposition 3.2 implies that the polynomial B_0 corresponding to this tuple is zero. Part 2 of Proposition 3.2 therefore implies that $(a, 0, d) \in S(XR(X))$ if and only

if $aR(aX) = dR(X)$. This condition is equivalent to $d = a^{p^i+1}$ for all $0 \leq i \leq h$ with $a_i \neq 0$, as can be readily seen by comparing coefficients in $aR(aX)$ and $dR(X)$. Part 2 now follows immediately.
3. Note that the order of H is prime to p. In particular, we have $H \cap P = \{1\}$. Part 3 follows since $\mathrm{Aut}^0(C_R)$ is generated by H and P.

□

For completeness we state the following theorem, which follows from [19], Satz 6 and Satz 7. (See also Theorem 3.1 of [15].) Since we study the automorphism group of C_R over the algebraically closed field k here, it is no restriction to assume that $R(X)$ is monic.

Theorem 4.4. *Let R be monic.*

1. *Assume that* $R(X) \notin \{X, X^p\}$. *Then* $\mathrm{Aut}(C_R) = \mathrm{Aut}^0(C_R)$.
2. *If* $R(X) = X^p$, *then* $\mathrm{Aut}(C_R) \cong \mathrm{PGU}_3(p)$.
3. *If* $R(X) = X$, *then* $\mathrm{Aut}(C_R) \simeq \mathrm{SL}_2(p)$.

For future reference we note the following result on the higher ramification groups of the point $\infty \in C_R$ in the cover $C_R \to C_R/\mathrm{Aut}^0(C_R)$. For the definition of the higher ramification groups and their basic properties we refer to [18], Chap. 4 or [20], Chap. 3.

Lemma 4.5. *Let R be an additive polynomial of degree* $h \geq 1$, *and* C_R *as given in (1).*

1. *The filtration of higher ramification groups in the lower numbering of* $\mathrm{Aut}^0(C_R)$ *is*

$$G = G_0 = \mathrm{Aut}^0(C_R) \supsetneq P = G_1 \supsetneq G_2 = \cdots = G_{1+p^h} = \langle \rho \rangle \supsetneq \{1\}.$$

2. *Let* $H \subset \mathrm{Aut}(C_R)$ *be any subgroup which contains* ρ. *Then* $g(C_R/H) = 0$.

Proof. To prove Part 1, write ν_∞ for the valuation at the unique point ∞ at infinity and choose a uniformizing parameter t at ∞. One easily computes that

$$\nu_\infty\left(\frac{\sigma(t) - t}{t}\right) = \begin{cases} 1 + p^h & \text{if } \sigma \in \langle \rho \rangle \setminus \{1\}, \\ 1 & \text{if } \sigma \in P \setminus \langle \rho \rangle. \end{cases}$$

This may also be deduced from the fact that the quotient of C_R by the subgroup generated by the Artin–Schreier automorphism $\rho(x, y) = (x, y + 1)$ has genus 0 ([16], Lemma 2.4).

Part 2 follows immediately from the fact that the function field of the curve $C_R/\langle \rho \rangle$ is $k(X)$. This can also be deduced from Part 1. □

5 Extraspecial Groups and the Structure of P

We now focus again on the subgroup P described in Part 1 of Theorem 4.3. Part 2 of Lemma 4.1 implies that the Sylow p-subgroup P of $\mathrm{Aut}^0(C_R)$ consists precisely of the automorphisms $\sigma_{1,b,c,1}(x,y) = (x+c, y+b+B_c(x))$. For brevity, we simplify their notation to

$$\sigma_{b,c} = \sigma_{1,b,c,1}.$$

The main result of the section, Theorem 5.3, states that P is an extraspecial group. For more details on extraspecial groups we refer the reader to Chap. III.13 of [9] and [22]. Recall that we assume that p is an odd prime. The classification of extraspecial 2-groups is different from that for odd primes.

Definition 5.1. A noncommutative p-group G is *extraspecial* if its center $Z(G)$ has order p and the quotient $G/Z(G)$ is elementary abelian.

We denote by $E(p^3)$ the unique nonabelian group of cardinality p^3 and exponent p. It can be given by generators and relations as follows:

$$E(p^3) = \langle x, y \mid x^p = y^p = [x,y]^p = 1, [x,y] \in Z(E(p^3))\rangle.$$

This group obviously is an extraspecial group.

The following lemma describes the commutation relation in P. The lemma contains the key steps to prove that P is an extraspecial group.

Lemma 5.2. *Assume that $h \geq 1$.*

1. *We have $[\sigma_{b_1,c_1}, \sigma_{b_2,c_2}] = \rho^{-\epsilon(c_1,c_2)}$, where*

$$\epsilon(c_1,c_2) = B_{c_1}(c_2) - B_{c_2}(c_1).$$

2. *We have $Z(P) = [P,P] = \langle\rho\rangle$. The quotient group $P/Z(P)$ is isomorphic to the space W defined in equation (3), where the isomorphism is induced by $\sigma_{b,c} \mapsto c$.*
3. *Any two non-commuting elements σ, σ' of P generate a normal subgroup $E_{\sigma,\sigma'} := \langle\sigma,\sigma'\rangle$ of P which is isomorphic to $E(p^3)$.*

Proof.

1. To prove Part 1, we compute that

$$\sigma_{b,c}^{-1}(x,y) = (x-c, y-b-B_c(x-c)).$$

We therefore have

$$\begin{aligned}\sigma_{b_1,c_1}\sigma_{b_2,c_2}\sigma_{b_1,c_1}^{-1}\sigma_{b_2,c_2}^{-1}(x,y) &= \sigma_{b_1,c_1}\sigma_{b_2,c_2}\sigma_{b_1,c_1}^{-1}(x-c_2, y-b_2-B_{c_2}(x-c_2))\\ &= \sigma_{b_1,c_1}\sigma_{b_2,c_2}(x-c_2-c_1, y-b_2-B_{c_2}(x-c_2)-b_1-B_{c_1}(x-c_2-c_1))\\ &= \sigma_{b_1,c_1}(x-c_1, y-B_{c_2}(x-c_2)-b_1-B_{c_1}(x-c_2-c_1)+B_{c_2}(x-c_2-c_1))\end{aligned}$$

$$= \sigma_{b_1,c_1}(x - c_1, y - b_1 - B_{c_1}(x - c_2 - c_1) - B_{c_2}(c_1))$$
$$= (x, y - B_{c_1}(x - c_2 - c_1) - B_{c_2}(c_1) + B_{c_1}(x - c_1))$$
$$= (x, y + B_{c_2}(c_1) - B_{c_1}(c_2)).$$

Since $\sigma_{b_1,c_1}\sigma_{b_2,c_2}\sigma_{b_1,c_1}^{-1}\sigma_{b_2,c_2}^{-1}$ certainly belongs to $\mathrm{Aut}^0(C_R)$, Part 1 of Lemma 4.1 implies that $\sigma_{b_1,c_1}\sigma_{b_2,c_2}\sigma_{b_1,c_1}^{-1}\sigma_{b_2,c_2}^{-1} = \sigma_{a,b,c,d}$ for some a, b, c, and d. From our computation above, $a = d = 1$ and $c = 0$. Since $c = 0$, by Part 2c of Proposition 3.2, $B_c(X) = 0$, which implies that $b = B_{c_2}(c_1) - B_{c_1}(c_2) \in \mathbb{F}_p$.

2. Part 1 shows that $[P, P] \subset \langle \rho \rangle$. Since P is noncommutative, we have equality. Because $\rho = \sigma_{1,0}$ and $B_0(X) = 0$ by Part 2c of Proposition 3.2, we have that for any $\sigma_{b,c}$,

$$\sigma_{b,c}\rho\sigma_{b,c}^{-1}\rho^{-1} = \rho^{B_c(0)} = 1,$$

since $B_c(X)$ is an additive polynomial and therefore has no constant term. Thus ρ commutes with every element of P, and $[P, P] = \langle \rho \rangle \subseteq Z(P)$.

To finish the proof of the first statement of Part 2, we now show that if $c_1 \neq 0$, then for each automorphism σ_{b_1,c_1} there exists an automorphism σ_{b_2,c_2} such that σ_{b_1,c_1} and σ_{b_2,c_2} do not commute. This shows that in fact $\langle \rho \rangle = Z(P)$.

Let $c_1 \in W \setminus \{0\}$. By Part 2 of Proposition 2.1 and Part 1 of Proposition 3.2, $(1, c_1, 1) \in S(XR(X))$ and by Part 2c of Proposition 3.2, $B_{c_1}(X)$ has degree p^{h-1}. Considering $c_2 =: C$ as a variable, the recursive formulas (7) and (8) for the coefficients b_i of B_C show that $\deg_C(b_i) \leq p^{h+i}$. We conclude that the degree of $\epsilon(c_1, C)$, when considered as polynomial in C, is at most p^{2h-1}. Since the cardinality of W is p^{2h}, it follows that there exists a $c_2 \in W$, and therefore $\sigma_{b_2,c_2} \in P$, such that $\epsilon(c_1, c_2) \neq 0$. We conclude that $Z(P) = \langle \rho \rangle$.

Since $\rho\sigma_{b,c} = \sigma_{b+1,c}$, it follows from Part 3 of Proposition 3.2 that the map

$$P \to W, \qquad \sigma_{b,c} \mapsto c$$

is a surjective group homomorphism with kernel $\langle \rho \rangle$.

3. Let $\sigma := \sigma_{b_1,c_1}, \sigma' := \sigma_{b_2,c_2} \in P$ be two non-commuting elements, and write $\epsilon = \epsilon(c_1, c_2)$. Part 1 implies that $\sigma\sigma' = \rho^{-\epsilon}\sigma'\sigma$. Since σ, σ', and ρ have order p (Part 2 of Lemma 4.1), it follows that σ and σ' generate a subgroup $E(\sigma, \sigma')$ of order p^3 of P, which contains $Z(P) = \langle \rho \rangle$. Since the exponent of this subgroup is p, it is isomorphic to $E(p^3)$.

For an arbitrary element $\sigma_{b,c} \in P$, Part 1 implies that $\sigma_{b,c}\sigma\sigma_{b,c}^{-1} \in \langle \rho, \sigma \rangle \subset E(\sigma, \sigma')$, and similarly for σ' replacing σ. Thus $E(\sigma, \sigma')$ is a normal subgroup, proving Part 3.

□

Theorem 5.3. *Assume that $h \geq 1$. Then the group P is an extraspecial group of exponent p.*

Proof. Since $h \geq 1$, Part 2 of Lemma 5.2 shows that P is an extraspecial group. Part 2 of Lemma 4.1 yields that P has exponent p. □

We now show that P is a central product of h copies of $E(p^3)$, i.e., P is isomorphic to the quotient of the direct product of h copies of $E(p^3)$, where the centers of each copy have been identified. These subgroups of P of order p^3 have been described in Part 3 of Lemma 5.2.

Corollary 5.4. *Assume that $h \geq 1$. Then P is a central product of h copies of $E(p^3)$.*

Proof. Theorem III.13.7.(c) of [9] states that P is the central product of h extraspecial groups P_i of order p^3. Since P has exponent p, it follows that the groups P_i have exponent p as well. Therefore $P_i \simeq E(p^3)$. □

We describe the decomposition of P as a central product from Corollary 5.4 explicitly; this description is in fact the proof given in Theorem III.13.7.(c) of [9]. The proof of Part 2 of Lemma 5.2 shows that $\epsilon(c_1, c_2)$ defines a non-degenerate symplectic pairing

$$W \times W \to \mathbb{F}_p, \qquad (c_1, c_2) \mapsto \epsilon(c_1, c_2).$$

We may choose a basis $(c_1, \ldots, c_h, c'_1, \ldots, c'_h)$ of W such that

$$\epsilon(c_i, c'_j) = \delta_{i,j},$$

where $\delta_{i,j}$ is the Kronecker function. In particular, it follows that $\langle c_1, \ldots, c_h \rangle \subset W$ is a maximal isotropic subspace of the bilinear form ϵ.

For every i, choose elements $\sigma_i, \sigma'_i \in P$ which map to c_i, c'_i, respectively, under the quotient map from Part 2 of Lemma 5.2. This corresponds to choosing an element b_i as in Part 3 of Proposition 3.2 for each i. Part 1 of Lemma 5.2 implies that σ_i does not commute with σ'_i, but commutes with σ_j, σ'_j for every $j \neq i$. Therefore $E_i = \langle \sigma_i, \sigma'_i \rangle$ is isomorphic to $E(p^3)$ (Part 3 of Lemma 5.2). It follows that P is the central product of the subgroups E_i.

We finish this section with a description of the maximal abelian subgroups of P. This will be used in Sect. 6 to obtain a decomposition of the Jacobian of C_R.

Proposition 5.5. *Let $h \geq 1$.*

1. *Every maximal abelian subgroup $\mathscr{A}$ of P is an elementary abelian group of order p^{h+1}, and is normal in P.*
2. *Let $\mathscr{A} \simeq (\mathbb{Z}/p\mathbb{Z})^{h+1}$ be a maximal abelian subgroup of P. For any subgroup $A = A_p \simeq (\mathbb{Z}/p\mathbb{Z})^h \subset \mathscr{A}$ with $A_p \cap Z(P) = \{1\}$ there exist subgroups $A_1, \ldots, A_{p-1}$ of $\mathscr{A}$ such that*

$$\mathscr{A} = Z(P) \cup A_1 \cup \cdots \cup A_p,$$

$$A_i \simeq (\mathbb{Z}/p\mathbb{Z})^h, \qquad A_i \cap Z(P) = \{1\}, \qquad A_i \cap A_j = \{1\} \text{ if } i \neq j.$$

3. *Any two subgroups A of $\mathscr{A}$ of order p^h which trivially intersect the center of P are conjugate inside P.*

Proof.

1. The statement that the maximal abelian subgroups $\mathscr{A}$ of P have order p^{h+1} is Theorem III.13.7.(e) of [9].
2. A maximal abelian subgroup $\mathscr{A}$ is the inverse image of a maximal isotropic subspace of W. Since P has exponent p, we conclude that $\mathscr{A} \simeq (\mathbb{Z}/p\mathbb{Z})^{h+1}$ is elementary abelian. Part 1 of Lemma 5.2 and the fact that $\mathscr{A}$ is the inverse image of a maximal isotropic subspace of W imply that $\mathscr{A}$ is a normal subgroup of P. This proves Part 1.

   Let $\mathscr{A} \subset P$ be a maximal abelian subgroup. Without loss of generality, we may assume that $\mathscr{A}$ corresponds to the maximal isotropic subspace generated by the basis elements $c_1, \ldots, c_h$ of W as described above. In this case we have $\mathscr{A} = \langle \rho, \sigma_1, \ldots, \sigma_h \rangle$ where σ_i maps to c_i under the map from Part 2 of Lemma 5.2. Define

   $$A_p := \langle \sigma_1, \ldots \sigma_h \rangle.$$

   This is a subgroup of $\mathscr{A}$ of order p^h such that $A_p \cap Z(P) = \{1\}$.

   We define $\tau = \sigma_{b,c_1'+\cdots+c_h'}$, where b is some solution of the equation

   $$b^p - b = (c_1' + \cdots + c_h')R(c_1' + \cdots + c_h')$$

   as specified in Remark 3.3. Let

   $$A_i = \tau^i A_p \tau^{-i}, \qquad i = 1, \ldots, p-1.$$

   By Part 2a of Proposition 2.1, $B_c(X)$ is additive in c. This implies that

   $$B_{c_1'+\cdots+c_h'}(X) = \sum_{i=1}^{h} B_{c_i'}(X).$$

   The choice of the basis c_i, c_i' of W, together with Part 1 of Lemma 5.2 implies therefore that

   $$\tau\sigma_i\tau^{-1} = \rho^{-\epsilon(c_i',c_i)}\sigma_i = \rho^{\epsilon(c_i,c_i')}\sigma_i = \rho\sigma_i.$$

   It follows that $A_i \cap Z(P) = \{1\}$ and $A_i \cap A_j = \{1\}$ if $i \neq j$. By counting, we see that each non-identity element of $\mathscr{A}$ is contained in exactly one A_i.
3. Let A, A' be two subgroups of $\mathscr{A}$ as in the statement of Part 3. Without loss of generality, we may assume that $A = A_p = \langle \sigma_1, \ldots \sigma_h \rangle$, as in the proof of Part 2. Then $A' = \langle \rho^{j_1}\sigma_1 \ldots, \rho^{j_h}\sigma_h \rangle$ for suitable $j_i \in \mathbb{F}_p$. Define $c = \sum_{i=1}^h j_i c_i \in W$ and choose b with $b^p - b = B_c(c)/2$. As in the proof of Part 2 it follows that $\tau := \sigma_{b,c}$ satisfies $\tau A \tau^{-1} = A'$.

□

6 Decomposition of the Jacobian of C_R

In this section we decompose the Jacobian of C_R over the splitting field $\mathbb{F}_q$ of the polynomial E. This decomposition allows us to reduce the calculation of the zeta function of C_R over $\mathbb{F}_q$ to that of a certain quotient curve. This quotient curve is computed in Sect. 7, and Sect. 8 combines these results to compute the zeta function of C_R over $\mathbb{F}_q$.

The decomposition result (Proposition 6.3) we prove below is based on the following general result of Kani–Rosen, Theorem B of [12]:

Theorem 6.1 (Kani-Rosen [12]). *Let C be a smooth projective curve defined over an algebraically closed field k, and G a (finite) subgroup of $\mathrm{Aut}_k(C)$ such that $G = H_1 \cup H_2 \cup \ldots \cup H_t$, where the subgroups $H_i \leq G$ satisfy $H_i \cap H_j = \{1\}$ for $i \neq j$. Then we have the isogeny relation*

$$\mathrm{Jac}(C)^{t-1} \times \mathrm{Jac}(C/G)^g \sim \mathrm{Jac}(C/H_1)^{h_1} \times \cdots \times \mathrm{Jac}(C/H_t)^{h_t},$$

where $g = \#G$, $h_i = \#H_i$, and $\mathrm{Jac}^n = \mathrm{Jac} \times \cdots \times \mathrm{Jac}$ (n times).

We apply Theorem 6.1 to a maximal abelian subgroup $\mathscr{A} \subset P$. Recall from Part 1 of Proposition 5.5 that $\mathscr{A}$ is an elementary abelian p-group of order p^{h+1} which contains the center $Z(P) = \langle \rho \rangle$ of P. Part 3 of Proposition 3.2 implies that all automorphisms in $\mathscr{A}$ are defined over $\mathbb{F}_q$.

Recall from Part 2 of Proposition 5.5 the existence of a decomposition

$$\mathscr{A} = A_0 \cup A_1 \cup \cdots \cup A_p, \tag{16}$$

where $A_0 = \langle \rho \rangle$ is the center of P and for $i \neq 0$ the A_i are elementary abelian p-groups of order p^h.

Each group A_i defines a quotient curve $\overline{C}_{A_i} := C_R/A_i$. Since all automorphisms in A_i are defined over $\mathbb{F}_q$, it follows that the quotient curve $\overline{C}_{A_i}$ together with the natural map $\pi_{A_i}\colon C_R \to \overline{C}_{A_i}$ may also be defined over $\mathbb{F}_q$. The following lemma implies that all curves $\overline{C}_{A_i}$ are isomorphic over $\mathbb{F}_q$:

Lemma 6.2. *Let $\mathscr{A}$ be a maximal abelian subgroup of P, and let A and A' be two subgroups of $\mathscr{A}$ of order p^h which have trivial intersection with the center of P. Then the curves C_R/A and C_R/A' are isomorphic over $\mathbb{F}_q$.*

Proof. Part 3 of Proposition 5.5 states that the subgroups A and A' are conjugate inside P. Namely, we have $A' = \tau A \tau^{-1}$ for an explicit element $\tau \in P$. The automorphism τ of C_R induces an isomorphism

$$\tau\colon C_R/A \to C_R/A'.$$

Since τ is defined over $\mathbb{F}_q$, this isomorphism is defined over $\mathbb{F}_q$ as well. □

We write $J_R := \mathrm{Jac}(C_R)$ for the Jacobian variety of C_R. Since C_R is defined over $\mathbb{F}_q$ and has an $\mathbb{F}_q$-rational point, the Jacobian variety J_R is also defined over $\mathbb{F}_q$. The map π_{A_i} induces $\mathbb{F}_q$-rational isogenies

$$\pi_{A_i,*}\colon J_R \to \mathrm{Jac}(\overline{C}_{A_i}), \qquad \pi_{A_i}^*\colon \mathrm{Jac}(\overline{C}_{A_i}) \to J_R. \tag{17}$$

The element

$$\varepsilon_{A_i} = \frac{1}{p^h}\pi_{A_i}^* \circ \pi_{A_i,*} \in \mathrm{End}^0(J_R) := \mathrm{End}(J_R) \otimes \mathbb{Q}$$

is an idempotent (Sect. 2 of [12]) and satisfies the property that $\varepsilon_{A_i}(J_R)$ is isogenous to $\mathrm{Jac}(\overline{C}_{A_i})$. Note that p^h is the degree of the map π_{A_i}.

In the following result we use these idempotents to decompose J_R. The same strategy was also used in Sect. 10 of [24] in the case that $p = 2$. In that source, Van der Geer and Van der Vlugt give a direct proof in their situation of the result of Kani–Rosen (Theorem 6.1) that we apply here.

Proposition 6.3. *There exists an $\mathbb{F}_q$-isogeny*

$$J_R \sim_{\mathbb{F}_q} \mathrm{Jac}(\overline{C}_{A_p})^{p^h}.$$

Proof. We apply Theorem 6.1 to the decomposition (16) of a maximal abelian subgroup $\mathscr{A}$ of P. This result shows the existence of a k-isogeny

$$J_R^p \times \mathrm{Jac}(C_R/\mathscr{A})^{p^{h+1}} \sim_k \mathrm{Jac}(\overline{C}_{A_0})^p \times \prod_{i=1}^{p} \mathrm{Jac}(\overline{C}_{A_i})^{p^h}. \tag{18}$$

The groups $\mathscr{A}$ and A_0 contain the Artin–Schreier element ρ; hence the curves $C_R/\mathscr{A}$ and $\overline{C}_{A_0}$ have genus zero (Part 2 of Lemma 4.5). Therefore the Jacobians of these curves are trivial and may be omitted from (18).

As before, let $\varepsilon_{A_i} \in \mathrm{End}(J_R)$ denote the idempotent corresponding to A_i. Theorem 2 of [12] states that the isogeny relation from (18) is equivalent to the relation

$$p\,\mathrm{Id} \sim p^h\Big(\sum_{i=1}^{p} \varepsilon_{A_i}\Big) \in \mathrm{End}^0(J_R).$$

Here, as defined on p. 312 of [12], the notation $a \sim b$ means that $\chi(a) = \chi(b)$ for all virtual characters of $\mathrm{End}^0(J_R)$. Since $\mathrm{End}^0(J_R)$ is a $\mathbb{Q}$-algebra, we may divide by p on both sides of this relation. Applying Theorem 2 of [12] once more yields the isogeny relation

$$J_R \sim_k \prod_{i=1}^{p} \mathrm{Jac}(\overline{C}_{A_i})^{p^{h-1}}. \tag{19}$$

We have already seen that the isogenies $\pi_{A_i}^*$ and $\pi_{A_i,*}$ are defined over $\mathbb{F}_q$. It follows that the isogeny (19) is defined over $\mathbb{F}_q$ as well (see also Remark 6 in Sect. 3 of [12]). Since the curves $\overline{C}_{A_i}$, and hence also their Jacobians, are isomorphic (Lemma 6.2), the statement of the proposition follows. □

7 Quotients of C_R by Elementary Abelian p-Groups

We consider again a maximal abelian subgroup $\mathscr{A} \simeq (\mathbb{Z}/p\mathbb{Z})^{h+1}$ of P and choose $A \subset \mathscr{A}$ with $A \simeq (\mathbb{Z}/p\mathbb{Z})^h$ and $A \cap Z(P) = \{1\}$. In this section we compute an $\mathbb{F}_q$-model of the quotient curve $\overline{C}_A = C_R/A$. Lemma 6.2 implies that the $\mathbb{F}_q$-isomorphism class of the quotient curve does not depend on the choice of the subgroup A.

Since $A \cap Z(P) = \{1\}$, Part 1 of Lemma 4.5 implies that the filtration of higher ramification groups in the lower numbering of A is

$$A = G_0 = G_1 \supsetneq G_2 = \{1\},$$

so the Riemann–Hurwitz formula yields

$$2g(C_R) - 2 = p^h(p-1) - 2 = (2g(\overline{C}_A) - 2)p^h + 2(p^h - 1).$$

We conclude that $g(\overline{C}_A) = (p-1)/2$.

Proposition 5.5 implies that the elements of A commute with ρ, since $\rho \in Z(P)$. It follows that $\overline{C}_A$ is an Artin–Schreier cover of the projective line branched at one point. Artin–Schreier theory implies therefore that $\overline{C}_A$ may be given by an Artin–Schreier equation

$$Y^p - Y = f_A(X),$$

where $f_A(X)$ is a polynomial of degree 2. Theorem 7.4 below implies that this polynomial $f_A(X)$ is in fact of the form $f_A(X) = a_A X^2$ for an explicit constant a_A. These curves are all isomorphic over the algebraically closed field k, but not over $\mathbb{F}_q$. The following lemma describes the different $\mathbb{F}_q$-models of the curves $Y^p - Y = eX^2$ for $e \in \mathbb{F}_q$.

Lemma 7.1. *For $e \in \mathbb{F}_q$, define the curve D_e by the affine equation*

$$Y^p - Y = eX^2. \tag{20}$$

Two curves D_{e_1} and D_{e_2} as in (20) are isomorphic over $\mathbb{F}_q$ if and only if e_1/e_2 is the product of a square in $\mathbb{F}_q^*$ with an element of $\mathbb{F}_p^*$. In particular, over $\overline{\mathbb{F}}_q$, any two of these curves are isomorphic.

Proof. Let D_{e_1} and D_{e_2} be curves of the form (20). Suppose there exists an $\mathbb{F}_q$-isomorphism $\varphi: D_{e_1} \to D_{e_2}$. We claim that there exists an $\mathbb{F}_q$-isomorphism which sends $\infty \in D_{e_1}$ to $\infty \in D_{e_2}$.

We first consider the case that $p > 3$, i.e., $g(D_{e_i}) \geq 2$. In this case, Proposition 3.3 of [15] states that there exists an automorphism σ of D_{e_1} over $\overline{\mathbb{F}}_q$ such that $\varphi \circ \sigma$ sends the point $\infty \in D_{e_1}$ to the point $\infty \in D_{e_2}$. To prove the claim it suffices to show that σ may be defined over $\mathbb{F}_q$.

To prove this, we follow the proof of Proposition 3.3 of [15] and use the fact that φ maps every point of D_{e_1} to a point of D_{e_2} with the same polar semigroup. Theorem 3.1.(a) of [15] implies that the only points of D_{e_1} with the same polar semigroup as ∞ are the points $Q_i := (0, i)$ with $i \in \mathbb{F}_p$. It follows that $\varphi^{-1}(\infty)$ is either ∞ or Q_i for some $i \in \mathbb{F}_p$. In the former case, there is nothing to show. If $\varphi^{-1}(\infty) = Q_i$, we may choose

$$\sigma(x, y) = \left(\frac{x}{y^{(p+1)/2}}, \frac{iy - 1}{y} \right).$$

Note that this is an automorphism of D_{e_1} which maps ∞ to Q_i. Moreover, σ is defined over the field of definition of D_{e_1}, and we are done.

We now prove the claim in the case that $p = 3$. In this case the curves D_{e_i} are elliptic curves. The inverse $\varphi^{-1}: D_{e_2} \to D_{e_1}$ of φ is also defined over $\mathbb{F}_q$. It follows that $Q := \varphi^{-1}(\infty) \in D_{e_1}(\mathbb{F}_q)$ is $\mathbb{F}_q$-rational. Then the translation $\tau_{Q-\infty}: P \mapsto P + Q - \infty$ is defined over $\mathbb{F}_q$ and sends the unique point $\infty \in D_{e_1}$ to Q. Precomposing φ with $\tau_{Q-\infty}$ gives an $\mathbb{F}_q$-isomorphism which sends $\infty \in D_{e_1}$ to $\infty \in D_{e_2}$.

Therefore, without loss of generality we let $\varphi: D_{e_1} \to D_{e_2}$ be an $\mathbb{F}_q$-isomorphism which sends the unique point of D_{e_1} at ∞ to the unique point of D_{e_2} at ∞. Any such automorphism can be written as $\varphi(x, y) = (\nu_0 x + \nu_1, \nu_2 y + \nu_3)$ with $\nu_i \in \mathbb{F}_q$ and $\nu_2 \nu_0 \neq 0$. The condition that φ maps D_{e_1} to D_{e_2} is equivalent to

$$\nu_2^p = \nu_2, \qquad \nu_2 e_1 = e_2 \nu_0^2, \tag{21}$$

$$0 = 2 e_2 \nu_0 \nu_1, \qquad \nu_3^p - \nu_3 = e_2 \nu_1^2. \tag{22}$$

It follows that $\nu_1 = 0$ and $\nu_2, \nu_3 \in \mathbb{F}_p$. The coefficient e_2 is given by

$$e_2 = \frac{\nu_2 e_1}{\nu_0^2}.$$

This proves the first assertion of the lemma. The second assertion is clear since any element of $\overline{\mathbb{F}}_q^*$ is a square in $\overline{\mathbb{F}}_q^*$. □

We now compute an $\mathbb{F}_q$-model of the curve C_R/A for $A \subset P$ an elementary abelian subgroup of cardinality p^h with $A \cap Z(P) = \{1\}$. We prove this by induction on h, following Sect. 13 of [24]. The following proposition is the key step in the inductive argument. It is a corrected version of Proposition 13.5 of [24], which extends to odd p Proposition 9.1 of [24] and is presented without proof. Indeed, the formula for the coordinate V of the quotient curve given in Proposition 13.5 of [24] contains an error that has been corrected here. We recall that $R(X)$ is an additive polynomial of degree p^h with leading coefficient $a_h \in \mathbb{F}_{p^r} \subseteq \mathbb{F}_q$.

Proposition 7.2. *Assume that $h \geq 1$, and let*

$$\sigma(x, y) := \sigma_{b,c}(x, y) = (x + c, y + b + B_c(x))$$

be an automorphism of C_R with $c \neq 0$ and $b = B_c(c)/2$. Then the quotient curve $C_R/\langle\sigma\rangle$ is isomorphic over $\mathbb{F}_q$ to the smooth projective curve given by an affine equation

$$V^p - V = \tilde{f}(U) = U\tilde{R}(U), \tag{23}$$

where $\tilde{R}(U) \in \mathbb{F}_q[U]$ is an additive polynomial of degree p^{h-1} with leading coefficient

$$\tilde{a} = \begin{cases} \frac{a_h}{c^{p-1}} & \text{if } h \neq 1, \\ \frac{a_h}{2c^{p-1}} & \text{if } h = 1. \end{cases}$$

Proof. In the proof c is fixed, therefore we write $B(X)$ for $B_c(X)$. We define new coordinates

$$U = X^p - c^{p-1}X, \qquad V = -Y + \Psi(X) = -Y + \gamma X^2 + \frac{X}{c}B(X), \tag{24}$$

where γ is defined by

$$\gamma = -\frac{B(c)}{2c^2}.$$

One easily checks that U and V are invariant under σ. The invariance of V under σ is equivalent to the property

$$\Psi(X + c) - \Psi(X) = B(X) + b.$$

Here we use the definition of b as $b = B(c)/2$. Since U and V generate a degree-p subfield of the function field of C_R and the automorphism σ has order p, U and V generate the function field of the quotient curve $C_R/\langle\sigma\rangle$.

From the definition of U and V above, one can see that the Artin–Schreier automorphism ρ induces an automorphism $\tilde{\rho}(U, V) = (U, V - 1)$ on the quotient curve $C_R/\langle\sigma\rangle$. It follows that the quotient curve is also given by an Artin–Schreier equation, which we may write as

$$V^p - V = -Y^p + Y + \Psi^p(X) - \Psi(X) = -XR(X) + \Psi^p(X) - \Psi(X). \tag{25}$$

It is clear that the right-hand side of (25) can be written as a polynomial $\tilde{f}(U)$ in U, since it is invariant under σ by construction. Since the constant term of Ψ is zero, the right-hand side has a zero at $X = 0$, so $\tilde{f}(U) \in U\mathbb{F}_q[U]$.

Recall that Part 1 of Proposition 3.2 established

$$B(X)^p - B(X) = cR(X) + XR(c). \tag{26}$$

This implies

$$XR(X) = \frac{X(B(X)^p - B(X))}{c} - \frac{X^2R(c)}{c}.$$

It follows that

$$-XR(X) + \Psi^p(X) - \Psi(X) = \frac{B(X)^p}{c^p}U + \gamma^p X^{2p} + X^2\left(\frac{R(c)}{c} - \gamma\right). \tag{27}$$

Using (26) one computes

$$\gamma^p X^{2p} + X^2\left(\frac{R(c)}{c} - \gamma\right) = \gamma^p U^2 - \frac{B(c)^p}{c^{p+1}}XU.$$

Define

$$\Theta(X) = \frac{B(X)^p}{c^p} - \frac{B(c)^p}{c^{p+1}}X.$$

Since Θ is invariant under σ, we may write $\Theta(X) = \theta(U)$ as a polynomial in U. Note that $\theta(0) = 0$ since $\Theta(0) = 0$. The additivity of the polynomials B and U in the variable X implies that the polynomial θ is additive in the variable U. It follows that we may write $\theta(U) = \sum_{i=0}^{h-1} \mu_i U^{p^i}$. From (9), we deduce that the leading coefficient of θ is

$$\mu_{h-1} = \frac{b_{h-1}^p}{c^p} = \frac{a_h}{c^{p-1}}.$$

Altogether, we find

$$V^p - V = \tilde{f}(U) = U\left(\theta(U) + \gamma^p U\right).$$

Setting $\tilde{R}(U) := \theta(U) + \gamma^p U$, we see that $\tilde{R}(U)$ is an additive polynomial in U. The statement about the leading coefficient of $\tilde{R}(U)$ follows from the definitions of θ and γ. □

Remark 7.3. We discuss a crucial difference between even and odd characteristic: Proposition 7.2 is a statement about the automorphisms $\sigma_{b,c}$ of order p which are not contained in the center of P. For p odd all elements of $P \setminus Z(P)$ have order p. This is not true for $p = 2$, as we already noted in Remark 4.2. Indeed all extraspecial 2-groups contain elements of order 4. The precise structure of the extraspecial group P in the case that $p = 2$ can be found in Theorem 4.1 of [24]. The automorphisms $\sigma_{b,c} \in P \setminus Z(P)$ of order 2 are easily recognized: they satisfy $c \neq 0$ but $B_c(c) = 0$. This observation considerably simplifies the computation in the proof of Proposition 7.2.

The distinction between elements of order 2 and 4 in $P \setminus Z(G)$ in characteristic 2 yields a decomposition of the polynomial E (Theorem 3.4 of [24]). There is no analogous result in odd characteristic.

Recall from Sect. 5 that every maximal abelian subgroup $\mathscr{A}$ of P is the inverse image of a maximal isotropic subspace $\overline{A}$ of W. For any such $\mathscr{A}$, let $\{c_1, \ldots, c_h\}$ be a basis of $\overline{A}$ as described prior to Proposition 5.5. Then every subgroup of $\mathscr{A}$ of order p^h that intersects $Z(P)$ trivially is generated by automorphisms of the form $\{\sigma_{b_1,c_1}, \ldots, \sigma_{b_h,c_h}\}$ where $b_i^p - b_i = c_i R(c_i)$ for $1 \leq i \leq h$. In fact, there is a one-to-one correspondence between such subgroups of $\mathscr{A}$ and sets of elements $\{b_1, \ldots, b_h\}$ satisfying $b_i^p - b_i = c_i R(c_i)$. By Remark 3.3 the elements in all these sets are of the form $b_i = B_{c_i}(c_i)/2 + i$ with $i \in \mathbb{F}_p$.

Theorem 7.4. *Assume $h \geq 0$. Let $\mathscr{A}$ be a maximal abelian subgroup of P. Any subgroup $A \subset \mathscr{A}$ of order p^h that intersects the center $Z(P)$ of P trivially gives rise to an $\mathbb{F}_q$-isomorphism of the quotient curve $\overline{C}_A$ onto the smooth projective curve given by the affine equation*

$$Y^p - Y = a_{\mathscr{A}} X^2.$$

Here

$$a_{\mathscr{A}} = \frac{a_h}{2} \prod_{c \in \overline{A} \setminus \{0\}} c,$$

for $h \geq 1$, where we recall that a_h is the leading coefficient of R and $\overline{A}$ is the maximal isotropic subspace of W that is the image of $\mathscr{A}$ under the quotient map $P \to W$. For $h = 0$, we let

$$a_{\mathscr{A}} = a_0.$$

Proof. We prove by induction on h that there exists a subgroup $A \subset \mathscr{A}$ with $A \simeq (\mathbb{Z}/p\mathbb{Z})^h$ and $Z(P) \cap A = \{1\}$ such that the quotient curve $\overline{C}_A = C_R/A$ is given over $\mathbb{F}_q$ by the equation stated in the theorem. The statement of the theorem follows from this using Lemma 6.2.

For $h = 0$ the statement is true by definition.

Assume that $h \geq 1$ and that the statement of the theorem holds for all additive polynomials $R(X)$ of degree p^{h-1}. Fix a basis $\{c_1, c_2, \ldots, c_h\}$ for the image of

$\mathscr{A}$ in W. We may choose $b_h = B_{c_h}(c_h)/2$. As in Sect. 5, we write $\sigma_h(x, y) = \sigma_{b_h,c_h}(x, y) = (x + c_h, y + b_h + B_{c_h}(x))$. Proposition 7.2 implies that the quotient curve $C_{h-1} := C_R/\langle\sigma_h\rangle$ is given by an Artin–Schreier equation

$$Y_{h-1}^p - Y_{h-1} = X_{h-1}R_{h-1}(X_{h-1}),$$

where R_{h-1} is an additive polynomial of degree p^{h-1}.

Since $\mathscr{A}$ is an abelian group, it follows that $\mathscr{A}_{h-1} := \mathscr{A}/\langle\sigma_h\rangle \simeq (\mathbb{Z}/p\mathbb{Z})^h$ is a maximal abelian subgroup of the Sylow p-subgroup P_{h-1} of $\mathrm{Aut}^0(C_{h-1})$. The definition of the coordinate X_{h-1} as $X^p - c_h^{p-1}X$ in the proof of Proposition 7.2 implies that $\mathscr{A}_{h-1}$ corresponds to the maximal isotropic subspace $\langle\bar{c}_1, \dots, \bar{c}_{h-1}\rangle$ of $W_{h-1} := W/\langle c_h, c_h'\rangle$, where $\bar{c}_i = c_i^p - c_h^{p-1}c_i$ and $c_h' \in W$ is an element with $\epsilon(c_i, c_h') = \delta_{i,h}$ as in Sect. 5.

The induction hypothesis implies that there exists a subgroup $A_{h-1} \subset \mathscr{A}_{h-1}$ with $A_{h-1} \simeq (\mathbb{Z}/p\mathbb{Z})^{h-1}$ and $A_{h-1} \cap Z(P_{h-1}) = \{1\}$ such that the quotient C_{h-1}/A_{h-1} is given by

$$Y_0^p - Y_0 = a_{\mathscr{A}_{h-1}}X_0^2.$$

We may choose b_i satisfying $b_i^p - b_i = c_iR(c_i)$ for $i = 1, \dots, h-1$ such that the images of $\sigma_{b_1,c_1}, \dots, \sigma_{b_{h-1},c_{h-1}}$ in $\mathscr{A}_{h-1}$ generate A_{h-1} (Remark 3.3). Put $\sigma_i = \sigma_{b_i,c_i}$ for $i = 1, \dots, h-1$. Then $A := \langle\sigma_1, \dots, \sigma_h\rangle$ satisfies

$$C_R/A \simeq_{\mathbb{F}_q} C_{h-1}/A_{h-1}.$$

This concludes the induction proof.

The statement about $a_{\mathscr{A}}$ follows immediately from the formula for the leading coefficient of the quotient curve given in Proposition 7.2. □

8 The Zeta Function of the Curve C_R

In this section, we describe the zeta function of the curve C_R over the splitting field $\mathbb{F}_q$ of the polynomial $E(X)$ defined in (2).

Let C be a curve defined over a finite field $\mathbb{F}_{p^s}$, and write $N_n = \#C(\mathbb{F}_{p^{sn}})$ for the number of points on C over any extension $\mathbb{F}_{p^{sn}}$ of $\mathbb{F}_{p^s}$. Recall that the *zeta function* of C, defined as

$$Z_C(T) = \exp\left(\sum_{n\geq 1}\frac{N_nT^n}{n}\right),$$

is a rational function with the following properties:

1. The zeta function may be written as

$$Z_C(T) = \frac{L_{C,\mathbb{F}_{p^s}}(T)}{(1-T)(1-p^sT)},$$

where $L_{C,\mathbb{F}_{p^s}}(T) \in \mathbb{Z}[T]$ is a polynomial of degree $2g(C)$ with constant term 1.
2. Write $L_{C,\mathbb{F}_{p^s}}(T) = \prod_{i=1}^{2g}(1-\alpha_i T)$ with $\alpha_i \in \mathbb{C}$. After suitably ordering the α_i, we have

$$\alpha_{2g-i} = \frac{p^s}{\alpha_i}, \qquad |\alpha_i| = p^{s/2}.$$

3. For each n, we have

$$N_n = \#C(\mathbb{F}_{p^{sn}}) = 1 + p^{sn} - \sum_{i=1}^{2g} \alpha_i^n.$$

4. If

$$L_{C,\mathbb{F}_{p^s}}(T) = \prod_{i=1}^{2g}(1-\alpha_i T)$$

as above, then for any $r \geq 0$, we have

$$L_{C,\mathbb{F}_{p^{rs}}}(T) = \prod_{i=1}^{2g}(1-\alpha_i^r T).$$

The numerator $L_{C,\mathbb{F}_{p^s}}(T)$ of the zeta function $Z_C(T)$ over $\mathbb{F}_{p^s}$ is called the *L-polynomial* of $C/\mathbb{F}_{p^s}$. If the field is clear from the context, we sometimes omit it from the notation and simply write $L_C(T)$.

Recall that the Hasse–Weil bound asserts that

$$|\#C(\mathbb{F}_{p^s}) - (p^s + 1)| \leq 2p^{s/2}g(C).$$

A curve $C/\mathbb{F}_{p^s}$ is called *maximal* if $\#C(\mathbb{F}_{p^s}) = p^s + 1 + 2p^{s/2}g(C)$ and *minimal* if $\#C(\mathbb{F}_{p^s}) = p^s + 1 - 2p^{s/2}g(C)$. Since the number of points on a curve must be an integer, if C is a maximal curve, then s must be even. Furthermore, using properties 2 and 3 above, it is clear that C is maximal if $\alpha_j = -p^{s/2}$ for each $1 \leq j \leq 2g(C)$, and C is minimal if $\alpha_j = p^{s/2}$ for each $1 \leq j \leq 2g(C)$.

Assume that s is even and that $\mathbb{F}_{p^s}$ is an extension of $\mathbb{F}_q$. In the notation of Proposition 2.6, we have $w_s = \dim_{\mathbb{F}_p} W = 2h$ (Corollary 2.3). Since the curve C_R has genus $p^h(p-1)/2$, Proposition 2.6 implies that C_R is either maximal or minimal in this case. Moreover, one easily sees that if either s is odd or $\mathbb{F}_{p^s}$ does not contain $\mathbb{F}_q$, then C_R is neither maximal nor minimal. The following proposition asserts that this almost determines the zeta function of C_R over $\mathbb{F}_q$. The statement is an extension to odd characteristic of Theorems 10.1 and 10.2 of [24]. Note that the statement for odd characteristic is simpler than that for characteristic 2.

Proposition 8.1. *Let $\mathbb{F}_{p^s}$ be an extension of $\mathbb{F}_q$, the splitting field of $E(X)$. Write $g = p^h(p-1)/2$ for the genus of C_R.*

1. *If s is even, the L-polynomial of C_R is*

$$L_{C_R}(T) = (1 \pm p^{s/2}T)^{2g}.$$

2. *If s is odd, the L-polynomial of C_R is*

$$L_{C_R}(T) = (1 \pm p^s T^2)^g.$$

Proof.

1. Let $\alpha_1, \ldots, \alpha_{2g}$ be the reciprocal zeros of the L-polynomial of C over $\mathbb{F}_{p^s}$, where we order the α_i such that $\alpha_i\alpha_{2g-i} = p^s$.
   We first assume that s is even. Since $\mathbb{F}_{p^s}$ is an extension of $\mathbb{F}_q$, we have

$$N_1 = \#C_R(\mathbb{F}_{p^s}) = 1 + p^s \pm 2gp^{s/2} = 1 + p^s - \sum_{i=1}^{2g} \alpha_i.$$

   Since $|\alpha_i| = p^{s/2}$ we conclude that

$$\alpha_1 = \cdots = \alpha_{2g} = \pm p^{s/2}.$$

   This proves Part 1.
2. We now assume that s is odd. Proposition 2.6 implies that

$$N_1 = \#C_R(\mathbb{F}_{p^s}) = 1 + p^s = 1 + p^s - \sum_{i=1}^{2g} \alpha_i. \tag{28}$$

   Since the reciprocal roots of the L-polynomial of C over $\mathbb{F}_{p^{2s}}$ are α_j^2, we conclude from Part 1 that either $\alpha_j^2 = p^s$ or $\alpha_j^2 = -p^s$ for all j.
   If $\alpha_j^2 = -p^s$ for all j, then $\alpha_j = \pm \mathrm{i} p^{s/2}$, where i is a primitive 4th root of unity. It follows that $\alpha_{2g-j} = p^s/\alpha_j = -\alpha_j$. Hence

$$(1 - \alpha_j T)(1 - \alpha_{2g-j}T) = 1 + p^s T^2.$$

   Assume now that $\alpha_j^2 = p^s$ for all j. In this case we have $\alpha_j = \pm p^{s/2}$ and $\alpha_{2g-j} = p^s/\alpha_j = \alpha_j$. Let $m = \#\{1 \le j \le g : \alpha_j = p^{s/2}\}$. It follows from (28) that

$$0 = \#C_R(\mathbb{F}_{p^s}) - (p^s + 1) = p^{s/2}(-2m + 2(g - m)).$$

We conclude that $2g = 4m$, i.e., $m = g/2$ (in particular, g is even). For the L-polynomial of C_R over $\mathbb{F}_{p^s}$ we find

$$L_{C_R}(T) = (1 - p^s T^2)^g,$$

as claimed in Part 2.

□

Remark 8.2.

1. The proof of Part 2 of Proposition 8.1 shows that the case $L_{C_R}(T) = (1 - p^s T^2)^g$ can only occur when g is even, i.e., if $p \equiv 1 \pmod 4$.
2. Assume that s is even. Then $\alpha_j = p^{s/2}$ or $\alpha_j = -p^{s/2}$ for all $1 \le j \le 2g$, and therefore C_R is either minimal or maximal. If C_R is minimal over $\mathbb{F}_{p^s}$, each $\alpha_j = p^{s/2}$. The curve C_R therefore remains minimal over each extension field $\mathbb{F}_{p^{sf}}$. If C_R is maximal over $\mathbb{F}_{p^s}$, each $\alpha_j = -p^{s/2}$. The reciprocal roots of the L-polynomial over $\mathbb{F}_{p^{sf}}$ are $\alpha_j^f = (-1)^f p^{sf/2}$. We conclude that C_R is maximal over $\mathbb{F}_{p^{sf}}$ if f is odd and minimal if f is even.

To determine the zeta function of C_R, it remains to decide when the different cases occur. The following result, which is an immediate corollary of Proposition 6.3, reduces this problem to the case $h = 0$.

Corollary 8.3. *Let $A \simeq (\mathbb{Z}/p\mathbb{Z})^h \subset P$ be a subgroup with $A \cap Z(P) = \{0\}$. Write $\overline{C}_A = C_R/A$. Then*

$$L_{C_R,\mathbb{F}_q}(T) = L_{\overline{C}_A,\mathbb{F}_q}(T)^{p^h}.$$

Proof. This is an immediate consequence of Proposition 6.3, since abelian varieties which are isogenous over $\mathbb{F}_q$ have the same zeta function over $\mathbb{F}_q$. This follows, for example, from the cohomological description of the zeta function in Sect. 1 of [13]. □

Recall from Theorem 7.4 that the curve $\overline{C}_A$ from Corollary 8.3 is a curve of genus $(p-1)/2$ given by an affine equation of the form

$$Y^p - Y = aX^2,$$

for some $a \in \mathbb{F}_q^*$. This corresponds to the case $h = 0$. All curves of this form are isomorphic over $\overline{\mathbb{F}}_q$, and the different $\mathbb{F}_q$-models are described in Lemma 7.1. The next result determines the L-polynomials of the curves $\overline{C}_A$. In the literature one finds many papers discussing the zeta function of similar curves using Gauss sums (for example, [6, 13, 27].) We give a self-contained treatment here based on the results of Sect. 2.

Theorem 8.4. *Consider the curve C_R over some extension of $\mathbb{F}_q$ and put $g = g(C_R)$. For $h \ge 0$ we put $a = a_{\mathscr{A}}$ with $a_{\mathscr{A}}$ as given in Theorem 7.4 for some choice of $\mathscr{A}$.*

1. *If* $p \equiv 1 \pmod 4$, *then the L-polynomial of* C_R *over* $\mathbb{F}_{p^s}$ *is given by*

$$L_{C_R,\mathbb{F}_{p^s}}(T) = \begin{cases} (1-p^sT^2)^g & \text{if } s \text{ is odd,} \\ (1-p^{s/2}T)^{2g} & \text{if } s \text{ is even and } a \text{ is a square in } \mathbb{F}_{p^s}^*, \\ (1+p^{s/2}T)^{2g} & \text{if } s \text{ is even and } a \text{ is a nonsquare in } \mathbb{F}_{p^s}^*. \end{cases}$$

2. *If* $p \equiv 3 \pmod 4$, *then the L-polynomial of* C_R *over* $\mathbb{F}_{p^s}$ *is given by*

$$L_{C_R,\mathbb{F}_{p^s}}(T) = \begin{cases} (1+p^sT^2)^g & \text{if } s \text{ is odd,} \\ (1-p^{s/2}T)^{2g} & \text{if } s \equiv 0 \pmod 4 \text{ and } a \text{ is a square in } \mathbb{F}_{p^s}^*, \\ (1+p^{s/2}T)^{2g} & \text{if } s \equiv 0 \pmod 4 \text{ and } a \text{ is a nonsquare in } \mathbb{F}_{p^s}^*, \\ (1+p^{s/2}T)^{2g} & \text{if } s \equiv 2 \pmod 4 \text{ and } a \text{ is a square in } \mathbb{F}_{p^s}^*, \\ (1-p^{s/2}T)^{2g} & \text{if } s \equiv 2 \pmod 4 \text{ and } a \text{ is a nonsquare in } \mathbb{F}_{p^s}^*. \end{cases}$$

Proof. Corollary 8.3 implies that it suffices to consider the case $h = 0$. To prove the theorem we may therefore assume that $R(X) = aX$. We label the corresponding curve D_a as we do in Lemma 7.1.

Case 1: The element a is a square in $\mathbb{F}_{p^s}^*$.

Then Lemma 7.1 implies that D_a is isomorphic over $\mathbb{F}_q$ to the curve D_1 given by the affine equation $Y^p - Y = X^2$. Since D_1 is defined over $\mathbb{F}_p$, we compute its L-polynomial over $\mathbb{F}_p$. The argument that we use here proceeds in the same manner as in the proof of Proposition 2.6. However, since both the polynomial $R(X)$ and the field are very simple, we do not need to consider the quadric Q considered in that proof explicitly.

As in the proof of Proposition 8.1, it suffices to determine the number N_2 of $\mathbb{F}_{p^2}$-rational points of the curve D_1. We have $p+1$ points with $x \in \{0, \infty\}$. As in the proof of Proposition 2.6, the $\mathbb{F}_{p^2}$-points with $x \neq 0, \infty$ correspond to squares $z = x^2$ with $\mathrm{Tr}_{\mathbb{F}_{p^2}/\mathbb{F}_p}(z) = 0$. Every such element z yields exactly $2p$ rational points. Since $\mathrm{Tr}_{\mathbb{F}_{p^2}/\mathbb{F}_p}(z) = z + z^p$, the nonzero elements of trace zero are exactly the elements with $z^{p-1} = -1$. Choosing an element $\zeta \in \mathbb{F}_{p^2}^*$ of order $2(p-1)$, we conclude that the nonzero elements with trace zero are

$$\ker(\mathrm{Tr}_{\mathbb{F}_{p^2}/\mathbb{F}_p}) \setminus \{0\} = \{\zeta^{2j+1} : j = 0, \ldots, p-2\}.$$

First suppose that $p \equiv 3 \pmod 4$. Then all the elements of $\ker(\mathrm{Tr}_{\mathbb{F}_{p^2}/\mathbb{F}_p})$ are squares in $\mathbb{F}_{p^2}$, so

$$\#D_1(\mathbb{F}_{p^2}) = 1 + p + (p-1)2p = 1 + p^2 + (p-1)p.$$

As in the proof of Proposition 8.1 it follows that $\alpha_j = \pm ip^{1/2} = -\alpha_{2g-j}$ for $1 \leq j \leq g$ after suitable relabeling. If s is even, then $\alpha_j^s = \alpha_{2g-j}^s = \mathrm{i}^s p^{s/2}$ and

$$(1-\alpha_j^s T)(1-\alpha_{2g-j}^s T) = 1 - 2\mathrm{i}^s p^{s/2} T + p^s T^2 = \begin{cases} (1-p^{s/2}T)^2 & \text{if } s \equiv 0 \pmod 4, \\ (1+p^{s/2}T)^2 & \text{if } s \equiv 2 \pmod 4. \end{cases}$$

If s is odd, then $\alpha_j^s = \pm \mathrm{i}^s p^{s/2} = -\alpha_{2g-j}^s$, and therefore

$$(1-\alpha_j^s T)(1-\alpha_{2g-j}^s T) = 1 + p^s T^2.$$

Now assume that $p \equiv 1 \pmod 4$. Then none of the elements of $\ker(\mathrm{Tr}_{\mathbb{F}_{p^2}/\mathbb{F}_p})$ are squares in $\mathbb{F}_{p^2}$, and we conclude that

$$\#D_1(\mathbb{F}_{p^2}) = 1 + p = 1 + p^2 - (p-1)p.$$

Again as in the proof of Proposition 8.1 it follows that, up to relabeling, $\alpha_j = p^{s/2} = \alpha_{2g-j}$ for $1 \leq j \leq g/2$, and $\alpha_j = -p^{s/2} = \alpha_{2g-j}$ for $g/2 + 1 \leq j \leq g$. (Note that g is even since $p \equiv 1 \pmod 4$.) We may therefore relabel again to ensure that $\alpha_j = p^{s/2} = -\alpha_{2g-j}$, for $1 \leq j \leq g$. With this new labeling, if s is even, then $\alpha_j^s = \alpha_{2g-j}^s = p^{s/2}$, and

$$(1-\alpha_j^s T)(1-\alpha_{j+g/2}^s T) = (1-p^{s/2}T)^2,$$

and if s is odd, then $\alpha_j^s = p^{s/2} = -\alpha_{2g-j}$ and

$$(1-\alpha_j^s T)(1-\alpha_{j+g/2}^s T) = (1-p^{s/2}T)(1+p^{s/2}T) = (1-p^s T^2).$$

This concludes Case 1.

Case 2: The element a is a nonsquare in $\mathbb{F}_{p^s}^*$ and s is odd.

Then the set $\{a\beta^2 : \beta \in \mathbb{F}_{p^s}^*\}$ contains $(p^s - 1)/2$ distinct elements, all of which are nonsquares. As a consequence, this set contains all nonsquares of $\mathbb{F}_{p^s}$. For s odd, the nonsquares in $\mathbb{F}_p^*$ are also nonsquares in $\mathbb{F}_{p^s}^*$, and therefore the set $\{a\beta^2 : \beta \in \mathbb{F}_{p^s}^*\}$ contains an element in $\mathbb{F}_p^*$. (In fact, this set contains all the nonsquares in $\mathbb{F}_p$.) Lemma 7.1 now implies that the curve D_a is isomorphic over $\mathbb{F}_q$ to the curve D_1, and the desired result follows therefore from Case 1.

Case 3: The element a is a nonsquare in $\mathbb{F}_{p^s}^*$ and s is even.

Here, we consider $M := \ker(\mathrm{Tr}_{\mathbb{F}_{p^s}/\mathbb{F}_p}) = \{z \in \mathbb{F}_{p^s} : \mathrm{Tr}_{\mathbb{F}_{p^s}/\mathbb{F}_p}(z) = 0\}$. Since the trace is surjective and $\mathbb{F}_p$-linear, the cardinality of M is p^{s-1}. We may write M as a disjoint union

$$M = \{0\} \cup M^{\mathrm{sq}} \cup M^{\mathrm{nsq}},$$

where M^{sq} (resp. M^{nsq}) are the elements of $M \setminus \{0\}$ which are squares (resp. nonsquares) in $\mathbb{F}_{p^s}^*$.

As in the proof of Case 1 we have

$$\#D_1(\mathbb{F}_{p^s}) = 1 + p + 2p\,\#M^{\mathrm{sq}},$$

and a similar argument gives

$$\#D_a(\mathbb{F}_{p^s}) = 1 + p + 2p\,\#M^{\mathrm{nsq}}.$$

From the expression for $\#D_1(\mathbb{F}_{p^s})$ computed in Case 1, it follows that

$$\#M^{\mathrm{sq}} = \begin{cases} \frac{p^{s-1}-1}{2} + \frac{(p-1)}{2}p^{(s-2)/2} & \text{if } p \equiv 3 \pmod 4 \text{ and } s \equiv 2 \pmod 4, \\ \frac{p^{s-1}-1}{2} - \frac{(p-1)}{2}p^{(s-2)/2} & \text{if } p \equiv 1 \pmod 4 \text{ or } s \equiv 0 \pmod 4. \end{cases}$$

Since $\#M^{\mathrm{nsq}} = \#M - 1 - \#M^{\mathrm{sq}} = p^{s-1} - 1 - \#M^{\mathrm{sq}}$, we conclude that

$$\#D_a(\mathbb{F}_{p^s}) = \begin{cases} 1 + p^s - (p-1)p^{s/2} & \text{if } p \equiv 3 \pmod 4 \text{ and } s \equiv 2 \pmod 4, \\ 1 + p^s + (p-1)p^{s/2} & \text{if } p \equiv 1 \pmod 4 \text{ or } s \equiv 0 \pmod 4. \end{cases}$$

The expressions for the L-polynomial now follow as in the previous cases. □

We finish this section by proving that all curves C_R are supersingular. This result is not new. Our proof just adds some details to Theorem 13.7 in [24]. An alternative proof is given by Blache, Corollary 3.7(ii) of [2].

Proposition 8.5. *The curve C_R is supersingular, i.e., its Jacobian is isogenous over $k = \overline{\mathbb{F}}_q$ to a product of supersingular elliptic curves.*

Proof. The curve C_R is supersingular if and only if all the slopes of the Newton polygon of the L-polynomial are $1/2$. (This follows, for example, from Theorem 2 of [23].) The statement of the proposition follows therefore from Theorem 8.4 . □

The reasoning of Van der Geer and Van der Vlugt for Theorem 13.7 of [24] is slightly different, since they do not compute the L-polynomial of C_R over $\mathbb{F}_q$. They argue that the Jacobian variety J_R of C_R is isogenous over k to p^h copies of the Jacobian of the curve D_1 with equation $Y^p - Y = X^2$. (This is a weaker version of Proposition 6.3.) They then use the fact that the curve D_1 is supersingular.

9 Examples

By work of Ihara [10], Stichtenoth and Xing [21], and Fuhrmann and Torres [8], we know that for q a power of a prime, a curve C which is maximal over $\mathbb{F}_{q^2}$ satisfies

$$g(C) \in \left[0, \frac{(q-1)^2}{4}\right] \cup \left\{\frac{q(q-1)}{2}\right\}.$$

Moreover, the Hermite curves are the only maximal curves of genus $(q(q-1))/2$ [17].

Recall from Sect. 8 that a curve C is maximal over $\mathbb{F}_{p^{2s}}$ if and only if its L-polynomial satisfies $L_{C,\mathbb{F}_{p^{2s}}} = (1+p^{2s}T)^{2g(C)}$. In our setting, Theorem 8.4 shows that for a curve C_R of the type considered in this paper and a defined as in Theorem 8.4, if $\mathbb{F}_{p^s}$ contains the splitting field $\mathbb{F}_q$ of $E(X)$, then C_R is maximal over $\mathbb{F}_{p^s}$ if and only if one of the following holds:

- s is even, a is a nonsquare in $\mathbb{F}_q^*$, and $p \equiv 1 \pmod 4$,
- $s \equiv 0 \pmod 4$, a is a nonsquare in $\mathbb{F}_q^*$, and $p \equiv 3 \pmod 4$, or
- $s \equiv 2 \pmod 4$, a is a square in $\mathbb{F}_q^*$, and $p \equiv 3 \pmod 4$.

In each case the negation of the condition on a guarantees that C_R is a minimal curve over $\mathbb{F}_{p^s}$.

In light of these facts, the only difficulty in generating examples of maximal and minimal curves lies in computing suitable elements a. In this section we present certain cases in which such a can be computed. We start with a discussion of the case $h = 0$, and then turn our attention to $R(X) = X^{p^h}$. For more results along the same lines we refer to [1, 3]. In [4] it is shown that all curves C_R that are maximal over the field $\mathbb{F}_{p^{2n}}$ are quotients of the Hermite curve H_{p^n} with affine equation $y^{p^n} - y = x^{p^n+1}$.

At the end of this section we briefly investigate isomorphisms between certain curves C_R and curves with defining equations

$$Y^p + Y = X^{p^h+1}.$$

Throughout this section, we let H_p denote the Hermite curve which is defined by the affine equation

$$Y^p + Y = X^{p+1}. \tag{29}$$

As mentioned above, this is a maximal curve over $\mathbb{F}_{p^2}$. The curve $Y^p + Y = X^2$ is a quotient of the Hermite curve H_p, and therefore this curve is maximal over $\mathbb{F}_{p^2}$. The following lemma determines when the twists

$$Y^p - Y = aX^2$$

of this curve are maximal. A similar result can also be found in Lemma 4.1 of [3].

Lemma 9.1. *Let $R(X) = aX \in \mathbb{F}_{p^{2s}}[X]$. Then C_R is maximal over $\mathbb{F}_{p^{2s}}$ if and only if one of the following conditions holds:*

1. *$p \equiv 1 \pmod 4$ and $a \in \mathbb{F}_{p^{2s}}^*$ is a nonsquare,*
2. *$p \equiv 3 \pmod 4$, s is even, and $a \in \mathbb{F}_{p^{2s}}^*$ is a nonsquare, or*
3. *$p \equiv 3 \pmod 4$, s is odd, and $a \in \mathbb{F}_{p^{2s}}^*$ is a square.*

Proof. In this case we have $E(X) = 2aX$, hence $\mathbb{F}_{p^{2s}}$ automatically contains the splitting field of E. The lemma therefore follows from Theorem 8.4. □

Remark 9.2. The database manYPoints [26] compiles records of curves with many points. The following two maximal curves fall in the range of genus and cardinality covered in the database, and have now been included in manYPoints. Previously, the database did not state any lower bound for the maximum number of points of a curve of genus 5 over $\mathbb{F}_{11^4}$ and a curve of genus 9 over $\mathbb{F}_{19^4}$.

1. In the case where $h = 0, p = 11$, and $s = 4$, let $a \in \mathbb{F}^*_{11^4}$ be a nonsquare. Then the curve
$$Y^{11} - Y = aX^2$$
is maximal over $\mathbb{F}_{11^4}$ and of genus 5.
2. In the case where $h = 0, p = 19$, and $s = 4$, let $a \in \mathbb{F}_{19^4}$ be a nonsquare. Then the curve
$$Y^{19} - Y = aX^2$$
is maximal over $\mathbb{F}_{19^4}$ and of genus 9.

The following proposition gives an example of a class of maximal curves with small genus compared to the size of their field of definition, in contrast to the Hermite curves which have large genus. A similar result for $p = 2$ can be found in Theorem 7.4 of [24]. A similar result with p replaced by an arbitrary prime power can be found in Proposition 4.6 of [3].

Proposition 9.3. *Let $h \geq 1$.*

1. *Let $R(X) = X^{p^h}$. Then $E(X) = X^{p^{2h}} + X$, which has splitting field $\mathbb{F}_q = \mathbb{F}_{p^{4h}}$. The curve C_R is minimal over $\mathbb{F}_q$.*
2. *Let $a_h \in \mathbb{F}^*_{p^{2h}}$ be an element with $a_h^{p^h-1} = -1$ and define $R(X) = a_h X^{p^h}$. Then $E(X) = a_h^{p^h}(X^{p^{2h}} - X)$, which has splitting field $\mathbb{F}_q = \mathbb{F}_{p^{2h}}$. The curve C_R is maximal over $\mathbb{F}_q$.*

Proof. We first prove the statement about the splitting field of $E(X)$ for both cases. Consider the additive polynomial $R(X) = a_h X^{p^h} \in \mathbb{F}_{p^s}[X]$ with $h \geq 1$. Then (2) shows that
$$E(X) = a_h^{p^h} X^{p^{2h}} + a_h X.$$

If $a_h = 1$, then E has splitting field $\mathbb{F}_q = \mathbb{F}_{p^{4h}}$. If $a_h \in \mathbb{F}^*_{p^{2h}}$ satisfies $a_h^{p^h-1} = -1$, then $E(X) = a_h^{p^h}(X^{p^{2h}} - X)$, which has splitting field $\mathbb{F}_q = \mathbb{F}_{p^{2h}}$. In both cases, we conclude from the explicit expression of E that
$$W = \{c \in \overline{\mathbb{F}}_p : c^{p^{2h}} = -a_h^{1-p^h} c\}.$$

For every $c \in W$, the formulas (7) and (8) imply that

$$B_c(X) = -\sum_{i=0}^{h-1} a_h^{p^i} c^{p^{h+i}} X^{p^i}.$$

We first consider the case where $a_h = 1$. Choose an element $c \in W \setminus \{0\}$, i.e., $c^{p^{2h}} = -c$, and define

$$\overline{A} = \{c\zeta : \zeta \in \mathbb{F}_{p^h}\} \subset W.$$

For any two ζ_j, ζ_k in $\mathbb{F}_{p^h}$, we have

$$B_{c\zeta_j}(c\zeta_k) = -\sum_{i=0}^{h-1} \zeta_j^{p^{h+i}} c^{p^{h+i}+p^i} \zeta_k^{p^i} = -\sum_{i=0}^{h-1} \zeta_j^{p^i} c^{p^{h+i}+p^i} \zeta_k^{p^{h+i}} = B_{c\zeta_k}(c\zeta_j),$$

since $\zeta^{p^h} = \zeta$ for any $\zeta \in \mathbb{F}_{p^h}$. Therefore the pairing from Part 1 of Lemma 5.2 satisfies

$$\epsilon(c\zeta_j, c\zeta_k) = B_{c\zeta_j}(c\zeta_k) - B_{c\zeta_k}(c\zeta_j) = 0 \quad \text{for any pair } (c\zeta_j, c\zeta_k) \in \overline{A}^2.$$

We conclude that $\overline{A} \subset W$ is a maximal isotropic subspace. Write $\mathscr{A} \subset P$ for the corresponding maximal abelian subgroup of P. Recall the constant from Theorem 7.4,

$$a_{\mathscr{A}} = \frac{a_h}{2} \prod_{\gamma \in \overline{A} \setminus \{0\}} \gamma,$$

when $h \geq 1$. Here the leading coefficient a_h of $R(X)$ is 1. The definition of $\overline{A}$ implies that

$$\prod_{\gamma \in \overline{A} \setminus \{0\}} \gamma = c^{p^h - 1} \prod_{\zeta \in \mathbb{F}_{p^h}^*} \zeta = -c^{p^h - 1}.$$

We conclude that $a_{\mathscr{A}} = -c^{p^h-1}/2$ is a square in $\mathbb{F}_q^*$, since $-1/2$ is a square in $\mathbb{F}_{p^2}^* \subset \mathbb{F}_q^*$. Theorem 8.4 now yields

$$L_{C_R, \mathbb{F}_q}(T) = (1 - \sqrt{q}T)^{2g}.$$

It follows that C_R is minimal over $\mathbb{F}_q$.

We now assume that $a_h \in \mathbb{F}_{p^{2h}}^*$ satisfies $a_h^{p^h} = -a_h$. In this case the splitting field of $E(X)$ is $\mathbb{F}_q = \mathbb{F}_{p^{2h}}$ as shown earlier. Choose a primitive $(p^{2h} - 1)$-st root of unity ζ. Then we may write $a_h = \zeta^{(2j+1)(p^h+1)/2}$ for some j. It follows that $a_h \in \mathbb{F}_q^*$ is a square if and only if $(p^h+1)/2$ is even. This is equivalent to $p \equiv 3 \pmod 4$ and h odd.

We choose $\overline{A} = \mathbb{F}_{p^h} \subset W = \mathbb{F}_{p^{2h}}$. For every $c, c' \in \overline{A}$, we have

$$B_c(c') = -\sum_{i=0}^{h-1}(a_h cc')^{p^i} = B_{c'}(c).$$

As in the proof of Part 1, we conclude that $\overline{A}$ is a maximal isotropic subspace for the pairing ϵ from Part 1 of Lemma 5.2. Since

$$\prod_{c\in\overline{A}\setminus\{0\}} c = -1,$$

we conclude that $a_{\mathscr{A}}$ is equivalent to a_h modulo squares in $\mathbb{F}_q^*$. (The argument is similar to that in the proof of Part 1.) We conclude that $a_{\mathscr{A}}$ is a square in $\mathbb{F}_q^*$ if and only if $p \equiv 3 \pmod 4$ and h is odd. Theorem 8.4 implies that C_R is a maximal curve over $\mathbb{F}_q$ in each of these cases. This proves Part 2. □

Remark 9.4. In their follow-up paper [25] to [24], Van der Geer and Van der Vlugt constructed further examples of maximal curves as a fiber product of the curves C_R. We have not considered this construction in the case of odd characteristic. We leave this as a subject for future research.

Example 9.5.

1. We consider the Hermite curve H_p given in (29), and the curve C_R given by

$$Y^p - Y = X^{p+1}.$$

We claim that the curves H_p and C_R are not isomorphic over $\mathbb{F}_{p^2}$. To see this, we show that $\#C_R(\mathbb{F}_{p^2}) = 1 + p \neq 1 + p^3 = \#H_p(\mathbb{F}_{p^2})$. This clearly implies that the two curves are not isomorphic over $\mathbb{F}_{p^2}$.

We note that

$$\psi : \mathbb{F}_{p^2}^* \to \mathbb{F}_{p^2}^*, \qquad x \mapsto x^{1+p}$$

is the restriction of the norm on $\mathbb{F}_{p^2}/\mathbb{F}_p$, so the image of ψ is $\mathbb{F}_p^*$. It follows that

$$\mathrm{Tr}_{\mathbb{F}_{p^2}/\mathbb{F}_p}(x^{1+p}) = 2x^{1+p} \neq 0 \quad \text{for all } x \in \mathbb{F}_{p^2}^*.$$

We conclude that the $\mathbb{F}_{p^2}$-rational points of C_R are the p points with $x = 0$ together with the unique point ∞. This proves the claim. (Exercise 6.7 in [20] asks to prove that H_p and C_R are isomorphic over $\mathbb{F}_{p^2}$ if $p \equiv 1$ (mod4). The above calculation shows that this does not hold.)

However, the Hermite curve H_p is isomorphic over $\mathbb{F}_{p^2}$ to the curve given by

$$C_{R'} : Y^p - Y = a_1 X^{p+1},$$

where $a_1 \in \mathbb{F}_{p^2}$ satisfies $a_1^{p-1} = -1$. The isomorphism is given by $\psi: C_{R'} \to H_p$, $(x, y) \mapsto (x, a_1^p y)$. This conforms with Part 2 of Proposition 9.3.

2. Let $a_h \in \mathbb{F}_{p^{2h}}^*$ be an element with $a_h^{p^h} = -a_h$ as in Part 2 of Proposition 9.3. Write $R(X) = a_h X^{p^h}$. Then $\psi: (x, y) \mapsto (x, a_h^{p^{2h-1}} y)$ defines an isomorphism between C_R and the curve given by

$$Y^p + Y = X^{p^h+1}.$$

Part 2 of Proposition 9.3 therefore implies that this curve is maximal over $\mathbb{F}_{p^{2h}}$. This can also be shown directly, for example, using Proposition 6.4.1 of [20].

Acknowledgements This research began at the Women in Numbers 3 workshop that took place April 20–25, 2014, at the Banff International Research Station (BIRS) in Banff, Alberta (Canada). We thank the organizers of this workshop as well as the hospitality of BIRS. We also thank Mike Zieve for pointing out some references to us.

IB is partially supported by DFG priority program SPP 1489, WH is partially supported by NSF grant DMS-1406066, and RS is supported by NSERC of Canada.

References

1. Anbar, N., Meidl, W.: Quadratic functions and maximal Artin–Schreier curves. Finite Fields Appl. **30**, 49–71 (2014)
2. Blache, R.: Valuation of exponential sums and the generic first slope for Artin–Schreier curves. J. Number Theory **132**(10), 2336–2352 (2012)
3. Çakçak, E., Özbudak, F.: Some Artin–Schreier type function fields over finite fields with prescribed genus and number of rational places. J. Pure Appl. Algebra **210**(1), 113–135 (2007)
4. Çakçak, E., Özbudak, F.: Curves related to Coulter's maximal curves. Finite Fields Appl. **14**(1), 209–220 (2008)
5. Cassels, J.W.S.: Rational Quadratic Forms. London Mathematical Society Monographs, vol. 13. Academic/[Harcourt Brace Jovanovich], London/New York (1978)
6. Coulter, R.S.: The number of rational points of a class of Artin–Schreier curves. Finite Fields Appl. **8**(4), 397–413 (2002)
7. Elkies, N.D.: Linearized algebra and finite groups of Lie type. I. Linear and symplectic groups. In: Applications of Curves Over Finite Fields (Seattle, WA, 1997). Contemporary Mathematics, vol. 245, pp. 77–107. American Mathematical Society, Providence, RI (1999)
8. Fuhrmann, R., Torres, F.: The genus of curves over finite fields with many rational points. Manuscripta Math. **89**(1), 103–106 (1996)
9. Huppert, B.: Endliche Gruppen. I. Die Grundlehren der Mathematischen Wissenschaften, vol. 134. Springer, Berlin/New York (1967)
10. Ihara, Y.: Some remarks on the number of rational points of algebraic curves over finite fields. J. Fac. Sci. Univ. Tokyo Sect. IA Math. **28**(3), 721–724 (1981)

11. Joly, J.R.: Équations et variétés algébriques sur un corps fini. Enseignement Math. (2) **19**, 1–117 (1973)
12. Kani, E., Rosen, M.: Idempotent relations and factors of Jacobians. Math. Ann. **284**(2), 307–327 (1989)
13. Katz, N.M.: Crystalline cohomology, Dieudonné modules, and Jacobi sums. In: Automorphic forms, representation theory and arithmetic (Bombay, 1979). Tata Institute of Fundamental Research Studies in Mathematics, vol. 10, pp. 165–246. Tata Institute of Fundamental Research, Bombay (1981)
14. Lehr, C., Matignon, M.: Automorphism groups for p-cyclic covers of the affine line. Preprint version of [15]. arXiv.math/0307031
15. Lehr, C., Matignon, M.: Automorphism groups for p-cyclic covers of the affine line. Compositio Math. **141**(5), 1213–1237 (2005)
16. Matignon, M., Rocher, M.: Smooth curves having a large automorphism p-group in characteristic $p > 0$. Algebra Number Theory **2**(8), 887–926 (2008)
17. Rück, H.G., Stichtenoth, H.: A characterization of Hermitian function fields over finite fields. J. Reine Angew. Math. **457**, 185–188 (1994)
18. Serre, J.P.: Corps locaux, deuxième edn. No. VIII in Publications de l'Université de Nancago. Hermann, Paris (1968)
19. Stichtenoth, H.: Über die Automorphismengruppe eines algebraischen Funktionenkörpers von Primzahlcharakteristik. II. Ein spezieller Typ von Funktionenkörpern. Arch. Math. (Basel) **24**, 615–631 (1973)
20. Stichtenoth, H.: Algebraic Function Fields and Codes. Graduate Texts in Mathematics, vol. 254, 2nd edn. Springer, Berlin (2009)
21. Stichtenoth, H., Xing, C.P.: The genus of maximal function fields over finite fields. Manuscripta Math. **86**(2), 217–224 (1995)
22. Suzuki, M.: Group Theory II. Grundlehren der Mathematischen Wissenschaften, vol. 248. Springer, New York (1986)
23. Tate, J.: Endomorphisms of abelian varieties over finite fields. Invent. Math. **2**, 134–144 (1966)
24. van der Geer, G., van der Vlugt, M.: Reed–Müller codes and supersingular curves I. Compositio Math. **84**, 333–367 (1992)
25. van der Geer, G., van der Vlugt, M.: How to construct curves over finite fields with many points. In: Catanese, F. (ed.) Arithmetic Geometry (Cortona 1994). Symposia Mathematica, pp. 169–189. Cambridge University Press, Cambridge (1997)
26. van der Geer, G., Howe, E., Lauter, K., Ritzenthaler, C.: Table of curves with many points. http://www.manypoints.org (2015)
27. Yui, N.: On the Jacobian variety of the Fermat curve. J. Algebra **65**(1), 1–35 (1980)

Hypergeometric Series, Truncated Hypergeometric Series, and Gaussian Hypergeometric Functions

Alyson Deines, Jenny G. Fuselier, Ling Long, Holly Swisher, and Fang-Ting Tu

Abstract In this paper, we investigate the relationships among hypergeometric series, truncated hypergeometric series, and Gaussian hypergeometric functions through some families of "hypergeometric" algebraic varieties that are higher dimensional analogues of Legendre curves.

Keywords Hypergeometric series • Gaussian hypergeometric functions • Algebraic varieties • Galois representations • Supercongruences

1 Introduction

1.1 *Motivation*

When a prime p satisfies $p \equiv 1 \pmod 6$, the p-adic gamma value $-\Gamma_p\left(\frac{1}{3}\right)^3$ is a quadratic algebraic number with absolute value $\sqrt{p}$ which can be written as a Jacobi sum. Thus, $\Gamma_p\left(\frac{1}{3}\right)^6$ is not a conjugate of $-\Gamma_p\left(\frac{1}{3}\right)^3$ in the sense of algebraic

---

A. Deines
Center for Communications Research, San Diego, CA 92121, USA
e-mail: aly.deines@gmail.com

J.G. Fuselier
Department of Mathematics and Computer Science, High Point University, High Point, NC 27268, USA
e-mail: jfuselie@highpoint.edu

L. Long (✉)
Louisiana State University, Baton Rouge, LA 70803, USA
e-mail: llong@math.lsu.edu

H. Swisher
Oregon State University, Corvallis, OR 97331, USA
e-mail: swisherh@math.oregonstate.edu

F.-T. Tu
National Center for Theoretical Sciences, Hsinchu 300, Taiwan, R.O.C.
e-mail: ft12am93g@gmail.com

© Springer International Publishing Switzerland 2016
E.E. Eischen et al. (eds.), *Directions in Number Theory*, Association for Women in Mathematics Series 3, DOI 10.1007/978-3-319-30976-7_5

numbers [13]. However, considering truncated hypergeometric series we have when $p \equiv 1 \pmod 6$,

$$ {}_3F_2\left[\begin{matrix} \frac{1}{3} & \frac{1}{3} & \frac{1}{3} \\ & 1 & 1 \end{matrix} ; 1\right]_{p-1} := \sum_{k=0}^{p-1} \binom{-\frac{1}{3}}{k}^3 \cdot (-1)^k \equiv \Gamma_p\left(\frac{1}{3}\right)^6 \pmod{p^3}, \tag{1} $$

which was shown by the third author and Ramakrishna in [31], while numerically we see that

$$ {}_3F_2\left[\begin{matrix} \frac{2}{3} & \frac{2}{3} & \frac{2}{3} \\ & 1 & 1 \end{matrix} ; 1\right]_{p-1} := \sum_{k=0}^{p-1} \binom{-\frac{2}{3}}{k}^3 \cdot (-1)^k \equiv -\Gamma_p\left(\frac{1}{3}\right)^3 \pmod{p^3}, \tag{2} $$

and we will show this holds modulo p^2 in this paper. By Dwork [16],

$$ \lim_{s\to\infty} {}_3F_2\left[\begin{matrix} \frac{1}{3} & \frac{1}{3} & \frac{1}{3} \\ & 1 & 1 \end{matrix} ; 1\right]_{p^s-1} \Big/ {}_3F_2\left[\begin{matrix} \frac{1}{3} & \frac{1}{3} & \frac{1}{3} \\ & 1 & 1 \end{matrix} ; 1\right]_{p^{s-1}-1} = \Gamma_p\left(\frac{1}{3}\right)^6, $$

while

$$ \lim_{s\to\infty} {}_3F_2\left[\begin{matrix} \frac{2}{3} & \frac{2}{3} & \frac{2}{3} \\ & 1 & 1 \end{matrix} ; 1\right]_{p^s-1} \Big/ {}_3F_2\left[\begin{matrix} \frac{2}{3} & \frac{2}{3} & \frac{2}{3} \\ & 1 & 1 \end{matrix} ; 1\right]_{p^{s-1}-1} = -\Gamma_p\left(\frac{1}{3}\right)^3. $$

When $p \equiv 5 \pmod 6$, Dwork in [16] showed that there is a similar congruence that involves both ${}_3F_2\left[\begin{matrix} \frac{1}{3} & \frac{1}{3} & \frac{1}{3} \\ & 1 & 1 \end{matrix} ; 1\right]_{p^s-1}$ and ${}_3F_2\left[\begin{matrix} \frac{2}{3} & \frac{2}{3} & \frac{2}{3} \\ & 1 & 1 \end{matrix} ; 1\right]_{p^{s-1}-1}$. It is tempting to think of the parameters $\frac{1}{3}$ and $\frac{2}{3}$ as "conjugates of some sort." Also, if one considers the finite field analogue of ${}_3F_2\left[\begin{matrix} \frac{1}{3} & \frac{1}{3} & \frac{1}{3} \\ & 1 & 1 \end{matrix} ; 1\right]$ due to Greene, what corresponds to $\frac{1}{3}$ is a cubic character, which is determined up to a conjugate. Putting these together, it appears that $-\Gamma\left(\frac{1}{3}\right)^3$ is some sort of "conjugate" of $\Gamma\left(\frac{1}{3}\right)^6$. One motivation of this paper is to investigate these seemingly contradicting phenomena via the relations between hypergeometric series, Gaussian hypergeometric functions, and truncated hypergeometric series. These objects correspond to periods, Galois representations, and unit roots (in the ordinary case) respectively.

In recent work [15], the authors use the perspective of Wolfart [43] and Archinard [3] to consider classical ${}_2F_1$-hypergeometric functions with rational parameters as periods of explicit generalized Legendre curves

$$ C_\lambda^{[N;i,j,k]} : y^N = x^i(1-x)^j(1-\lambda x)^k. $$

In [15], the main players are hypergeometric series and Gaussian hypergeometric functions. The authors use Gaussian ${}_2F_1$-hypergeometric functions to count points of $C_\lambda^{[N;i,j,k]}$ over finite fields and hence compute the corresponding Galois representations. This arithmetic information together with the periods of $C_\lambda^{[N;i,j,k]}$ in terms of hypergeometric values yields information about the decomposition of the Jacobian variety $J_\lambda^{[N;i,j,k]}$ constructed from the desingularization of $C_\lambda^{[N;i,j,k]}$. When $\gcd(i,j,k)$ is coprime to N and $N \nmid i+j+k$, then $J_\lambda^{[N;i,j,k]}$ has a degree $2\varphi(N)$ "primitive" factor J_λ^{new}, where φ is the Euler phi function. The authors prove the following theorem:

Theorem 1 ([15]). *Let $N = 3, 4, 6$ and $1 \le i, j, k < N$ with $\gcd(i,j,k)$ coprime to N and $N \nmid i+j+k$. Then for each $\lambda \in \overline{\mathbb{Q}}$, the endomorphism algebra of J_λ^{new} contains a four-dimensional algebra over $\mathbb{Q}$ if and only if*

$$B\left(\frac{N-i}{N}, \frac{N-j}{N}\right) \Big/ B\left(\frac{k}{N}, \frac{2N-i-j-k}{N}\right) \in \overline{\mathbb{Q}},$$

where $B(a,b) = \frac{\Gamma(a)\Gamma(b)}{\Gamma(a+b)}$, and $\Gamma(\cdot)$ is the Gamma function.

The second motivation for this paper is to explore the following higher dimensional analogues of Legendre curves:

$$C_{n,\lambda}: \quad y^n = (x_1x_2\cdots x_{n-1})^{n-1}(1-x_1)\cdots(1-x_{n-1})(x_1 - \lambda x_2x_3\cdots x_{n-1}).$$

In particular, the curves $C_{2,\lambda}$ are known as Legendre curves. Up to a scalar multiple, the hypergeometric series

$${}_nF_{n-1}\left[\begin{matrix} \frac{i}{n} & \frac{i}{n} & \ldots & \frac{i}{n} \\ & 1 & \ldots & 1 \end{matrix}; \lambda\right]$$

for any $1 \le j \le n-1$, when convergent, can be realized as a period of $C_{n,\lambda}$.

1.2 Results

Our first theorem shows that the number of rational points on $C_{n,\lambda}$ over finite fields $\mathbb{F}_q$ can be expressed in terms of Gaussian hypergeometric functions. For a definition of Gaussian hypergeometric functions please see Sect. 2.3.[1] Let $\widehat{\mathbb{F}_q^\times}$ denote the group of all multiplicative characters on $\mathbb{F}_q^\times$.

Theorem 2. *Let $q = p^e \equiv 1 \pmod{n}$ be a prime power. Let η_n be a primitive order n character and ε the trivial multiplicative character in $\widehat{\mathbb{F}_q^\times}$. Then*

[1]The subscript q for a Gaussian hypergeometric function records the size of the corresponding finite field and should not be confused with the subscript for truncated hypergeometric series which records the location of truncation.

$$\#C_{n,\lambda}(\mathbb{F}_q) = 1 + q^{n-1} + q^{n-1}\sum_{i=1}^{n-1} {}_nF_{n-1}\left(\begin{matrix}\eta_n^i, \eta_n^i, \cdots, \eta_n^i \\ \varepsilon, \cdots, \varepsilon,\end{matrix}; \lambda\right)_q.$$

Meanwhile, we are also interested in knowing how to use information from truncated hypergeometric series to obtain information about the Galois representations and hence local zeta functions of $C_{n,\lambda}$. For instance, we have the following conjecture based on numerical evidence for the case $\lambda = 1$:

Conjecture 3. *Let $n \geq 3$ be a positive integer, and p be prime such that $p \equiv 1 \pmod{n}$. Then*

$${}_nF_{n-1}\left[\begin{matrix}\frac{n-1}{n} & \frac{n-1}{n} & \cdots & \frac{n-1}{n} \\ & 1 & \cdots & 1\end{matrix}; 1\right]_{p-1} := \sum_{k=0}^{p-1}\binom{\frac{1-n}{n}}{k}^n (-1)^{kn} \equiv -\Gamma_p\left(\frac{1}{n}\right)^n \pmod{p^3}.$$

Using the Gross–Koblitz formula [22], recalled in Sect. 2.4, we have for a prime $p \equiv 1 \pmod{n}$,

$$J(\eta_n, \eta_n)J(\eta_n, \eta_n^2)\cdots J(\eta_n, \eta_n^{n-2}) = (-1)^{n-2+\frac{1+(n-1)p}{n}}\Gamma_p\left(\frac{1}{n}\right)^n,$$

where $J(\cdot, \cdot)$ denotes the standard Jacobi sum. We see that $(-1)^{n-2+\frac{1+(n-1)p}{n}} = 1$ when n is odd and η_n is an order n character of $\mathbb{F}_p^\times$ such that $\eta_n(x) \equiv x^{\frac{p-1}{n}} \pmod{p}$ for all $x \in \mathbb{F}_p$. From the perspective of Grössencharacters (Hecke characters) (see Weil [40]), this product of Jacobi sums is associated with a linear representation χ of the Galois group $\mathrm{Gal}(\overline{\mathbb{Q}}/\mathbb{Q}(e^{2\pi i/n}))$. We would like to explore whether the Galois representation arising from $C_{n,1}$ contains a factor that is related to χ. When $n = 3, 4$, the answer is positive. In proving these results the work of Greene [21] and McCarthy [32] on finite field analogues of classical hypergeometric evaluation formulas plays an essential role.

Ahlgren and Ono [1] show that for any odd prime p,

$$p^3 \cdot {}_4F_3\left(\begin{matrix}\eta_4^2, \eta_4^2, \eta_4^2, \eta_4^2 \\ \varepsilon, \varepsilon, \varepsilon\end{matrix}; 1\right)_p = -a(p) - p, \tag{3}$$

where $a(p)$ is the pth coefficient of the weight 4 Hecke cuspidal eigenform

$$\eta(2z)^4\eta(4z)^4,$$

with $\eta(z)$ being the Dedekind eta function. Here, we show the following:

Theorem 4. *Let η_2, η_3, η_4 denote characters of order 2, 3, 4, respectively, in $\widehat{\mathbb{F}_q^\times}$.*

1. Let $q \equiv 1 \pmod{3}$ be a prime power. Then

$$q^2 \cdot {}_3F_2\left(\begin{matrix}\eta_3, \eta_3, \eta_3 \\ \varepsilon, \varepsilon\end{matrix}; 1\right)_q = J(\eta_3, \eta_3)^2 - J(\eta_3^2, \eta_3^2).$$

2. *Let $q \equiv 1 \pmod 4$ be a prime power. Then*

$$q^3 \cdot {}_4F_3\left(\begin{matrix}\eta_4, \eta_4, \eta_4, \eta_4 \\ \varepsilon, \varepsilon, \varepsilon\end{matrix}; 1\right)_q = J(\eta_4, \eta_2)^3 + qJ(\eta_4, \eta_2) - J(\overline{\eta_4}, \eta_2)^2.$$

Here we observe $J(\overline{\eta_4}, \eta_2)^2 = \eta_4(-1)J(\overline{\eta_4}, \overline{\eta_4})J(\overline{\eta_4}, \overline{\eta_4}^2)$. To prove Theorem 4 we use the work of Greene [21] and McCarthy [32], except in case (2) when $q \equiv 5 \pmod 8$, in which we use Grössencharacters and representation theory. The reason we do this is because a key ingredient of our proof is Theorem 1.6 of McCarthy [32], for which we assume η_4 is a square, i.e., $q \equiv 1 \pmod 8$. Combining this with the theory of Galois representations, we can reach our conclusion when $q \equiv 5 \pmod 8$. We wish to point out that the above results can be interpreted in terms of Galois representations.[2] Result (1) describes the trace of the Frobenius element at q in $\mathrm{Gal}(\overline{\mathbb{Q}}/\mathbb{Q}(\sqrt{-3}))$ under a two-dimensional Galois representation arising from the second étale cohomology of $C_{3,1}$ in terms of Jacobi sums (and hence Grössencharacters); while (2) describes a three-dimensional Galois representation of $\mathrm{Gal}(\overline{\mathbb{Q}}/\mathbb{Q}(\sqrt{-1}))$ arising from the third étale cohomology of $C_{4,1}$ in terms of Jacobi sums. Both cases are exceptional. Consequently we can describe the local zeta function of $C_{3,1}$ and $C_{4,1}$ completely. For instance, when $p \equiv 1 \pmod 3$ is prime, by the Hasse–Davenport relation for Jacobi sums (see [23]), the local zeta function of $C_{3,1}$ over $\mathbb{F}_p$ is

$$Z_{C_{3,1}}(T,p) = \frac{1}{(1-T)(1+(\alpha_p+\overline{\alpha}_p)T+pT^2)(1-p^2T)(1-(\alpha_p^2+\overline{\alpha}_p^2)T+p^2T^2)}$$

where $\alpha_p = J(\eta_3, \eta_3)$. Note that the factor $(1 + (\alpha_p + \overline{\alpha}_p)T + pT^2)$ appearing in the denominator has roots of absolute value $1/\sqrt{p}$; meanwhile following Weil's conjecture (see [23]) such a term should appear in the numerator instead. We believe the discrepancy is due to the fact that we are not computing using the smooth model of $C_{n,\lambda}$ as no resolution of singularities is involved so far. Similarly, we have for any prime $p \equiv 1 \pmod 4$

[2] In a different language, our results correspond to the explicit descriptions of some mixed weight hypergeometric motives arising from exponential sums which are initiated by Katz [24], and are explicitly formulated and implemented by a group of mathematicians including Beukers, Cohen, Rodriguez-Villegas and others (from private communication with H. Cohen and F. Rodriguez-Villegas). Here we can use the explicit algebraic varieties to compute the Galois representations directly. A different algebraic model for the algebraic varieties is given in the following recent preprint [9].

$$Z_{C_{4,1}}(T,p) =$$
$$(1+(\beta_p^3+\overline{\beta}_p^3)T+p^3T^2)(1+(\beta_p+\overline{\beta})pT+p^3T^2)$$
$$\cdot\frac{(1-(\beta_p^2+\overline{\beta}_p^2)T+p^2T^2)(1-a(p)T+p^3T^2)(1-pT)}{(1-T)(1-p^3T)},$$

where $a(p)$ as in (3) and $\beta_p = J(\eta_4, \eta_2)$. The factor corresponding to

$$y^2 = (x_1x_2x_3)^3(1-x_1)(1-x_2)(1-x_3)(x_1-x_2x_3)$$

is

$$Z_{C_{4,1}^{old}}(T,p) = \frac{(1-a(p)T+p^3T^2)(1-pT)}{(1-T)(1-p^3T)},$$

and new primitive portion is

$$Z_{C_{4,1}^{new}}(T,p) =$$
$$(1+(\beta_p^3+\overline{\beta}_p^3)T+p^3T^2)(1+(\beta_p+\overline{\beta})pT+p^3T^2)(1-(\beta_p^2+\overline{\beta}_p^2)T+p^2T^2).$$

Part (1) of Theorem 4 explains why $-\Gamma_p(\frac{1}{3})^3$ appears to be a conjugate of $\Gamma_p(\frac{1}{3})^6$. There are two ways to specify a cubic character in $\widehat{\mathbb{F}_p^\times}$ when $p \equiv 1 \pmod 3$, i.e., $\eta_3(x) \equiv x^{(p-1)/3} \pmod p$ for all $x \in \mathbb{F}_p$ or $\eta_3(x) \equiv x^{2(p-1)/3} \pmod p$. Either way gives an embedding of

$$p^2 \cdot {}_3F_2\left(\begin{matrix}\eta_3, \eta_3, \eta_3 \\ \varepsilon, \varepsilon\end{matrix}; 1\right)_p$$

to $\mathbb{Z}_p$. Then the image of the Gaussian hypergeometric function is congruent to $-\Gamma_p(\frac{1}{3})^3$ or $\Gamma_p(\frac{1}{3})^6$, respectively, via the Gross–Koblitz formula [22, 37]. Using this formula, we also prove the following result which relates Gaussian hypergeometric functions to truncated hypergeometric series:

Lemma 1. *Let r, n, j be positive integers with $1 \leq j < n$. Let $p \equiv 1 \pmod n$ be prime and $\eta_n \in \widehat{\mathbb{F}_p^\times}$ such that $\eta_n(x) \equiv x^{j(p-1)/n} \pmod p$ for each $x \in \mathbb{F}_p$. Then,*

$$p^{r-1} \cdot {}_rF_{r-1}\left(\begin{matrix}\eta_n, \eta_n, \cdots, \eta_n \\ \varepsilon, \cdots, \varepsilon\end{matrix}; x\right)_p \equiv$$
$$(-1)^{r+1} \cdot {}_rF_{r-1}\left[\begin{matrix}\frac{n-j}{n} & \frac{n-j}{n} & \cdots & \frac{n-j}{n} \\ & 1 & \cdots & 1\end{matrix}; \frac{1}{x}\right]_{(p-1)\left(\frac{n-j}{n}\right)}$$
$$+ (-1)^{r+1+\frac{(p-1)}{n}jr}\left(x^{(p-1)\frac{n-j}{n}} - x^{\frac{p-1}{n}j}\right) \pmod p;$$

$$p^{r-1} \cdot {}_rF_{r-1}\left(\begin{matrix} \overline{\eta_n}, \overline{\eta_n}, \cdots, \overline{\eta_n} \\ \varepsilon, \cdots, \varepsilon \end{matrix}; x\right)_p \equiv$$

$$(-1)^{r+1} \cdot p^r \, {}_{r+1}F_r \left[\begin{matrix} 1 & 1 & \cdots & 1 \\ & \frac{2n-j}{n} & \cdots & \frac{2n-j}{n} \end{matrix}; \frac{1}{x}\right]_{p-1} \pmod{p}.$$

Thus, (1) and (2) hold modulo p. It is shown in [31] that (1) holds modulo p^3. These kinds of stronger congruences are known as *supercongruences* as they are stronger than what the theory of formal groups can predict. We will establish a few here. In particular, we prove the claim that Conjecture 3 holds modulo p^2.

Theorem 5. *Conjecture 3 holds modulo p^2. Namely, for $n \geq 3$, and $p \equiv 1 \pmod{n}$ prime,*

$${}_nF_{n-1}\left[\begin{matrix} \frac{n-1}{n} & \frac{n-1}{n} & \cdots & \frac{n-1}{n} \\ & 1 & \cdots & 1 \end{matrix}; 1\right]_{p-1} := \sum_{k=0}^{p-1} \binom{\frac{1-n}{n}}{k}^n (-1)^{kn} \equiv -\Gamma_p\left(\frac{1}{n}\right)^n \pmod{p^2}.$$

Remark. Theorem 5 also holds for $n = 2$, due to Mortenson [35].

We note that in [33, Definition 1.4], McCarthy defines a new function ${}_nG_n[\cdots]$ in terms of sums and ratios of p-adic Gama functions. Recently, the second author and McCarthy produced families of congruences between these ${}_nG_n$ functions and truncated hypergeometric series [19]. New identities for these functions have also recently been obtained by McCarthy et al. [7] and it is possible they could be used to prove Conjecture 3 in full.

For the truncated hypergeometric series related to $C_{4,1}$ we have another result.

Theorem 6. *For each prime $p \equiv 1 \pmod{4}$,*

$${}_4F_3\left[\begin{matrix} \frac{1}{4} & \frac{1}{4} & \frac{1}{4} & \frac{1}{4} \\ & 1 & 1 & 1 \end{matrix}; 1\right]_{p-1} = \sum_{k=0}^{p-1} \binom{-\frac{1}{4}}{k}^4 \equiv (-1)^{\frac{p-1}{4}} \Gamma_p\left(\frac{1}{2}\right) \Gamma_p\left(\frac{1}{4}\right)^6 \pmod{p^4}.$$

Corresponding to Ahlgren and Ono's result (3), Kilbourn [25] shows the supercongruence

$${}_4F_3\left[\begin{matrix} \frac{1}{2} & \frac{1}{2} & \frac{1}{2} & \frac{1}{2} \\ & 1 & 1 & 1 \end{matrix}; 1\right]_{p-1} := \sum_{k=0}^{p-1} \binom{-\frac{1}{2}}{k}^4 \equiv a(p) \pmod{p^3},$$

where $a(p)$ is defined as in (3).

Supercongruences are not only intellectually appealing, but they are also of very practical use for our computations. For instance, Theorem 6 corresponds to the properties of the third étale cohomology group of $C_{4,1}$ as mentioned earlier. By the Hasse-Weil bounds for them, which are constant multiples of p and $p^{3/2}$, respectively, the supercongruence results allowed us to compute the traces of Frobenius without any ambiguity, from which we were able to nail down the local zeta functions of $C_{3,1}$ and $C_{4,1}$ and discover Theorem 4 numerically before proving it. There are a variety of different techniques for proving such results and each has its own strength. See [11] for another Women in Numbers (WIN) project on supercongruences, which was motivated by the work of Zudilin [44] and his conjectures. We prove Theorems 5 and 6 by deforming truncated hypergeometric series using hypergeometric evaluation identities (several of them are due to Whipple [2, 41]) together with p-adic analysis via harmonic sums and p-adic Gamma functions. This technique is originated in [10] and is later formulated explicitly in [31].

1.3 Outline of this Paper

Section 2 contains some background material. In Sect. 3, we consider the familiar setting of Legendre curves. This section serves as a showcase of our techniques without getting into too much technicality. We prove Theorem 2 and Lemma 1 in Sect. 4. Section 5 is devoted to proving the results on Gaussian hypergeometric functions in Theorem 4. In Sect. 6, we prove Theorem 5, based on an idea of Zudilin (private communication), and then prove Theorem 6. Sections 4, 5, and 6 are technical in nature. In Sect. 7 we end with some remarks including a few conjectures based on our numerical data computed using `Sage`.

2 Preliminaries

2.1 Generalized Hypergeometric Series and Truncation

For a positive integer r, and α_i, $\beta_i \in \mathbb{C}$ with $\beta_i \notin \{\dots,-3,-2,-1\}$, the (generalized) hypergeometric series ${}_rF_{r-1}$ is defined by

$${}_rF_{r-1}\left[\begin{matrix}\alpha_1 & \alpha_2 & \dots & \alpha_r \\ & \beta_1 & \dots & \beta_{r-1}\end{matrix} ; \lambda\right] := \sum_{k=0}^{\infty} \frac{(\alpha_1)_k(\alpha_2)_k \dots (\alpha_r)_k}{(\beta_1)_k \dots (\beta_{r-1})_k} \cdot \frac{\lambda^k}{k!}$$

where $(a)_0 := 1$ and $(a)_k := a(a+1)\cdots(a+k-1)$ are rising factorials. This series converges for $|\lambda| < 1$.

When we truncate the above sum at $k = m$, we use the subscript notation

$$ {}_rF_{r-1}\left[\begin{matrix}\alpha_1\ \alpha_2\ \dots\ \alpha_r \\ \beta_1\ \dots\ \beta_{r-1}\end{matrix};\lambda\right]_m := \sum_{k=0}^{m} \frac{(\alpha_1)_k(\alpha_2)_k\dots(\alpha_r)_k}{(\beta_1)_k\dots(\beta_{r-1})_k}\cdot\frac{\lambda^k}{k!}. $$

We note that the books by Slater [38], Bailey [5], and Andrews et al. [2] are excellent sources for information on classical hypergeometric series.

The following gives an alternate truncation for hypergeometric series modulo powers of primes:

Lemma 2. *Let $n \geq 2$ be a positive integer, j an integer $1 \leq j < n$, and $p \equiv 1 \pmod n$ prime. Then for $x \in \mathbb{Z}_p$,*

$$ {}_rF_{r-1}\left[\begin{matrix}\frac{j}{n}\ \frac{j}{n}\ \dots\ \frac{j}{n} \\ 1\ \dots\ 1\end{matrix};x\right]_{\frac{j}{n}(p-1)} \equiv {}_rF_{r-1}\left[\begin{matrix}\frac{j}{n}\ \frac{j}{n}\ \dots\ \frac{j}{n} \\ 1\ \dots\ 1\end{matrix};x\right]_{p-1} \pmod{p^r}. $$

Proof. The lemma follows from the fact that when $\frac{j(p-1)}{n} + 1 \leq k \leq (p-1)$, the rising factorial $\left(\frac{j}{n}\right)_k \in p\mathbb{Z}_p$, since it contains the factor $p\left(\frac{j}{n}\right)$, while $(1)_k$ is not divisible by p. □

2.2 *Euler's Integral Formula and Higher Generalization*

When $\mathrm{Re}(\beta_1) > \mathrm{Re}(\alpha_2) > 0$, Euler's integral representation for ${}_2F_1$ [2] states that

$$ {}_2F_1\left[\begin{matrix}\alpha_1\ \alpha_2 \\ \beta_1\end{matrix};\lambda\right] = \frac{\Gamma(\beta_1)}{\Gamma(\alpha_2)\Gamma(\beta_1-\alpha_2)}\int_0^1 x^{\alpha_2-1}(1-x)^{\beta_1-\alpha_2-1}(1-\lambda x)^{-\alpha_1}dx. $$

More generally, one has that when $\mathrm{Re}\,(\beta_r) > \mathrm{Re}\,(\alpha_{r+1}) > 0$ (see [2, (2.2.2)])

$$ {}_{r+1}F_r\left[\begin{matrix}\alpha_1\ \alpha_2\ \dots\ \alpha_{r+1} \\ \beta_1\ \dots\ \beta_r\end{matrix};\lambda\right] = \frac{\Gamma(\beta_r)}{\Gamma(\alpha_{r+1})\Gamma(\beta_r-\alpha_{r+1})} \cdot\int_0^1 x^{\alpha_{r+1}-1}(1-x)^{\beta_r-\alpha_{r+1}-1}\,{}_rF_{r-1}\left[\begin{matrix}\alpha_1\ \alpha_2\ \dots\ \alpha_r \\ \beta_1\ \dots\ \beta_{r-1}\end{matrix};\lambda x\right]dx. \tag{4} $$

From the above two integral formulas, one can derive that for each $1 \leq j \leq n-1$ the series ${}_nF_{n-1}\left[\begin{matrix}\frac{j}{n}\ \frac{j}{n}\ \dots\ \frac{j}{n} \\ 1\ \dots\ 1\end{matrix};\lambda\right]$, with a suitable beta quotient factor, is a period of $C_{n,\lambda}$.

2.3 Gaussian Hypergeometric Functions

Let p be prime and let $q = p^e$. We extend any character $\chi \in \widehat{\mathbb{F}_q^\times}$ to all of $\mathbb{F}_q$ by setting $\chi(0) = 0$, including the trivial character ε, so that $\varepsilon(0) = 0$. For $A, B \in \widehat{\mathbb{F}_q^\times}$, let $J(A, B) := \sum_{x \in \mathbb{F}_q} A(x)B(1-x)$ denote the Jacobi sum and define

$$\binom{A}{B} := \frac{B(-1)}{q} J(A, \overline{B}) = \frac{B(-1)}{q} \sum_{x \in \mathbb{F}_q} A(x)\overline{B}(1-x).$$

In [21], Greene defines a finite field analogue of hypergeometric series called Gaussian hypergeometric functions, defined below.

Definition 1 ([21] Definition 3.10). If n is a positive integer, $x \in \mathbb{F}_q$, and $A_0, A_1, \ldots, A_n, B_1, B_2, \ldots, B_n \in \widehat{\mathbb{F}_q^\times}$, then

$$_{n+1}F_n \left(\begin{matrix} A_0, & A_1, & \ldots, & A_n \\ & B_1, & \ldots, & B_n \end{matrix} ; x \right)_q := \frac{q}{q-1} \sum_{\chi \in \widehat{\mathbb{F}_q^\times}} \binom{A_0\chi}{\chi} \binom{A_1\chi}{B_1\chi} \cdots \binom{A_n\chi}{B_n\chi} \chi(x).$$

Greene showcases a variety of identities satisfied by his Gaussian hypergeometric functions, many of which provide direct analogues for transformations of classical hypergeometric series. For example, he provides a finite field analogue of (4), shown below.

Theorem 7 (Greene [21]). *For characters $A_0, A_1, \ldots, A_n, B_1, \ldots, B_n$ in $\widehat{\mathbb{F}_q^\times}$, and $x \in \mathbb{F}_q$,*

$$_{n+1}F_n \left(\begin{matrix} A_0, & A_1, & \ldots, & A_n \\ & B_1, & \ldots, & B_n \end{matrix} ; x \right)_q =$$

$$\frac{A_n B_n(-1)}{q} \cdot \sum_y {}_nF_{n-1} \left(\begin{matrix} A_0, & A_1, & \ldots, & A_{n-1} \\ & B_1, & \ldots, & B_{n-1} \end{matrix} ; xy \right)_q \cdot A_n(y)\overline{A_n}B_n(1-y).$$

To extend Greene's program, McCarthy provides a modification of Greene's functions below. We let $g(\chi) = \sum_{x \in \mathbb{F}_q} \chi(x)\zeta_p^{\mathrm{Tr}(x)}$ denote the Gauss sum of χ, and Tr the usual trace map form $\mathbb{F}_q$ to $\mathbb{F}_p$.

Definition 2 ([32]). For characters $A_0, A_1, \ldots, A_n, B_1, \ldots, B_n$ in $\widehat{\mathbb{F}_q^\times}$,

$$
{}_{n+1}F_n\left(\begin{matrix} A_0, A_1, \ldots, A_n \\ B_1, \ldots, B_n \end{matrix}; x\right)_q^* :=
$$

$$
\frac{1}{q-1}\sum_{\chi\in\widehat{\mathbb{F}_q^\times}}\prod_{i=0}^{n}\frac{g(A_i\chi)}{g(A_i)}\prod_{j=1}^{n}\frac{g(\overline{B_j\chi})}{g(\overline{B_j})}g(\overline{\chi})\chi(-1)^{n+1}\chi(x).
$$

McCarthy makes explicit how the two hypergeometric functions are related, via the following:

Proposition 8 (McCarthy [32]). *If $A_0 \neq \varepsilon$ and $A_i \neq B_i$ for each $1 \leq i \leq n$, then*

$$
{}_{n+1}F_n\left(\begin{matrix} A_0, A_1, \ldots, A_n \\ B_1, \ldots, B_n \end{matrix}; x\right)_q^* = \left[\prod_{i=1}^{n}\binom{A_i}{B_i}^{-1}\right]{}_{n+1}F_n\left(\begin{matrix} A_0, A_1, \ldots, A_n \\ B_1, \ldots, B_n \end{matrix}; x\right)_q.
$$

McCarthy uses this hypergeometric function to provide analogues to classical formulas of Dixon, Kummer, and Whipple for well-poised classical hypergeometric series [32]. For example, consider Whipple's classical transformation below:

Theorem 9 (Whipple [41]). *If one of $1+\frac{1}{2}a-b$, c, d, e is a negative integer, then*

$$
{}_5F_4\left[\begin{matrix} a & b & c & d & e \\ & 1+a-b & 1+a-c & 1+a-d & 1+a-e \end{matrix}; 1\right]
$$

$$
= \frac{\Gamma(1+a-c)\Gamma(1+a-d)\Gamma(1+a-e)\Gamma(1+a-c-d-e)}{\Gamma(1+a)\Gamma(1+a-d-e)\Gamma(1+a-c-d)\Gamma(1+a-c-e)}
$$

$$
\cdot\, {}_4F_3\left[\begin{matrix} 1+\frac{1}{2}a-b & c & d & e \\ & 1+\frac{1}{2}a & c+d+e-a & 1+a-b \end{matrix}; 1\right].
$$

McCarthy provides a finite field analogue to this result using his hypergeometric series.

Theorem 10 (McCarthy, Theorem 1.6 of [32]). *Let $A, B, C, D, E \in \widehat{\mathbb{F}_q^\times}$ such that, when A is a square, $A \neq \varepsilon$, $B \neq \varepsilon$, $B^2 \neq A$, $CD \neq A$, $CE \neq A$, $DE \neq A$, and $CDE \neq A$. Then, if A is not a square,*

$$
{}_5F_4\left(\begin{matrix} A, & B, & C, & D, & E \\ & A\overline{B}, & A\overline{C}, & A\overline{D} & A\overline{E} \end{matrix}; 1\right)_q^* = 0,
$$

and if A is a square,

$$
{}_5F_4\begin{pmatrix} A, & B, & C, & D, & E \\ & A\overline{B}, & A\overline{C}, & A\overline{D} & A\overline{E} \end{pmatrix}; 1\Big)_q^* =
$$

$$
\frac{g(\overline{A})g(\overline{A}DE)g(\overline{A}CD)g(\overline{A}CE)}{g(\overline{A}C)g(\overline{A}D)g(\overline{A}E)g(\overline{A}CDE)} \sum_{R^2=A} {}_4F_3\begin{pmatrix} R\overline{B}, & C, & D, & E \\ & R & \overline{A}CDE, & A\overline{B} \end{pmatrix}; 1\Big)_q^*
$$

$$
+ \frac{g(\overline{A}DE)g(\overline{A}CD)g(\overline{A}CE)q}{g(C)g(D)g(E)g(\overline{A}C)g(\overline{A}D)g(\overline{A}E)} {}_2F_1\begin{pmatrix} A, & B \\ & A\overline{B} \end{pmatrix}; -1\Big)_q^* .
$$

Gaussian hypergeometric functions have been used to count points on different types of varieties over $\mathbb{F}_q$ and they are related to coefficients of various modular forms [1, 6, 17, 18, 26, 27, 36, 39]. We use Greene's hypergeometric functions to count points on $C_{n,\lambda}$ in Sect. 4.1. We make use of McCarthy's hypergeometric function, as well as the previous theorem, Theorem 10, in the proof of Theorem 4 in Sect. 5. Values of McCarthy's normalized version of the hypergeometric function over finite fields have also been shown to be related to eigenvalues of Siegel modular forms of higher degree [34].

2.4 p-adic Gamma Functions and the Gross–Koblitz Formula

We first recall the p-adic Γ-function. The p-adic Γ-function is defined for $n \in \mathbb{N}$ by

$$
\Gamma_p(n) := (-1)^n \prod_{\substack{0<j<n \\ p \nmid j}} j,
$$

and extends to $x \in \mathbb{Z}_p$ by defining $\Gamma_p(0) := 1$, and for $x \neq 0$,

$$
\Gamma_p(x) := \lim_{n \to x} \Gamma_p(n),
$$

where n runs through any sequence of positive integers p-adically approaching x.

We recall some basic properties for $\Gamma_p(\cdot)$ which will be useful later (see Theorem 14 of [31]).

Proposition 11. *Let $x \in \mathbb{Z}_p$. We have the following facts about Γ_p:*

(a) $\Gamma_p(0) = 1$,
(b) $\Gamma_p(x+1)/\Gamma_p(x) = -x$ *unless* $x \in p\mathbb{Z}_p$ *in which case the quotient takes value* -1,
(c) $\Gamma_p(x)\Gamma_p(1-x) = (-1)^{a_0(x)}$ *where* $a_0(x)$ *is the least positive residue of* x *modulo* p,
(d) Given $p > 11$, *there exist* $G_1(x), G_2(x) \in \mathbb{Z}_p$ *such that for any* $m \in \mathbb{Z}_p$,

$$
\Gamma_p(x+mp) \equiv \Gamma_p(x)\left[1 + G_1(x)mp + G_2(x)\frac{(mp)^2}{2} + G_3(x)\frac{(mp)^3}{6}\right] \pmod{p^4}.
$$

(e) $G_1(x) = G_1(1-x)$ *and* $G_2(x) + G_2(1-x) = 2G_1(x)^2$,
(f) If $x \equiv y \pmod{p^n}$, *then* $\Gamma_p(x) \equiv \Gamma_p(y) \pmod{p^n}$.

We note that (c) above implies in particular that for any integer $n > 1$ and prime $p \equiv 1 \pmod{n}$,

$$\Gamma_p\left(\frac{1}{n}\right)\Gamma_p\left(1-\frac{1}{n}\right) = (-1)^{\frac{1+(n-1)p}{n}}.$$

Thus when n is odd, $\Gamma_p(\frac{1}{n})\Gamma_p(1-\frac{1}{n}) = -1$.

We now recall the Gross–Koblitz formula [22, 37] in the case of $\mathbb{F}_p$. Let

$$\varphi : \mathbb{F}_p^\times \to \mathbb{Z}_p^\times$$

be the Teichmüller character such that $\varphi(x) \equiv x \pmod{p}$. The Gross–Koblitz formula states that the Gauss sum $g(\varphi^{-j})$ defined using the Dwork exponential as the additive character satisfies

$$g\left(\varphi^{-j}\right) = -\pi_p^j \Gamma_p\left(\frac{j}{p-1}\right), \tag{5}$$

where $0 \le j \le p-2$, and $\pi_p \in \mathbb{C}_p$ is a root of $x^{p-1} + p = 0$.

3 In the Setting of Legendre Curves

We first briefly explain the relationships between Gaussian hypergeometric functions, hypergeometric series, and truncated hypergeometric series using the Legendre curve

$$C_{2,\lambda} : y^2 = x_1(1-x_1)(x_1-\lambda),$$

which is isomorphic to the more familiar form $y^2 = x(x-1)(x-\lambda)$ over $\mathbb{Q}(\sqrt{-1})$. It is well known that one period of $C_{2,\lambda}$ is given by

$$\pi \cdot {}_2F_1\left[\begin{matrix}\frac{1}{2} & \frac{1}{2} \\ & 1\end{matrix} ; \lambda\right].$$

Assume $\lambda \in \mathbb{Q}$ and $\eta_2 \in \widehat{\mathbb{F}_p^\times}$ is of order 2. It follows from the Taniyama–Shimura–Wiles theorem [42] that for good primes p,

$$a_p(\lambda) = p+1-\#C_{2,\lambda}(\mathbb{F}_p) = -\sum_{x_1 \in \mathbb{F}_p} \eta_2(x_1(1-x_1)(x_1-\lambda)) = -p \cdot {}_2F_1\left(\begin{matrix}\eta_2, & \eta_2 \\ & \varepsilon\end{matrix} ; \lambda\right)_p$$

is the pth coefficient of a weight 2 cuspidal Hecke eigenform that can be computed from a compatible family of two-dimensional ℓ-adic Galois representations of $G_{\mathbb{Q}} := \mathrm{Gal}(\overline{\mathbb{Q}}/\mathbb{Q})$ constructed from the Tate module of $C_{2,\lambda}$ via L-series. This gives a correspondence between the ${}_2F_1$ Gaussian hypergeometric functions and the Galois representations arising from $C_{2,\lambda}$.

When $a_p(\lambda)$ is not divisible by p, i.e., p is ordinary for $C_{2,\lambda}$, then a result of Dwork [16] says that

$$\lim_{s\to\infty} {}_2F_1\left[\begin{matrix}\frac{1}{2} & \frac{1}{2} \\ & 1\end{matrix}; \hat{\lambda}\right]_{p^s-1} \Big/ {}_2F_1\left[\begin{matrix}\frac{1}{2} & \frac{1}{2} \\ & 1\end{matrix}; \hat{\lambda}\right]_{p^{s-1}-1} \tag{6}$$

is the unit root of $T^2 - a_p(\lambda)T + p$, where $\hat{\lambda} = \varphi(\lambda)$ is the image of λ under the Teichmüller character.

Since $\lambda = 1$ is a singular case, we study the case when $\lambda = -1$, for which the corresponding Legendre curve admits complex multiplication. Let $p \equiv 1 \pmod 4$ be prime, then by Greene [21, (4.11)],

$$p \cdot {}_2F_1\left(\begin{matrix}\eta_2, & \eta_2 \\ & \varepsilon\end{matrix}; -1\right)_p = J(\eta_4, \eta_2) + J(\overline{\eta_4}, \eta_2),$$

where η_4 is a primitive order 4 character of $\mathbb{F}_p^\times$.[3] In the perspective of Dwork, the value (6) agrees with the unit root of $T^2 + (J(\eta_4, \eta_2) + J(\overline{\eta_4}, \eta_2))T + p$, which is $\frac{\Gamma_p(\frac{1}{2})\Gamma_p(\frac{1}{4})}{\Gamma_p(\frac{3}{4})}$ by the Gross–Koblitz formula. Using Lemma 1, we have the following:

Proposition 12. *For each prime $p \equiv 1 \pmod 4$*

$${}_2F_1\left[\begin{matrix}\frac{1}{2} & \frac{1}{2} \\ & 1\end{matrix}; -1\right]_{\frac{p-1}{2}} \equiv \frac{\Gamma_p\left(\frac{1}{2}\right)\Gamma_p\left(\frac{1}{4}\right)}{\Gamma_p\left(\frac{3}{4}\right)} = -\frac{\Gamma_p\left(\frac{1}{4}\right)}{\Gamma_p\left(\frac{1}{2}\right)\Gamma_p\left(\frac{3}{4}\right)} \pmod p.$$

Proof. By Lemmas 2 and 1, we obtain that

$${}_2F_1\left[\begin{matrix}\frac{1}{2} & \frac{1}{2} \\ & 1\end{matrix}; -1\right]_{\frac{p-1}{2}} \equiv {}_2F_1\left[\begin{matrix}\frac{1}{2} & \frac{1}{2} \\ & 1\end{matrix}; -1\right]_{p-1} \equiv -p \cdot {}_2F_1\left(\begin{matrix}\eta_2, & \eta_2 \\ & \varepsilon\end{matrix}; -1\right)_p \pmod p.$$

It remains to show

$$p \cdot {}_2F_1\left(\begin{matrix}\eta_2, & \eta_2 \\ & \varepsilon\end{matrix}; -1\right)_p \equiv -\frac{\Gamma_p\left(\frac{1}{2}\right)\Gamma_p\left(\frac{1}{4}\right)}{\Gamma_p\left(\frac{3}{4}\right)} \pmod p. \tag{7}$$

By the relations

$$p \cdot {}_2F_1\left(\begin{matrix}\eta_2, & \eta_2 \\ & \varepsilon\end{matrix}; -1\right)_p = J(\eta_4, \eta_2) + J(\overline{\eta_4}, \eta_2) = \frac{g(\eta_2)\left(g(\eta_4)^2 + g(\overline{\eta_4})^2\right)}{g(\overline{\eta_4})g(\eta_4)},$$

[3] When $p \equiv 3 \pmod 4$, ${}_2F_1\left(\begin{matrix}\eta_2, & \eta_2 \\ & \varepsilon\end{matrix}; -1\right)_p = 0.$

using the Gross–Koblitz formula, we see that

$$p \cdot {}_2F_1\left(\begin{matrix}\eta_2, \eta_2 \\ \varepsilon\end{matrix}; -1\right)_p = \frac{-\pi_p^{\frac{p-1}{2}} \Gamma_p\left(\frac{1}{2}\right)\left(\pi_p^{3\frac{p-1}{2}} \Gamma_p\left(\frac{3}{4}\right)^2 + \pi_p^{\frac{p-1}{2}} \Gamma_p\left(\frac{1}{4}\right)^2\right)}{\pi_p^{p-1} \Gamma_p\left(\frac{1}{4}\right) \Gamma_p\left(\frac{3}{4}\right)}$$
$$= -\frac{\Gamma_p\left(\frac{1}{2}\right)\left(-p\Gamma_p\left(\frac{3}{4}\right)^2 + \Gamma_p\left(\frac{1}{4}\right)^2\right)}{\Gamma_p\left(\frac{1}{4}\right)\Gamma_p\left(\frac{3}{4}\right)},$$

which yields the result. □

Using a different technique via hypergeometric evaluation identities, one can prove the following stronger result. We will also outline this strategy here (for details, see [31]). First we deform the truncated hypergeometric series p-adically so that it becomes a whole family of terminating series that can be written as a quotient of Pochhammer symbols $(a)_k$ via appropriate hypergeometric evaluation formulas. We further rewrite the quotient of Pochhammer symbols as a quotient of p-adic Gamma values using the functional equation of p-adic Gamma functions. Then we use harmonic sums to analyze the terminating series on one side and use the Taylor expansion of p-adic Gamma functions on the other side. Now picking suitable members in the deformed family, we compare both sides to get a linear system which allows us to conclude the desired congruence.

Proposition 13. *For any prime* $p \equiv 1 \pmod 4$,

$$ {}_2F_1\left[\begin{matrix}\frac{1}{2} & \frac{1}{2} \\ & 1\end{matrix}; -1\right]_{\frac{p-1}{2}} \equiv -\frac{\Gamma_p(\frac{1}{4})}{\Gamma_p(\frac{1}{2})\Gamma_p(\frac{3}{4})} \pmod{p^2}.$$

Remark 1. Proposition 13 was first obtained by Coster and van Hamme [14] using a refined version of formal group laws.

Proof. We first recall a theorem of Kummer (see Corollary 3.1.2 of [2]) which says that whenever b is a negative integer,

$$ {}_2F_1\left[\begin{matrix}a & b \\ & a-b+1\end{matrix}; -1\right] = \frac{\Gamma(a-b+1)\Gamma(a/2+1)}{\Gamma(a+1)\Gamma(a/2-b+1)} = \frac{(a+1)_{-b}}{(a/2+1)_{-b}}. \tag{8}$$

We now estimate the left-hand side of the proposition statement, modulo p^2. Observe that for any positive integer $1 \le k \le \frac{p-1}{2}$, and $x, y \in \mathbb{Z}_p$,

$$\left(\frac{1}{2}+xp\right)_k \equiv \left(\frac{1}{2}\right)_k (1 + 2xpH_k^{(odd)}) \pmod{p^2},$$
$$(1+yp)_k \equiv (1)_k(1+ypH_k) \pmod{p^2},$$

where $H_k^{(odd)} := \sum_{j=1}^{k} \frac{1}{2j-1}$ and $H_k := \sum_{j=1}^{k} \frac{1}{j}$ are harmonic sums. Thus we have for any $x_1, x_2, y \in \mathbb{Z}_p$,

$$
{}_2F_1\left[\begin{matrix} \frac{1}{2}+x_1p & \frac{1}{2}+x_2p \\ & 1+yp \end{matrix} ; -1\right]_{\frac{p-1}{2}} \equiv
$$

$$
{}_2F_1\left[\begin{matrix} \frac{1}{2} & \frac{1}{2} \\ & 1 \end{matrix} ; -1\right]_{\frac{p-1}{2}} + (x_1+x_2)Ap - yBp \pmod{p^2}, \tag{9}
$$

where $A = \sum_{k=0}^{\frac{p-1}{2}} \left(\frac{(\frac{1}{2})_k^2}{k!^2}\right)(-1)^k \cdot 2H_k^{(odd)}$ and $B = \sum_{k=0}^{\frac{p-1}{2}} \left(\frac{(\frac{1}{2})_k^2}{k!^2}\right)(-1)^k H_k$.

If $b = \frac{1}{2} + x_2p$ is a negative integer, then by (8) and the above analysis

$$
{}_2F_1\left[\begin{matrix} \frac{1}{2} & \frac{1}{2} \\ & 1 \end{matrix} ; -1\right]_{\frac{p-1}{2}} + (x_1+x_2)Ap - (x_1-x_2)Bp \equiv \frac{\left(\frac{3}{2}+x_1p\right)_{-b}}{\left(\frac{5}{4}+\frac{x_1p}{2}\right)_{-b}} \pmod{p^2}. \tag{10}
$$

We now estimate the right-hand side of the proposition, which can be also written in terms of the Pochhammer symbols. One can use Lemma 17 of [31] to convert it to a quotient of p-adic Gamma function values. For example, if we let $x_1 = \frac{1}{2}$, $x_2 = -\frac{1}{2}$ in (10) (thus $b = \frac{1-p}{2}$), the right-hand side becomes

$$
\frac{\left(\frac{3+p}{2}\right)_{\frac{p-1}{2}}}{\left(\frac{5+p}{4}\right)_{\frac{p-1}{2}}} = -\frac{\Gamma_p(p)\Gamma_p(\frac{1}{4}+\frac{p}{4})}{\Gamma_p(\frac{1}{2}+\frac{p}{2})\Gamma_p(\frac{3}{4}+\frac{3p}{4})}.
$$

By Proposition 11,

$$
\Gamma_p(\alpha+mp) \equiv \Gamma_p(\alpha)[1+G_1(\alpha)mp] \pmod{p^2},
$$

and $G_1(\alpha) = G_1(1-\alpha)$. Thus,

$$
-\frac{\Gamma_p(p)\Gamma_p(\frac{1}{4}+\frac{p}{4})}{\Gamma_p(\frac{1}{2}+\frac{p}{2})\Gamma_p(\frac{3}{4}+\frac{3p}{4})} \equiv
$$

$$
-\frac{\Gamma_p(\frac{1}{4})}{\Gamma_p(\frac{1}{2})\Gamma_p(\frac{3}{4})}\left[1+G_1(0)p - G_1\left(\frac{1}{2}\right)\frac{p}{2} - G_1\left(\frac{1}{4}\right)\frac{p}{2}\right] \pmod{p^2}.
$$

Equating both sides in (10) gives that

$$
{}_2F_1\left[\begin{matrix}\frac{1}{2} & \frac{1}{2} \\ & 1\end{matrix}; -1\right]_{\frac{p-1}{2}} - Bp \equiv
$$

$$
-\frac{\Gamma_p(\frac{1}{4})}{\Gamma_p(\frac{1}{2})\Gamma_p(\frac{3}{4})}\left[1 + G_1(0)p - G_1\left(\frac{1}{2}\right)\frac{p}{2} - G_1\left(\frac{1}{4}\right)\frac{p}{2}\right] \pmod{p^2}.
$$

Similarly, letting $x_1 = -3/2, x_2 = -1/2$, we get

$$
{}_2F_1\left[\begin{matrix}\frac{1}{2} & \frac{1}{2} \\ & 1\end{matrix}; -1\right]_{\frac{p-1}{2}} - 2A + Bp \equiv
$$

$$
-\frac{\Gamma_p(\frac{1}{4})}{\Gamma_p(\frac{1}{2})\Gamma_p(\frac{3}{4})}\left[1 - G_1(0)p + G_1\left(\frac{1}{2}\right)\frac{3p}{2} - G_1\left(\frac{1}{4}\right)\frac{p}{2}\right] \pmod{p^2}.
$$

Letting $x_1 = -1/2, x_2 = -3/2$ (here $b = \frac{1-3p}{2}$ but the series terminates at the $\frac{p-1}{2}$th term as $a = -\frac{p-1}{2}$),

$$
{}_2F_1\left[\begin{matrix}\frac{1}{2} & \frac{1}{2} \\ & 1\end{matrix}; -1\right]_{\frac{p-1}{2}} - 2A - Bp \equiv
$$

$$
-\frac{\Gamma_p(\frac{1}{4})}{\Gamma_p(\frac{1}{2})\Gamma_p(\frac{3}{4})}\left[1 + G_1(0)p + G_1\left(\frac{1}{2}\right)\frac{p}{2} - G_1\left(\frac{1}{4}\right)\frac{3p}{2}\right] \pmod{p^2}.
$$

By summing the first two and last two congruences and comparing, we arrive at the proposition, in light of Lemma 2. □

Based on Lemma 1, we observe the following numerically which is a companion form of the congruence above:

Conjecture 14. *For any prime* $p \equiv 1 \pmod 4$,

$$
-\frac{\Gamma_p(\frac{1}{4})}{\Gamma_p(\frac{1}{2})\Gamma_p(\frac{3}{4})} \equiv \sum_{k=0}^{p-1}\left(\frac{(\frac{1}{2})_k}{k!}\right)^2(-1)^k \stackrel{?}{\equiv}
$$

$$
\sum_{k=0}^{p-1}\left(\frac{k!}{(\frac{3}{2})_k}\right)^2 p^2(-1)^k \equiv \sum_{k=\frac{p-1}{2}}^{p-1}\left(\frac{k!}{(\frac{3}{2})_k}\right)^2 p^2(-1)^k \pmod{p^2}.
$$

4 Proofs of Theorem 2 and Lemma 1

4.1 Proof of Theorem 2

We first prove Theorem 2 which counts points on $C_{n,\lambda}$ over finite fields $\mathbb{F}_q$ in terms of Gaussian hypergeometric functions. We begin with a lemma.

Lemma 3. *Let $n \geq 2$ be an integer and let q be a prime power with $q = p^e \equiv 1 \pmod{n}$. Suppose $\eta_n \in \widehat{\mathbb{F}_q^{\times}}$ is a character of order n and ε denotes the trivial character. If $k \in \{1, \ldots, n-1\}$, then*

$$\sum_{x_i \in \mathbb{F}_q} \eta_n^k((x_1 \cdots x_{n-1})^{n-1}(1-x_1)\cdots(1-x_{n-1})(x_1 - \lambda x_2 \cdots x_{n-1}))$$
$$= q^{n-1} \cdot {}_nF_{n-1}\left(\begin{matrix} \eta_n^{n-k}, & \eta_n^{n-k}, & \ldots, & \eta_n^{n-k} \\ & \varepsilon, & \ldots, & \varepsilon \end{matrix} ; \lambda\right)_q .$$

Proof. We apply Theorem 7 $(n-3)$ times, noting that $\overline{\eta_n^{n-k}} = \eta_n^k$. This gives

$${}_nF_{n-1}\left(\begin{matrix} \eta_n^{n-k}, & \eta_n^{n-k}, & \ldots, & \eta_n^{n-k} \\ & \varepsilon, & \ldots, & \varepsilon \end{matrix} ; \lambda\right)_q$$
$$= \frac{(\eta_n^{n-k}(-1))^{n-3}}{q^{n-3}} \sum_{x_2,\ldots,x_{n-2} \in \mathbb{F}_q} {}_3F_2\left(\begin{matrix} \eta_n^{n-k}, & \eta_n^{n-k}, & \eta_n^{n-k} \\ & \varepsilon, & \varepsilon \end{matrix} ; \lambda x_2 \cdots x_{n-2}\right)_q$$
$$\cdot \eta_n^{n-k}(x_2 \cdots x_{n-2}) \eta_n^k((1-x_2)\cdots(1-x_{n-2})).$$

We next apply Corollary 3.14(ii) of [21] to the ${}_3F_2$ function to obtain

$${}_nF_{n-1}\left(\begin{matrix} \eta_n^{n-k}, & \eta_n^{n-k}, & \ldots, & \eta_n^{n-k} \\ & \varepsilon, & \ldots, & \varepsilon \end{matrix} ; \lambda\right)_q = \frac{(\eta_n^{n-k}(-1))^{n-2}}{q^{n-1}}$$
$$\cdot \sum_{x_1,\ldots,x_{n-1} \in \mathbb{F}_q} \eta_n^{n-k}(x_1 \cdots x_{n-1}) \eta_n^k((1-x_1)\cdots(1-x_{n-1})(x_1 - \lambda x_2 \cdots x_{n-1})).$$

Since $(\eta_n^{n-k}(-1))^{n-2} = \eta_n^{(n-k)(n-2)}(-1) = 1$ and

$$\eta_n^k((x_1 \cdots x_{n-1})^{n-1}) = \eta_n^{kn-k}(x_1 \cdots x_{n-1}) = \eta_n^{n-k}(x_1 \cdots x_{n-1}),$$

we have the result. □

We are now able to prove Theorem 2.

Proof of Theorem 2. For convenience, we denote

$$f(x_1,\ldots,x_{n-1},\lambda) = (x_1\cdots x_{n-1})^{n-1}(1-x_1)\cdots(1-x_{n-1})(x_1-\lambda x_2\cdots x_{n-1}).$$

Then

$$\begin{aligned}
\#C_{n,\lambda}(\mathbb{F}_q) &= 1+\sum_{x_i\in\mathbb{F}_q}\#\{y\in\mathbb{F}_q : y^n = f(x_1,\ldots,x_{n-1},\lambda)\}\\
&= 1+\sum_{\substack{x_i\in\mathbb{F}_q\\ f(x_1,\ldots,x_{n-1},\lambda)\neq 0}}\#\{y\in\mathbb{F}_q : y^n = f(x_1,\ldots,x_{n-1},\lambda)\}\\
&\qquad + \#\{(x_1,\ldots,x_{n-1})\in\mathbb{F}_q^{n-1} : f(x_1,\ldots,x_{n-1},\lambda)=0\}.
\end{aligned}$$

Using Proposition 8.1.5 in [23], we rewrite the first sum to see

$$\begin{aligned}
\#C_{n,\lambda}(\mathbb{F}_q) &= 1+\sum_{x_i\in\mathbb{F}_q}\sum_{i=0}^{n-1}\eta_n^i(f(x_1,\ldots,x_{n-1},\lambda))\\
&\qquad + \#\{(x_1,\ldots,x_{n-1})\in\mathbb{F}_q^{n-1} : f(x_1,\ldots,x_{n-1},\lambda)=0\}\\
&= 1+q^{n-1}+\sum_{x_i\in\mathbb{F}_q}\sum_{i=1}^{n-1}\eta_n^i(f(x_1,\ldots,x_{n-1},\lambda)),
\end{aligned}$$

since

$$\varepsilon(f(x_1,\ldots,x_{n-1},\lambda)) + \#\{(x_1,\ldots,x_{n-1})\in\mathbb{F}_q^{n-1} : f(x_1,\ldots,x_{n-1},\lambda)=0\} = q^{n-1}.$$

Finally, we have

$$\begin{aligned}
\#C_{n,\lambda}(\mathbb{F}_q) &= 1+q^{n-1}+\sum_{i=1}^{n-1}\sum_{x_i\in\mathbb{F}_q}\eta_n^i(f(x_1,\ldots,x_{n-1},\lambda))\\
&= 1+q^{n-1}+q^{n-1}\sum_{i=1}^{n-1} {}_nF_{n-1}\left(\begin{matrix}\eta_n^{n-i}, & \eta_n^{n-i}, & \ldots, & \eta_n^{n-i}\\ & \varepsilon, & \ldots, & \varepsilon\end{matrix}\;;\lambda\right)_q
\end{aligned}$$

by Lemma 3, which gives the result. □

4.2 Proof of Lemma 1

To prove Lemma 1, which relates certain Gaussian hypergeometric functions to truncated hypergeometric series, we first give a lemma, which analyzes the Gaussian hypergeometric functions modulo a power of p.

Lemma 4. *Let n be a positive integer, and $p \equiv 1 \pmod{n}$ prime. Let φ denote the Teichmüller character, and η_n a character of order n on $\mathbb{F}_p$ corresponding to $\varphi^{\frac{p-1}{n}j}$, for some $0 < j < p-1$. Then*

$$p^{r-1}{}_rF_{r-1}\left(\begin{matrix}\overline{\eta_n}, \overline{\eta_n}, \cdots, \overline{\eta_n} \\ \varepsilon, \cdots, \varepsilon\end{matrix}; x\right)_p \equiv$$

$$\frac{1}{p-1}\left(\sum_{k=(p-1)(\frac{n-j}{n})}^{p-2}\left(\frac{\Gamma_p\left(\frac{k}{p-1}-\frac{n-j}{n}\right)}{\Gamma_p\left(\frac{k}{p-1}\right)\Gamma_p\left(\frac{j}{n}\right)}\right)^r \overline{\varphi}^k(x)+(-1)^r\right) \pmod{p^r}, \tag{11}$$

$$p^{r-1}{}_rF_{r-1}\left(\begin{matrix}\eta_n, \eta_n, \cdots, \eta_n \\ \varepsilon, \cdots, \varepsilon\end{matrix}; x\right)_p \equiv$$

$$\frac{(-1)^r}{p-1}\left(\sum_{k=0}^{(p-1)(\frac{n-j}{n})-1}\frac{\Gamma_p\left(\frac{k}{p-1}\right)^r\Gamma_p\left(\frac{j}{n}\right)^r}{\Gamma_p\left(\frac{j}{n}+\frac{k}{p-1}\right)^r}\overline{\varphi}^k(x)+\overline{\eta_n}((-1)^r x)\right) \pmod{p^r}. \tag{12}$$

Proof. For $x \in \mathbb{F}_p^\times$, it follows from the definition that

$${}_rF_{r-1}\left(\begin{matrix}\eta_n, \eta_n, \cdots, \eta_n \\ \varepsilon, \cdots, \varepsilon\end{matrix}; x\right)_p =$$

$$\frac{p}{p-1}\sum_{\chi\in\widehat{\mathbb{F}_p^\times}}\binom{\eta_n\chi}{\chi}^r\chi(x) = \frac{p^{1-r}}{p-1}\sum_\chi J(\eta_n\chi, \overline{\chi})^r\chi^r(-1)\chi(x),$$

and also that

$$J(\overline{\eta_n}\chi, \overline{\chi}) = \begin{cases}\frac{\chi(-1)pg(\overline{\eta_n}\chi)}{g(\overline{\eta_n})g(\chi)}, & \text{if } \chi \neq \varepsilon, \\ -1, & \text{if } \chi = \varepsilon,\end{cases}$$

$$\text{while, } J(\eta_n\chi, \overline{\chi}) = \begin{cases}\frac{\chi(-1)g(\overline{\eta_n})g(\overline{\chi})}{g(\overline{\eta_n}\chi)}, & \text{if } \chi \neq \overline{\eta_n}, \\ -1, & \text{if } \chi = \overline{\eta_n}.\end{cases}$$

Thus, we have that

$$\sum_\chi J(\overline{\eta_n}\chi, \overline{\chi})^r\chi^r(-1)\chi(x) = p^r\sum_{\chi\neq\varepsilon}\frac{g(\overline{\eta_n}\chi)^r}{g(\overline{\eta_n})^r g(\chi)^r}\chi(x)+(-1)^r$$

$$= p^r \sum_{k=1}^{(p-1)(\frac{n-j}{n})-1} \frac{g(\overline{\varphi}^{\frac{p-1}{n}j+k})^r}{g(\overline{\varphi}^{\frac{p-1}{n}j})^r g(\overline{\varphi}^k)^r} \overline{\varphi}^k(x)$$

$$+ p^r \sum_{k=(p-1)(\frac{n-j}{n})}^{p-2} \frac{g(\overline{\varphi}^{k-(p-1)\frac{n-j}{n}})^r}{g(\overline{\varphi}^{\frac{p-1}{n}j})^r g(\overline{\varphi}^k)^r} \overline{\varphi}^k(x) + (-1)^r$$

$$= p^r \sum_{k=1}^{(p-1)(\frac{n-j}{n})-1} (-1)^r \left(\frac{\Gamma_p\left(\frac{j}{n} + \frac{k}{p-1}\right)}{\Gamma_p\left(\frac{k}{p-1}\right)\Gamma_p\left(\frac{j}{n}\right)} \right)^r \overline{\varphi}^k(x)$$

$$+ \sum_{k=(p-1)(\frac{n-j}{n})}^{p-2} \left(\frac{\Gamma_p\left(\frac{k}{p-1} - \frac{n-j}{n}\right)}{\Gamma_p\left(\frac{k}{p-1}\right)\Gamma_p\left(\frac{j}{n}\right)} \right)^r \overline{\varphi}^k(x) + (-1)^r.$$

Similarly,

$$\sum_{\chi} J(\eta_n\chi, \overline{\chi})^r \chi^r(-1)\chi(x) = \sum_{\chi \neq \overline{\eta_n}} \frac{g(\overline{\eta_n}\chi)^r}{g(\overline{\eta_n})^r g(\overline{\chi})^r} \chi(x) + (-1)^r \overline{\eta_n}^r(-1)\overline{\eta_n}(x)$$

$$= (-1)^r \sum_{k=0}^{(p-1)(\frac{n-j}{n})-1} \left(\frac{\Gamma_p\left(\frac{k}{p-1}\right)\Gamma_p\left(\frac{j}{n}\right)}{\Gamma_p\left(\frac{j}{n} + \frac{k}{p-1}\right)} \right)^r \overline{\varphi}^k(x) + (-1)^r \overline{\eta_n}^r(-1)\overline{\eta_n}(x)$$

$$+ p^r \sum_{k=(p-1)(\frac{n-j}{n})+1}^{p-2} \left(\frac{\Gamma_p\left(\frac{k}{p-1}\right)\Gamma_p\left(\frac{j}{n}\right)}{\Gamma_p\left(\frac{k}{p-1} - \frac{n-j}{n}\right)} \right)^r \overline{\varphi}^k(x).$$

□

We are now able to prove Lemma 1.

Proof of Lemma 1. For the first congruence, we use (12). By Proposition 11, when $k < p$ we have that $\Gamma_p(-k) = \Gamma_p(-k)/\Gamma_p(0) = 1/k!$ and $\Gamma_p(\frac{k}{p-1}) \equiv \Gamma_p(-k) \pmod{p}$. Similarly,

$$\frac{\Gamma_p(\frac{j}{n})}{\Gamma_p(\frac{j}{n} + \frac{k}{p-1})} \equiv \frac{\Gamma_p(\frac{j}{n})}{\Gamma_p(\frac{j}{n} - k)} \pmod{p}$$

and

$$\frac{\Gamma_p(\frac{j}{n})}{\Gamma_p(\frac{j}{n} - k)} = \left(1 - \frac{j}{n}\right) \cdots \left(k - \frac{j}{n}\right) = \left(1 - \frac{j}{n}\right)_k.$$

When $k = (p-1)\left(\frac{n-j}{n}\right)$, we have $\overline{\eta_n}(\pm(-1)^r) \equiv \frac{(1-\frac{j}{n})_k^r}{k!^r}(\pm 1)^k \pmod p$. Thus the first claim follows.

For the second claim, we consider (11) and use a similar argument. Notice that

$$\frac{\Gamma_p(\frac{k}{p-1} - \frac{n-j}{n})}{\Gamma_p(\frac{j}{n})} \equiv \frac{\Gamma_p(-k - \frac{n-j}{n})}{\Gamma_p(\frac{j}{n})} = \frac{-p(\frac{n-j}{n})}{(1-\frac{j}{n})(2-\frac{j}{n})\cdots(k-\frac{j}{n})} \pmod p,$$

and there is no $-p(\frac{n-j}{n})$ in the numerator since for $(p-1)(\frac{n-j}{n}) \le k \le p-2$ the denominator, $(1-\frac{j}{n})_{k+1}$, will contain a multiple of p, which is $p(\frac{n-j}{n})$. This term will not show up in the quotient of p-adic Gamma values and the corresponding term is -1 by the functional equation of $\Gamma_p(\cdot)$. Thus

$$\frac{\Gamma_p(\frac{k}{p-1} - \frac{n-j}{n})}{\Gamma_p(\frac{j}{n})} \equiv -\frac{p}{(2-\frac{j}{n})_k} \pmod p,$$

which concludes the proof of the second claim. □

5 The Proof of Theorem 4

The following proposition establishes case (1) of Theorem 4:

Proposition 15. *Let $q = p^e \equiv 1 \pmod 3$ be a prime power and let η_3 be a character of order 3 in $\widehat{\mathbb{F}_q^\times}$. Then*

$$q^2 \cdot {}_3F_2\left(\begin{matrix}\eta_3, \eta_3, \eta_3 \\ \varepsilon, \ \varepsilon\end{matrix}; 1\right)_q = J(\eta_3, \eta_3)^2 - J(\eta_3^2, \eta_3^2).$$

Proof. Beginning with Theorem 4.35 in [21], we have

$$q^2 \cdot {}_3F_2\left(\begin{matrix}\eta_3, \eta_3, \eta_3 \\ \varepsilon, \ \varepsilon\end{matrix}; 1\right)_q = q^2\eta_3^2(-1)\binom{\eta_3}{\eta_3^2}\binom{\eta_3}{\eta_3^2} - q\eta_3(-1)\binom{\eta_3^2}{\eta_3}$$

Then, since $\binom{\eta_3}{\eta_3^2} = \frac{1}{q}J(\eta_3, \eta_3)$ and $\binom{\eta_3^2}{\eta_3} = \frac{\eta_3(-1)}{q}J(\eta_3^2, \eta_3^2)$, we get the result. □

We now restate an equivalent form of case (2) of Theorem 4.

Theorem 16. *Let $q = p^e \equiv 1 \pmod 4$ be a prime power and let η_4 and η_2 be characters of order 4 and 2, respectively, in $\widehat{\mathbb{F}_q^\times}$. Then*

$$\sum_{x,y,z\in\mathbb{F}_q} \eta_4(x^3y^3z^3(1-x)(1-y)(1-z)(x-yz)) = J(\overline{\eta_4},\eta_2)^3 + qJ(\overline{\eta_4},\eta_2) - J(\eta_4,\eta_2)^2.$$

Equivalently, we have

$$q^3 \cdot {}_4F_3\left(\begin{matrix}\overline{\eta_4}, & \overline{\eta_4}, & \overline{\eta_4}, & \overline{\eta_4} \\ & \varepsilon, & \varepsilon, & \varepsilon\end{matrix};1\right)_q = J(\overline{\eta_4},\eta_2)^3 + qJ(\overline{\eta_4},\eta_2) - J(\eta_4,\eta_2)^2.$$

As mentioned in the introduction, this result says that a three-dimensional Galois representation of $G_{\mathbb{Q}(\sqrt{-1})} := \mathrm{Gal}(\overline{\mathbb{Q}}/\mathbb{Q}(\sqrt{-1}))$ arising from $C_{4,1}$ is isomorphic to a direct sum of three Grössencharacters. The proof contains two steps using totally different approaches. We first establish the case for $q \equiv 1 \pmod 8$ using results of Greene and McCarthy. This is equivalent to proving the two Galois representations are isomorphic when they are restricted to the subgroup $G_{\mathbb{Q}(\sqrt{-1},\sqrt{2})} := \mathrm{Gal}(\overline{\mathbb{Q}}/\mathbb{Q}(\sqrt{-1},\sqrt{2}))$. In the second part of the proof, we use representation theory to draw the final conclusion. We now start with the case when $q = p^e \equiv 1 \pmod 8$ is a prime power, we prove this case via the series of results below. The lemma below evaluates two modified Gaussian hypergeometric functions.

Lemma 5. *Let $q = p^e \equiv 1 \pmod 8$ be a prime power. Then*

$${}_4F_3\left(\begin{matrix}\overline{\eta_8}, & \eta_4, & \eta_4, & \eta_8 \\ & \eta_8, & \eta_8^3, & \varepsilon\end{matrix};1\right)_q^* =$$
$$\frac{1}{J(\eta_4,\overline{\eta_8}^3)}\left(J(\eta_8,\overline{\eta}_4) - \eta_8(-1)J(\overline{\eta_8},\overline{\eta_8})J(\eta_2,\eta_4) + \frac{J(\eta_8,\overline{\eta}_4)^3}{q}\right),$$

$${}_4F_3\left(\begin{matrix}\eta_8^3, & \eta_4, & \eta_4, & \eta_8 \\ & \overline{\eta_8}^3, & \eta_8^3, & \varepsilon\end{matrix};1\right)_q^* =$$
$$\frac{1}{J(\eta_4,\eta_8^3)}\left(J(\eta_8^3,\overline{\eta}_4) - \eta_8(-1)J(\eta_8,\overline{\eta_8}^3)J(\eta_2,\eta_4) + \frac{J(\overline{\eta}_8,\overline{\eta}_4)J(\overline{\eta}_4,\eta_8^3)^2}{q}\right).$$

Proof. Firstly, we obtain that

$${}_4F_3\left(\begin{matrix}\overline{\eta_8}, & \eta_4, & \eta_4, & \eta_8 \\ & \eta_8, & \eta_8^3, & \varepsilon\end{matrix};1\right)_q^* = -q\left(\binom{\eta_4}{\eta_8}\binom{\eta_4}{\eta_8^3}\right)^{-1} {}_4F_3\left(\begin{matrix}\overline{\eta_8}, & \eta_4, & \eta_4, & \eta_8 \\ & \eta_8, & \eta_8^3, & \varepsilon\end{matrix};1\right)_q.$$

Using the transformations Theorem 3.15(iv), Theorem 4.35 in [21], and the fact $\eta_4(-1) = 1$ when $q \equiv 1 \pmod 8$, we have

$$
\begin{aligned}
{}_4F_3\left(\begin{matrix}\overline{\eta_8}, \eta_4, \eta_4, \eta_8\\ \eta_8, \eta_8^3, \varepsilon\end{matrix}; 1\right)_q &= {}_4F_3\left(\begin{matrix}\eta_8, \eta_4, \eta_4, \overline{\eta}_8\\ \eta_8, \eta_8^3, \varepsilon\end{matrix}; 1\right)_q \\
&= \binom{\eta_4}{\eta_8} {}_3F_2\left(\begin{matrix}\eta_4, \eta_4, \overline{\eta}_8\\ \eta_8^3, \varepsilon\end{matrix}; 1\right)_q - \frac{\eta_8(-1)}{q}\binom{\eta_8}{\eta_4}\binom{\overline{\eta}_4}{\overline{\eta}_8} \\
&= \binom{\eta_4}{\eta_8}\left(\eta_8(-1)\binom{\eta_8}{\overline{\eta}_8}\binom{\eta_4}{\eta_2} - \frac{\eta_8(-1)}{q}\binom{\eta_8}{\eta_4}\right) - \frac{1}{q}\binom{\eta_8}{\eta_4}^2.
\end{aligned}
$$

Therefore, we can conclude that

$$
\begin{aligned}
&{}_4F_3\left(\begin{matrix}\overline{\eta_8}, \eta_4, \eta_4, \eta_8\\ \eta_8, \eta_8^3, \varepsilon\end{matrix}; 1\right)_q^* \\
&= \frac{1}{J(\eta_4, \overline{\eta_8}^3)}\left(J(\eta_8, \overline{\eta}_4) - \eta_8(-1)J(\overline{\eta_8}, \overline{\eta_8})J(\eta_2, \eta_4) + \frac{J(\eta_8, \overline{\eta}_4)^3}{q}\right).
\end{aligned}
$$

Likewise, we have the equality

$$
\begin{aligned}
&{}_4F_3\left(\begin{matrix}\eta_8^3, \eta_4, \eta_4, \eta_8\\ \overline{\eta_8}^3, \eta_8^3, \varepsilon\end{matrix}; 1\right)_q^* = \\
&\frac{1}{J(\eta_4, \eta_8^3)}\left(J(\eta_8^3, \overline{\eta}_4) - \eta_8(-1)J(\eta_8, \overline{\eta_8}^3)J(\eta_2, \eta_4) + \frac{J(\overline{\eta}_8, \overline{\eta}_4)J(\overline{\eta}_4, \eta_8^3)^2}{q}\right).
\end{aligned}
$$

□

Next, we relate our target to a modified Gaussian ${}_5F_4$ hypergeometric function.

Proposition 17. *Let $q = p^e \equiv 1 \pmod 8$ be a prime power, and η_8 a character of order 8 in $\widehat{\mathbb{F}_q^\times}$ with $\eta_8^2 = \eta_4$. Then*

$$
q^4 \cdot {}_4F_3\left(\begin{matrix}\eta_4, \eta_4, \eta_4, \eta_4\\ \varepsilon, \varepsilon, \varepsilon\end{matrix}; 1\right)_q = J(\eta_8, \eta_8)^4 - q \cdot {}_5F_4\left(\begin{matrix}\eta_4, \eta_4, \eta_4, \eta_4, \eta_8\\ \varepsilon, \varepsilon, \varepsilon, \eta_8\end{matrix}; 1\right)_q^*.
$$

Proof. Comparing the definitions of finite field hypergeometric functions given by Greene and McCarthy, one can find

$$
\begin{aligned}
&{}_5F_4\left(\begin{matrix}\eta_4, \eta_4, \eta_4, \eta_4, \eta_8\\ \varepsilon, \varepsilon, \varepsilon, \eta_8\end{matrix}; 1\right)_q \\
&= \frac{1}{q^3}{}_5F_4\left(\begin{matrix}\eta_4, \eta_4, \eta_4, \eta_4, \eta_8\\ \varepsilon, \varepsilon, \varepsilon, \eta_8\end{matrix}; 1\right)_q^* + \left(1 - \frac{1}{q}\right){}_4F_3\left(\begin{matrix}\eta_4, \eta_4, \eta_4, \eta_4\\ \varepsilon, \varepsilon, \varepsilon\end{matrix}; 1\right)_q.
\end{aligned}
$$

On the other hand, by [21, Theorem 3.15(ii)], the Greene's ${}_5F_4$ function is equal to

$$-\frac{1}{q}{}_4F_3\left(\begin{matrix}\eta_4, \eta_4, \eta_4, \eta_4\\ \varepsilon, \ \varepsilon, \ \varepsilon\end{matrix};1\right)_q+\frac{J(\eta_8,\eta_8)^4}{q^4}.$$

These lead to the desired result. □

We now evaluate the ${}_5F_4$ modified Gaussian hypergeometric function using Theorem 10 which is due to McCarthy.

Proposition 18. *Let $q = p^e \equiv 1 \pmod 8$ be a prime power. Then*

$${}_5F_4\left(\begin{matrix}\eta_4, \eta_4, \eta_4, \eta_4, \eta_8\\ \varepsilon, \ \varepsilon, \ \varepsilon, \ \eta_8\end{matrix};1\right)_q^* = \frac{J(\eta_8,\eta_8)^4}{q} - qJ(\eta_4,\eta_2) - J(\eta_2,\eta_4)^3 + J(\eta_2,\overline{\eta}_4)^2.$$

Proof. According to Theorem 10, we can deduce that

$${}_5F_4\left(\begin{matrix}\eta_4, \eta_4, \eta_4, \eta_4, \eta_8\\ \varepsilon, \ \varepsilon, \ \varepsilon, \ \eta_8\end{matrix};1\right)_q^* = \frac{q}{J(\overline{\eta}_8,\eta_4)}{}_2F_1\left(\begin{matrix}\eta_4, \eta_4\\ \varepsilon\end{matrix};-1\right)_q^*$$
$$+q\frac{J(\eta_8,\eta_8)}{J(\eta_8^3,\overline{\eta}_8)}\left({}_4F_3\left(\begin{matrix}\overline{\eta_8} \ \eta_4 \ \eta_4 \ \eta_8\\ \eta_8, \eta_8^3, \ \varepsilon\end{matrix};1\right)_q^* + {}_4F_3\left(\begin{matrix}\eta_8^3, \ \eta_4, \ \eta_4, \eta_8\\ \overline{\eta_8}^3, \eta_8^3, \ \varepsilon\end{matrix};1\right)_q^*\right).$$

We now apply Lemma 5 and the following fact which arises from Theorem 1.9 of [32]:

$${}_2F_1\left(\begin{matrix}\eta_4 \ \eta_4\\ \varepsilon\end{matrix};1\right)_q^* = -J(\eta_8,\overline{\eta}_4) - J(\overline{\eta}_4,\overline{\eta}_8^3).$$

To simplify the formulas, we recall the facts that if $q = p^e \equiv 1 \pmod 8$, we have $\eta_2(2) = \eta_4(-1) = 1$ and the identities

$$J(\eta_8,\eta_4) = \overline{\eta}_8^3(-4)J(\eta_2,\eta_4),$$
$$J(\overline{\eta}_4,\eta_8^3) = \eta_8(-1)J(\overline{\eta}_8,\eta_8^3) = \overline{\eta}_8(-4)J(\eta_2,\overline{\eta}_4),$$
$$J(\eta_8,\overline{\eta}_4) = J(\eta_4,\eta_8^3) = \eta_8(-1)J(\eta_8^3,\eta_8^3) = \eta_8(-1)J(\eta_8,\eta_8).$$

For more details on Jacobi sums, please see [8, Chap. 3]. □

We now conclude with the second case using representation theory. We first describe the Grössencharacters corresponding to the Jacobi sums

$$-J(\overline{\eta_4},\eta_2)^3 - qJ(\overline{\eta_4},\eta_2) + J(\eta_4,\eta_2)^2 =$$
$$-J(\overline{\eta_4},\eta_2)^3 - J(\overline{\eta_4},\eta_2)J(\overline{\eta_4},\eta_2)^2 + J(\eta_4,\eta_2)^2$$

and then we use properties of induced representations to draw the final conclusion.

Proof (The proof of the case when prime powers $q \equiv 5 \pmod 8$.).

For each prime ideal $\mathfrak{p}$ prime to (4) in $\mathbb{Z}[\sqrt{-1}]$, let q be the norm of $\mathfrak{p}$. Then $q \equiv 1 \mod 4$, and let $\psi_\mathfrak{p}$ be a homomorphism of $\mathbb{Z}[\sqrt{-1}]/\mathfrak{p}$ to the order 4 multiplicative group $\langle\sqrt{-1}\rangle$ such that $\psi_\mathfrak{p}(x) \equiv x^{\frac{q-1}{4}} \mod \mathfrak{p}$ for each $x \in \mathbb{Z}[\sqrt{-1}]$. The map that assigns $-\sum_{x \mod \mathfrak{p}} \psi_\mathfrak{p}(x)\psi_\mathfrak{p}^2(1-x) = -J(\psi_\mathfrak{p}, \psi_\mathfrak{p}^2)$ to $\mathfrak{p}$ extends to a Hecke (or Grössencharacter) character ψ of $G_{\mathbb{Q}(\sqrt{-1})}$ (see [40] by Weil). In particular, ψ is of conductor $\left((1+\sqrt{-1})^4\right) = (4)$ and infinity-type $[1,0]$, which is corresponding to the elliptic curve with complex multiplication which has conductor 64. Explicitly, for any $a+b\sqrt{-1} \in \mathbb{Z}[\sqrt{-1}]$, $\psi(a+b\sqrt{-1}) = (-1)^b(a+b\sqrt{-1})\chi_1(a+b\sqrt{-1})$, where

$$\chi_1(a+b\sqrt{-1}) = \begin{cases} (-1)^{\frac{a+b-1}{2}}\sqrt{-1}, & \text{if } a \equiv 0 \pmod 2, b \equiv 1 \pmod 2, \\ (-1)^{\frac{a+b-1}{2}}, & \text{if } a \equiv 1 \pmod 2, b \equiv 0 \pmod 2, \\ 0, & \text{otherwise.} \end{cases}$$

Here, we remark that the unit group of $\mathbb{Z}[\sqrt{-1}]/(4)$ has order 8 and it is generated by $\sqrt{-1}, -1+2\sqrt{-1}$. The Dirichlet character χ_1 takes values $\chi_1(\sqrt{-1}) = \sqrt{-1}$, $\chi_1(-1+2\sqrt{-1}) = 1$.

By class field theory, ψ corresponds to a character χ of $G_{\mathbb{Q}(\sqrt{-1})}$. Similar to the discussion in [15], for each Frobenius class $\mathrm{Fr}_q \in G_{\mathbb{Q}(\sqrt{-1},\sqrt{2})}$ with $q \equiv 1 \pmod 4$,

$$\sum_{x,y,z\in\mathbb{F}_q} \Big[-\eta_4\left(x^3y^3z^3(1-x)(1-y)(1-z)(x-yz)\right) - \overline{\eta_4}\left(x^3y^3z^3(1-x)(1-y)(1-z)(x-yz)\right)\Big]$$

coincides with the trace of $\mathrm{Fr}_\mathfrak{p}$ under the six-dimensional semisimple representation

$$\rho := \mathrm{Ind}_{G_{\mathbb{Q}(\sqrt{-1})}}^{G_\mathbb{Q}} \left(\overline{\chi}^3 \oplus (\overline{\chi}^2 \otimes \chi) \oplus \chi^2\right).$$

Moreover, $\rho|_{G_{\mathbb{Q}(\sqrt{-1})}} = \sigma \oplus \bar{\sigma}$, with the restriction $\sigma|_{G_{\mathbb{Q}(\sqrt{-1},\sqrt{2})}}$ being isomorphic to the restriction of $\overline{\chi}^3 \oplus (\overline{\chi}^2 \otimes \chi) \oplus \chi^2$ to $G_{\mathbb{Q}(\sqrt{-1},\sqrt{2})}$. As $G_{\mathbb{Q}(\sqrt{-1},\sqrt{2})}$ is an index-2 subgroup of $G_{\mathbb{Q}(\sqrt{-1})}$, by Clifford's result [12], σ is also direct sum of the form

$$(\overline{\chi}^3 \otimes \varphi^{n_1}) \oplus (\overline{\chi}^2 \otimes \chi \otimes \varphi^{n_2}) \oplus (\chi^2 \otimes \varphi^{n_3})$$

where φ is the order 2 character of $G_{\mathbb{Q}(\sqrt{-1})}$ with kernel $G_{\mathbb{Q}(\sqrt{-1},\sqrt{2})}$, $n_1, n_2, n_3 \in \{0,1\}$. From computing a few primes $p \equiv 5 \pmod 8$, we determine that each $n_i = 0$ and the claim for $p \equiv 5 \pmod 8$ thus follows. □

6 Supercongruences

We first prove Theorem 5 using the technique outlined before Proposition 13. The initial idea of the proof is due to Zudilin, and uses the following particular case of the Karlsson–Minton formula [20, Eq. (1.9.3)]: for any non-negative integers $m_1, \ldots, m_n$,

$$
{}_{n+1}F_n\left[\begin{matrix} -(m_1+\cdots+m_n) & b_1+m_1 & \ldots & b_n+m_n \\ & b_1 & \ldots & b_n \end{matrix} ; 1\right] = (-1)^{m_1+\cdots+m_n}\frac{(m_1+\cdots+m_n)!}{(b_1)_{m_1}\cdots(b_n)_{m_n}}. \tag{13}
$$

Note that when $n=2$, we can derive a different proof using a formula of Dixon. We present the proof below as it applies to all n.

Proof of Theorem 5. Let $n \geq 3$, and $p \equiv 1 \pmod{n}$ be prime, which has to be odd. Set $m = \frac{p-1}{n}$, and let y be any integer. Letting $b_1 = 1+yp$, $b_2 = \cdots = b_n = 1$, and $m_1 = \cdots = m_n = m$ in (13), we get that

$$
{}_{n+1}F_n\left[\begin{matrix} 1-p & 1+m+yp & 1+m & \cdots & 1+m \\ & 1+yp & 1 & \cdots & 1 \end{matrix} ; 1\right] = \frac{(-1)^{p-1}(p-1)!}{(1+yp)_m(m!)^{n-1}} = \frac{(p-1)!}{(1+yp)_m(m!)^{n-1}}. \tag{14}
$$

Now we compare the left-hand side with the right-hand side of Theorem 5. We observe that

$$
(1-p)_k \equiv (1)_k - p\sum_{i=1}^{k}\frac{(1)_k}{i} \pmod{p^2},
$$

$$
(1+yp)_k \equiv (1)_k + yp\sum_{i=1}^{k}\frac{(1)_k}{i} \pmod{p^2},
$$

and so

$$
\frac{(1-p)_k}{(1+yp)_k} \equiv 1-(1+y)\left[\sum_{i=1}^{k} i^{-1}\right]p \pmod{p^2}. \tag{15}
$$

Also,

$$
(1+m+yp)_k = \left(\left(1-\frac{1}{n}\right)+\left(\frac{1}{n}+y\right)p\right)_k \equiv
$$

$$\left(1-\frac{1}{n}\right)_k+\left(\frac{1}{n}+y\right)\left[\sum_{i=1}^{k}\frac{(1-\frac{1}{n})_k}{(i-\frac{1}{n})}\right]p \pmod{p^2},$$

$$(1+m)_k^{n-1}\equiv\left(1-\frac{1}{n}\right)_k^{n-1}+\left(\frac{n-1}{n}\right)\left(1-\frac{1}{n}\right)_k^{n-2}\left[\sum_{i=1}^{k}\frac{(1-\frac{1}{n})_k}{(i-\frac{1}{n})}\right]p \pmod{p^2},$$

and so

$$(1+m+yp)_k(1+m)_k^{n-1}\equiv$$
$$\left(1-\frac{1}{n}\right)_k^n\left(1+(1+y)\left[\sum_{i=1}^{k}\left(i-\frac{1}{n}\right)^{-1}\right]p\right) \pmod{p^2}. \qquad (16)$$

Thus (15) and (16) give that the left-hand side of (14) can be written modulo p^2 as

$$\sum_{k=0}^{p-1}\frac{(1-\frac{1}{n})_k^n}{(1)_k^{n-1}}\left(1+(1+y)\left[\sum_{i=1}^{k}\left(\left(i-\frac{1}{n}\right)^{-1}-i^{-1}\right)\right]p\right) \pmod{p^2}.$$

Using harmonic sums we conclude that there exists $A\in\mathbb{Z}_p$ such that

$$_{n+1}F_n\left[\begin{matrix}1-p & 1+m+yp & 1+m & \cdots & 1+m\\ & 1+yp & 1 & \cdots & 1\end{matrix};1\right]$$
$$\equiv {}_nF_{n-1}\left[\begin{matrix}\frac{n-1}{n} & \frac{n-1}{n} & \cdots & \frac{n-1}{n}\\ & 1 & \cdots & 1\end{matrix};1\right]_{p-1}\cdot(1-A(y+1)p) \pmod{p^2}. \qquad (17)$$

Using part c) of Proposition 11, we see that $a_0(\frac{1}{n})=\frac{1+(n-1)p}{n}=p-m$, and by part b),

$$\Gamma_p(-m)=1/m! \text{ and } \Gamma_p(-p)=-1/(p-1)!.$$

Also

$$\frac{1}{(1+yp)_m}=\frac{(-1)^m\Gamma_p(1+yp)}{\Gamma_p(1+yp+\frac{p-1}{n})}=\frac{-(-1)^m\Gamma_p(yp)}{\Gamma_p(1-\frac{1}{n}+(\frac{1}{n}+yp))}=$$
$$\Gamma_p\left(\frac{1}{n}-\left(\frac{1}{n}+y\right)p\right)\Gamma_p(yp).$$

We thus have

$$\frac{(p-1)!}{(1+yp)_m(m!)^{n-1}} = -\frac{\Gamma_p(\frac{1-p}{n})^{n-1}\Gamma_p(\frac{1}{n}-(\frac{1}{n}+y)p)\Gamma_p(yp)}{\Gamma_p(-p)}$$
$$\equiv -\Gamma_p\left(\frac{1}{n}\right)^n\left(1+\left(G_1(0)-G_1\left(\frac{1}{n}\right)\right)(1+y)p\right) \pmod{p^2}. \tag{18}$$

Finally, letting $y = -1$ in (17) and (18), we see the desired congruence modulo p^2. □

We now prove Theorem 6.

Proof of Theorem 6. Let p be a prime such that $p \equiv 1 \pmod 4$. We will use the following formula of Dougall (see Theorem 3.5.1 of [2]) which says that if $2a+1 = b+c+d+e-m$, then

$$\begin{aligned} &{}_7F_6\left[\begin{matrix} a & a/2+1 & b & c & d & e & -m \\ & a/2 & 1+a-b & 1+a-c & 1+a-d & 1+a-e & 1+a+m \end{matrix}; 1\right] \\ &= \frac{(1+a)_m(1+a-b-c)_m(1+a-b-d)_m(1+a-c-d)_m}{(1+a-b)_m(1+a-c)_m(1+a-d)_m(1+a-b-c-d)_m}. \end{aligned} \tag{19}$$

Letting $a = 1/4, b = 5/8, c = 1/8, d = (1+pu)/4, e = (1+(1-u)p)/4, m = (p-1)/4$, we have $2a+1 = b+c+d+e-m$ and thus we can use (19). We first observe that the left-hand side reduces to

$${}_4F_3\left[\begin{matrix} \frac{1}{4} & \frac{1+pu}{4} & \frac{1+(1-u)p}{4} & \frac{1-p}{4} \\ & 1-\frac{pu}{4} & 1+\frac{(u-1)p}{4} & 1+\frac{p}{4} \end{matrix}; 1\right]$$

after deleting three matching pairs of upper and lower parameters corresponding to $a/2+1$, b, and c. Writing the right-hand side in terms of Gamma functions, we thus get that

$$\begin{aligned} &{}_4F_3\left[\begin{matrix} \frac{1}{4} & \frac{1+pu}{4} & \frac{1+(1-u)p}{4} & \frac{1-p}{4} \\ & 1-\frac{pu}{4} & 1+\frac{(u-1)p}{4} & 1+\frac{p}{4} \end{matrix}; 1\right] \\ &= \frac{\Gamma(1+\frac{p}{4})\Gamma(\frac{1+p}{4})\Gamma(\frac{1}{8}+\frac{(1-u)p}{4})\Gamma(\frac{5}{8}+\frac{(1-u)p}{4})\Gamma(\frac{5}{8})\Gamma(\frac{9}{8})\Gamma(1-\frac{pu}{4})\Gamma(\frac{1-pu}{4})}{\Gamma(\frac{5}{4})\Gamma(\frac{1}{2})\Gamma(\frac{3}{8}-\frac{pu}{4})\Gamma(\frac{7}{8}-\frac{pu}{4})\Gamma(\frac{3}{8}+\frac{p}{4})\Gamma(\frac{7}{8}+\frac{p}{4})\Gamma(\frac{3+(1-u)p}{4})\Gamma(\frac{(1-u)p}{4})}. \end{aligned}$$

Using the duplication formula $\Gamma(z)\Gamma(z+\frac{1}{2}) = 2^{1-2z}\Gamma(\frac{1}{2})\Gamma(2z)$, we have

$$\frac{\Gamma(\frac{1}{8}+\frac{(1-u)p}{4})\Gamma(\frac{5}{8}+\frac{(1-u)p}{4})\Gamma(\frac{5}{8})\Gamma(\frac{9}{8})}{\Gamma(\frac{3}{8}-\frac{pu}{4})\Gamma(\frac{7}{8}-\frac{pu}{4})\Gamma(\frac{3}{8}+\frac{p}{4})\Gamma(\frac{7}{8}+\frac{p}{4})} = \frac{\Gamma(\frac{1}{4}+\frac{(1-u)p}{2})\Gamma(\frac{5}{4})}{\Gamma(\frac{3}{4}-\frac{pu}{2})\Gamma(\frac{3}{4}+\frac{p}{2})}.$$

Using Proposition 11, we can thus rewrite the right-hand side as

$$\frac{\Gamma_p(\frac{p}{4})\Gamma_p(\frac{1+p}{4})\Gamma_p(\frac{1}{4}+\frac{(1-u)p}{2})\Gamma_p(-\frac{pu}{4})\Gamma_p(\frac{1-pu}{4})}{\Gamma_p(\frac{1}{2})\Gamma_p(\frac{3}{4}-\frac{pu}{2})\Gamma_p(\frac{3}{4}+\frac{p}{2})\Gamma_p(\frac{3+(1-u)p}{4})\Gamma_p(\frac{(1-u)p}{4})}$$
$$\equiv (-1)^{\frac{p-1}{4}}\Gamma_p\left(\frac{1}{2}\right)\Gamma_p\left(\frac{1}{4}\right)^6\cdot\left(1-\frac{5(u^2-u+1)(G_1(\frac{1}{4})^2-G_2(\frac{1}{4}))}{16}p^2\right.$$
$$\left.+\frac{u(u-1)(G_1(0)^3-G_3(0)-21G_2(\frac{1}{4})G_1(\frac{1}{4})+7G_3(\frac{1}{4})+14G_1(\frac{1}{4})^3)}{128}p^3\right).$$

Meanwhile, we expand the left-hand side using harmonic sums as in Proposition 13. So there exist $a_{k,i}, b_{k,i} \in \mathbb{Z}_p$ such that modulo p^4 we have

$${}_4F_3\left[\begin{matrix}\frac{1}{4} & \frac{1+pu}{4} & \frac{1+(1-u)p}{4} & \frac{1-p}{4} \\ & 1-\frac{pu}{4} & 1+\frac{(u-1)p}{4} & 1+\frac{p}{4}\end{matrix};1\right]=$$
$$\sum_{k=0}^{\frac{p-1}{4}}\left(\frac{(\frac{1}{4})_k^4}{k!^4}\cdot\frac{(1+a_{k,1}\frac{pu}{4}+a_{k,2}(\frac{pu}{4})^2+a_{k,3}(\frac{pu}{4})^3)}{(1+b_{k,1}\frac{-pu}{4}+b_{k,2}(\frac{-pu}{4})^2+b_{k,3}(\frac{-pu}{4})^3)}\cdot\right.$$
$$\left.\frac{(1+a_{k,1}\frac{(1-u)p}{4}+a_{k,2}(\frac{(1-u)p}{4})^2+a_{k,3}(\frac{(1-u)p}{4})^3)(1+a_{k,1}\frac{-p}{4}+a_{k,2}(\frac{-p}{4})^2+a_{k,3}(\frac{-p}{4})^3)}{(1+b_{k,1}\frac{(u-1)p}{4}+b_{k,2}(\frac{(u-1)p}{4})^2+b_{k,3}(\frac{(u-1)p}{4})^3)(1+b_{k,1}\frac{p}{4}+b_{k,2}(\frac{p}{4})^2+b_{k,3}(\frac{p}{4})^3)}\right).$$

Note that if we collect coefficients of p, p^2, p^3 we get 0,

$$-(u^2-u+1)\sum_{k=0}^{\frac{p-1}{4}}\left(\frac{(\frac{1}{4})_k^4}{k!^4}\right)\frac{(a_{k,1}^2+2b_{k,2}-2a_{k,2}-b_{k,1}^2)}{16},$$

and

$$u(u-1)\sum_{k=0}^{\frac{p-1}{4}}\left(\frac{(\frac{1}{4})_k^4}{k!^4}\right)\frac{(3b_{k,3}+3a_{k,3}+b_{k,1}^3-3a_{k,1}a_{k,2}-3b_{k,1}b_{k,2}+a_{k,1}^3)}{64},$$

respectively.[4]

By collecting terms, we see there are $A_i, B_i \in \mathbb{Z}_p$ such that for all $u \in \mathbb{Z}_p$

[4]Comparing both sides, when we pick $u = -\zeta_3$ where ζ_3 be a primitive cubic root, we can derive that the claim of the theorem holds modulo p^3.

$$
{}_4F_3\left[\begin{matrix}\frac{1}{4} & \frac{1}{4} & \frac{1}{4} & \frac{1}{4} \\ & 1 & 1 & 1\end{matrix}\ ;\ 1\right]_{\frac{p-1}{4}} [1 + A_2(u^2 - u + 1)p^2 + A_3(u^2 - u)p^3]
$$

$$
\equiv (-1)^{\frac{p-1}{4}} \Gamma_p\left(\frac{1}{2}\right) \Gamma_p\left(\frac{1}{4}\right)^6 [1 + B_2(u^2 - u + 1)p^2 + B_3(u^2 - u)p^3] \pmod{p^4}.
$$

The fact that this congruence holds for all choices of $u \in \mathbb{Z}_p$ gives a lot of information by comparing coefficients. For example, from considering the cases when $u = 1$ and $u = -1$ we obtain by subtraction the two congruences

$$
{}_4F_3\left[\begin{matrix}\frac{1}{4} & \frac{1}{4} & \frac{1}{4} & \frac{1}{4} \\ & 1 & 1 & 1\end{matrix}\ ;\ 1\right]_{\frac{p-1}{4}} [1 + A_2p^2] \equiv
$$

$$
(-1)^{\frac{p-1}{4}} \Gamma_p\left(\frac{1}{2}\right) \Gamma_p\left(\frac{1}{4}\right)^6 [1 + B_2p^2] \pmod{p^4}, \tag{20}
$$

and

$$
{}_4F_3\left[\begin{matrix}\frac{1}{4} & \frac{1}{4} & \frac{1}{4} & \frac{1}{4} \\ & 1 & 1 & 1\end{matrix}\ ;\ 1\right]_{\frac{p-1}{4}} [A_2p^2 + A_3p^3] \equiv
$$

$$
(-1)^{\frac{p-1}{4}} \Gamma_p\left(\frac{1}{2}\right) \Gamma_p\left(\frac{1}{4}\right)^6 [B_2p^2 + B_3p^3] \pmod{p^4}. \tag{21}
$$

From (20), we see that to establish the claim of the theorem, it suffices to show that

$$
{}_4F_3\left[\begin{matrix}\frac{1}{4} & \frac{1}{4} & \frac{1}{4} & \frac{1}{4} \\ & 1 & 1 & 1\end{matrix}\ ;\ 1\right]_{\frac{p-1}{4}} \cdot A_2 \equiv (-1)^{\frac{p-1}{4}} \Gamma_p\left(\frac{1}{2}\right) \Gamma_p\left(\frac{1}{4}\right)^6 \cdot B_2 \pmod{p^2}. \tag{22}
$$

We can already see that (22) holds modulo p from considering (21) modulo p^3.

To see that we can further refine this equation to obtain the congruence modulo p^2, we argue in a similar way as above by setting up a series of congruences. We first let $a = (1 + xp)/4$, $b = a/2$, $c = (a + 1)/2$, $d = 1/4$, $e = (1 + (x + 1)p)/4$, $m = (1 - p)/4$ in (19). Then we can expand both sides of (19) as we have done above. Now by comparing the coefficients of $1, x, x^2$, and x^3, respectively, we obtain four congruences like (20) above, each modulo p^4. We then further let $a = (1-p)/4$, and since $p \equiv 1 \pmod 4$, we must have that one of $a/2, (a + 1)/2$ is a negative integer. We set this to be $-m$ and then choose b, c, d, e so that the upper parameters will be, up to permutation, a, $1+a/2$, $a/2$, $(a+1)/2$, $1/4$, $(1+(x-1)p)/4$, $(1-xp)/4$ in (19). This allows us get another three congruences modulo p^4 by comparing the coefficients of 1, x, x^2 on both sides of (19). The desired (22) then follows from these congruences. □

7 Remarks

7.1 Truncated Hypergeometric Series and Noncongruence Modular Forms

We first recall the following hypergeometric series transformation (see [2, (2.4.12)]):

$$ {}_3F_2\left[\begin{matrix} -m & a & b \\ & d & e \end{matrix} ; 1\right] = \frac{(e-a)_m}{(e)_m}\, {}_3F_2\left[\begin{matrix} -m & a & d-b \\ & d & a+1-m-e \end{matrix} ; 1\right], \tag{23} $$

which holds when $-m$ is a negative integer and both sides converge. Given an integer $n \geq 2$ and a prime $p \equiv 1 \pmod{n}$, letting first $m = \frac{p-1}{n}$, $a = \frac{n-1+p}{n}$, $b = \frac{1}{n}$, $d = e = 1$, and then $m = \frac{(n-1)(p-1)}{3}$, $a = \frac{1+(n-1)p}{n}$, $b = \frac{1}{n}$, $d = e = 1$, respectively, we derive the following supercongruence:

$$ (-1)^{\frac{p-1}{n}}\, {}_3F_2\left[\begin{matrix} \frac{n-1}{n} & \frac{n-1}{n} & \frac{1}{n} \\ & 1 & 1 \end{matrix} ; 1\right]_{p-1} \equiv {}_3F_2\left[\begin{matrix} \frac{1}{n} & \frac{1}{n} & \frac{n-1}{n} \\ & 1 & 1 \end{matrix} ; 1\right]_{p-1} \pmod{p^2}. \tag{24} $$

This was first observed and proved by McCarthy in a private communication via a different approach using the work of Mortenson [35] for the case of $n = 3$. Moreover, we would like to mention the following conjecture to demonstrate that truncated hypergeometric series arise in many different settings including the theory of noncongruence modular forms:

Conjecture 19. *For any integer $n > 1$ and prime $p \equiv 1 \pmod{n}$,*

$$ {}_3F_2\left[\begin{matrix} \frac{1}{n} & \frac{1}{n} & \frac{n-1}{n} \\ & 1 & 1 \end{matrix} ; 1\right]_{p-1} \equiv a_p(f_n(z)) \pmod{p^2}, $$

where $a_p(f_n(z))$ is the pth coefficient of $f_n(z) = \sqrt[n]{E_1(z)^{n-1}E_2(z)}$ when expanded in terms of the local uniformizer $e^{2\pi iz/5n}$, and $E_1(z)$ and $E_2(z)$ are two explicit level 5 weight 3 Eisenstein series with coefficients in $\mathbb{Z}$ (see (17) and (18) of [29] or [28, §3] for their expansions).

In a series of papers [4, 29, 30], the third author and her collaborators studied the properties of these functions $f_n(z)$ which are weight 3 cusp forms for some finite index subgroups of $SL_2(\mathbb{Z})$ that contain no principal congruence subgroups. For $n = 3, 4, 6$, the pth coefficients of $f_n(z)$ are shown to be related to the coefficients of classical Hecke or Hilbert modular forms.

In terms of Gaussian hypergeometric functions, we have when $p \equiv 1 \pmod{n}$ and η_n any order n character of $\mathbb{F}_p^\times$ then by work of Greene [21]

$$ {}_3F_2\left(\begin{matrix} \eta_n, \eta_n, \overline{\eta_n} \\ \varepsilon, \ \varepsilon \end{matrix}; 1\right)_p = \eta_n(-1)\, {}_3F_2\left(\begin{matrix} \eta_n, \overline{\eta_n}, \overline{\eta_n} \\ \varepsilon, \ \varepsilon \end{matrix}; 1\right)_p. \tag{25} $$

7.2 Other Observations

We conclude with some patterns observed from numerical data. For each prime $p \equiv 1 \pmod 5$, it appears that

$$ {}_5F_4\left[\begin{matrix} \frac{2}{5} & \frac{2}{5} & \frac{2}{5} & \frac{2}{5} & \frac{2}{5} \\ & 1 & 1 & 1 & 1 \end{matrix}; 1\right]_{p-1} \overset{?}{\equiv} -\Gamma_p\left(\frac{1}{5}\right)^5 \Gamma_p\left(\frac{2}{5}\right)^5 \pmod{p^5}. \tag{26} $$

Using the strategy of the proof of Theorem 5 and Dougall's formula (19), one can obtain (26) modulo p^4. By the Gross–Koblitz formula, the p-adic Gamma value agrees with $J(\eta_n^2, \eta_n^2)^3 J(\eta_n, \eta_n)$ when we choose the right order n character. Meanwhile, by Conjecture 3, we sense the presence of another Jacobi sum factor $-J(\eta_n, \eta_n)J(\eta_n, \eta_n^2)J(\eta_n, \eta_n^3)$. It will be interesting to know whether we can reconstruct the local zeta function of $C_{5,1}$ as we have done for $C_{3,1}$ and $C_{4,1}$. We leave this task to interested readers.

We conclude with a few more observations. Motivated by Lemma 1, we numerically observed the following supercongruences. Each corresponds to a supercongruence mentioned earlier.

1. For any prime $p \equiv 1 \pmod 3$,

$$ \sum_{k=0}^{p-1} \left(p\frac{k!}{(\frac{5}{3})_k}\right)^3 \equiv \sum_{k=\frac{2(p-1)}{3}}^{p-1} \left(p\frac{k!}{(\frac{5}{3})_k}\right)^3 \overset{?}{\equiv} \Gamma_p\left(\frac{1}{3}\right)^6 \pmod{p^3}. $$

2. For any prime $p \equiv 1 \pmod 4$,

$$ \sum_{k=0}^{p-1} \left(p\frac{k!}{(\frac{7}{4})_k}\right)^4 \equiv \sum_{k=\frac{3(p-1)}{4}}^{p-1} \left(p\frac{k!}{(\frac{7}{4})_k}\right)^4 \overset{?}{\equiv} (-1)^{\frac{p-1}{4}} \Gamma_p\left(\frac{1}{2}\right) \Gamma_p\left(\frac{1}{4}\right)^6 \pmod{p^4}. $$

3. For any prime $p \equiv 1 \pmod 5$,

$$ \sum_{k=0}^{p-1} \left(p\frac{k!}{(\frac{8}{5})_k}\right)^5 \equiv \sum_{k=\frac{3(p-1)}{5}}^{p-1} \left(p\frac{k!}{(\frac{8}{5})_k}\right)^5 \overset{?}{\equiv} -\Gamma_p\left(\frac{1}{5}\right)^5 \Gamma_p\left(\frac{2}{5}\right)^5 \pmod{p^5}. $$

4. For an integer $n > 2$, and any prime $p \equiv 1 \pmod n$,

$$\sum_{k=0}^{p-1}\left(p\frac{k!}{(\frac{1}{n}+1)_k}\right)^n \equiv \sum_{k=\frac{(p-1)}{n}}^{p-1}\left(p\frac{k!}{(\frac{1}{n}+1)_k}\right)^n \overset{?}{\equiv} -\Gamma_p\left(\frac{1}{n}\right)^n \pmod{p^3}.$$

Acknowledgements We warmly thank the Banff International Research Station (BIRS) and Women in Numbers 3 BIRS 2014 (14w5009) workshop for the opportunity to initiate this collaboration. Thanks to the National Center for Theoretical Sciences in Taiwan for supporting the travel of Fang-Ting Tu to visit Ling Long. Long was supported by NSF DMS1303292. We are indebted to Wadim Zudilin for his insightful suggestions and sharing his ideas. We also thank Jerome W. Hoffman for his interest and valuable comments; Rupam Barman, Jesús Guillera, and Ravi Ramakrishna for helpful discussions. Further thanks to the referees for their careful readings and helpful comments. We used `Magma` and `Sage` for our computations related to this project and `Sage Math Cloud` to collaborate.

References

1. Ahlgren, S., Ono, K.: A Gaussian hypergeometric series evaluation and Apéry number congruences. J. Reine Angew. Math. **518**, 187–212 (2000)
2. Andrews, G.E., Askey, R., Roy, R.: Special Functions. Encyclopedia of Mathematics and its Applications, vol. 71. Cambridge University Press, Cambridge (1999)
3. Archinard, N.: Hypergeometric abelian varieties. Can. J. Math. **55**(5), 897–932 (2003)
4. Atkin, A.O.L., Li, W.-C.W., Long, L.: On Atkin and Swinnerton-Dyer congruence relations. II. Math. Ann. **340**(2), 335–358 (2008)
5. Bailey, W.N.: Generalized Hypergeometric Series. Cambridge Tracts in Mathematics and Mathematical Physics, vol. 32, pp. pp. v+108. Stechert-Hafner, Inc., New York (1964)
6. Barman, R., Kalita, G.: Hypergeometric functions and a family of algebraic curves. Ramanujan J. **28**(2), 175–185 (2012)
7. Barman, R., Saikia, N., McCarthy, D.: Summation identities and special values of hypergeometric series in the p-adic setting. J. Number Theory **153**, 63–84 (2015)
8. Berndt, B.C., Evans, R.J., Williams, K.S.: Gauss and Jacobi Sums. Canadian Mathematical Society Series of Monographs and Advanced Texts. A Wiley-Interscience Publication. Wiley, New York (1998)
9. Beukers, F., Cohen, H., Mellit, A.: Finite hypergeometric functions. Mathematics - Number Theory, 11F46, 11T24 (2015). http://adsabs.harvard.edu/abs/2015arXiv150502900B. ArXiv e-prints: 1505.02900. Provided by the SAO/NASA Astrophysics Data System
10. Chan, K.K., Long, L., Zudilin, V.V.: A supercongruence motivated by the Legendre family of elliptic curves. Mat. Zametki **88**(4), 620–624 (2010)
11. Chisholm, S., Deines, A., Long, L., Nebe, G., Swisher, H.: $p-$adic analogues of Ramanujan type formulas for $1/\pi$. Mathematics **1**(1), 9–30 (2013)
12. Clifford, A.H.: Representations induced in an invariant subgroup. Ann. Math. (2) **38**(3), 533–550 (1937)
13. Cohen, H.: Number Theory: Volume II. Analytic and Modern Tools. Graduate Texts in Mathematics, vol. 240. Springer, New York (2007)
14. Coster, M.J., Van Hamme, L.: Supercongruences of Atkin and Swinnerton-Dyer type for Legendre polynomials. J. Number Theory **38**(3), 265–286 (1991)
15. Deines, A., Fuselier, J.G., Long, L., Swisher, H., Tu, F.T.: Generalized Legendre curves and quaternionic multiplication. J. Number Theory **161**, 175–203 (2016)
16. Dwork, B.: p-adic cycles. Publ. Math. l'IHÉS. **37**(1), 27–115 (1969)
17. Frechette, S., Ono, K., Papanikolas, M.: Gaussian hypergeometric functions and traces of Hecke operators. Int. Math. Res. Not. **60**, 3233–3262 (2004)

18. Fuselier, J.G.: Hypergeometric functions over $\mathbb{F}_p$ and relations to elliptic curves and modular forms. Proc. Am. Math. Soc., **138**(1), 109–123 (2010)
19. Fuselier, J.G., McCarthy, D.: Hypergeometric type identities in the p-adic setting and modular forms. Proc. Am. Math. Soc. **144**(4), 1493–1508 (2016). doi:10.1090/proc/12837
20. Gasper, G., Rahman, M.: Basic Hypergeometric Series, 2nd edn. Encyclopedia of Mathematics and its Applications, vol. 96. Cambridge University Press, Cambridge (2004) With a foreword by Richard Askey
21. Greene, J.: Hypergeometric functions over finite fields. Trans. Am. Math. Soc. **301**(1), 77–101 (1987)
22. Gross, B.H., Koblitz, N.: Gauss sums and the p-adic y-function. Ann. Math. 109(3), 569–581 (1979). doi:10.2307/1971226
23. Ireland, K., Rosen, M.: A Classical Introduction to Modern Number Theory, 2nd edn. Graduate Texts in Mathematics, vol. 84. Springer, New York (1990)
24. Katz, N.M.: Exponential Sums and Differential Equations, vol. 124. Princeton University Press, Princeton (1990)
25. Kilbourn, T.: An extension of the Apéry number supercongruence. Acta Arith. **123**(4), 335–348 (2006)
26. Koike, M.: Hypergeometric series over finite fields and Apéry numbers. Hiroshima Math. J. **22**(3), 461–467 (1992)
27. Lennon, C.: Gaussian hypergeometric evaluations of traces of Frobenius for elliptic curves. Proc. Am. Math. Soc. **139**(6), 1931–1938 (2011)
28. Li, W.-C.W., Long, L.: Atkin and Swinnerton-Dyer congruences and noncongruence modular forms. IMS Kôkyûroku Bessatsu **B51**, 269–299 (2014). arXiv:1303.6228
29. Li, W.-C.W., Long, L., Yang, Z.: On Atkin-Swinnerton-Dyer congruence relations. J. Number Theory **113**(1), 117–148 (2005)
30. Long, L.: On Atkin and Swinnerton-Dyer congruence relations. III. J. Number Theory **128**(8), 2413–2429 (2008)
31. Long, L., Ramakrishna, R.: Some supercongruences occurring in truncated hypergeometric series. Adv. Math. **290**, 773–808 (2016). doi:10.1016/j.aim.2015.11.043
32. McCarthy, D.: Transformations of well-poised hypergeometric functions over finite fields. Finite Fields Appl. **18**(6), 1133–1147 (2012)
33. McCarthy, D.: The trace of Frobenius of elliptic curves and the p-adic gamma function. Pac. J. Math. **261**(1), 219–236 (2013)
34. McCarthy, D., Papanikolas, M.A.: A finite field hypergeometric function associated to eigenvalues of a siegel eigenform. Int. J. Number Theory **11**(8), 2431–2450
35. Mortenson, E.: A supercongruence conjecture of Rodriguez-Villegas for a certain truncated hypergeometric function. J. Number Theory **99**(1), 139–147 (2003)
36. Ono, K.: Values of Gaussian hypergeometric series. Trans. Am. Math. Soc. **350**(3), 1205–1223 (1998)
37. Robert, A.M.: The Gross-Koblitz formula revisited. Rendiconti del Seminario Matematico della Università di Padova **105**, 157–170 (2001)
38. Slater, L.J.: Generalized Hypergeometric Functions, pp. xiii+273. Cambridge University Press, Cambridge (1966)
39. Vega, M.V.: Hypergeometric functions over finite fields and their relations to algebraic curves. Int. J. Number Theory **7**(8), 2171–2195 (2011)
40. Weil, A.: Jacobi sums as "Grössencharaktere". Trans. Am. Math. Soc. **73**, 487–495 (1952)
41. Whipple, F.J.W.: On well-poised series, generalized hypergeometric series having parameters in pairs, each pair with the same sum. Proc. Lond. Math. Soc. **s2–24**(1), 247–263 (1926)
42. Wiles, A.: Modular elliptic curves and Fermat's last theorem. Ann. Math. Sec., **141**(3), 443–551 (1995) DOI: 10.2307/2118559, http://dx.doi.org/10.2307/2118559
43. Wiles, A.: Modular elliptic curves and FermatâĂŹs last theorem. Ann. Math. 141(3), 443–551 (1995). doi:10.2307/2118559
44. Zudilin, W.: Ramanujan-type supercongruences. J. Number Theory **129**(8), 1848–1857 (2009)

A Generalization of S. Zhang's Local Gross–Zagier Formula for GL_2

Kathrin Maurischat

Abstract S. Zhang's local Gross–Zagier formulae for GL_2 can be interpreted as a fundamental lemma for some relative trace formulae. From this point of view we prove the existence of the corresponding local transfer. Further we construct universally defined geometric operators which realize the behavior of Hecke operators on the analytic side. We use them to give a proof of the local Gross–Zagier formula for GL_2. We work locally and throughout computationally.

Keywords Local Gross–Zagier formula • Relative trace formula • Hecke operators

2010 *Mathematics Subject Classification.* 11F70, 11G18, 11G50.

1 Introduction

The Gross–Zagier formula [4] is a relation between a Heegner point of discriminant D on the moduli space $X_0(N)$ and the Rankin–Selberg L-function attached to a newform f of weight 2 and level N and a character χ of the class group of $K = \mathbb{Q}(\sqrt{D})$, in case D is squarefree and prime to N:

$$L'(f, \chi, s = \frac{1}{2}) = const \cdot \hat{h}(c_{\chi,f}) \,.$$

Here $\hat{h}$ is the height function on $\mathrm{Jac}(X_0(N))$ and $c_{\chi,f}$ is a component of the Heegner class depending on χ and f. S. Zhang [12, 13] switches the point of view to a local one. CM-points are studied on the corresponding Shimura curve, modula r forms

K. Maurischat (✉)
Mathematisches Institut, Universität Heidelberg, Im Neuenheimer Feld 288, 69120 Heidelberg, Germany
e-mail: maurischat@mathi.uni-heidelberg.de

© Springer International Publishing Switzerland 2016

E.E. Eischen et al. (eds.), *Directions in Number Theory*, Association for Women in Mathematics Series 3, DOI 10.1007/978-3-319-30976-7_6

are automorphic representations. The height pairing of CM-cycles is replaced by a geometric pairing of Schwartz functions $\phi, \psi \in \mathcal{S}(\chi, G(\mathbb{A}_{\mathbb{F},f}))$,

$$< \phi, \psi > = \sum_{\gamma} m_\gamma < \phi, \psi >_\gamma . \tag{1}$$

Here G is an inner form of PGL_2, and T is the maximal subtorus given by the quadratic extension $\mathbb{K}/\mathbb{F}$ of the totally real field $\mathbb{F}$. The sum is over double cosets γ of $T\backslash G/T$. The multiplicities m_γ carry heavy arithmetic input. They are global data determined by intersection numbers. The coefficients $< \phi, \psi >_\gamma$ are adèlic integrals, given by their local components. Here a first parallel with a (relative) trace formula,

$$\sum_{\gamma} t_\gamma O_\gamma(\phi),$$

becomes visible. The sum is over double cosets γ again. The Tamagawa numbers t_γ are global data, and the orbital integrals $O_\gamma(\phi)$ are computed by their local factors. The local components of the coefficients above are (for nondegenerate γ) given by

$$< \phi, \psi >_\gamma = \int_{T\backslash G} \int_T \phi(t^{-1}\gamma t y)\, dt\, \bar{\psi}(y)\, dy . \tag{2}$$

These expressions are close to orbital integrals on G relative to T (as in Jacquet's work on relative trace formula [6]). They can even be read as orbital integrals for (γ, id) on $G \times G$ relative to $T \times G$, the action given by $(\gamma, \delta) \cdot (t, g) = (t^{-1}\gamma t, \delta g)$ (for $t \in T$, $g, \gamma, \delta \in G$), of the function $(\gamma, \delta) \mapsto \phi(\gamma\delta)\psi(\delta)$. As this is of no further use here, we call them *local linking numbers* according to their origin [12].

S. Zhang [12] invents a kernel function for the L-function which satisfies a functional equation similar to that for L. The local Fourier coefficients of the kernel are given by products of Whittaker newforms for the theta series $\Pi(\chi)$ and the Eisenstein series Π_E occurring in the Rankin convolution. They do not depend on the cusp form anymore. (See Sect. 2 for concrete definitions.) We generalize these products to get invariant linear forms on the isobaric sum $\Pi(\chi) \boxplus \Pi_E$ defined by evaluating functions in the Kirillov models,

$$(W_\chi, W_E) \mapsto W_\chi(\eta) W_E(\xi) ,$$

for $\xi, \eta = 1 - \xi \in F \setminus \{0, 1\}$, $W_\chi \in \mathcal{K}(\Pi(\chi))$, $W_E \in \mathcal{K}(\Pi_E)$. Let $\mathcal{W}$ be the space of distributions on $\Pi(\chi) \boxplus \Pi_E$ defined by these evaluations at ξ.

Let $\phi \in \mathcal{S}(\chi, G)$ be essentially the characteristic function of the maximal compact subgroup. Then the local Gross–Zagier formula for GL_2 ([12, Lemma 4.3.1], resp., Theorem 6.4 below) essentially states that for the newforms $W_{\chi,\mathrm{new}}$, $W_{E,\mathrm{new}}$ we have an equality

$$\mathbf{T}_b \left(W_{\chi,\mathrm{new}}(\eta) W_{E,\mathrm{new}}(\xi)\right) = |b|^{-1} |\xi\eta|^{\frac{1}{2}} < \tilde{\mathbf{T}}_b \phi, \phi >_{\gamma=\gamma(\xi)} .$$

Here $\mathbf{T}_b$ is a Hecke operator indexed by $b \in F^\times$, and $\tilde{\mathbf{T}}_b\phi$ is a special transform of ϕ. S. Zhang et al. prove local Gross–Zagier formulae with no level constraints [10] on more general Shimura curves [11].

In the language of trace formula, this is a fundamental lemma for the comparison of relative trace formulae. W. Zhang [14, 15] gives a general relative trace formula approach to the Gross–Zagier problem on unitary Shimura varieties. He formulates an arithmetic fundamental lemma in terms of unitary Rapoport–Zink spaces, proving it for small degrees. Gross–Zagier fits in the case of degree 2.

We discuss some local aspects in comparing relative trace formulae in the GL_2 case. In trace formula theory, it is a nontrivial problem to find enough local test vectors on each side at almost all places which can be compared. In the first part of the paper, we solve this problem in the case above, i.e., establish a transfer. We choose a parametrization of the double cosets $\gamma = \gamma(\xi)$ by the projective line, $\xi \in \mathbb{P}^1(F)$, and characterize the expansion of the local linking numbers with respect to this variable (Propositions 3.1, 3.2, and 3.5). This is very close to Jacquet's characterization of orbital integrals [6]. The space of distributions built by evaluating local linking numbers at ξ multiplied with the factor $|\xi\eta|^{\frac{1}{2}}$ will be denoted by $\tilde{\mathcal{L}}$. On the other hand, the expansion in ξ of the space $\mathcal{W}$ of distributions on $\Pi(\chi) \boxplus \Pi_E$ can be described by the theory of automorphic forms (Propositions 2.9 and 2.10). The transfer result is:

Theorem 1.1. *The spaces $\tilde{\mathcal{L}}$ and $\mathcal{W}$ have identical ξ-expansion.*

The second part of the paper is concerned with more quantitative aspects. We construct operators on the geometric side which realize the behavior of Hecke operators on the analytic side. The existence of such operators is not surprising but to have an explicit shape of them is appealing for several aspects. It provides a tool to produce more identities like the fundamental lemma out of given ones, i.e., it is a first step towards a general matching of orbital integrals. Moreover it makes the behavior around degenerate elements more visible. (Which in general forces a stabilization process.) Lastly we use these general geometric Hecke operators to rephrase S. Zhang's local Gross–Zagier formula for GL_2 and give a shorter proof.

On the analytic side, the Hecke operators $\mathbf{T}_b$ are essentially given by translations by $b \in F^\times$. In case of a split torus T they produce logarithmic and, in case of a quadratic character χ, even double logarithmic singularities as $b \to 0$,

$$\mathbf{T}_b\left(W_\chi(\eta)W_E(\xi)\right) = |b|^{-1}|\xi\eta|^{\frac{1}{2}}\chi_1(b\eta)(c_1 v(b\eta) + c_2)(c_3 v(b\xi) + c_4),$$

if $\chi_1^2 = 1$ (Proposition 5.1). The geometric Hecke operators are constructed in a simple manner to realize this pole behavior. The first natural guess is to translate the local linking numbers as well,

$$< \phi, \begin{pmatrix} b & 0 \\ 0 & 1 \end{pmatrix} \psi >_{\gamma(\xi)} .$$

These translations are studied in Sect. 4. It turns out (Theorems 4.1 and 4.3) that they do not suffice, as they do not produce the double logarithmic term $v(b)^2$. In Sect. 5 we construct an operator $\mathbf{S}_b$, which essentially is a weighted sum of translations by elements of valuations at most $v(b)$. This operator has good properties (Propositions 5.3 and 5.4), and we get:

Theorem 1.2. *The local linking numbers* $|b|^{-1}|\xi\eta|^{\frac{1}{2}}\mathbf{S}_b < \phi, \psi >_{\gamma(\xi)}$ *and the Whittaker products* $\mathbf{T}_b\left(W_\chi(\eta)W_E(\xi)\right)$ *have the same asymptotics in b.*

Accordingly, we formulate and prove the local Gross–Zagier formula in terms of $\mathbf{S}_b$ (Theorem 6.5).

Concerning concrete calculations, the case of a compact torus T is much easier than that of a noncompact one. This is due to the inner integral of the local linking numbers having compact support in the first case. In view of the noncompact torus we have to reduce ourselves to an arbitrary but fixed ξ to describe the asymptotics in the translation variable b. Anyway the calculations for the translation (Theorem 4.3) take about one hundred pages of $\wp$-adic integration in [9]. We sketch the outline of the proof in Sect. 4. Due to this difficulty, the results on Hecke operators are of asymptotic nature.

2 Terminology and Preparation

2.1 Geometry

The geometric setting is that of S. Zhang [12]

2.1.1 Global Data

Let $\mathbb{F}$ be a totally real algebraic number field and let $\mathbb{K}$ be an imaginary quadratic extension of $\mathbb{F}$. Further, let $\mathbb{D}$ be a division quaternion algebra over $\mathbb{F}$ which contains $\mathbb{K}$ and splits at the archimedean places. Let $\mathbf{G}$ denote the inner form of the projective group PGL_2 over $\mathbb{F}$ which is given by the multiplicative group $\mathbb{D}^\times$,

$$\mathbf{G}(\mathbb{F}) = \mathbb{F}^\times\backslash\mathbb{D}^\times.$$

Let $\mathbf{T}$ be the maximal torus of $\mathbf{G}$ given by $\mathbb{K}^\times$, i.e., $\mathbf{T}(\mathbb{F}) = \mathbb{F}^\times\backslash\mathbb{K}^\times$. Let $\mathbb{A}_\mathbb{F}$ (resp., $\mathbb{A}_\mathbb{K}$) be the adèles of $\mathbb{F}$ (resp., $\mathbb{K}$) and let $\mathbb{A}_{\mathbb{F},f}$ be the subset of finite adèles. On $\mathbf{T}(\mathbb{F})\backslash\mathbf{G}(\mathbb{A}_{\mathbb{F},f})$ there is an action of $\mathbf{T}(\mathbb{A}_{\mathbb{F},f})$ from the left and an action of $\mathbf{G}(\mathbb{A}_{\mathbb{F},f})$ from the right. The factor space $\mathbf{T}(\mathbb{F})\backslash\mathbf{G}(\mathbb{A}_{\mathbb{F},f})$ can be viewed as the set of CM-points of the Shimura variety defined by the inverse system of

$$\mathrm{Sh}_K := \mathbf{G}(\mathbb{F})^+\backslash\mathcal{H}_1^n \times \mathbf{G}(\mathbb{A}_{\mathbb{F},f})/K\,,$$

where K runs though sufficiently small compact open subgroups of $\mathbf{G}(\mathbb{A}_{\mathbb{F}f})$, $\mathcal{H}_1$ is the upper halfplane, and n is the number of the infinite places of $\mathbb{F}$. The CM-points are embedded in Sh_K by mapping the coset of $g \in \mathbf{G}(\mathbb{A}_{\mathbb{F}f})$ to the coset of (z, g), where $z \in \mathcal{H}_1^n$ is fixed by $\mathbf{T}$.

Let $\mathcal{S}(\mathbf{T}(\mathbb{F})\backslash\mathbf{G}(\mathbb{A}_{\mathbb{F}f}))$ be the Schwartz space, i.e., the space of complex valued functions on $\mathbf{T}(\mathbb{F})\backslash\mathbf{G}(\mathbb{A}_{\mathbb{F}f})$ which are locally constant and of compact support. A character of $\mathbf{T}$ is a character χ of $\mathbf{T}(\mathbb{F})\backslash\mathbf{T}(\mathbb{A}_{\mathbb{F}f})$, that is a character of $\mathbb{A}_{\mathbb{K}f}^\times/\mathbb{K}^\times$ trivial on $\mathbb{A}_{\mathbb{F}f}^\times/\mathbb{F}^\times$. Especially, $\chi = \prod \chi_v$ is the product of its local unitary components. We have

$$\mathcal{S}(\mathbf{T}(\mathbb{F})\backslash\mathbf{G}(\mathbb{A}_{\mathbb{F}f})) = \oplus_\chi \mathcal{S}(\chi, \mathbf{T}(\mathbb{F})\backslash\mathbf{G}(\mathbb{A}_{\mathbb{F}f})),$$

where $\mathcal{S}(\chi, \mathbf{T}(\mathbb{F})\backslash\mathbf{G}(\mathbb{A}_{\mathbb{F}f}))$ is the subspace of those functions ϕ transforming under $\mathbf{T}(\mathbb{A}_{\mathbb{F}f})$ by χ, i.e., for $t \in \mathbf{T}(\mathbb{A}_{\mathbb{F}f})$ and $g \in \mathbf{G}(\mathbb{A}_{\mathbb{F}f})$: $\phi(tg) = \chi(t)\phi(g)$. Any such summand is the product of its local components,

$$\mathcal{S}(\chi, \mathbf{T}(\mathbb{F})\backslash\mathbf{G}(\mathbb{A}_{\mathbb{F}f})) = \otimes_v \mathcal{S}(\chi_v, \mathbf{G}(\mathbb{A}_{\mathbb{F}_v})).$$

A pairing on $\mathcal{S}(\chi, \mathbf{T}(\mathbb{F})\backslash\mathbf{G}(\mathbb{A}_{\mathbb{F}f}))$ can be defined as follows. For functions ϕ, ψ in $\mathcal{S}(\chi, \mathbf{T}(\mathbb{F})\backslash\mathbf{G}(\mathbb{A}_{\mathbb{F}f}))$ and a double coset $[\gamma] \in \mathbf{T}(\mathbb{F})\backslash\mathbf{G}(\mathbb{F})/\mathbf{T}(\mathbb{F})$ define the **linking number**

$$< \phi, \psi >_\gamma := \int_{\mathbf{T}_\gamma(\mathbb{F})\backslash\mathbf{G}(\mathbb{A}_{\mathbb{F}f})} \phi(\gamma y)\bar{\psi}(y)\, dy, \tag{3}$$

where $\mathbf{T}_\gamma = \gamma^{-1}\mathbf{T}\gamma \cap \mathbf{T}$. For γ normalizing $\mathbf{T}$ we have $\mathbf{T}_\gamma = \mathbf{T}$. Otherwise $\mathbf{T}_\gamma$ is trivial. We call γ nondegenerate, if it belongs to the latter case. Here dy denotes the quotient measure of nontrivial Haar measures on $\mathbf{G}$ and $\mathbf{T}$. Further, let

$$m : \mathbf{T}(\mathbb{F})\backslash\mathbf{G}(\mathbb{F})/\mathbf{T}(\mathbb{F}) \to \mathbb{C}$$

be a multiplicity function. Then

$$< \phi, \psi > := \sum_{[\gamma]} m([\gamma]) < \phi, \psi >_\gamma$$

defines a sesquilinear pairing on $\mathcal{S}(\chi, \mathbf{T}(\mathbb{F})\backslash\mathbf{G}(\mathbb{A}_{\mathbb{F}f}))$. Determining the multiplicity function is an essential global problem, which was solved by S. Zhang for (local) Gross–Zagier in terms of intersection numbers. Concerning the parallels with trace formula, they take over the role of Tamagawa numbers. The coefficients $< \phi, \psi >_\gamma$ are the data linking global height pairings on curves and local approaches.

2.1.2 Local Data

In studying the local components of the linking numbers (3), we restrict to the nondegenerate case. First notice that

$$< \phi, \psi >_\gamma = \int_{\mathbf{T}(\mathbb{A}_{\mathbb{F},f})\backslash\mathbf{G}(\mathbb{A}_{\mathbb{F},f})} \int_{\mathbf{T}(\mathbb{A}_{\mathbb{F},f})} \phi(t^{-1}\gamma t y)\, dt\, \bar{\psi}(y)\, dy. \tag{4}$$

Assume $\phi = \prod_v \phi_v$ and $\psi = \prod_v \psi_v$ are products of local components. Then

$$\int_{\mathbf{T}(\mathbb{A}_{\mathbb{F},f})} \phi(t^{-1}\gamma t y)\, dt \;=\; \prod_v \int_{\mathbf{T}(F_v)} \phi_v(t_v^{-1}\gamma_v t_v y_v)\, dt_v$$

as well as $< \phi, \psi >_\gamma = \prod_v < \phi, \psi >_{\gamma,v}$, where

$$< \phi, \psi >_{\gamma,v} := \int_{\mathbf{T}(\mathbb{F}_v)\backslash\mathbf{G}(\mathbb{F}_v)} \int_{\mathbf{T}(\mathbb{F}_v)} \phi_v(t_v^{-1}\gamma_v t_v y_v)\, dt_v\, \bar{\psi}_v(y_v)\, dy_v. \tag{5}$$

Observe that $< \phi, \psi >_{\gamma,v}$ depends on the choice γ while $< \phi, \psi >_\gamma$ does not. An appropriate local definition is given below (Definition 2.1).

As all the following is local, we simplify notation: Let F denote a localization of $\mathbb{F}$ at a finite place not dividing 2. Let K be the quadratic extension of F given by localizing $\mathbb{K}$. K is either a field, $K = F(\sqrt{A})$, or a split algebra $K = F \oplus F$. For $t \in K$, let $\bar{t}$ denote the Galois conjugate of t (resp., $\overline{(x,y)} = (y,x)$ in the split case). The local ring of F (resp., K) is $\mathbf{o}_F$ (resp., $\mathbf{o}_K$). It contains the maximal ideal $\wp_F$ (resp., $\wp_K$, where in the split case $\wp_K := \wp_F \oplus \wp_F$). Let π_F be a uniformizer for $\mathbf{o}_F$. If it can't be mixed up, we write $\wp$ (resp., π) for $\wp_F$ (resp., π_F). The residue class field of F has characteristic p and q elements. Further, let ω be the quadratic character of $F^\times$ given by the extension K/F, that is, $\omega(x) = -1$ if x is not in the image of the norm of K/F. Let $D := \mathbb{D}(F)$, $T := \mathbf{T}(F)$, and $G := \mathbf{G}(F)$. There exists $\epsilon \in D^\times$ such that for all $t \in K$ we have $\epsilon t = \bar{t}\epsilon$ and

$$D = K + \epsilon K.$$

Let $c := \epsilon^2 \in F^\times$. Let N denote the reduced norm on D. Restricted to K it equals the norm of K/F. For $\gamma_1, \gamma_2 \in K$ we have

$$\mathrm{N}(\gamma_1 + \epsilon\gamma_2) = \mathrm{N}(\gamma_1) - c\,\mathrm{N}(\gamma_2)\,.$$

D splits exactly in case $c \in \mathrm{N}(K^\times)$. We parametrize the double cosets $[\gamma] \in T\backslash G/T$ by the projective line:

Definition 2.1. Let $P : T\backslash G/T \to \mathbb{P}^1(F)$ be given by

$$P(\gamma_1 + \epsilon\gamma_2) \;:=\; \frac{c\,\mathrm{N}(\gamma_2)}{\mathrm{N}(\gamma_1)}$$

for $\gamma_1 + \epsilon\gamma_2 \in D^\times$ as above.

This is well defined: $P(t(\gamma_1+\epsilon\gamma_2)t') = P(\gamma_1+\epsilon\gamma_2)$ for all $t, t' \in K^\times$. The non-empty fibers of P not belonging to 0 or ∞ are exactly the nondegenerate double cosets. In case that K/F is a field extension, P is injective with range $c\,\mathrm{N}(K^\times) \cup \{0, \infty\}$. In case K/F split, the range of P is $F^\times \backslash \{1\} \cup \{0, \infty\}$ and the fibers of $F^\times \backslash \{1\}$ are single double cosets [6]. This is one possible parametrization, another is $\xi := \frac{P}{P-1}$.

Lemma 2.2 ([12], Chap. 4). *Let $\gamma \in D^\times$. In the double coset $T\gamma T$ of G there exists one and only one T-conjugacy class of trace zero.*

Now the local components $< \phi, \psi >_\gamma$ of the linking numbers can be declared precisely:

Definition 2.3. Let $\phi, \psi \in \mathcal{S}(\chi, G)$. For $x \in F^\times$ define the **local linking number**

$$< \phi, \psi >_x := < \phi, \psi >_{\gamma(x)}$$

if there is a trace zero preimage $\gamma(x) \in D^\times$ of x under P. If there is no preimage, let $< \phi, \psi >_x := 0$. Thus, for $x \in c\mathrm{N} := c\,\mathrm{N}(K^\times)$

$$< \phi, \psi >_x = \int_{T\backslash G} \int_T \phi(t^{-1}\gamma(x)ty)\, dt\, \bar{\psi}(y)\, dy.$$

By unimodularity of the Haar measure on T, this definition is independent of the choice of the element $\gamma(x)$ of trace zero. In all the following we make a general natural assumption on the character χ:

Hypothesis 2.4. *The conductors of χ and ω are coprime.*

The conductor $f(\chi) < \mathbf{o}_K$ of χ may be viewed as an ideal of $\mathbf{o}_F$: If $K = F \oplus F$, then $\chi = (\chi_1, \chi_1^{-1})$ for a character χ_1 of $F^\times$ and $f(\chi) = f(\chi_1)$. If K/F is a ramified field extension, then χ is unramified, thus $f(\chi) \cap \mathbf{o}_F = \mathbf{o}_F$. If K/F is an unramified field extension, then $f(\chi) = \pi^{c(\chi)}\mathbf{o}_K$, where π is a uniformizing element for K as well as F. That is, $f(\chi) \cap \mathbf{o}_F = \pi^{c(\chi)}\mathbf{o}_F$. There are some simple properties of χ following from the Hypothesis 2.4.

Lemma 2.5. *Assume 2.4. The following are equivalent:*

(a) χ is quadratic.
(b) χ factorizes via the norm.

Corollary 2.6. *Assume 2.4. If K/F is a ramified field extension, then χ is a quadratic character. If K/F is an unramified field extension and χ is unramified, then $\chi = 1$.*

We use compatible Haar measures: Let da be a nontrivial additive Haar measure on F. Then the measure $d^\times a$ of the multiplicative group $F^\times$ is normalized by

$$\mathrm{vol}^\times(\mathbf{o}_F^\times) = (1 - q^{-1})\,\mathrm{vol}(\mathbf{o}_F^\times)\,,$$

where vol (resp., $\mathrm{vol}^\times$) is the volume associated with da (resp., $d^\times a$). The measure on $T\backslash G$ is the quotient measure induced of those on G and T.

2.2 Automorphic Forms

The central object on the analytic side is the Rankin–Selberg convolution of two automorphic representations. Gross–Zagier formulae describe the central order of its L-function.

Let Π_1 be a cuspidal representation of $\mathrm{GL}_2(\mathbb{A}_\mathbb{F})$ with trivial central character (i.e., an irreducible component of the discrete spectrum of the right translation on $L^2(\mathrm{GL}_2(\mathbb{F})\backslash \mathrm{GL}_2(\mathbb{A}_\mathbb{F})), 1))$ and conductor N. Further, let $\Pi(\chi)$ be the irreducible component belonging to χ of the Weil representation of $\mathrm{GL}_2(\mathbb{A}_\mathbb{F})$ for the norm form of $\mathbb{K}/\mathbb{F}$ (e.g., [2, Sect. 7]). It has conductor $f(\chi)^2 f(\omega)$ and central character ω. The Rankin–Selberg convolution of Π_1 and $\Pi(\chi)$ produces [5] the Mellin transform

$$\Psi(s, W_1, W_2, \Phi) = \int_{Z(F)N(F)\backslash \mathrm{GL}_2(F)} W_1(g)W_2(eg)f_\Phi(s, \omega, g)\, dg$$

for Whittaker functions W_1 of Π_1 (resp., W_2 of $\Pi(\chi)$) for an arbitrary nontrivial character of F. Here $e := \begin{pmatrix} -1 & 0 \\ 0 & 1 \end{pmatrix}$. For a function $\Phi \in \mathcal{S}(F^2)$ let

$$f_\Phi(s, \omega, g) = |\det g|^s \int_{F^\times} \Phi\left((0, t)g\right) |t|^{2s} \omega(t)\, d^\times t\,.$$

f_Φ belongs to the principal series $\Pi(|\cdot|^{s-\frac{1}{2}}, \omega|\cdot|^{\frac{1}{2}-s})$. There is an adèlic analogue. Analytical continuation of Ψ leads to the L-function, the greatest common divisor of all Ψ. It is defined by newforms ϕ for Π_1 and θ_χ of $\Pi(\chi)$ as well as a special form E of $\Pi_E := \Pi(|\cdot|^{s-\frac{1}{2}}, \omega|\cdot|^{\frac{1}{2}-s})$:

$$\begin{aligned} L(s, \Pi_1 \times \Pi(\chi)) &= \int_{Z(\mathbb{A}_\mathbb{F})\,\mathrm{GL}_2(\mathbb{F})\backslash \mathrm{GL}_2(\mathbb{A}_\mathbb{F})} \phi(g)\theta_\chi(g)E(s, g)\, dg \\ &= \int_{Z(\mathbb{A}_\mathbb{F})\,\mathrm{GL}_2(\mathbb{F})\backslash \mathrm{GL}_2(\mathbb{A}_\mathbb{F})} W_\phi(g)W_{\theta_\chi}(g)f_E(s, \omega, g)\, dg, \end{aligned}$$

where W_ϕ, etc., denotes the associated Whittaker function. For places where $c(\chi)^2 c(\omega) \leq v(N)$, the form E (resp., W_E) is the newform of the Eisenstein series. As Π_1 and $\Pi(\chi)$ are self-dual, the functional equation is

$$L(s, \Pi_1 \times \Pi(\chi)) \;=\; \epsilon(s, \Pi_1 \times \Pi(\chi))L(1 - s, \Pi_1 \times \Pi(\chi))\,.$$

In [12, Chap. 1.4] an integral kernel $\Xi(s, g)$ is constructed which has a functional equation analogous to that of L and for which

$$L(s, \Pi_1 \times \Pi(\chi)) = \int_{Z(\mathbb{A}_\mathbb{F})\,\mathrm{GL}_2(\mathbb{F})\backslash \mathrm{GL}_2(\mathbb{A}_\mathbb{F})} \phi(g)\Xi(s, g)\, dg.$$

We do not report the construction of this kernel, but we remark that the kernel depends on the newform of the theta series $\Pi(\chi)$ as well as the Eisenstein series Π_E, but not on the choice of Π_1. Its local nonconstant Fourier coefficients are defined by

$$W(s,\xi,\eta,g) := W_\theta(\begin{pmatrix}\eta & 0\\ 0 & 1\end{pmatrix} g)W_E(s,\begin{pmatrix}\xi & 0\\ 0 & 1\end{pmatrix} g). \tag{6}$$

Here $\eta := 1-\xi$. These Fourier coefficients are exactly those analytic functions which are compared to special local linking numbers in the local Gross–Zagier formula [12, Lemma 4.3.1]. We get rid of the restriction to newforms in (6) by reading it in the Kirillov models of the representations: Starting from the Whittaker model $\mathcal{W}(\Pi,\psi)$ of an irreducible admissible representation Π for an additive character ψ, the Kirillov space $\mathcal{K}(\Pi)$ is given by

$$\begin{aligned}\mathcal{W}(\Pi,\psi) &\to \mathcal{K}(\Pi),\\ W &\mapsto k : (a \mapsto W\begin{pmatrix}a & 0\\ 0 & 1\end{pmatrix}).\end{aligned}$$

Proposition 2.7 ([3], I.36). *Let Π be an infinite dimensional irreducible admissible representation of $\mathrm{GL}_2(F)$. The Kirillov space $\mathcal{K}(\Pi)$ is generated by the Schwartz space $\mathcal{S}(F^\times)$ along with the following stalks around zero:*

(a) If Π is supercuspidal, this stalk is zero.
(b) If $\Pi = \Pi(\mu_1,\mu_2)$ is a principle series representation, then it is given by representatives of the form

- $\left(|a|^{\frac{1}{2}}c_1\mu_1(a) + |a|^{\frac{1}{2}}c_2\mu_2(a)\right)\mathbf{1}_{\wp^n}(a)$, *if* $\mu_1 \neq \mu_2$,
- $|a|^{\frac{1}{2}}\mu_1(a)\,(c_1 + c_2 v(x))\,\mathbf{1}_{\wp^n}(a)$, *if* $\mu_1 = \mu_2$.

Here $c_1, c_2 \in \mathbb{C}$.
(c) If $\Pi = \Pi(\mu_1,\mu_2)$ is special, it is given by representatives

- $|a|^{\frac{1}{2}}\mu_1(a)\mathbf{1}_{\wp^n}(a)$, *if* $\mu_1\mu_2^{-1} = |\cdot|$,
- $|a|^{\frac{1}{2}}\mu_2(a)\mathbf{1}_{\wp^n}(a)$, *if* $\mu_1\mu_2^{-1} = |\cdot|^{-1}$.

Definition 2.8. Let $\Pi(\chi)$ be the theta series and Π_E be the Eisenstein series at the central place $s = \frac{1}{2}$. The products

$$W(\xi,\eta) = W_\theta(\eta)W_E(\xi)$$

of Kirillov functions $W_\theta \in \mathcal{K}(\Pi(\chi))$ and $W_E \in \mathcal{K}(\Pi(1,\omega))$ are called **Whittaker products**. As $\eta = 1-\xi$, they define linear forms on the isobaric sum $\Pi(\chi) \boxplus \Pi_E$. We denote the corresponding space of distributions by $\mathcal{W}$.

Being a component of a Weil representation, the theta series $\Pi(\chi)$ is completely described ([7, Sect. 1], [2, Sect. 7]). Adèlically, it is a Hilbert modular form of conductor $f(\chi)^2 f(\omega)$ and of weight $(1,\dots,1)$ at the infinite places. If $K = F \oplus F$ is split, then $\chi = (\chi_1, \chi_1^{-1})$ and $\Pi(\chi) = \Pi(\chi_1, \omega\chi_1^{-1}) = \Pi(\chi_1, \chi_1^{-1})$ is a principle series representation. If K/F is a field extension and χ does not factorize via the norm, then $\Pi(\chi)$ is supercuspidal. While if $\chi = \chi_1 \circ \mathrm{N}$, it is the principle series representation $\Pi(\chi_1, \chi_1^{-1}\omega) = \Pi(\chi_1, \chi_1\omega)$, as $\chi_1^2 = 1$ by Lemma 2.5. Thus, by Proposition 2.7:

Proposition 2.9. *Let $\Pi(\chi)$ be the theta series and let $\mathcal{K}(\Pi(\chi))$ be its Kirillov space. It is a function space in one variable η generated by $\mathcal{S}(F^\times)$ along with the following stalks around zero:*

- *The zero function, if K/F is a field extension and $\chi \neq 1$.*
- *$|\eta|^{\frac{1}{2}}\chi_1(\eta)\,(a_1 + a_2\omega(\eta))$, if K/F is a field extension and $\chi^2 = 1$.*
- *$|\eta|^{\frac{1}{2}}\left(a_1\chi_1(\eta) + a_2\chi_1^{-1}(\eta)\right)$, if K/F is split and $\chi_1^2 \neq 1$,*
- *$|\eta|^{\frac{1}{2}}\chi_1(\eta)\,(a_1 + a_2 v(\eta))$, if K/F is split and $\chi_1^2 = 1$.*

We collect some properties of principal series. For an automorphic form $f \in \Pi(\mu_1|\cdot|^{s-\frac{1}{2}}, \mu_2|\cdot|^{\frac{1}{2}-s})$ there is $\Phi \in \mathcal{S}(F^2)$ such that

$$f(s,g) = \mu_1(\det g)|\det g|^s \int_{F^\times} \Phi\left((0,t)g\right)(\mu_1\mu_2^{-1})(t)|t|^{2s}\, d^\times t. \tag{7}$$

Conversely, any $\Phi \in \mathcal{S}(F^2)$ defines a form $f_\Phi \in \Pi(|\cdot|^{s-\frac{1}{2}}, \omega|\cdot|^{\frac{1}{2}-s})$ in that way [1, Chap. 3.7]. The Whittaker function belonging to f (in a Whittaker model with unramified character ψ) is given by the first Fourier coefficient

$$W_f(s,g,\psi) = \int_F f(s, \begin{pmatrix} 0 & -1 \\ 1 & 0 \end{pmatrix} \begin{pmatrix} 1 & x \\ 0 & 1 \end{pmatrix} g)\bar{\psi}(x)\, dx.$$

Read off in the Kirillov model, the form for $s = \frac{1}{2}$ is given by evaluation at $g = \begin{pmatrix} a & 0 \\ 0 & 1 \end{pmatrix}$, thus

$$W_f(a) := W_f(\frac{1}{2}, \begin{pmatrix} a & 0 \\ 0 & 1 \end{pmatrix}, \psi).$$

For unramified μ_i the newform is obtained by choosing

$$\phi(x,y) = \mathbf{1}_{\mathfrak{o}_F}(x)\mathbf{1}_{\mathfrak{o}_F}(y)$$

in (7). Thus,

$$\begin{aligned} W_{\text{new}}(a) &= \mu_1(a)|a|^{\frac{1}{2}} \int_F \int_{F^\times} \mathbf{1}_{\mathbf{o}_F}(at)\mathbf{1}_{\mathbf{o}_F}(xt)\mu_1\mu_2^{-1}(t)|t|\, d^\times t\, \bar{\psi}(x)\, dx \\ &= \mu_1(a)|a|^{\frac{1}{2}} \mathbf{1}_{\mathbf{o}_F}(a) \operatorname{vol}(\mathbf{o}_F) \operatorname{vol}^\times(\mathbf{o}_F^\times) \sum_{j=-v(a)}^{0} \mu_1\mu_2^{-1}(\pi^j) \\ &= |a|^{\frac{1}{2}} \mathbf{1}_{\mathbf{o}_F}(a) \operatorname{vol}(\mathbf{o}_F) \operatorname{vol}^\times(\mathbf{o}_F^\times) \begin{cases} \frac{\mu_1(a\pi)-\mu_2(a\pi)}{\mu_1(\pi)-\mu_2(\pi)}, \text{ if } \mu_1 \neq \mu_2 \\ \mu_1(a)(v(a)+1), \text{ if } \mu_1 = \mu_2 \end{cases}. \end{aligned} \tag{8}$$

By Proposition 2.7 we have:

Proposition 2.10. *At $s = \frac{1}{2}$ the Eisenstein series Π_E is the principle series representation $\Pi(1, \omega)$. Its Kirillov space as a function space in the variable ξ is generated by $\mathcal{S}(F^\times)$ along with the following stalks around zero:*

- $|\xi|^{\frac{1}{2}}\,(a_1 + a_2\omega(\xi))$, *if K/F is a field extension,*
- $|\xi|^{\frac{1}{2}}\,(a_1 + a_2 v(\xi))$, *if K/F is split.*

For a finite set S of places of $\mathbb{F}$, let $\hat{\mathbf{o}}_{\mathbb{F}}^S := \prod_{v\notin S} \mathbf{o}_{\mathbb{F}_v}$ and $\mathbb{A}_S := \prod_{v\in S} \mathbb{F}_v \cdot \hat{\mathbf{o}}_{\mathbb{F}}^S$. We recall a property of Hecke operators.

Proposition 2.11 ([12], Chap. 2.4). *Let μ be a character of $\mathbb{A}^\times/\mathbb{F}^\times$. Let $\phi \in L^2(\mathrm{GL}_2(\mathbb{F})\backslash \mathrm{GL}_2(\mathbb{A}), \mu)$, and let W_ϕ be the Whittaker function of ϕ in some Whittaker model. Let S be the finite set of infinite places and of those finite places v for which ϕ_v is not invariant under the maximal compact subgroup $\mathrm{GL}_2(\mathbf{o}_{\mathbb{F}_v})$. For $b \in \hat{\mathbf{o}}_{\mathbb{F}}^S \cap \mathbb{A}^\times$ define*

$$H(b) := \left\{g \in M_2(\hat{\mathbf{o}}_{\mathbb{F}}^S) \mid \det(g)\hat{\mathbf{o}}_{\mathbb{F}}^S = b\hat{\mathbf{o}}_{\mathbb{F}}^S\right\}.$$

Then the following Hecke operator $\mathbf{T}_b$ is well defined for $g \in \mathrm{GL}_2(\mathbb{A}_S)$:

$$\mathbf{T}_b W_\phi(g) := \int_{H(b)} W_\phi(gh)\, dh.$$

If $y \in \hat{\mathbf{o}}_{\mathbb{F}}^S$ and $(b, y_f) = 1$, then

$$\mathbf{T}_b W_\phi(g \begin{pmatrix} y & 0 \\ 0 & 1 \end{pmatrix}) = |b|^{-1} W_\phi(g \begin{pmatrix} yb & 0 \\ 0 & 1 \end{pmatrix}).$$

That is, the action of the Hecke operator $\mathbf{T}_b$ on some Whittaker product is essentially translation by b:

$$\mathbf{T}_b W(\xi, \eta) = |b|^{-2} W(b\xi, b\eta). \tag{9}$$

3 Expansion of Local Linking Numbers

Let $\mathcal{L}$ denote the space of distributions defined by the local linking numbers $<\phi,\psi>_x$ for $\phi,\psi \in \mathcal{S}(\chi,G)$. We describe the expansion in $x \in F^\times$ of $\mathcal{L}$. The characterizing properties are close to those satisfied by the orbital integrals of [6], and Propositions 3.1 and 3.2 are influenced by the methods there. We distinguish whether the torus T is compact or not.

Proposition 3.1. *Let $K = F(\sqrt{A})$ be a field extension and let ω be the associated quadratic character. Let $\phi,\psi \in \mathcal{S}(\chi,G)$. The local linking number $<\phi,\psi>_x$ is a function of $x \in F^\times$ with the following properties:*

(a) It is zero on the complement of $c\mathrm{N}$.
(b) It is zero on a neighborhood of $1 \in F^\times$.
(c) There is a locally constant function A_1 on a neighborhood U of 0 depending on ϕ such that for all $0 \neq x \in U$: $<\phi,\psi>_x = A_1(x)(1+\omega(cx))$.
(d) There is an open set V containing zero such that for all $x^{-1} \in V \cap c\mathrm{N}$

$$<\phi,\psi>_x = \delta(\chi^2=1)\chi_1(\frac{A}{c})\chi_1(x)\int_{T\backslash G}\phi(\epsilon y)\bar{\psi}(y)\,dy.$$

Here the character χ_1 of $F^\times$ is given by $\chi = \chi_1 \circ \mathrm{N}$ if $\chi^2 = 1$. Especially, the local linking number vanishes in a neighborhood of infinity if $\chi^2 \neq 1$.

Proposition 3.2. *Let $K = F \oplus F$ be a split algebra. Let $\chi = (\chi_1, \chi_1^{-1})$ and let $\phi,\psi \in \mathcal{S}(\chi,G)$. The local linking number $<\phi,\psi>_x$ is a function of $x \in F^\times$ with the following properties:*

(a) It is zero on a neighborhood of $1 \in F^\times$.
(b) It is locally constant on $F^\times$.
(c) There is an open set $U \ni 0$ and locally constant functions A_1, A_2 on U depending on ϕ and ψ such that for $0 \neq x \in U$: $<\phi,\psi>_x = A_1(x) + A_2(x)v(x)$.
(d) There is an open set V containing zero and locally constant functions B_1, B_2 on V depending on ϕ and ψ such that for $x^{-1} \in V$:

$$<\phi,\psi>_x = \begin{cases} \chi_1(x)(B_1(x^{-1}) + B_2(x^{-1})v(x)), & \text{if } \chi_1^2 = 1 \\ \chi_1(x)B_1(x^{-1}) + \chi_1^{-1}(x)B_2(x^{-1}), & \text{if } \chi_1^2 \neq 1 \end{cases}.$$

For $\chi_1^2 = 1$, the function B_2 is nonzero only if $\mathrm{id} \in \operatorname{supp}\phi(\operatorname{supp}\psi)^{-1}$.

We need two lemmas.

Lemma 3.3. *Let $\phi \in \mathcal{S}(\chi,G)$.*

(a) For each $y \in G$ there is an open set $V \ni y$ such that for all $g \in \operatorname{supp}(\phi)y^{-1}$ and all $\tilde{y} \in V$

$$\phi(g\tilde{y}) = \phi(gy).$$

(b) Let $C \subset G$ be compact. For each $g \in G$ there is an open set $U \ni g$ such that for all $\tilde{g} \in U$ and all $y \in TC$

$$\int_T \phi(t^{-1}\tilde{g}ty)\, dt = \int_T \phi(t^{-1}gty)\, dt. \tag{10}$$

Proof of Lemma 3.3. (a) It is enough to prove the statement for $y = \mathrm{id}$. As ϕ is locally constant, for every $g \in G$ there is an open set $U_g \ni \mathrm{id}$ with $\phi(gU_g) = \phi(g)$. Let $C \subset G$ be compact such that $\operatorname{supp}\phi = TC$. We cover C by finitely many gU_g and choose U to be the intersection of those U_g. Then $\phi(gU) = \phi(g)$ for all $g \in TC$.

(b) It is enough to prove the statement for $y \in C$ rather than $y \in TC$, as a factor $s \in T$ just changes the integral by a factor $\chi(s)$. By (a) there is an open set $V_y \ni y$ such that $\phi(t^{-1}gt\tilde{y}) = \phi(t^{-1}gty)$ for $\tilde{y} \in V_y$ and $t^{-1}gt \in \operatorname{supp}(\phi)y^{-1}$. Take finitely many $y \in C$ such that the V_y cover C. It is enough to find open sets $U_y \ni g$ for these y so that Eq. (10) is fulfilled. Then $\cap U_y$ is an open set such that (10) is satisfied for all $y \in TC$. Write $g = g_1 + \epsilon g_2$ and describe a neighborhood U_y of g by $k_1, k_2 > 0$ depending on y and the obstructions $|\tilde{g}_i - g_i| < k_i$, $i = 1, 2$, for $\tilde{g}$ lying in U_y. Write $t^{-1}\tilde{g}t = g_1 + \epsilon g_2 t\bar{t}^{-1} + (\tilde{g}_1 - g_1) + \epsilon(\tilde{g}_2 - g_2)t\bar{t}^{-1}$. As ϕ is locally constant, we may choose k_1, k_2 depending on y such that

$$\phi(t^{-1}\tilde{g}t) = \phi((g_1 + \epsilon g_2 t\bar{t}^{-1})y) = \phi(t^{-1}gty).$$

These constants are independent from t as $|(\tilde{g}_2 - g_2)t\bar{t}^{-1}| = |\tilde{g}_2 - g_2|$. □

Lemma 3.4. *Let $\phi \in \mathcal{S}(F \oplus F)$.*

(a) There are $A_1, A_2 \in \mathcal{S}(F)$ such that

$$\int_{F^\times} \phi(a^{-1}y, a)\, d^\times a = A_1(y) + A_2(y)v(y).$$

(b) Let η be a nontrivial (finite) character of $F^\times$. There are $B_1, B_2 \in \mathcal{S}(F)$ and $m \in \mathbb{Z}$ such that for $0 \neq y \in \wp^m$

$$\int_{F^\times} \phi(a^{-1}y, a)\eta(a)\, d^\times a = B_1(y) + B_2(y)\eta(y).$$

Proof of Lemma 3.4. (a) Any $\phi \in \mathcal{S}(F \oplus F)$ is a finite linear combination of the following elementary functions: $\mathbf{1}_{\wp^n}(a)\mathbf{1}_{\wp^n}(b)$, $\mathbf{1}_{x+\wp^n}(a)\mathbf{1}_{\wp^n}(b)$, $\mathbf{1}_{\wp^n}(a)\mathbf{1}_{z+\wp^n}(b)$, $\mathbf{1}_{x+\wp^n}(a)\mathbf{1}_{z+\wp^n}(b)$ for suitable $n \in \mathbb{Z}$ and $v(x), v(z) > n$. It is enough to prove the statement for these functions. We get

$$\int_{F^\times} \mathbf{1}_{\wp^n}(a^{-1}y)\mathbf{1}_{\wp^n}(a)\, d^\times a = \mathbf{1}_{\wp^{2n}}(y)v(y\pi^{-2n+1}) \operatorname{vol}^\times(\mathbf{o}_F^\times).$$

Thus, if $0 \in \operatorname{supp}\phi$, the integral has a pole at $y = 0$, otherwise it hasn't:

$$\int_{F^\times} \mathbf{1}_{x+\wp^n}(a^{-1}y)\mathbf{1}_{\wp^n}(a)\, d^\times a = \mathbf{1}_{\wp^{v(x)+n}}(y)\operatorname{vol}^\times(1+\wp^{n-v(x)}),$$

$$\int_{F^\times} \mathbf{1}_{\wp^n}(a^{-1}y)\mathbf{1}_{z+\wp^n}(a)\, d^\times a = \mathbf{1}_{\wp^{v(z)+n}}(y)\operatorname{vol}^\times(1+\wp^{n-v(z)})$$

and

$$\int_{F^\times} \mathbf{1}_{x+\wp^n}(a^{-1}y)\mathbf{1}_{z+\wp^n}(a)\, d^\times a = \mathbf{1}_{xz(1+\wp^m)}(y)\operatorname{vol}^\times(1+\wp^m),$$

where $m := n - \min\{v(x), v(z)\}$.

(b) Similar computations to those of (a). □

Proof of Proposition 3.1. (a) is clear by definition.

(b) Assume $1 \in c\mathrm{N}$, otherwise this property is trivial. We have to show that for all γ with $P(\gamma) \in U$, where U is a sufficiently small neighborhood of 1,

$$\int_{T\backslash G}\int_T \phi(t^{-1}\gamma t y)\, dt \bar{\psi}(y)\, dy = 0.$$

We show that the inner integral is zero. Let $C \subset G$ be compact such that $\operatorname{supp}\phi \subset TC$. Then ϕ vanishes outside of TCT. It is enough to show that there is $k > 0$ such that $|P(\gamma)-1| > k$ holds for all $\gamma \in TCT$. Assume there isn't such k. Let $(\gamma_i)_i$ be a sequence such that $P(\gamma_i)$ tends to 1. Multiplying by elements of T and enlarging C occasionally (this is possible as T is compact), we assume $\gamma_i = 1 + \epsilon t_i = z_i c_i$, where $t_i \in T$, $c_i \in C$, $z_i \in Z$. Then $P(\gamma_i) = ct_i\bar{t}_i = 1 + a_i$, where $a_i \to 0$. We have $\det\gamma_i = 1 - ct_i\bar{t}_i = -a_i$ as well as $\det\gamma_i = z_i^2\det c_i$. As C is compact, $(z_i)_i$ is forced to tend to zero. This implies $\gamma_i \to 0$ contradicting $\gamma_i = 1 + \epsilon t_i$.

(c) A coset $\gamma \in F^\times\backslash D^\times$ of trace zero has a representative of the form $\gamma = \sqrt{A}+\epsilon\gamma_2$ (by abuse of notation). Thus,

$$<\phi,\psi>_x = \int_{T\backslash G}\int_T \phi((\sqrt{A}+\epsilon\gamma_2 t\bar{t}^{-1})y)\, dt\, \bar{\psi}(y)\, dy.$$

As $\phi \in \mathcal{S}(\chi, G)$, there exists an ideal $\wp_K^m$ of K such that for all $y \in G$ and all $l \in \wp_K^m$ one has $\phi((\sqrt{A}+\epsilon l)y) = \phi(\sqrt{A}y)$. Let $x = P(\gamma)$ be near zero, i.e., x belongs to an ideal U of F given by the obstruction that $\frac{cl\bar{l}}{-A} \in U$ implies $l \in \wp_K^m$. For such x we have

$$<\phi,\psi>_x = \operatorname{vol}_T(T)\chi(\sqrt{A})\int_{T\backslash G} \phi(y)\bar{\psi}(y)\, dy.$$

So if $\langle\cdot,\cdot\rangle$ denotes the L^2-scalar product, we have

$$<\phi,\psi>_x = \frac{1}{2}\,\mathrm{vol}_T(T)\chi(\sqrt{A})\langle\phi,\psi\rangle(1+\omega(cx)).$$

(d) Let $\gamma = \sqrt{A} + \epsilon\gamma_2$ denote a trace zero preimage of x under P. Then

$$\int_T \phi(t^{-1}\gamma t y)\,dt \;=\; \chi(\gamma_2)\int_T \phi((\sqrt{A}\gamma_2^{-1} + t^{-1}\bar{t}\epsilon)y)\,dt.$$

By Lemma 3.3 there exists $k>0$ such that for $|\gamma_2|>k$ and for $y\in\operatorname{supp}\psi$ we have $\phi((\sqrt{A}\gamma_2^{-1}+t^{-1}\bar{t}\epsilon)y) = \phi(t^{-1}\bar{t}\epsilon y)$. Thus, for $|x| > |cA^{-1}|k^2$,

$$<\phi,\psi>_x = \chi(\gamma_2)\int_T \chi(t^{-1}\bar{t})\,dt\int_{T\backslash G}\phi(\epsilon y)\bar{\psi}(y)\,dy.$$

As $\chi(t^{-1}\bar{t})$ defines the trivial character of T if and only if $\chi^2=1$, the statement follows by noticing that in this case $\chi(\gamma_2) = \chi_1(\frac{Ax}{c})$. □

Proof of Proposition 3.2. There is an isomorphism from $D^\times$ to $\mathrm{GL}_2(F)$ given by embedding $K^\times$ diagonally and sending ϵ to $\begin{pmatrix}0&1\\1&0\end{pmatrix}$. Then P is given by

$$P\begin{pmatrix}a&b\\c&d\end{pmatrix} = \frac{bc}{ad}.$$

The only value not contained in the image of P is 1. A preimage of $x\neq 1$ of trace zero is

$$\gamma(x) \;=\; \begin{pmatrix}-1&x\\-1&1\end{pmatrix}.$$

(a) We show that for $\phi\in\mathcal{S}(\chi,G)$ there is a constant $k>0$ such that for all $\gamma\in\operatorname{supp}\phi$: $|P(\gamma)-1|>k$. By Bruhat–Tits decomposition, $G=\mathrm{PGL}_2(F) = TNN'\cup TNwN$, where N is the group of unipotent upper triangular matrices, N' its transpose, and $w=\begin{pmatrix}0&-1\\1&0\end{pmatrix}$. Thus, there is $c>0$ such that

$$\operatorname{supp}\phi\subset T\left\{\begin{pmatrix}1&u\\0&1\end{pmatrix}\begin{pmatrix}1&0\\v&1\end{pmatrix}\;\middle|\;|u|<c,|v|<c\right\}$$
$$\bigcup T\left\{\begin{pmatrix}1&u\\0&1\end{pmatrix}w\begin{pmatrix}1&v\\0&1\end{pmatrix}\;\middle|\;|u|<c,|v|<c\right\}.$$

On the first set P we have $P = \frac{uv}{1+uv}$. On the second one we have $P = \frac{uv-1}{uv}$. Thus, for all $\gamma \in \operatorname{supp}\phi$ we have $|P(\gamma) - 1| \geq \min\{1, c^{-2}\}$. Next we show that there is a constant $k > 0$ such that $|P(\gamma) - 1| > k$ for all $\gamma y \in \operatorname{supp}\phi$ for all $y \in \operatorname{supp}\psi$. This implies that $< \phi, \psi >_x = 0$ in the neighborhood $|x - 1| < k$ of 1. There is such a constant k_y for any $y \in \operatorname{supp}\psi$. By Lemma 3.3(a) this constant is valid for all $\tilde{y}$ in a neighborhood V_y. Modulo T the support of ψ can be covered by finitely many V_y. The minimum of the associated k_y is the global constant we claimed.

(b) By Lemma 3.3(b), there is for every $x \in F^\times \backslash \{1\}$ a neighborhood U_x such that for all $y \in \operatorname{supp}\psi$ the inner integral

$$\int_T \phi(t^{-1}\gamma(\tilde{x})ty)\, dt$$

is constant in $\tilde{x} \in U_x$. Even more the local linking number is locally constant on $F^\times \backslash \{1\}$. By (a) it is locally constant in $x = 1$ as well.

For (c) and (d) we regard the inner integral separately first. For representatives we have

$$\begin{aligned} t^{-1}\gamma(x)t &= \begin{pmatrix} a^{-1} & 0 \\ 0 & 1 \end{pmatrix} \begin{pmatrix} -1 & x \\ -1 & 1 \end{pmatrix} \begin{pmatrix} a & 0 \\ 0 & 1 \end{pmatrix} \\ &= \begin{pmatrix} (x-1) & 0 \\ 0 & 1 \end{pmatrix} \begin{pmatrix} 1 & \frac{x}{a(x-1)} \\ 0 & 1 \end{pmatrix} \begin{pmatrix} 1 & 0 \\ -a & 1 \end{pmatrix} \in K^\times NN' \\ &= \begin{pmatrix} \frac{1-x}{a} & 0 \\ 0 & -a \end{pmatrix} \begin{pmatrix} 1 & \frac{a}{x-1} \\ 0 & 1 \end{pmatrix} w \begin{pmatrix} 1 & -a^{-1} \\ 0 & 1 \end{pmatrix} \in K^\times NwN. \end{aligned}$$

As $\operatorname{supp}\phi$ is compact modulo T, the intersections $\operatorname{supp}\phi \cap NN'$ and $\operatorname{supp}\phi \cap NwN$ are compact. We write ϕ^y for the right translation of ϕ by y. Then ϕ^y is a sum $\phi^y = \phi_1^y + \phi_2^y$, $\phi_i^y \in \mathcal{S}(\chi, G)$, with $\operatorname{supp}\phi_1^y \subset TNN'$ and $\operatorname{supp}\phi_2^y \subset TNwN$. Using the transformation under T by χ, we can actually regard ϕ_i^y, $i = 1, 2$, as functions on $F \oplus F$ identifying N with F. Thus, $\phi_i^y \in \mathcal{S}(F \oplus F)$. Then we get

$$\begin{aligned} \int_T \phi(t^{-1}\gamma(x)ty)\, dt &= \chi_1(x-1) \int_{F^\times} \phi_1^y(\frac{x}{a(x-1)}, -a)\, d^\times a \\ &+ \chi_1(1-x) \int_{F^\times} \chi_1(a^{-2})\phi_2^y(\frac{a}{x-1}, -a^{-1})\, d^\times a. \quad (11) \end{aligned}$$

(c) We have $\chi_1(x-1) = \chi_1(-1)$ if $x \in \wp^{c(\chi_1)}$, where $c(\chi_1)$ is the leader of χ_1. By Lemma 3.4, the first integral of (11) for small x equals

$$A_1(\frac{x}{x-1}) + A_2(\frac{x}{x-1})v(\frac{x}{x-1}),$$

where A_1, A_2 are locally constant functions on a neighborhood of zero depending on y. Then $\tilde{A}_i(x) := A_i(\frac{x}{x-1})$ are locally constant functions on a neighborhood U_1 of zero as well. The second integral of (11) is constant on a neighborhood U_2 of $x = 0$ depending on y, as ϕ_2^y is locally constant for $(x-1)^{-1} \to -1$. Thus, the complete inner integral can be expressed on $U_y := \wp^{c(\chi_1)} \cap U_1 \cap U_2$ as

$$A_y(x) := \tilde{A}_1(x) + \tilde{A}_2(x)v(x) + B.$$

By Lemma 3.3(a), there is a neighborhood V_y of y where the inner integral is constant. Take V_y that small that ψ is constant there, too, and cover supp ψ modulo T by finitely many such V_y, $y \in I$, for some finite set I. The local linking number for $x \in U = \cap_{y \in I} U_y$ now is computed as

$$< \phi, \psi >_x = \sum_{y \in I} \mathrm{vol}_{T\backslash G}(TV_y)\bar{\psi}(y)A_y(x).$$

That is, there are locally constant functions B_1, B_2 on U such that for $x \in U$

$$< \phi, \psi >_x = B_1(x) + B_2(x)v(x).$$

(d) Let $x^{-1} \in \wp^{c(\chi_1)}$. Then $\chi_1(x-1) = \chi(x)$. As ϕ_1^y is locally constant, the first integral of (11) equals a locally constant function $A_1(x^{-1})$ for x^{-1} in a neighborhood U_1 of zero depending on y. For the second integral, we distinguish whether $\chi_1^2 = 1$ or not. Let $\eta := \chi_1^2 \neq 1$. Applying Lemma 3.4(b), we get locally constant functions A_2, A_3 on a neighborhood U_2 of zero depending on y such that the second integral equals $A_2(x^{-1}) + \chi_1^2(x^{-1})A_3(x^{-1})$. Thus, for fixed y the inner integral for $x^{-1} \in U_y = U_1 \cap U_2 \cap \wp^{c(\chi_1)}$ is

$$A_y(x) := \int_T \phi^y(t^{-1}\gamma(c)t)\, dt = \chi_1(x)\big(A_1(x^{-1}) + A_2(x^{-1}) + A_3(x^{-1})\chi_1^{-1}(x)\big).$$

Proceeding as in (c), we get the assertion. Let $\chi_1^2 = 1$. By Lemma 3.4(a), we have locally constant functions A_2, A_3 on a neighborhood U_2 of zero such that for $x^{-1} \in U$ the second integral of (11) is given by $A_2(x^{-1}) + A_2(x^{-1})v(x)$. Thus, for $x^{-1} \in U_y := U_1 \cap U_2 \cap \wp^{c(\chi_1)}$ the inner integral is given by

$$A_y(x) := \chi_1(x)\big(A_1(x^{-1}) + A_2(x^{-1}) + A_3(x^{-1})v(x)\big).$$

The term $A_3(x^{-1})v(x)$ by Lemma 3.4(a) is obtained from functions $\phi_2^y(a,b)$ having the shape $\mathbf{1}_{\wp^n}(a)\mathbf{1}_{\wp^n}(b)$ around zero. Those function can only occur if y is contained in supp ϕ. Again proceeding as in part (c), the local linking number for x^{-1} in a sufficiently small neighborhood U of zero is

$$< \phi, \psi >_x = \chi_1(x)\big(B_1(x^{-1}) + B_2(x^{-1})v(x)\big),$$

where B_1, B_2 are locally constant on U and B_2 doesn't vanish only if id $\in$ (supp ϕ)(supp ψ)$^{-1}$. □

The above properties of the local linking numbers describe them completely:

Proposition 3.5. *The properties (a) to (d) of Proposition 3.1, resp., 3.2 characterize* $\mathcal{L}$*: Given a function* H *on* $F^\times$ *satisfying these properties, there are* $\phi, \psi \in \mathcal{S}(\chi, G)$ *such that* $H(x) = <\phi, \psi>_x$.

We first describe the construction in general before going into detail in the case of a field extension K/F. The case of a split algebra $K = F \oplus F$ will be omitted, as it is similar and straightforward after the case of a field extension. A complete proof can be found in [9, Chap. 2]. We choose a description of a function H satisfying the properties (a) to (d),

$$H(x) = \mathbf{1}_{c\mathrm{N}}(x)\big(A_0(x)\mathbf{1}_{V_0}(x) + A_1(x)\mathbf{1}_{V_1}(x) + \sum_{j=2}^{M} H(x_j)\mathbf{1}_{V_j}(x)\big),$$

where $V_j = x_j(1 + \wp_F^{n_j}), j = 2, \ldots, M$, are open sets in $F^\times$ on which H is constant. Similarly,

$$V_0 = \wp_F^{n_0} \quad \text{resp.} \quad V_1 = F\backslash\wp^{-n_1}$$

are neighborhoods of 0 (resp., ∞) where H is characterized by A_0 (resp., A_1) according to property (c) (resp., (d)). We may assume $n_j > 0$ for $j = 0, \ldots, M$ and $V_i \cap V_j = \emptyset$ for $i \neq j$. Then we construct a function ψ and functions ϕ_j, $j = 0, \ldots, M$, in $\mathcal{S}(\chi, G)$ such that $\operatorname{supp}\phi_i \cap \operatorname{supp}\phi_j = \emptyset$ if $i \neq j$ and such that

$$<\phi_j, \psi>_x = H(x_j)\mathbf{1}_{V_j}(x) \quad \text{resp.} \quad <\phi_j, \psi>_x = A_j(x)\mathbf{1}_{V_j}(x).$$

There is essentially one possibility to construct such functions in $\mathcal{S}(\chi, G)$: Take a compact open subset C of G which is **fundamental** for χ, that is, if $t \in T$ and $c \in C$ as well as $tc \in C$, then $\chi(t) = 1$. Then the function $\phi = \chi \cdot \mathbf{1}_C$ given by $\phi(tg) = \chi(t)\mathbf{1}_C(g)$ is well defined in $\mathcal{S}(\chi, G)$ with support TC. The function ψ is then chosen as $\psi = \chi \cdot \mathbf{1}_U$, where U is a compact open subgroup of G that small that for $j = 0, \ldots, M$

$$P(P^{-1}(V_j)U) = V_j \cap c\mathrm{N}.$$

For $j \geq 2$ we take C_j compact such that C_jU is fundamental and $P(C_jU) = V_j$ and define $\phi_j := H(x_j) \cdot \chi \cdot \mathbf{1}_{C_jU}$. The stalks of zero and infinity are constructed similarly. As the local linking numbers are linear in the first component and as the supports of the ϕ_j are disjoint by construction, we get

$$H(x) = <\sum_{j=0}^{M} \phi_j, \psi>_x .$$

Proof of Proposition 3.5 in the case K a field. Let $K = F(\sqrt{A})$. Let the function H satisfying (a) to (d) of Proposition 3.1 be given by

$$H(x) = \mathbf{1}_{c\mathrm{N}}(x)\big(A_0(x)\mathbf{1}_{V_0}(x) + A_1(x)\mathbf{1}_{V_1}(x) + \sum_{j=2}^{M} H(x_j)\mathbf{1}_{V_j}(x)\big),$$

where

$$\begin{aligned}
&V_0 = \wp^{n_0} \text{ and } A_0(x) = a_0,\\
&V_1 = F\backslash\wp^{-n_1} \text{ and } A_1(x) = \begin{cases} \chi_1(x)a_1, & \text{if } \chi^2 = 1 \\ 0, & \text{if } \chi^2 \neq 1 \end{cases},\\
&V_j = x_j(1+\wp^{n_j}) \text{ for } j = 2, \dots, M,
\end{aligned}$$

with $a_0, a_1, H(x_j) \in \mathbb{C}$, and $n_j > 0$ for $j = 0, \dots, M$. We further assume

$$n_0 - v(\frac{c}{A}) > 0, \quad n_1 + v(\frac{c}{A}) > 0 \text{ and both even,}$$

as well as $V_i \cap V_j = \emptyset$ for $i \neq j$. Let

$$\begin{aligned}
\tilde{n}_0 &= \begin{cases} \frac{1}{2}(n_0 - v(\frac{c}{A})), & \text{if } K/F \text{ unramified} \\ n_0 - v(\frac{c}{A}), & \text{if } K/F \text{ ramified} \end{cases},\\
\tilde{n}_1 &= \begin{cases} \frac{1}{2}(n_1 + v(\frac{c}{A})), & \text{if } K/F \text{ unramified} \\ n_1 + v(\frac{c}{A}), & \text{if } K/F \text{ ramified} \end{cases},
\end{aligned}$$

as well as for $j = 2, \dots, M$

$$\tilde{n}_j = \begin{cases} n_j, & \text{if } K/F \text{ unramified} \\ 2n_j, & \text{if } K/F \text{ ramified} \end{cases}.$$

Then $\mathrm{N}(1 + \wp_K^{\tilde{n}_j}) = 1 + \wp_F^{n_j}, j \geq 2$. Here $\wp_K$ is the prime ideal of K. Define

$$U := 1 + \wp_K^k + \epsilon\wp_K^m,$$

where $k > 0$ and $m > 0$ are chosen such that

$$\begin{aligned}
&k \geq c(\chi), \quad m \geq c(\chi)\\
&k \geq \tilde{n}_j, \quad m \geq \tilde{n}_j + 1, \text{ for } j = 0, \dots, M,\\
&m \geq c(\chi) + 1 - \frac{1}{2}v(x_j), \text{ for } j = 2, \dots, M,\\
&m \geq \tilde{n}_j + 1 + \frac{1}{2}|v(x_j)|, \text{ for } j = 2, \dots, M.
\end{aligned} \tag{12}$$

As $k, m > 0$ and $k, m \geq c(\chi)$, U is fundamental. Define

$$\psi := \chi \cdot \mathbf{1}_U.$$

To realize the stalks for $x_j, j \geq 2$, let $\sqrt{A} + \epsilon\gamma_j$ be a preimage of x_j,

$$P(\sqrt{A} + \epsilon\gamma_j) = \frac{c\,\mathrm{N}(\gamma_j)}{-A} = x_j.$$

The preimage of V_j is

$$P^{-1}(V_j) = T\big(\sqrt{A} + \epsilon\gamma_j(1 + \wp_K^{\tilde{n}_j})\big)T = T\big(\sqrt{A} + \epsilon\gamma_j(1 + \wp_K^{\tilde{n}_j})\,\mathrm{N}_K^1\big).$$

Let $C_j := \sqrt{A} + \epsilon\gamma_j(1 + \wp_K^{\tilde{n}_j})\,\mathrm{N}_K^1$. The compact open set

$$C_jU = \sqrt{A}(1 + \wp_K^k) + c\bar{\gamma}_j\wp_K^m + \epsilon\big(\gamma_j(1 + \wp_K^k + \wp_K^{\tilde{n}_j})\,\mathrm{N}_K^1 + \sqrt{A}\wp_K^m\big).$$

is fundamental, due to the choices (12): We have to check that if $t \in T$, $c \in C_j$, and $tc \in C_jU$, then $\chi(t) = 1$ (observe that U is a group). Let

$$tc = t\sqrt{A} + \epsilon\bar{t}\gamma_j(1 + \pi_K^{\tilde{n}_j}c_1)l \in C_jU.$$

The first component forces $t \in 1 + \wp_K^k + \frac{c}{A}\bar{\gamma}_j\wp_K^m$, for which $\chi(t) = 1$, by (12). For the image of C_jU we find again by (12)

$$P(C_jU) = \frac{c\,\mathrm{N}(\gamma_j)\,\mathrm{N}(1 + \wp_K^k + \wp_K^{\tilde{n}_j} + \wp_K^m\frac{\sqrt{A}}{\gamma_j})}{-A\,\mathrm{N}(1 + \wp_K^k + \frac{c}{\sqrt{A}}\bar{\gamma}_j\wp_K^m)} = V_j.$$

So the functions $\phi_j := \chi \cdot \mathbf{1}_{C_jU} \in \mathcal{S}(\chi, G)$ are well defined. We compute

$$< \phi_j, \psi >_x = \int_{T\backslash G}\int_T \phi_j(t^{-1}\gamma(x)ty)\,dt\,\bar{\psi}(y)\,dy.$$

The integrand doesn't vanish only if there is $s \in K^\times$ such that

$$st^{-1}\gamma(x)t = s\sqrt{A} + \epsilon\bar{s}\gamma_2(x)t\bar{t}^{-1} \in C_jU.$$

The first component implies $s \in 1 + \wp_K^{\tilde{n}_j}$. The second one implies $\gamma_2(x) \in \gamma_j(1 + \wp_K^{\tilde{n}_j})\,\mathrm{N}_K^1$, which is equivalent to $x \in V_j$. In this case we take $s = 1$ and get

$$< \phi_j, \psi >_x = \mathbf{1}_{V_j}(x)\int_{T\backslash G}\int_T 1\,dt\,\bar{\psi}(y)\,dy = \mathbf{1}_{V_j}(x)\,\mathrm{vol}_T(T)\,\mathrm{vol}_G(U).$$

Normalizing $\tilde{\phi}_j := \frac{H(x_j)}{\mathrm{vol}_T(T)\,\mathrm{vol}_G(U)}\phi_j$, we get $H|_{V_j}(x) =< \tilde{\phi}_j, \psi >_x$.

For the stalk at zero, we find $P(C_0) = \wp_F^{n_0} \cap c\,\mathrm{N}$, where $C_0 := \sqrt{A} + \epsilon\wp_K^{\tilde{n}_0}$. The preimage $P^{-1}(V_0)$ equals $TC_0T = TC_0$. The open and compact set C_0U is easily seen to be fundamental and to satisfy $P(C_0U) = V_0 \cap cN$. Define $\phi_0 := \chi \cdot \mathbf{1}_{C_0U}$ and compute the local linking number $< \phi_0, \psi >_x$. It doesn't vanish only if there is $s \in K^\times$ such that

$$st^{-1}\gamma(x)t \;=\; s\sqrt{A} + \epsilon\bar{s}\gamma_2(x)t\bar{t}^{-1} \;\in\; C_0U.$$

This forces $\gamma_2(x) \in \wp_K^{\tilde{n}_0}$. Then we take $s = 1$ and get

$$< \phi_0, \psi >_x = \mathbf{1}_{V_0 \cap c\mathrm{N}}(x)\,\mathrm{vol}_T(T)\,\mathrm{vol}_G(U).$$

That is, $H|_{V_0}(x) = a_0 =< \frac{a_0}{\mathrm{vol}_T(T)\,\mathrm{vol}_G(U)}\phi_0, \psi >_x$. It remains to construct the stalk at infinity in case $\chi^2 = 1$. Thus, $\chi = \chi_1 \circ \mathrm{N}$. The preimage of $V_1 = F\backslash\wp_F^{-n_1}$ is given by

$$P^{-1}(V_1) \;=\; T\big(\sqrt{A} + \epsilon(\wp_K^{\tilde{n}_1})^{-1}\big)T = T\big(\sqrt{A}\wp_K^{\tilde{n}_1} + \epsilon\,\mathrm{N}_K^1\big).$$

Take $C_1 = \sqrt{A}\wp_K^{\tilde{n}_1} + \epsilon\,\mathrm{N}_K^1$ to get a fundamental compact open set

$$C_1U \;=\; \sqrt{A}\wp_K^{\tilde{n}_1} + c\wp_K^{m} + \epsilon\big(\mathrm{N}_K^1(1 + \wp_K^k) + \sqrt{A}\wp_K^{m+\tilde{n}_1}\big),$$

By the choices (12) we get $P(C_1U) = V_1 \cap c\mathrm{N}$. Taking $\phi_1 := \chi \cdot \mathbf{1}_{C_1U}$ this time, we get $H|_{V_1}(x) = \frac{a_1}{\mathrm{vol}_T(T)\,\mathrm{vol}_G(U)} < \phi_1, \psi >_x$. □

We use the parametrization $\xi = \frac{x}{x-1}$. The properties of the local linking numbers (Propositions 3.1 and 3.2) transform accordingly. Let $\tilde{\mathcal{L}}$ be the space of distributions made up by evaluating the multiple $|\xi\eta|^{\frac{1}{2}} < \phi, \psi >_{\gamma(\xi)}$ of local linking numbers at $\xi \in F^\times$ for $\phi, \psi \in \mathcal{S}(\chi, G)$. This is the space of test vectors of the geometric side, while the space $\mathcal{W}$ of analytic test vectors is given by evaluation of Whittaker products. We have the following transfer:

Theorem 3.6. *Assume $\omega(-\xi\eta) = 1$ if D is split, resp., $\omega(-\xi\eta) = -1$ if D is a division algebra. The spaces of test vectors $\tilde{\mathcal{L}}$ and $\mathcal{W}$ have identical ξ-expansion.*

Proof of Theorem 3.6. The space $\mathcal{W}$ is characterized by Propositions 2.9 and 2.10. Comparing it with $\tilde{\mathcal{L}}$ (Propositions 3.1, resp., 3.2) yields the claim. For example, by Proposition 2.9 for K/F split and $\chi_1^2 \neq 1$, the Whittaker products for $\xi \to 1$ ($\eta \to 0$) are given by

$$|\xi\eta|^{\frac{1}{2}}\left(a_1\chi_1(\eta) + a_2\chi_1^{-1}(\eta)\right),$$

which corresponds to Proposition 3.2(d). For $\xi \to 0$ ($\eta \to 1$) we apply Proposition 2.10: The Whittaker products have the shape $|\xi\eta|^{\frac{1}{2}}\,(a_1 + a_2 v(\xi))$. This is property (c) of Proposition 3.2. Away from $\xi \to 1$ and $\xi \to 0$, the Whittaker products are locally constant with compact support. This is equivalent to (a) and (b) of Proposition 3.2. □

4 Translated Linking Numbers

In the remaining, the quaternion algebra D is assumed to be split, that is, $G = F^\times \backslash D^\times$ is isomorphic to the projective group $\mathrm{PGL}_2(F)$. The aim is to give an operator on the local linking numbers realizing the Hecke operator on the analytic side. As the analytic Hecke operator essentially is given by translation by $b \in F^\times$ (Proposition 2.11), the first candidate for this study is the translation by b,

$$< \phi, \begin{pmatrix} b & 0 \\ 0 & 1 \end{pmatrix} \psi >_x = \int_{T\backslash G} \int_T \phi(t^{-1}\gamma(x)ty)\, dt\, \bar{\psi}(y \begin{pmatrix} b & 0 \\ 0 & 1 \end{pmatrix})\, dy.$$

Let

$$I_\phi(y) = \int_T \phi(t^{-1}\gamma(x)ty)\, dt \tag{13}$$

be its inner integral. Here the difference between the case of a compact torus and that of a noncompact one becomes crucial. Fixing x and viewing the translated linking number as a function of b alone, we describe the behavior in the compact case in a few lines. In the noncompact case our computational approach makes up one hundred pages. Referring to [9], this case is only sketched.

4.1 *The Compact Case*

Let $K = F(\sqrt{A})$ be a field extension of F. So T is compact. As functions $\phi \in \mathcal{S}(\chi, G)$ have compact support modulo T, the set $T\gamma(x)T \cdot \operatorname{supp}\phi$ is compact. Left translation by $t' \in T$ yields $I_\phi(t'y) = \chi(t')I_\phi(y)$. Thus, the inner integral I_ϕ itself is an element of $\mathcal{S}(\chi, G)$. Choose the following isomorphism of $D^\times = (K + \epsilon K)^\times$ with $\mathrm{GL}_2(F)$:

$$\epsilon \mapsto \begin{pmatrix} 0 & -A \\ 1 & 0 \end{pmatrix},$$

$$K^\times \ni t = a + b\sqrt{A} \mapsto \begin{pmatrix} a & bA \\ b & a \end{pmatrix}.$$

Let $M = \left\{ \begin{pmatrix} y_1 & y_2 \\ 0 & 1 \end{pmatrix} \mid y_1 \in F^\times, y_2 \in F \right\}$ be the mirabolic subgroup of the standard Borel group. It carries the right invariant Haar measure $d^\times y_1\, dy_2$. As the map $K^\times \times M \to \mathrm{GL}_2(F)$, $(t, m) \mapsto t \cdot m$, is a homeomorphism [8, Sect. 2.2], we may normalize the quotient measure dy on $T\backslash G$ such that $dy = d^\times y_1\, dy_2$. We identify $\phi \in \mathcal{S}(\chi, G)$ with a function in $\mathcal{S}(F^\times \times F)$,

$$\phi(y_1, y_2) := \phi\begin{pmatrix} y_1 & y_2 \\ 0 & 1 \end{pmatrix}.$$

ϕ being locally constant with compact support, there are finitely many points $(z_1, z_2) \in F^\times \times F$ and $m > 0$ such that

$$\phi(y_1, y_2) = \sum_{(z_1, z_2)} \phi(z_1, z_2)\mathbf{1}_{z_1(1+\wp^m)}(y_1)\mathbf{1}_{z_2+\wp^m}(y_2).$$

Applying this for I_ϕ and ψ,

$$I_\phi(y_1, y_2) = \sum_{(z_1, z_2)} I_\phi(z_1, z_2)\mathbf{1}_{z_1(1+\wp^m)}(y_1)\mathbf{1}_{z_2+\wp^m}(y_2),$$

$$\psi(y_1, y_2) = \sum_{(w_1, w_2)} \psi(w_1, w_2)\mathbf{1}_{w_1(1+\wp^m)}(y_1)\mathbf{1}_{w_2+\wp^m}(y_2),$$

we compute the translated local linking number

$$\begin{aligned} < \phi, \begin{pmatrix} b & 0 \\ 0 & 1 \end{pmatrix} \psi >_x &= \int_{T\backslash G} I_\phi(y)\bar{\psi}(y\begin{pmatrix} b & 0 \\ 0 & 1 \end{pmatrix})\, dy \\ &= \sum_{(z_1, z_2), (w_1, w_2)} I_\phi(z_1, z_2)\bar{\psi}(w_1, w_2)\mathbf{1}_{z_2+\wp^m}(w_2)\mathbf{1}_{\frac{w_1}{z_1}(1+\wp^m)}(b) \\ &\quad \cdot \mathrm{vol}^\times(1+\wp^m)\,\mathrm{vol}(\wp^m). \end{aligned}$$

We have proved:

Theorem 4.1. *Let T be compact. For fixed x, the translated local linking number $< \phi, \begin{pmatrix} b & 0 \\ 0 & 1 \end{pmatrix} \psi >_x$ is a locally constant function of $b \in F^\times$ with compact support.*

We give an explicit example used later on.

Example 4.2 ([9], Bsp. 4.8). Let K/F be an unramified field extension and let $\chi = 1$. Then $\phi = \chi \cdot \mathbf{1}_{\mathrm{GL}_2(\mathbf{o}_F)}$ is well defined in $\mathcal{S}(\chi, G)$ and

$$< \phi, \begin{pmatrix} b & 0 \\ 0 & 1 \end{pmatrix} \phi >_x \cdot \mathrm{vol}^{-1} =$$

$$\mathbf{1}_{\mathrm{N}\backslash(1+\wp)}(x)\mathbf{1}_{\mathbf{o}_F^\times}(b) + \mathbf{1}_{1+\wp}(x)\big(\mathbf{1}_{(1-x)\mathbf{o}_F^\times}(b) + \mathbf{1}_{(1-x)^{-1}\mathbf{o}_F^\times}(b)\big)q^{-v(1-x)},$$

where $\mathrm{vol} := \mathrm{vol}_T(T)\,\mathrm{vol}^\times(\mathbf{o}_F^\times)\,\mathrm{vol}(\mathbf{o}_F)$.

4.2 The Noncompact Case

Let $K = F \oplus F$ be a split algebra. The character χ is of the form $\chi = (\chi_1, \chi_1^{-1})$ for a character χ_1 of $F^\times$. As in the proof of Proposition 3.2, $G = TNN' \cup TNwN$. Both of these open subsets are invariant under right translation by $\begin{pmatrix} b & 0 \\ 0 & 1 \end{pmatrix}$. Choose coset representatives for $T\backslash TNN'$ of the form

$$y = \begin{pmatrix} 1 & y_2 \\ 0 & 1 \end{pmatrix} \begin{pmatrix} 1 & 0 \\ y_3 & 1 \end{pmatrix}$$

as well as coset representatives for $T\backslash TNwN$ of the form

$$y = \begin{pmatrix} 1 & y_1 \\ 0 & 1 \end{pmatrix} w \begin{pmatrix} 1 & 0 \\ y_4 & 1 \end{pmatrix}.$$

Any function $\psi \in \mathcal{S}(\chi, G)$ can be split into a sum $\psi = \psi_1 + \psi_2$, $\psi_i \in \mathcal{S}(\chi, G)$, with $\operatorname{supp} \psi_1 \subset TNN'$ (resp., $\operatorname{supp} \psi_2 \subset TNwN$). The function ψ_1 can be viewed as an element of $\mathcal{S}(F^2)$ in the variable (y_2, y_3). Choose the quotient measure dy on $T\backslash TNN'$ such that $dy = dy_2\, dy_3$ for fixed Haar measure dy_i on F. Proceed analogously for ψ_2. For fixed x the inner integral I_ϕ (13) is a locally constant function in y. Its support is not compact anymore, but I_ϕ is the locally constant limit of Schwartz functions. This is the reason for this case being that more elaborate than the case of a compact torus. The shape of the translated linking numbers is given by the following theorem.

Theorem 4.3. *Let T be a noncompact torus. For fixed x, the translated local linking number $< \phi, \begin{pmatrix} b & 0 \\ 0 & 1 \end{pmatrix} \psi >_x$ is a function in $b \in F^\times$ of the form*

$$\chi_1^{-1}(b)\Big(\mathbf{1}_{\wp^n}(b)|b|(a_{+,1}v(b) + a_{+,2}) + A(b) + \mathbf{1}_{\wp^n}(b^{-1})|b|^{-1}(a_{-,1}v(b) + a_{-,2})\Big)$$
$$+ \chi_1(b)\Big(\mathbf{1}_{\wp^n}(b)|b|(c_{+,1}v(b) + c_{+,2}) + C(b) + \mathbf{1}_{\wp^n}(b^{-1})|b|^{-1}(c_{-,1}v(b) + c_{-,2})\Big),$$

with suitable constants $a_{\pm,i}, c_{\pm,i} \in \mathbb{C}$, integral $n > 0$, and functions $A, C \in \mathcal{S}(F^\times)$.

Sketch of proof of Theorem 4.3. This is done by brute force computations in [9] Chapter 8. We will outline the reduction to $\wp$-adic integration here. We choose the functions ϕ, ψ locally as simple as possible: $z \in \operatorname{supp} \phi$ belongs to TNN' or $TNwN$. We restrict to $z \in TNN'$, the other case is done similarly. There is a representative

$$\begin{pmatrix} 1 + z_2 z_3 & z_2 \\ z_3 & 1 \end{pmatrix}$$

of z modulo T and an open set

$$U_z = \begin{pmatrix} 1+z_2z_3 & z_2 \\ z_3 & 1 \end{pmatrix} + \begin{pmatrix} \wp^m & \wp^m \\ \wp^m & \wp^m \end{pmatrix}$$

such that $\phi|_{U_z} = \phi(z)$. Choosing m that large that U_z is fundamental, ϕ locally has the shape $\phi_z := \chi \cdot \mathbf{1}_{U_z}$ up to some multiplicative constant. For the exterior function ψ proceed similarly. It is enough to determine the behavior of the translated local linking numbers for functions of this type, i.e.,

$$\int_{T\backslash G}\int_T \phi_z(t^{-1}\gamma(x)ty)\, dt\, \bar{\psi}_{\tilde{z}}(y\begin{pmatrix} b & 0 \\ 0 & 1 \end{pmatrix})\, dy.$$

According to whether z_2 or z_3 is zero or not, and $\operatorname{supp}\psi \subset TNN'$ or $\operatorname{supp}\psi \subset TNwN$, there are $2^3 = 8$ types of integrals to be done [9, Chaps. 5.2 and 8]. For later use we include an explicit example.

Example 4.4 ([9], Bsp. 5.2). Let T be noncompact. Let $\chi = (\chi_1, \chi_1)$, where χ_1 is unramified and quadratic. Then $\phi = \chi \cdot \mathbf{1}_{\mathrm{GL}_2(\mathbf{o}_F)}$ is well defined in $\mathcal{S}(\chi, G)$. The translated local linking number $< \phi, \begin{pmatrix} b & 0 \\ 0 & 1 \end{pmatrix} \phi >_x$ is given by

$$\begin{aligned}
&\chi_1(1-x)\chi_1(b)\,\mathrm{vol}^\times(\mathbf{o}_F^\times)\,\mathrm{vol}(\mathbf{o}_F)^2 \cdot \\
&\Bigg[\mathbf{1}_{F^\times\backslash(1+\wp)}(x)\Bigg(\mathbf{1}_{\mathbf{o}_F^\times}(b)\big(|v(x)|+1\big)(1+q^{-1}) + \mathbf{1}_\wp(b)|b|\big(4v(b)+2|v(x)|\big) \\
&\qquad\qquad + \mathbf{1}_\wp(b^{-1})|b^{-1}|\big(-4v(b)+2|v(x)|\big)\Bigg) \\
&+\ \mathbf{1}_{1+\wp}(x)\Bigg(\mathbf{1}_{\wp^{v(1-x)+1}}(b)|b|\big(4v(b)-4v(1-x)\big) \\
&\qquad\qquad + \mathbf{1}_{v(1-x)\mathbf{o}_F^\times}(b)|b| + \mathbf{1}_{v(1-x)\mathbf{o}_F^\times}(b^{-1})|b^{-1}| \\
&\qquad\qquad + \mathbf{1}_{\wp^{v(1-x)+1}}(b^{-1})|b^{-1}|\big(-4v(b)-4v(1-x)\big)\Bigg)\Bigg].
\end{aligned}$$

5 A Geometric Hecke Operator

We construct operators on the local linking numbers that realize the asymptotics ($b \to 0$) of the Hecke operators on Whittaker products. The asymptotics of the second is as follows.

Proposition 5.1. *The Whittaker products $W(b\xi, b\eta)$ have the following behavior for $b \to 0$ and fixed $\xi = \frac{x}{x-1}$, $\eta = 1 - \xi$.*

(a) In case of a compact Torus T and χ not factorizing via the norm,

$$W(b\xi, b\eta) = 0.$$

In case of a compact Torus T and $\chi = \chi_1 \circ \mathrm{N}$,

$$W(b\xi, b\eta) = |b||\xi\eta|^{\frac{1}{2}} \chi_1(b\eta)\,(c_1 + c_2\omega(b\xi))\,\left(c_3 \mathbf{1}_{\wp^m \cap (1-x)\mathrm{N}}(b) + c_4 \mathbf{1}_{\wp^m \cap (1-x)z\mathrm{N}}(b)\right),$$

where $z \in F^\times \setminus \mathrm{N}$.

(b) In case of a noncompact Torus T,

$$W(b\xi, b\eta) = \begin{cases} |b||\xi\eta|^{\frac{1}{2}}\left(c_1\chi_1(b\eta) + c_2\chi_1^{-1}(b\eta)\right)(c_3 v(b\xi) + c_4)), & \text{if } \chi_1^2 \neq 1 \\ |b||\xi\eta|^{\frac{1}{2}}\chi_1(b\eta)\,(c_1 v(b\eta) + c_2)\,(c_3 v(b\xi) + c_4), & \text{if } \chi_1^2 = 1 \end{cases}.$$

Here, $c_i \in \mathbb{C}$, $i = 1, \ldots, 4$.

Proof of Proposition 5.1. For $b \to 0$ both arguments $b\xi$ and $b\eta$ tend to zero. The stated behaviors are collected from Propositions 2.9 and 2.10. □

Notice that the translation by b of the local linking numbers underlies this asymptotics (Theorems 4.1 and 4.3), but it does not realize the leading terms in case χ is quadratic. In case of a noncompact torus T, the leading term is $v(b)^2$, while translation only produces $v(b)$. In case of a compact torus, the translated linking numbers have compact support, while the Hecke operator on Whittaker products has not. In the following, we make the additional "completely unramified" assumption which is satisfied at almost all places.

Hypothesis 5.2. *D is a split algebra. K/F is an unramified extension (split or nonsplit) contained in D. The character χ is unramified.*

For a noncompact torus T the translated local linking numbers (Theorem 4.3) split into sums of the form

$$< \phi, \begin{pmatrix} \beta & 0 \\ 0 & 1 \end{pmatrix} \psi >_x = < \phi, \begin{pmatrix} \beta & 0 \\ 0 & 1 \end{pmatrix} \psi >_x^+ + < \phi, \begin{pmatrix} \beta & 0 \\ 0 & 1 \end{pmatrix} \psi >_x^-,$$

where

$$< \phi, \begin{pmatrix} \beta & 0 \\ 0 & 1 \end{pmatrix} \psi >_x^\pm := \chi_1^{\pm 1}(\beta) \cdot \tag{14}$$

$$\left(\mathbf{1}_{\wp^n}|\beta|(c_{\pm,1}v(\beta) + c_{\pm,2}) + C_\pm(\beta) + \mathbf{1}_{\wp^n}(\beta^{-1})|\beta|^{-1}(d_{\pm,1}v(\beta) + d_{\pm,2})\right)$$

are the summands belonging to $\chi_1^{\mp 1}$, respectively. In here, the constants $c_{\pm,i}, d_{\pm,i}$, and $C_\pm \in \mathcal{S}(F^\times)$ as well as $n > 0$ depend on ϕ, ψ, and x. If χ_1 is a quadratic character, these two summands coincide. To give an operator fitting all cases, define in case of a compact torus

$$< \phi, \begin{pmatrix} \beta & 0 \\ 0 & 1 \end{pmatrix} \psi >_x^\pm := < \phi, \begin{pmatrix} \beta & 0 \\ 0 & 1 \end{pmatrix} \psi >_x .$$

For $v(b) \geq 0$ define the operator $\mathbf{S}_b$ to be

$$\mathbf{S}_b := \frac{1}{4} \left(\mathbf{S}_b^+ + \mathbf{S}_b^- \right), \tag{15}$$

where

$$\mathbf{S}_b^\pm < \phi, \psi >_x := \sum_{s=0,1} \sum_{i=0}^{v(b)} \frac{\chi_1^{\mp 1}(\pi)^{i(-1)^s} \omega(b(1-x))^{i+s}}{|\pi^{v(b)-i}|} \cdot < \phi, \begin{pmatrix} \pi^{(-1)^s(v(b)-i)} & 0 \\ 0 & 1 \end{pmatrix} \psi >_x^\pm .$$

This "Hecke operator" is not unique. For example, the summand for $s = 0$ has the same properties as $\mathbf{S}_b$ itself. The crucial point is that an averaging sum occurs. The operator $\mathbf{S}_b$ is chosen such that this sum includes negative exponents $-v(b) + i$ as well. This kind of symmetry will make the results on the local Gross–Zagier formula look smoothly (Sect. 6.2).

Proposition 5.3. *Let T be a compact torus. Then the operator $\mathbf{S}_b$ reduces to*

$$\mathbf{S}_b < \phi, \psi >_x = \frac{1}{2} \sum_{s=0,1} \sum_{i=0}^{v(b)} \frac{\omega(b(1-x))^{i+s}}{|\pi^{v(b)-i}|} < \phi, \begin{pmatrix} \pi^{(-1)^s(v(b)-i)} & 0 \\ 0 & 1 \end{pmatrix} \psi >_x .$$

Let $x \in c\mathrm{N}$ be fixed. For $\phi, \psi \in \mathcal{S}(\chi, G)$ there are constants $c_1, c_2 \in \mathbb{C}$ and $n \in \mathbb{N}$ such that for $v(b) \geq n$

$$\mathbf{S}_b < \phi, \psi >_x = c_1 \mathbf{1}_{\wp^n \cap (1-x)\mathrm{N}}(b) + c_2 \mathbf{1}_{\wp^n \cap (1-x) z \mathrm{N}}(b).$$

Proposition 5.4. *Let T be a noncompact torus. The operators $\mathbf{S}_b^\pm$ reduce to*

$$\mathbf{S}_b^\pm < \phi, \psi >_x = \sum_{s=0,1} \sum_{i=0}^{v(b)} \frac{\chi_1^{\mp 1}(\pi)^{i(-1)^s}}{|\pi^{v(b)-i}|} < \phi, \begin{pmatrix} \pi^{(-1)^s(v(b)-i)} & 0 \\ 0 & 1 \end{pmatrix} \psi >_x^\pm .$$

Let $x \in F^\times$ be fixed. For $\phi, \psi \in \mathcal{S}(\chi, G)$ there are constants $c_0, \ldots, c_3 \in \mathbb{C}$ and $n \in \mathbb{N}$ such that for $v(b) \geq n$

$$\mathbf{S}_b < \phi, \psi >_x = \begin{cases} \chi_1^{-1}(b)\big(c_3 v(b) + c_2\big) + \chi_1(b)\big(c_1 v(b) + c_0\big), & \text{if } \chi_1^2 \neq 1 \\ \chi_1(b)\big(c_2 v(b)^2 + c_1 v(b) + c_0\big), & \text{if } \chi_1^2 = 1 \end{cases}.$$

Theorem 5.5. *For fixed x, the local linking numbers $|b|^{-1}|\xi\eta|^{\frac{1}{2}}\mathbf{S}_b < \phi, \psi >_x$ and the Whittaker products $\mathbf{T}_b W(\xi, \eta)$ have the same asymptotics in b.*

Proof of Theorem 5.5. Recall that $\mathbf{T}_b W(\xi, \eta) = |b|^{-2} W(b\xi, b\eta)$. In case T compact, combine Propositions 5.3 and 5.1(a) for $\chi = 1$. In case T noncompact, combine Propositions 5.4 and 5.1(b). □

Proof of Proposition 5.3. For T compact Assumption 5.2 induces $\chi = 1$ by Corollary 2.6. By Theorem 4.1, the translated linking number can be written as

$$< \phi, \begin{pmatrix} \beta & 0 \\ 0 & 1 \end{pmatrix} \psi >_x = \sum_i d_{a_i} |a_i|^{\operatorname{sign} v(a_i)} \mathbf{1}_{a_i(1+\wp^m)}(\beta),$$

for finitely many $a_i \in F^\times$, $d_{a_i} \in \mathbb{C}$, and some $m > 0$, where the sets $a_i(1 + \wp^m)$ are pairwise disjoint. We may assume that in this sum all π^l, $-\max_i |v(a_i)| \leq l \leq \max_i |v(a_i)|$, occur. Let $n := \max_i |v(a_i)| + 1$. Then, for $v(b) \geq n$,

$$\begin{aligned}
\mathbf{S}_b < \phi, \psi >_x &= \frac{1}{2} \sum_{i=0}^{v(b)} \Bigg(\omega(b(1-x))^i \sum_{l=-n+1}^{n-1} \frac{d_{\pi^l} |\pi^l|^{\operatorname{sign}(l)}}{|\pi^{v(b)-i}|} \mathbf{1}_{\pi^l(1+\wp^m)}(\pi^{v(b)-i}) \\
&\quad + \omega(b(1-x))^{i+1} \sum_{l=-n+1}^{n-1} \frac{d_{\pi^l} |\pi^l|^{\operatorname{sign}(l)}}{|\pi^{v(b)-i}|} \mathbf{1}_{\pi^l(1+\wp^m)}(\pi^{i-v(b)}) \Bigg) \\
&= \frac{1}{2} \sum_{l=0}^{n-1} \omega(b(1-x))^{v(b)+l} d_{\pi^l} + \frac{1}{2} \sum_{l=-n+1}^{0} \omega(b(1-x))^{v(b)+l+1} d_{\pi^l} \\
&= c_1 \mathbf{1}_{\wp^n \cap (1-x)\mathrm{N}}(b) + c_2 \mathbf{1}_{\wp^n \cap (1-x) z \mathrm{N}}(b),
\end{aligned}$$

where $c_1 := \frac{1}{2} \sum_{l=0}^{n-1} (d_{\pi^l} + d_{\pi^{-l}})$ and $c_2 := \frac{1}{2} \sum_{l=0}^{n-1} (-1)^l (d_{\pi^l} - d_{\pi^{-l}})$. Notice, that for $b(1-x) \in z\mathrm{N}$ one has $\omega(b(1-x))^{v(b)} = (-1)^{v(b)} = -\omega(1-x)$. □

Proof of Proposition 5.4. T is noncompact, so $\omega = 1$. First we prove this asymptotics for the part $\mathbf{T}_b^-$ of $\mathbf{S}_b$ belonging to $\mathbf{S}_b^-$ and $s = 0$,

$$\mathbf{T}_b^- < \phi, \psi >_x := \sum_{i=0}^{v(b)} \frac{\chi_1(\pi)^i}{|\pi^{v(b)-i}|} < \phi, \begin{pmatrix} \pi^{v(b)-i} & 0 \\ 0 & 1 \end{pmatrix} \psi >_x^- .$$

Let $n > 0$ be the integer of (14). Let $v(b) \geq n$. In the formula for $\mathbf{T}_b^-$, we distinguish the summands whether $v(b) - i < n$ or not. If $v(b) - i < n$, then

$$< \phi, \begin{pmatrix} \pi^{v(b)-i} & 0 \\ 0 & 1 \end{pmatrix} \psi >_x^- = \chi_1^{-1}(\pi^{v(b)-i}) C_-(\pi^{v(b)-i}).$$

The function $\tilde{C}_-$ defined by

$$\tilde{C}_-(\beta) := \frac{\chi_1^{-2}(\beta)}{|\beta|} C_-(\beta)$$

belongs to $\mathcal{S}(F^\times)$. The part of $\mathbf{T}_b^-$ made up by summands satisfying $v(b) - i < n$ is now simplified to

$$\sum_{i=v(b)-n+1}^{v(b)} \frac{\chi_1(\pi)^i}{|\pi^{v(b)-i}|} < \phi, \begin{pmatrix} \pi^{v(b)-i} & 0 \\ 0 & 1 \end{pmatrix} \psi >_x^-$$

$$= \sum_{i=v(b)-n+1}^{v(b)} \chi_1(b) \tilde{C}_-(\pi^{v(b)-i}) = \chi_1(b) \sum_{l=0}^{n-1} \tilde{C}_-(\pi^l).$$

In here, the last sum is independent of b. Thus, this part of $\mathbf{T}_b^-$ satisfies the claim. In the remaining part

$$\mathbf{T}(i \leq v(b) - n) := \sum_{i=0}^{v(b)-n} \frac{\chi_1(\pi)^i}{|\pi^{v(b)-i}|} < \phi, \begin{pmatrix} \pi^{v(b)-i} & 0 \\ 0 & 1 \end{pmatrix} \psi >_x^-$$

all the translated local linking numbers occurring can be written as

$$< \phi, \begin{pmatrix} \pi^{v(b)-i} & 0 \\ 0 & 1 \end{pmatrix} \psi >_x^- = \chi_1^{-1}(\pi^{v(b)-i}) |\pi^{v(b)-i}| \left(c_{-,1}(v(b) - i) + c_{-,2}\right).$$

So

$$\mathbf{T}(i \leq v(b) - n) = \chi_1^{-1}(b) \sum_{i=0}^{v(b)-n} \chi_1(\pi)^{2i} \left(c_{-,1}(v(b) - i) + c_{-,2}\right).$$

In case $\chi_1^2 = 1$, we have

$$\mathbf{T}(i \leq v(b) - n) = \chi_1(b)(v(b) - n + 1) \left(c_{-,2} + \frac{1}{2} c_{-,1}(v(b) + n)\right),$$

which owns the claimed asymptotics. If $\chi_1^2 \neq 1$, enlarge n such that $\chi_1^n = 1$. The remaining part of $\mathbf{T}_b^-$ then is

$$\mathbf{T}(i \leq v(b) - n) = (c_{-,1}v(b) + c_{-,2})\frac{\chi_1(b\pi) - \chi_1^{-1}(b\pi)}{\chi_1(\pi) - \chi_1^{-1}(\pi)}$$
$$- c_{-,1}\frac{\chi_1(b\pi)(v(b) - n + 1)}{\chi_1(\pi) - \chi_1^{-1}(\pi)} + c_{-,1}\frac{\chi_1(b\pi^2) - 1}{(\chi_1(\pi) - \chi_1^{-1}(\pi))^2}.$$

Thus, the claim is satisfied in case $\chi_1^2 \neq 1$. The other parts of $\mathbf{S}_b$ satisfy the claimed asymptotics as well: If $\mathbf{T}_b^+$ denotes the part of $\mathbf{S}_b$ belonging to $\mathbf{S}_b^+$ and $s = 0$, then the statement for $\mathbf{T}_b^+$ follows from the proof for $\mathbf{T}_b^-$ replacing there χ_1^{-1} by χ_1, C_- by C_+, and $c_{-,i}$ by $c_{+,i}$, where the constants are given by (14). For $s = 1$ notice that

$$\chi_1(\pi)^{i(-1)^s}\chi_1^{-1}(\pi^{(-1)^s(v(b)-i)}) = \chi_1(b)\chi_1(\pi)^{-2i}.$$

So the claim follows from the proof for $s = 0$ if there we substitute χ_1 by χ_1^{-1} as well as $c_{\pm,i}$ by $d_{\pm,i}$ of (14). □

6 Local Gross–Zagier Formula

We report on Zhang's local Gross–Zagier formulae for GL_2 [12] using our notations in order to compare them directly with the results given by the operator $\mathbf{S}_b$. We include short proofs of S. Zhang's results.

6.1 S. Zhang's Local Gross–Zagier Formula

The local Gross–Zagier formula compares the Whittaker products of local newforms with a local linking number belonging to a very special function ϕ [12, Chap. 4.1],

$$\phi = \chi \cdot \mathbf{1}_{R^\times},$$

where $R^\times$ is the unit group of a carefully chosen order R in D. Almost everywhere, especially under Hypothesis 5.2, $R^\times = \mathrm{GL}_2(\mathbf{o}_F)$ and the function ϕ is well defined. The specially chosen local linking number then is

$$< \tilde{\mathbf{T}}_b\phi, \phi >_x,$$

where the geometric Hecke operator $\tilde{\mathbf{T}}_b$ is defined as follows [12, 4.1.22 et sqq.]. Let

$$H(b) := \{g \in M_2(\mathbf{o}_F) \mid v(\det g) = v(b)\}.$$

Then

$$\tilde{\mathbf{T}}_b\phi(g) := \int_{H(b)} \phi(hg)\, dh.$$

This operator is well defined on $\phi = \chi \cdot \mathbf{1}_{\mathrm{GL}_2(\mathbf{o}_F)}$, but not generally on $\mathcal{S}(\chi, G)$. In our construction of the universal operator $\mathbf{S}_b$ we followed the idea that $\tilde{\mathbf{T}}_b$ reflects summation over translates by coset representatives, as

$$H(b) = \bigcup \begin{pmatrix} y_1 & 0 \\ 0 & y_3 \end{pmatrix} \begin{pmatrix} 1 & y_2 \\ 0 & 1 \end{pmatrix} \mathrm{GL}_2(\mathbf{o}_F),$$

where the union is over representatives $(y_1, y_3) \in \mathbf{o}_F \times \mathbf{o}_F$ with $v(y_1 y_3) = v(b)$ and $y_2 \in \wp^{-v(y_1)} \backslash \mathbf{o}_F$.

Lemma 6.1 ([12], Lemma 4.2.2). *Let K/F be a field extension and assume Hypothesis 5.2. Let $\phi = \chi \cdot \mathbf{1}_{\mathrm{GL}_2(\mathbf{o}_F)}$. Then*

$$< \tilde{\mathbf{T}}_b\phi, \phi >_x = \mathrm{vol}(\mathrm{GL}_2(\mathbf{o}_F))^2 \,\mathrm{vol}_T(T) \mathbf{1}_{\mathrm{N}}(x) \mathbf{1}_{\frac{1-x}{x}(\mathbf{o}_F \cap \mathrm{N})}(b) \mathbf{1}_{(1-x)(\mathbf{o}_F \cap \mathrm{N})}(b).$$

Lemma 6.2 ([12], Lemma 4.2.3). *Let K/F be split, let $\chi = (\chi_1, \chi_1^{-1})$ be an unramified character, and let $\phi = \chi \cdot \mathbf{1}_{\mathrm{GL}_2(\mathbf{o}_F)}$. In case $\chi_1^2 \neq 1$,*

$$\begin{aligned} < \tilde{\mathbf{T}}_b\phi, \phi >_x = {} & \frac{\chi_1(b(1-x)^{-1}\pi) - \chi_1^{-1}(b(1-x)^{-1}\pi)}{\chi_1(\pi) - \chi_1^{-1}(\pi)} \mathrm{vol}(\mathrm{GL}_2(\mathbf{o}_F))^2 \,\mathrm{vol}^\times(\mathbf{o}_F^\times) \\ & \cdot \mathbf{1}_{\frac{1-x}{x}\mathbf{o}_F \cap (1-x)\mathbf{o}_F}(b) \mathbf{1}_{F^\times}(x) \left(v(b) + v(\frac{x}{1-x}) + 1 \right). \end{aligned}$$

In case $\chi_1^2 = 1$,

$$\begin{aligned} < \tilde{\mathbf{T}}_b\phi, \phi >_x = {} & \chi_1(b(1-x)) \,\mathrm{vol}(\mathrm{GL}_2(\mathbf{o}_F))^2 \,\mathrm{vol}^\times(\mathbf{o}_F^\times) \mathbf{1}_{\frac{1-x}{x}\mathbf{o}_F \cap (1-x)\mathbf{o}_F}(b) \\ & \cdot \mathbf{1}_{F^\times}(x) \big(v(b) - v(1-x) + 1 \big) \big(v(b) + v(\frac{x}{1-x}) + 1 \big). \end{aligned}$$

For the proofs of Lemmas 6.1 and 6.2 we follow a hint by Uwe Weselmann. Write

$$\phi(x) = \sum_{\tau \in T(F)/T(\mathbf{o}_F)} \chi(\tau) \mathbf{1}_{\tau \,\mathrm{GL}_2(\mathbf{o}_F)}(x).$$

For the Hecke operator we find

$$\tilde{\mathbf{T}}_b \mathbf{1}_{\tau \,\mathrm{GL}_2(\mathbf{o}_F)}(x) = \mathrm{vol}(\mathrm{GL}_2(\mathbf{o}_F)) \mathbf{1}_{\tau b^{-1} H(b)}(x),$$

as $b^{-1}H(b) = \{h \in \mathrm{GL}_2(F) \mid h^{-1} \in H(b)\}$. As the Hecke operator is invariant under right translations, $\tilde{\mathbf{T}}_b\phi(xy) = \tilde{\mathbf{T}}_b\phi(x)$ for $y \in \mathrm{GL}_2(\mathbf{o}_F)$, we get

$$< \tilde{\mathbf{T}}_b\phi, \phi >_x = \mathrm{vol}(\mathrm{GL}_2(\mathbf{o}_F))^2 \sum_\tau \chi(\tau) \int_T \mathbf{1}_{\tau b^{-1}H(b)}(t^{-1}\gamma(x)t)\, dt. \tag{16}$$

This formula is evaluated in the different cases for K/F.

Proof of Lemma 6.1. Let $K = F(\sqrt{A})$, where $v(A) = 0$. Choose a trace zero $\gamma(x) = \sqrt{A} + \epsilon(\gamma_1 + \gamma_2\sqrt{A})$, where $\mathrm{N}(\gamma_1 + \gamma_2\sqrt{A}) = x$. The conditions for the integrands of (16) not to vanish are

$$\begin{aligned} \tau^{-1}b\sqrt{A} &\in \mathbf{o}_K \\ \tau^{-1}b\bar{t}^{-1}t(\gamma_1 + \gamma_2\sqrt{A}) &\in \mathbf{o}_K \\ \det(t^{-1}\gamma(x)t) = A(x-1) &\in b^{-1}\,\mathrm{N}(\tau)\mathbf{o}_F^\times. \end{aligned}$$

They are equivalent to $|\mathrm{N}(\tau)| = |b(1-x)|$ and $|b| \leq \min\{|\frac{1-x}{x}|, |1-x|\}$. There is at most one coset $\tau \in T(F)/T(\mathbf{o}_F)$ satisfying this, and this coset exists only if $b \in (1-x)\,\mathrm{N}$. Thus,

$$\begin{aligned} < \tilde{\mathbf{T}}_b\phi, \phi >_x = {} & \mathrm{vol}(\mathrm{GL}_2(\mathbf{o}_F))^2\, \mathrm{vol}_T(T) \\ & \cdot \left(\mathbf{1}_{\mathrm{N}\setminus(1+\wp)}(x)\mathbf{1}_{\mathbf{o}_F\cap\mathrm{N}}(b) + \mathbf{1}_{1+\wp}(x)\mathbf{1}_{(1-x)(\mathbf{o}_F\cap\mathrm{N})}(b)\right), \end{aligned}$$

which equals the claimed result. □

Proof of Lemma 6.2. Choose $\gamma(x) = \begin{pmatrix} -1 & x \\ -1 & 1 \end{pmatrix}$ of trace zero, and set $\tau = (\tau_1, \tau_2) \in K^\times/\mathbf{o}_K^\times$ as well as $t = (a, 1) \in T$. The conditions for an integrand of (16) not to vanish are

$$\begin{aligned} (-\tau_1^{-1}b, \tau_2^{-1}b) &\in \mathbf{o}_K, \\ (-\tau_1^{-1}a^{-1}bx, \tau_2^{-1}ab) &\in \mathbf{o}_K, \\ \det(t^{-1}\gamma(x)t) = x - 1 &\in \mathrm{N}(\tau)b^{-1}\mathbf{o}_K^\times. \end{aligned}$$

So only if $v(\tau_2) = -v(\tau_1) + v(b) + v(1-x)$ satisfies $v(1-x) \leq v(\tau_2) \leq v(b)$, the integral does not vanish. Then the scope of integration is given by $-v(b) + v(\tau_2) \leq v(a) \leq v(\tau_2) + v(x) - v(1-x)$ and the integral equals

$$\mathrm{vol}^\times(\mathbf{o}_F^\times)\big(v(b) + v(x) - v(1-x) + 1\big)\mathbf{1}_{\mathbf{o}_F\cap\wp^{v(1-x)-v(x)}}(b).$$

Evaluating $\chi(\tau)$ we get $\chi(\tau) = \chi_1(b(1-x))\chi_1^{-2}(\tau_2)$, as χ is unramified. Summing up the terms of (16) yields the lemma. □

The other constituents of the local Gross–Zagier formulae are the Whittaker products of newforms for both the theta series $\Pi(\chi)$ and the Eisenstein series $\Pi(1,\omega)$ at $s = \frac{1}{2}$. By Hypothesis 5.2, the theta series equals $\Pi(\chi_1, \chi_1^{-1})$ if K/F splits, and it equals $\Pi(1,\omega)$ if K/F is a field extension. Thus, all occurring representations are principal series and the newforms read in the Kirillov model are given by (8). In case of a field extension we get

$$W_{\theta,new}(a) = W_{E,new}(a) = \operatorname{vol}(\mathbf{o}_F)\operatorname{vol}^\times(\mathbf{o}_F^\times)\cdot |a|^{\frac{1}{2}}\mathbf{1}_{\mathbf{o}_F\cap \mathrm{N}}(a)\,.$$

In case K/F splits we get

$$W_{\theta,new}(a) = \operatorname{vol}(\mathbf{o}_F)\operatorname{vol}^\times(\mathbf{o}_F^\times)\cdot |a|^{\frac{1}{2}}\mathbf{1}_{\mathbf{o}_F}(a)\begin{cases}\frac{\chi_1(a\pi)-\chi_1^{-1}(a\pi)}{\chi_1(\pi)-\chi_1^{-1}(\pi)}, & \text{if } \chi_1^2\neq 1\\ \chi_1(a)(v(a)+1), & \text{if } \chi_1^2 = 1\end{cases},$$

while

$$W_{E,new}(a) = \operatorname{vol}(\mathbf{o}_F)\operatorname{vol}^\times(\mathbf{o}_F^\times)\cdot |a|^{\frac{1}{2}}\mathbf{1}_{\mathbf{o}_F}(a)(v(a)+1)\,.$$

Summing up, we get the following lemma. Recall $\xi = \frac{x}{x-1}$ and $\eta = 1-\xi$.

Lemma 6.3 ([12], Lemma 3.4.1). *Assume Hypothesis 5.2. Then the Whittaker products for the newforms of theta series and Eisenstein series have the following form up to the factor* $\operatorname{vol}(\mathbf{o}_F)^2\operatorname{vol}^\times(\mathbf{o}_F^\times)^2$. *If K/F is a field extension, then*

$$\begin{aligned}W_{\theta,new}(b\eta)W_{E,new}(b\xi) &= |\xi\eta|^{\frac{1}{2}}|b|\mathbf{1}_{\mathbf{o}_F}(b\xi)\mathbf{1}_{\mathbf{o}_F}(b\eta)\\ &= |\xi\eta|^{\frac{1}{2}}|b|\mathbf{1}_{\frac{1-x}{x}(\mathbf{o}_F\cap\mathrm{N})}(b)\mathbf{1}_{(1-x)(\mathbf{o}_F\cap\mathrm{N})}(b).\end{aligned}$$

If K/F splits and χ is quadratic, then

$$\begin{aligned}&W_{\theta,new}(b\eta)W_{E,new}(b\xi)\\ &= |\xi\eta|^{\frac{1}{2}}|b|\mathbf{1}_{\mathbf{o}_F}(b\xi)\mathbf{1}_{\mathbf{o}_F}(b\eta)\chi_1(b\eta)\,(v(b\xi)+1)\,(v(b\eta)+1)\\ &= |\xi\eta|^{\frac{1}{2}}|b|\mathbf{1}_{\frac{1-x}{x}\mathbf{o}_F\cap(1-x)\mathbf{o}_F}(b)\chi_1(b(1-x))\big(v(b)+v(\frac{x}{1-x})+1\big)\cdot\\ &\qquad\cdot\big(v(b)-v(1-x)+1\big).\end{aligned}$$

If K/F splits and χ is not quadratic, then

$$\begin{aligned}&W_{\theta,new}(b\eta)W_{E,new}(b\xi)\\ &= |\xi\eta|^{\frac{1}{2}}|b|\mathbf{1}_{\mathbf{o}_F}(b\xi)\mathbf{1}_{\mathbf{o}_F}(b\eta)\,(v(b\xi)+1)\,\frac{\chi_1(b\eta\pi)-\chi_1^{-1}(b\eta\pi)}{\chi_1(\pi)-\chi_1^{-1}(\pi)}\end{aligned}$$

$$= |\xi\eta|^{\frac{1}{2}}|b|\mathbf{1}_{\frac{1-x}{x}\mathbf{o}_F\cap(1-x)\mathbf{o}_F}(b)\big(v(b)+v(\frac{x}{1-x})+1\big)\cdot$$
$$\cdot\frac{\chi_1(b(1-x)^{-1}\pi)-\chi_1^{-1}(b(1-x)^{-1}\pi)}{\chi_1(\pi)-\chi_1^{-1}(\pi)}.$$

Comparing Lemma 6.1, resp., 6.2 with Lemma 6.3 we get S. Zhang's local Gross–Zagier formula:

Theorem 6.4 ([12], Lemma 4.3.1). *Assume Hypothesis 5.2. Let $W_{\theta,new}$, resp., $W_{E,new}$ be the newform for the theta series, resp., Eisenstein series. Let $\phi = \chi\cdot\mathbf{1}_{\mathrm{GL}_2(\mathbf{o}_F)}$. Then up to a factor of volumes,*

$$W_{\theta,new}(b\eta)W_{E,new}(b\xi) = |\xi\eta|^{\frac{1}{2}}|b| < \tilde{\mathbf{T}}_b\phi,\phi >_{x=\frac{\xi}{\xi-1}}.$$

6.2 Reformulation of Local Gross–Zagier

We re-prove S. Zhang's local Gross–Zagier formula in terms of $\mathbf{S}_b$:

Theorem 6.5. *Assume Hypothesis 5.2 and assume $\chi_1^2 = 1$ in case K/F splits. Let $W_{\theta,new}$, resp., $W_{E,new}$ be the newform for the theta series, resp., Eisenstein series. Let $\phi = \chi\cdot\mathbf{1}_{\mathrm{GL}_2(\mathbf{o}_F)}$. Then up to a factor of volumes,*

$$W_{\theta,new}(b\eta)W_{E,new}(b\xi) = |\xi\eta|^{\frac{1}{2}}|b|\mathbf{S}_b < \phi,\phi >_x + O(v(b)),$$

where in case K/F a field extension the term of $O(v(b))$ is actually zero, while in case K/F split the term of $O(v(b))$ can be given precisely by collecting terms in the proof of Example 4.4.

Proof of Theorem 6.5. Compare the Whittaker products for newforms given in Lemma 6.3 with the action of the operator $\mathbf{S}_b$ on the local linking number belonging to ϕ, given by Lemma 6.6, resp., 6.7 below. □

Lemma 6.6. *Let K/F be a field extension. Assume Hypothesis 5.2. Let $\phi = \chi\cdot\mathbf{1}_{\mathrm{GL}_2(\mathbf{o}_F)}$. Then up the factor $\mathrm{vol}_T(T)\,\mathrm{vol}^\times(\mathbf{o}_F^\times)\,\mathrm{vol}(\mathbf{o}_F)$,*

$$\mathbf{S}_b < \phi,\phi >_x = \mathbf{1}_{\mathrm{N}}(x)\mathbf{1}_{\frac{1-x}{x}(\mathbf{o}_F\cap\mathrm{N})}(b)\mathbf{1}_{(1-x)(\mathbf{o}_F\cap\mathrm{N})}(b).$$

Proof of Lemma 6.6. The translated local linking number is that of Example 4.2. We compute the action of $\mathbf{S}_b$ given by Proposition 5.3. If $x\in\mathrm{N}\setminus(1+\wp)$, then up to the factor $\mathrm{vol}_T(T)\,\mathrm{vol}^\times(\mathbf{o}_F^\times)\,\mathrm{vol}(\mathbf{o}_F)$,

$$\mathbf{S}_b < \phi,\phi >_x = \frac{1}{2}\left(\omega(b(1-x))^{v(b)} + \omega(b(1-x))^{v(b)+1}\right) = \mathbf{1}_{\mathrm{N}}(b).$$

If $x \in 1 + \wp$, then again up to the factor of volumes

$$\begin{aligned}\mathbf{S}_b < \phi, \phi >_x &= \frac{1}{2}\mathbf{1}_{\wp^{v(1-x)}}(b)\omega(b(1-x))^{v(b)-v(1-x)}\left(1+\omega(b(1-x))\right)\\ &= \mathbf{1}_{\wp^{v(1-x)}\cap(1-x)\mathrm{N}}(b).\end{aligned}$$

□

In case K/F we restrict ourselves to the case $\chi_1^2 = 1$.

Lemma 6.7. *Let K/F be split and assume Hypothesis 5.2 as well as $\chi_1^2 = 1$. Let $\phi = \chi \cdot \mathbf{1}_{\mathrm{GL}_2(\mathbf{o}_F)}$. Then up the factor* $\mathrm{vol}^\times(\mathbf{o}_F^\times)\,\mathrm{vol}(\mathbf{o}_F)^2$,

$$\begin{aligned}\mathbf{S}_b < \phi, \phi >_x = {} & \chi_1(b(1-x))\cdot \\ & \Big[\mathbf{1}_{F^\times\backslash(1+\wp)}(x)\Big(2v(b)^2 + 2(|v(x)|+1)v(b) + (1+q^{-1})(|v(x)|+1)\Big) \\ & + \mathbf{1}_{1+\wp}(x)\mathbf{1}_{\wp^{v(1-x)}}(b)\Big(2\big(v(b)-v(1-x)+1\big)\big(v(b)-v(1-x)\big)+1\Big)\Big].\end{aligned}$$

Proof of Lemma 6.7. The operator $\mathbf{S}_b$ is given by Proposition 5.4. The translated local linking number is given by Example 4.4. As χ_1 is quadratic, $\mathbf{S}_b = \frac{1}{2}\mathbf{S}_b^+$. For $x \in 1 + \wp$ we compute

$$\begin{aligned}&\mathbf{S}_b < \phi, \phi >_x \\ &= \chi_1(b(1-x))\mathbf{1}_{\wp^{v(1-x)}}(b)\left(1 + \sum_{i=0}^{v(b)-v(1-x)-1} 4(v(b)-i-v(1-x))\right) \\ &= \chi_1(b(1-x))\mathbf{1}_{\wp^{v(1-x)}}(b)\Big(2(v(b)-v(1-x)+1)(v(b)-v(1-x))+1\Big),\end{aligned}$$

while for $x \in F^\times\backslash(1+\wp)$,

$$\begin{aligned}&\mathbf{S}_b < \phi, \phi >_x \\ &= \chi_1(b(1-x))\left[(|v(x)|+1)(1+q^{-1}) + \sum_{i=0}^{v(b)-1}\left(4(v(b)-i)+2|v(x)|\right)\right] \\ &= \chi_1(b(1-x))\Big(2v(b)^2 + (1+|v(x)|)(2v(b)+1+q^{-1})\Big).\end{aligned}$$

□

Acknowledgements We thank Rainer Weissauer for challenging and supervising, as well as Uwe Weselmann for many useful comments, and Lynne Walling for mentoring. This work was partially supported by the European Social Fund.

References

1. Bump, D.: Automorphic Forms and Representations. Cambridge Studies in Advanced Mathematics, vol. 55. Cambridge University Press, Cambridge (1996)
2. Gelbart, S.S.: Automorphic Forms on Adele Groups. Princeton University Press and University of Tokyo Press, Princeton (1975)
3. Godement, R.: Notes on Jacquet-Langlands' Theory. The Institute for Advanced Study, Princeton (1970)
4. Gross, B.H., Zagier, D.B.: Heegner points and derivatives of L-series. Invent. Math. **84**(2), 225–320 (1986)
5. Jacquet, H.: Automorphic Forms on GL(2), Part II. Lecture Notes in Mathematics, vol. 278. Springer, Berlin (1972)
6. Jacquet, H.: Sur un résultat de Waldspurger. Ann. Sci. Ec. Norm. Sup. **4**(19), 185–229 (1986)
7. Jacquet, H., Langlands, R.P.: Automorphic Forms on GL(2). Springer Lecture Notes, vol. 114. Springer, Berlin (1970)
8. Jacquet, H., Zagier, D.: Eisenstein series and the Selberg trace formula II. Trans. Am. Math. Soc. **300**(1), 1–48 (1987)
9. Maurischat, K.: Eine Verallgemeinerung der lokalen Gross-Zagier-Formeln von Zhang. Dissertation, Heidelberg University, Heidelberg (2008). Available at: http://archiv.ub.uni-heidelberg.de/volltextserver/volltexte/2008/8420/pdf/maurischat_diss.pdf.
10. Yuan, X., Zhang, S.-W., Zhang, W.: Heights of CM points I. Gross-Zagier formula. Preprint (2009)
11. Yuan, X., Zhang, S.-W., Zhang, W.: The Gross-Zagier Formula on Shimura Curves. Annals of Mathematics Studies, vol. 184. Princeton University Press, Princeton (2013)
12. Zhang, S.-W.: Gross-Zagier formula for GL_2. Asian J. Math. **5**(2), 183–290 (2001)
13. Zhang, S.-W.: Gross-Zagier formula for GL_2, II. In: Heegner Points and Rankin L-Series. Mathematical Sciences Research Institute Publications, vol. 49, pp. 191–214. Cambridge University Press, Cambridge (2004)
14. Zhang, W.: On arithmetic fundamental lemmas. Invent. Math. **188**, 197–252 (2012)
15. Zhang, W.: Gross-Zagier formula and arithmetic fundamental lemma. In: Fifth International Congress of Chinese Mathematicians, Part 1, 2. AMS/IP Studies in Advanced Mathematics, vol. 51, pp. 447–459. AMS, Providence (2012)

p-Adic q-Expansion Principles on Unitary Shimura Varieties

Ana Caraiani, Ellen Eischen, Jessica Fintzen, Elena Mantovan, and Ila Varma

Abstract We formulate and prove certain vanishing theorems for p-adic automorphic forms on unitary groups of arbitrary signature. The p-adic q-expansion principle for p-adic modular forms on the Igusa tower says that if the coefficients of (sufficiently many of) the q-expansions of a p-adic modular form f are zero, then f vanishes everywhere on the Igusa tower. There is no p-adic q-expansion principle for unitary groups of arbitrary signature in the literature. By replacing q-expansions with Serre–Tate expansions (expansions in terms of Serre–Tate deformation coordinates) and replacing modular forms with automorphic forms on unitary groups of arbitrary signature, we prove an analogue of the p-adic q-expansion principle.

Ana Caraiani's research is partially supported by NSF Postdoctoral Fellowship DMS-1204465 and NSF Grant DMS-1501064.

Ellen Eischen's research is partially supported by NSF Grants NSF DMS-1249384 and DMS-1559609.

Jessica Fintzen's research is partially supported by the Studienstiftung des deutschen Volkes.

Elena Mantovan's research is partially supported by NSF Grant DMS-1001077.

Ila Varma's research is partially supported by a National Defense Science and Engineering Fellowship.

A. Caraiani • I. Varma
Department of Mathematics, Princeton University, Fine Hall, Washington Road, Princeton, NJ 08544-1000, USA
e-mail: caraiani@princeton.edu; ivarma@math.princeton.edu

E. Eischen (✉)
Department of Mathematics, University of Oregon, Fenton Hall, Eugene, OR 97403, USA
e-mail: eeischen@uoregon.edu

J. Fintzen
Department of Mathematics, Harvard University, One Oxford Street, Cambridge, MA 02138, USA
e-mail: fintzen@math.harvard.edu

E. Mantovan
Department of Mathematics, CalTech, Pasadena, CA 91125, USA
e-mail: mantovanelena@gmail.com

© Springer International Publishing Switzerland 2016

E.E. Eischen et al. (eds.), *Directions in Number Theory*, Association for Women in Mathematics Series 3, DOI 10.1007/978-3-319-30976-7_7

More precisely, we show that if the coefficients of (sufficiently many of) the Serre–Tate expansions of a p-adic automorphic form f on the Igusa tower (over a unitary Shimura variety) are zero, then f vanishes identically on the Igusa tower.

This paper also contains a substantial expository component. In particular, the expository component serves as a complement to Hida's extensive work on p-adic automorphic forms.

1 Introduction

The purpose of this paper is twofold: to provide an expository guide to the theory of p-adic automorphic forms on unitary groups and to formulate and prove certain vanishing theorems for these p-adic automorphic forms, which are analogous to the p-adic q-expansion principle for modular forms.

In the case of modular forms, which are automorphic forms for $GL_2/\mathbb{Q}$, the q-expansion principle is important for constructing families of p-adic modular forms and for explicitly computing the Hecke operators acting on (p-adic) modular forms. In turn, the algebraic q-expansion principle relies on the geometric interpretation of (p-adic) modular forms and on the underlying geometry of the moduli spaces they live on.

Automorphic forms for GL_n, when $n > 2$, do not have a natural interpretation in terms of algebraic geometry, because the locally symmetric spaces of GL_n do not have the structure of algebraic varieties. The locally symmetric spaces for unitary groups, however, do have the structure of Shimura varieties, and their cohomology realizes systems of Hecke eigenvalues coming from GL_n (either directly since unitary groups are outer forms of GL_n or through congruences—via p-adic interpolation). This is why unitary groups have been key in trying to extend results in the Langlands program from GL_2 to GL_n in recent years [13, 29, 32]. This is also why unitary groups provide a natural context in which to define and study p-adic automorphic forms geometrically.

The first part of our paper discusses unitary Shimura varieties, their moduli interpretation, and the geometry of their integral models. This leads to the geometric definition of p-adic automorphic forms on unitary groups. This is a vast area of research and many different aspects could be highlighted, but we focus on providing an expository account of H. Hida's extensive work in this area, including [14, 15]. In the second part of our paper, we formulate and prove certain analogues of the q-expansion principle in this context. We expect that these vanishing theorems will play a key role in constructing families of p-adic automorphic forms on unitary groups of arbitrary signature. We discuss these types of theorems in more depth below.

1.1 Vanishing Theorems

1.1.1 q-Expansion Principles for Modular (And Hilbert Modular) Forms

We start with some overview and motivation. We then review the different incarnations of the q-expansion principle for modular forms. (Note that q-expansion principles—and the Serre–Tate expansion principle discussed later in this paper—are instances of the principle of analytic continuation, which says that an analytic function on a connected domain is completely determined by its restriction to any non-empty open subset and, in particular, is determined by its Taylor expansion around any point.)

Let $\mathcal{H} = \{z \in \mathbb{C} | \mathrm{Im}\, z > 0\}$ be the complex upper half plane. The upper half plane can be identified with the *symmetric space* for the group $SL_2/\mathbb{Q}$:

$$\mathcal{H} \simeq SL_2(\mathbb{R})/SO_2(\mathbb{R})$$

and it has a natural action of $SL_2(\mathbb{Z})$ by Möbius transformations, which is equivariant for this identification. Given a congruence subgroup of $SL_2(\mathbb{Z})$, such as

$$\Gamma_1(N) := \left\{ g \in SL_2(\mathbb{Z}) | g \equiv \begin{pmatrix} 1 & * \\ 0 & 1 \end{pmatrix} \pmod{N} \right\},$$

we can form the associated *locally symmetric space*

$$Y_1(N) := \mathcal{H}/\Gamma_1(N).$$

This construction generalizes from SL_2 to any reductive group over $\mathbb{Q}$, such as GL_n or a unitary group. Moreover, the Betti cohomology of the associated locally symmetric spaces (thought of simply as real manifolds) can be related to automorphic representations of the reductive group.

In the case of SL_2, something special happens: $Y_1(N)$ is not merely a real manifold, but it has a natural complex structure, inherited from the complex structure on $\mathcal{H}$. A modular form of weight k and level N is a holomorphic function on $\mathcal{H}$ satisfying certain symmetries under the action of $\Gamma_1(N)$ by Möbius transformations:

$$f\left(\frac{az+b}{cz+d}\right) = (cz+d)^k f(z)$$

and also satisfying a growth condition. Since $\begin{pmatrix} 1 & 1 \\ 0 & 1 \end{pmatrix} \in \Gamma_1(N)$, we have

$$f(z+1) = f(z),$$

which means that f has a Fourier expansion, which ends up looking like

$$f(z) = \sum_{n=0}^{\infty} a_n q^n, \text{ where } q = e^{2\pi i z}.$$

We can think of the q-expansion as the Taylor expansion of f around the missing point $z = i\infty$ of $Y_1(N)$. The *analytic q-expansion principle* says that the Fourier expansion uniquely determines the modular form f.

There is also an algebraic q-expansion principle, which comes from the fact that modular forms have an interpretation in terms of algebraic geometry. Again, this is something special for SL_2 (and other groups that admit Shimura varieties): the Riemann surface $Y_1(N)$ has a natural moduli interpretation (parametrizing elliptic curves) and therefore comes from an algebraic curve defined over $\mathbb{Q}$. The symmetries that the holomorphic function f is required to satisfy make f into a section of a line bundle ω^k on this curve $Y_1(N)$. This curve is not projective: it will miss a finite number of cusps, one of which corresponds to the point $i\infty$ on the compactification of $\mathcal{H}/\Gamma_1(N)$ as a Riemann surface. If we call this cusp ∞, then the coordinate q in the Fourier expansion of f can be identified with a canonical coordinate in a formal neighborhood of ∞. The *algebraic* q-expansion of f can be identified with the localization of f at the cusp ∞. (The line bundle ω^k is also canonically trivialized.) The *algebraic q-expansion principle* says that a modular form of weight k is uniquely determined by its q-expansion. This principle follows from the fact that a section of a line bundle which vanishes in a formal neighborhood of a point on an irreducible curve must vanish everywhere on that curve.

The p-adic interpolation of modular forms is crucial for understanding their connection with Galois representations: for example, constructing Galois representations in weight 1 by congruences and proving modularity via the Taylor–Wiles patching method. This leads to the natural question of how to formalize the notion of p-adic interpolation and how to define a p-adic modular form.

One option is geometric: it only works for groups that have Shimura varieties and uses the geometry of their integral models. In the case of modular forms, this approach goes back to Katz [19, 20]: p-adic modular forms can be thought of as sections of the trivial line bundle over the *Igusa tower*, which can be constructed over the *ordinary locus*. The Igusa tower has the property that it simultaneously trivializes all the line bundles ω^k, corresponding to different weight modular forms k. In a very rough sense, this construction can be thought of as a p-adic analogue of the upper half plane $\mathcal{H}$. The advantage of this approach is that the underlying geometry provides more tools for studying p-adic modular forms (for example, the *Hasse invariant* is such a tool) and answering questions such as when a p-adic modular form is classical.

The p-adic q-expansion principle says that a p-adic modular form is uniquely determined by its q-expansion. This principle relies on the irreducibility of the Igusa tower. This principle has been extremely important for further studying p-adic modular forms. Applications to the construction of p-adic L-functions are mentioned in Sect. 1.2. The p-adic q-expansion principle is also a crucial ingredient in the work of Buzzard and Taylor on the icosahedral Artin conjecture [5] and in generalizations of this type of argument to Hilbert modular varieties. A key aspect of this application is the fact that, for GL_2, q-expansions of Hecke eigenforms are closely related to Hecke eigenvalues and, therefore, to Galois representations. This is a connection that is not yet understood for other groups, such as unitary groups.

1.1.2 Principles for Other Groups, Including Unitary Groups

In the case of unitary groups or symplectic groups, which admit Shimura varieties, the story described above largely generalizes. Their locally symmetric spaces have an algebraic structure, admit a moduli interpretation, and have integral models. We describe these models in Sect. 2. One can define automorphic forms in a way similar to how one defines modular forms, and they have an algebro-geometric interpretation as sections of certain vector bundles. We give more details on this in Sect. 3. It is also possible to talk about p-adic automorphic forms by constructing a (higher-dimensional) Igusa tower over the ordinary locus. These notions are made precise in Sect. 4. However, it is not clear what the best analogue of the p-adic q-expansion principle would be, in this level of generality.

There are q-expansion principles, or partial results in this direction, in a number of cases. For Siegel modular forms, i.e., automorphic forms on symplectic groups, there is an algebraic q-expansion principle in [6]. By [14, Corollary 8.17], there is a p-adic q-expansion principle for Siegel modular forms. By [23, Proposition 7.1.2.14], there is an algebraic q-expansion principle for scalar-valued automorphic forms on unitary groups of signature (a, a) for any positive integer a. As mentioned in the last paragraph of [14], there is a p-adic q-expansion principle for automorphic forms on unitary groups of signature (n, n). The proofs of all of these q-expansion principles rely on the existence of cusps, whose formal neighborhoods have canonical coordinates and on the irreducibility of the underlying moduli space (i.e., a Shimura variety or Igusa tower).

For automorphic forms on unitary groups of signature (a, b) with $a \neq b$, the underlying geometry of the associated moduli spaces prevents the existence of a q-expansion principle, because these spaces have no cusps (whose formal neighborhoods have canonical coordinates). In the case of automorphic forms on unitary groups of signature (a, b), when the corresponding Shimura varieties are non-compact, the usual q-expansion is replaced by a Fourier–Jacobi expansion, a generalization of the Fourier expansion, in which the coefficients are themselves functions formed from theta-functions. Nevertheless, there is an algebraic *Fourier–Jacobi principle* for unitary groups [23, Proposition 7.1.2.14]. (This Fourier–Jacobi principle gives the algebraic q-expansion principle for unitary groups of signature (a, a).)

While it is natural to ask for a "p-adic Fourier–Jacobi expansion principle" for unitary groups of arbitrary signature, a slightly different—but analogous—principle, a "Serre–Tate expansion principle" follows more naturally from the existing literature. The main result of this paper is the formulation and proof of the Serre–Tate expansion principle (in Theorem 5.14). Algebraic q-expansions and algebraic Fourier–Jacobi expansions are expansions of a modular (or automorphic) form at the boundary of the Shimura variety. On the other hand, a Serre–Tate expansion is the expansion of a modular form at an ordinary CM point. There is a canonical choice of coordinates for the local ring at the ordinary point; these are called *Serre–Tate deformation coordinates*. Roughly speaking our main result (stated precisely in Theorem 5.14) says that given suitable conditions on the prime p (namely, when p splits completely in the reflex field), if f is an automorphic form on a unitary group and for each irreducible component C of the associated Igusa tower, a Serre–Tate expansion of f at some CM point in C is 0, then f vanishes identically on the Igusa tower. The proof relies on Hida's description of the geometry of the Igusa tower. The key point is, again, the irreducibility of the Igusa tower.

1.2 Anticipated Applications

As noted above, the use of q-expansion principles in the construction of p-adic families of modular forms is well-established. In [20], Katz used the q-expansion principle for Hilbert modular forms to study congruences between values of different Hilbert modular forms, which led to the construction of certain p-adic families of Hilbert modular forms. Similarly, in [9], the second author used the q-expansion principle to construct p-adic families of automorphic forms on unitary groups of signature (a, a) for all positive integers a. Katz's p-adic families of Hilbert modular forms are the main ingredient in his construction of p-adic L-functions for CM fields [20]. Analogously, the second author constructed the p-adic families of automorphic forms in [9] to complete a step in the construction of p-adic L-functions (for unitary groups) proposed in [12].

We plan to use the Serre–Tate expansion principle in Theorem 5.14 analogously to how the q-expansion principle is used in contexts in which q-expansions exist. More precisely, in a joint paper in preparation, we are using the Serre–Tate expansion principle introduced in this paper to construct p-adic families of automorphic forms on unitary groups of signature (a, b) with $a \neq b$. As explained in [10], the lack of such a principle in the literature was an obstacle faced by the second author in her effort to extend her results on p-adic families of automorphic forms to unitary groups of arbitrary signature. This paper eliminates that obstacle and fills in a hole in the literature. We also are using the expository portion of this paper as part of the foundation for our construction of these families.

One advantage of expansions around CM points over q-expansions is that they can be used for compact as well as non-compact Shimura varieties. The Serre–Tate expansion has been used before by Hida (for example, to define his idempotent in [14]) and also appears in work of Brooks [3] (for Shimura curves) and Burungale and Hida [4] (for Hilbert modular varieties) with applications to special values of p-adic L-functions.

In a more speculative direction, we note the potential for applications to homotopy theory. Certain p-adic families of modular forms, studied in terms of their q-expansions, were used to defined an invariant (the *Witten genus*) in homotopy theory [1, 16, 17]. The Witten genus is a p-adic modular form valued invariant that occurs in the theory of *topological modular forms*. Recently, there have been attempts to construct an analogue of the Witten genus in the theory of *topological automorphic forms*, where there is conjecturally an invariant taking values in the space of p-adic automorphic forms on unitary groups of signature $(1, n)$ [2]. Vanishing theorems analogous to the q-expansion principle will likely play an analogously important role in this context.

1.3 Structure of the Paper

We now provide a brief overview of the paper. Section 2 introduces Shimura varieties for unitary groups and the associated moduli problem. We work with these Shimura varieties throughout most of the paper. Section 3 reviews the theory of classical automorphic forms on unitary groups, from several perspectives. Section refpadic-section introduces Hida's geometric theory of p-adic automorphic forms (i.e., over the ordinary locus). This section includes details about the Igusa tower, as well as the space of p-adic automorphic forms (defined as global sections of the structure sheaf over the Igusa tower). Section 5 covers the main results of this paper, namely the *Serre–Tate expansion principle*, an analogue of the q-expansion principle, for p-adic automorphic forms on unitary groups of arbitrary signature. We are using this result in a paper in preparation that constructs families of p-adic automorphic forms on unitary groups of arbitrary signature. Finally, Sect. 6 discusses how Serre–Tate expansions behave with respect to pullbacks, as an example of the kind of application we have in mind to computational aspects of p-adic automorphic forms on unitary groups.

1.4 Notation and Conventions

We now establish some notation and conventions that we will use throughout the paper.

First, we establish some notation for fields. Fix a totally real number field K^+ and an imaginary quadratic extension F of $\mathbb{Q}$. Define K to be the composition of K^+ and F. Let c denote complex conjugation on K, i.e., the generator of $\mathrm{Gal}(K/K^+)$. We denote by Σ the set of complex embeddings of K^+, and we denote by Σ_K the set of complex embeddings of K. We typically use τ to denote an element of Σ, and for each $\tau \in \Sigma$, we fix an extension $\tilde{\tau}$ of τ to K, i.e., $\tilde{\tau}$ is an element of Σ_K. A reflex field will be denoted by E (with subscripts to denote different reflex fields when there is more than one reflex field appearing in the same context). Given a local or global field L, we denote the ring of integers in L by $\mathcal{O}_L$. We write $\mathbb{A}$ to denote the adeles over $\mathbb{Q}$, we write $\mathbb{A}^\infty$ to denote the adeles away from the archimedean places, and we write $\mathbb{A}^{\infty,p}$ to denote the adeles away from the archimedean places and p.

Fix a rational prime p that splits as $p = w \cdot w^c$ in the imaginary quadratic extension $F/\mathbb{Q}$. We make this assumption in order to ensure that our unitary group at p is a product of (restrictions of scalars of) general linear groups. Instead, we could assume that every place of K^+ above p splits in the quadratic extension K/K^+ and choose a CM type for K. In addition, we restrict our attention to the case when the prime p is *unramified* in K. This ensures that the Shimura varieties we consider have smooth integral models over $\mathcal{O}_{E,(p)}$ (where $\mathcal{O}_E$ is the ring of integers in the reflex field E of these Shimura varieties) when no level structure at p is imposed.

To help the reader keep track of each setting, we adhere to the following conventions for fonts used to denote schemes, integral models, and formal completions throughout the paper. Schemes over $\mathbb{Q}$ are in normal font, their integral models are in mathcal font, and their formal completions are in mathfrak font.

2 Unitary Shimura Varieties

In this section, we introduce unitary Shimura varieties. In Sect. 2.1, we introduce PEL data and conventions for unitary groups, with which we work throughout the paper. Section 2.2 introduces the PEL moduli problem, and Sect. 2.3 specializes to the setting over $\mathbb{C}$. In our exposition, we follow [22, 23].

2.1 PEL Data and Unitary Groups

The following definition of the PEL datum follows [23, Sect. 1.2], and it is an integral version of the datum in [22, Sect. 4].

By a *PEL datum*, we mean a tuple $(K, c, L, \langle , \rangle, h)$ consisting of

- the CM field K equipped with the involution c introduced in Sect. 1.4,
- an $\mathcal{O}_K$-lattice L, i.e., a finitely generated free $\mathbb{Z}$-module with an action of $\mathcal{O}_K$,
- a non-degenerate Hermitian pairing $\langle \cdot, \cdot \rangle : L \times L \to \mathbb{Z}$ satisfying $\langle k \cdot v_1, v_2 \rangle = \langle v_1, k^c \cdot v_2 \rangle$ for all $v_1, v_2 \in L$ and $k \in \mathcal{O}_k$,
- an $\mathbb{R}$-algebra endomorphism

$$h : \mathbb{C} \to \mathrm{End}_{\mathcal{O}_K \otimes_{\mathbb{Z}} \mathbb{R}}(L \otimes_{\mathbb{Z}} \mathbb{R})$$

  such that $(v_1, v_2) \mapsto \langle v_1, h(i) \cdot v_2 \rangle$ is symmetric and positive definite and such that $\langle h(z)v_1, v_2 \rangle = \langle v_1, h(\overline{z})v_2 \rangle$.

Furthermore, for considering the moduli problem over a p-adic ring and for defining p-adic automorphic forms, we require

- $L_p := L \otimes_{\mathbb{Z}} \mathbb{Z}_p$ is self-dual under the alternating Hermitian pairing $\langle \cdot, \cdot \rangle_p$ on $L \otimes_{\mathbb{Z}} \mathbb{Q}_p$.

To the PEL datum $(K, c, L, \langle , \rangle, h)$, we associate algebraic groups $GU = GU(L, \langle , \rangle)$, $U = U(L, \langle , \rangle)$, and $SU = SU(L, \langle , \rangle)$ defined over $\mathbb{Z}$, whose R-points (for any $\mathbb{Z}$-algebra R) are given by

$$\begin{aligned} GU(R) &:= \{(g, \nu) \in \mathrm{End}_{\mathcal{O}_K \otimes_{\mathbb{Z}} R}(L \otimes_{\mathbb{Z}} R) \times R^{\times} \mid \langle g \cdot v_1, g \cdot v_2 \rangle = \nu \langle v_1, v_2 \rangle\} \\ U(R) &:= \{g \in \mathrm{End}_{\mathcal{O}_K \otimes_{\mathbb{Z}} R}(L \otimes_{\mathbb{Z}} R) \mid \langle g \cdot v_1, g \cdot v_2 \rangle = \langle v_1, v_2 \rangle\} \\ SU(R) &:= \{g \in U(R) \mid \det g = 1\}. \end{aligned}$$

Note that ν is called a *similitude factor*. In the following, for $R = \mathbb{Q}$ or $\mathbb{R}$, we also write

$$GU_+(R) := \{(g, \nu) \in GU(R) \mid \nu > 0\}.$$

Moreover, given a PEL datum $(K, c, L, \langle , \rangle, h)$, we define the $\mathbb{R}$-vector space with an action of K

$$V := L \otimes_{\mathbb{Z}} \mathbb{R}.$$

Then $h_{\mathbb{C}} = h \times_{\mathbb{R}} \mathbb{C}$ gives rise to a decomposition $V \otimes_{\mathbb{R}} \mathbb{C} = V_1 \oplus V_2$ (where $h(z) \times 1$ acts by z on V_1 and by $\bar{z}$ on V_2). We have decompositions $V_1 = \oplus_{\tau \in \Sigma_K} V_{1,\tau}$ and $V_2 = \oplus_{\tau \in \Sigma_K} V_{2,\tau}$ induced from the decomposition of $K \otimes_{\mathbb{Q}} \mathbb{C} = \oplus_{\tau \in \Sigma_K} \mathbb{C}$ where only the τth $\mathbb{C}$ acts nontrivially on $V_{1,\tau} \oplus V_{2,\tau}$, acting via the standard action on $V_{1,\tau}$. As defined in [23, Definition 1.2.5.2], the *signature* of $(V, \langle , \rangle, h)$ is the tuple of pairs $(a_{+\tau}, a_{-\tau})_{\tau \in \Sigma_K}$ such that $a_{+\tau} = \dim_{\mathbb{C}} V_{1,\tau}$ and $a_{-\tau} = \dim_{\mathbb{C}} V_{2,\tau}$ for all $\tau \in \Sigma_K$. Let

$$n = a_{+\tau} + a_{-\tau}.$$

Note that n is independent of τ and furthermore $a_{\pm\tau} = a_{\mp\tau^c}$.

In order to define automorphic forms of nonscalar weight in Sects. 3 and 4, we define the algebraic group over $\mathbb{Z}$

$$H := \prod_{\tau \in \Sigma} \mathrm{GL}_{a_{+\tilde{\tau}}} \times \mathrm{GL}_{a_{-\tilde{\tau}}},$$

where $\tilde{\tau} \in \Sigma_K$ is a previously fixed lift of $\tau \in \Sigma$. Note that $H_{\mathbb{C}}$ can be identified with the Levi subgroup of $U(\mathbb{C})$ that preserves the decomposition $V_{\mathbb{C}} = V_1 \oplus V_2$. This identification also works over $\mathbb{Z}_p$, as we will see in Sect. 4.1.

Moreover, using the decomposition $\mathbb{C} \otimes \mathbb{C} = \mathbb{C} \oplus \mathbb{C}$ of $\mathbb{R}$-modules (where $z \in \mathbb{C}$ acts on the first summand by z and on the second summand by $\bar{z}$) we define $\mu : \mathbb{C} \to V_{\mathbb{C}}$ by $z \mapsto h_{\mathbb{C}}(z, 1)$. (Compare [27, Sect. 12].) Then the *reflex field* is defined to be the field of definition of the $GU(\mathbb{C})$-conjugacy class of μ (or equivalently as the conjugacy class of V_1). Henceforth, we denote the reflex field by E (note that $E \subset \mathbb{C}$).

2.2 PEL Moduli Problem

The goal of this section is to introduce PEL-type unitary Shimura varieties from a moduli-theoretic perspective. We will restrict our attention to cases where these Shimura varieties have no level structure and good reduction at p. For more details, see [22, Sect. 5] or [23, Sect. 1.4].

We now define a moduli problem for abelian varieties equipped with extra structures (more precisely, polarizations, endomorphisms, and level structure) and which will be representable by unitary Shimura varieties that have integral models. For each open compact subgroup $\mathcal{U} \subset GU(\mathbb{A}^{\infty})$, consider the moduli problem

$$(S, s) \mapsto \{(A, i, \lambda, \alpha)\}$$

which assigns to every connected, locally noetherian scheme S over E together with a geometric point s of S the set of tuples (A, i, λ, α), where

- A is an abelian variety over S of dimension $g := [K^+ : \mathbb{Q}] \cdot n$,
- $i : K \hookrightarrow \mathrm{End}^0(A) := (\mathrm{End}(A)) \otimes_{\mathbb{Z}} \mathbb{Q}$ is an embedding of $\mathbb{Q}$-algebras,
- $\lambda : A \to A^\vee$ (where $A^\vee$ denotes the dual abelian variety) is a polarization satisfying $\lambda \circ i(k^c) = i(k)^\vee \circ \lambda$ for all $k \in K$,
- α is a $\pi_1(S, s)$-invariant $\mathcal{U}$-orbit of $K \otimes_{\mathbb{Q}} \mathbb{A}^{\infty}$-equivariant isomorphisms

$$L \otimes_{\mathbb{Z}} \mathbb{A}^{\infty} \xrightarrow{\sim} V_f A_s,$$

  which takes the Hermitian pairing $\langle \cdot, \cdot \rangle$ on L to an $(\mathbb{A}^{\infty})^{\times}$-multiple of the λ-Weil pairing on the rational (adelic) Tate module $V_f A_s$.

Note that Lie A is a locally free $\mathcal{O}_S$-module of rank g and has an induced action of K via i. The tuple (A, i, λ, α) must satisfy Kottwitz's *determinant condition*:

$$\det(K|V_1) = \det_{\mathcal{O}_S}(K|\mathrm{Lie}\, A).$$

Here, by $\det(K|V_1)$ we denote the element in $E[K^\vee] = \mathrm{Sym}(K^\vee) \otimes_{\mathbb{Q}} E$, for $K^\vee$ the $\mathbb{Q}$-vector space dual to K, defined by $k \mapsto \det_{\mathbb{C}}(k|V_1)$, for all $k \in K$. By definition of the reflex field, $\det(K|V_1) \in E[K^\vee] \hookrightarrow \mathbb{C}[K^\vee]$. Similarly, $\det_{\mathcal{O}_S}(K|\mathrm{Lie}\, A)$ denotes the element in $\mathcal{O}_S[K^\vee] = \mathrm{Sym}(K^\vee) \otimes_{\mathbb{Q}} \mathcal{O}_S(S)$ defined by $k \mapsto \det_{\mathcal{O}_S}(k|\mathrm{Lie}\, A)$, for all $k \in K$. The determinant condition is an equality of elements in $\mathcal{O}_S[K^\vee]$, after taking the image of $\det(K|V_1)$ under the structure homomorphism of E to $\mathcal{O}_S(S)$.

Two tuples (A, i, λ, α) and $(A', i', \lambda', \alpha')$ are equivalent if there exists an isogeny $A \to A'$ taking i to i', λ to a rational multiple of λ', and α to α'. We note that the definition is independent of the choice of geometric point s of S. We can extend the definition to non-connected schemes by choosing a geometric point for each connected component.

If the compact open subgroup $\mathcal{U}$ is neat (in particular, if it is sufficiently small) as defined in [23, Definition 1.4.1.8], then this moduli problem is representable by a smooth, quasi-projective scheme $M_{\mathcal{U}}/E$.

From now on, assume that $\mathcal{U} = \mathcal{U}^p\mathcal{U}_p$ is neat and that $\mathcal{U}_p \subset GU(\mathbb{Q}_p)$ is hyperspecial. We can construct an integral model of $M_{\mathcal{U}}$ by considering an integral version of the above moduli problem. To a pair (S, s), where S is now a scheme over $\mathcal{O}_{E,(p)}$, we assign the set of tuples $(A, i, \lambda, \alpha^p)$, where

- A is an abelian variety over S of dimension g,
- $i : \mathcal{O}_{K,(p)} \hookrightarrow (\mathrm{End}(A)) \otimes_{\mathbb{Z}} \mathbb{Z}_{(p)}$ is an embedding of $\mathbb{Z}_{(p)}$-algebras
- $\lambda : A \to A^\vee$ is a prime-to-p polarization satisfying $\lambda \circ i(k^c) = i(k)^\vee \circ \lambda$ for all $k \in \mathcal{O}_K$,
- α^p is a $\pi_1(S, s)$-invariant $\mathcal{U}^p$-orbit of $K \otimes_{\mathbb{Q}} \mathbb{A}^{\infty,p}$-equivariant isomorphisms

$$L \otimes_{\mathbb{Z}} \mathbb{A}^{\infty,p} \xrightarrow{\sim} V_f^p A_s,$$

  which takes the Hermitian pairing $\langle \cdot, \cdot \rangle$ on L to an $(\mathbb{A}^{\infty,p})^\times$-multiple of the λ-Weil pairing on $V_f^p A_s$ (the Tate module away from p).

In addition, the tuple (A, i, λ, α) must satisfy Kottwitz's *determinant condition*:

$$\det(\mathcal{O}_K|V_1) = \det_{\mathcal{O}_S}(\mathcal{O}_K|\mathrm{Lie}A).$$

Here, the determinant condition is an equality of elements in $\mathcal{O}_S[\mathcal{O}_K^\vee]$, after taking the image of $\det(\mathcal{O}_K|V_1) \in (\mathcal{O}_{E,(p)})[\mathcal{O}_K^\vee]$ under the structure homomorphism of $\mathcal{O}_{E,(p)}$ to $\mathcal{O}_S$, for $\mathcal{O}_K^\vee$ the dual $\mathbb{Z}$-module of $\mathcal{O}_K$.

Two tuples (A, i, λ, α) and $(A', i', \lambda', \alpha')$ are equivalent if there exists a prime-to-p isogeny $A \to A'$ taking i to i', λ to a prime-to-p rational multiple of λ', and α to α'.

This moduli problem is representable by a smooth, quasi-projective scheme $\mathcal{M}_{\mathcal{U}}$ over $\mathcal{O}_{E,(p)}$. (See, for example, page 391 of [22] for a discussion of representability and smoothness. The representability is reduced to the Siegel case, proved in [28], while the smoothness follows from Grothendieck–Messing deformation theory.) We have a canonical identification

$$M_{\mathcal{U}} = \mathcal{M}_{\mathcal{U}} \times_{\mathrm{Spec}\,(\mathcal{O}_{E,(p)})} \mathrm{Spec}\, E,$$

which can be checked directly on the level of moduli problems. As the level $\mathcal{U}^p$ varies, the inverse system of Shimura varieties $\mathcal{M}_{\mathcal{U}}$ has a natural action of $GU(\mathbb{A}^{\infty,p})$. (More precisely, $g \in GU(\mathbb{A}^{\infty,p})$ acts by precomposing the level structure α with it.) Since our interest is in the p-adic theory, we will fix and suppress the level $\mathcal{U}$ starting from Sect. 4.

2.3 *Abelian Varieties and Shimura Varieties over* $\mathbb{C}$

In this section, we specialize to working over $\mathbb{C}$. Our goal is to sketch how the set $M_{\mathcal{U}}(\mathbb{C})$ of complex points of $M_{\mathcal{U}}$ is naturally identified with the set of points of a finite union of locally symmetric complex varieties corresponding to (GU, h). (For more details, see [22, Sect. 8].)

We remark that what we show is merely a bijection of sets. Proving that the Shimura varieties corresponding to (GU, h) are moduli spaces of abelian varieties over $\mathbb{C}$ would also require matching the complex structures on the two sides.

2.3.1 Abelian Varieties over $\mathbb{C}$

Recall that the $\mathbb{C}$-points of an abelian variety $A/\mathbb{C}$ are of the form $V(A)/\Lambda$, where Λ is a $\mathbb{Z}$-lattice in a complex vector space $V(A)$. Any abelian variety over $\mathbb{C}$ admits a polarization; since $V(A)/\Lambda$ comes from a complex abelian variety A, it is also polarizable, i.e., there exists a non-degenerate, positive definite Hermitian form

$$\lambda_{\mathbb{C}} : V(A) \times V(A) \to \mathbb{C} \quad \text{s.t. } \lambda_{\mathbb{C}}(\Lambda, \Lambda) \subset \mathbb{Z}.$$

We call each such Hermitian form $\lambda_{\mathbb{C}}$ a *polarization* of $V(A)/\Lambda$. It may be better to think of a polarization as an alternating form $\lambda_{\mathbb{R}} : V(A) \times V(A) \to \mathbb{R}$ satisfying $\lambda_{\mathbb{R}}(iu, iv) = \lambda_{\mathbb{R}}(u, v)$ for all $u, v \in V(A)$ and

$$\lambda_{\mathbb{C}}(u, v) = \lambda_{\mathbb{R}}(u, iv) + i\lambda_{\mathbb{R}}(u, v).$$

It is enough to characterize a pair $(A, \lambda_{\mathbb{C}})$, where A is an abelian variety of dimension g, by considering the following triple:

1. the free $\mathbb{Z}$-module $\Lambda = H_1(A, \mathbb{Z})$ of rank $2g$
2. the $\mathbb{R}$-algebra homomorphism $\mathbb{C} \to \mathrm{End}_{\mathbb{R}}(\Lambda \otimes \mathbb{R}) = \mathrm{End}_{\mathbb{R}}(H_1(A, \mathbb{R})) = \mathrm{End}_{\mathbb{R}}(\mathrm{Lie}\, A)$ describing the complex structure on $\mathrm{Lie}\, A$ (so $V(A) := \Lambda \otimes \mathbb{R}$, endowed with this complex structure)
3. the alternating form on $\Lambda = H_1(A, \mathbb{Z})$ induced by $\lambda_{\mathbb{C}}$ denoted by $\langle \cdot, \cdot \rangle$ after identifying $A(\mathbb{C}) \cong \mathrm{Lie}\, A(\mathbb{C})/H_1(A, \mathbb{Z})$

2.3.2 Shimura Varieties over $\mathbb{C}$

Recall that, associated with the PEL datum, we have the $\mathbb{R}$-vector space $V = L\otimes_{\mathbb{Z}}\mathbb{R}$, which is endowed with an action of K and the complex structure defined by h. (Note that the complex structure depends uniquely on $h(i) \in \mathrm{End}_{K\otimes_{\mathbb{Q}}\mathbb{R}}(V)$.)

Let $\mathfrak{h}$ denote the set of elements $I \in \mathrm{End}_{K\otimes_{\mathbb{Q}}\mathbb{R}}(V)$ which satisfy

1. $I^2 = -1$
2. $I^c = -I$
3. $(w, v) \mapsto \langle w, Iv\rangle$ is a positive or negative definite form on V
4. the $K \otimes_{\mathbb{Q}} \mathbb{C}$-structures on V defined by I and $h(i)$ are isomorphic.

In [22, Lemmas 4.1 and 4.2], Kottwitz shows that the set $\mathfrak{h}$ is equal to $GU(\mathbb{R})/C_h$, for C_h the stabilizer of $h(i)$ in $GU(\mathbb{R})$, and that it can be identified with a finite union of copies of the symmetric domain for the identity component of $GU(\mathbb{R})$.

For $\mathcal{U} \subset GU(\mathbb{A}^\infty)$ a neat open compact subgroup, we define the quotient

$$X_{\mathcal{U}} = GU(\mathbb{Q})\backslash(GU(\mathbb{A}^\infty)/\mathcal{U} \times \mathfrak{h}). \tag{1}$$

We sketch how the set of complex points of $M_{\mathcal{U}}$ corresponds to a disjoint union of finitely many copies of $X_{\mathcal{U}}$.

By definition, the set $M_{\mathcal{U}}(\mathbb{C})$ parametrizes equivalence classes of tuples (A, i, λ, α) where

1. A is an abelian variety over $\mathbb{C}$,
2. $i : K \hookrightarrow \mathrm{End}^0(A)$ is an embedding of $\mathbb{Q}$-algebras,
3. $\lambda : A \to A^\vee$ is a polarization satisfying $\lambda \circ i(b^c) = i(b)^\vee \circ \lambda$ for all $b \in K$,
4. α is a $\mathcal{U}$-orbit of isomorphisms of skew-Hermitian K-vector spaces (in the sense of Kottwitz, i.e., preserving the pairing only up to scalar) $L \otimes_{\mathbb{Z}} \mathbb{A}^\infty \cong H_1(A, \mathbb{A}^\infty)$.

In addition, the tuple (A, i, λ, α) must satisfy the determinant condition.

Every equivalence class of tuples (A, i, λ, α) satisfying the above restrictions gives rise to an element $GU(\mathbb{A}^\infty)/\mathcal{U}\times\mathfrak{h}$ as follows. The existence of the equivalence class of isomorphisms α implies that the skew-Hermitian K-vector spaces $L_{\mathbb{Q}} := L \otimes_{\mathbb{Z}} \mathbb{Q}$ and $H = H_1(A, \mathbb{Q})$ are isomorphic over any finite place of $\mathbb{Q}$. We conclude in particular that $H_1(A, \mathbb{Q})$ and $L_{\mathbb{Q}}$ have the same dimension over K. By [22, Lemma 4.2], $H_{\mathbb{R}}$ and $V = L_{\mathbb{Q}} \otimes_{\mathbb{Q}} \mathbb{R}$ are isomorphic as skew-Hermitian K-vector spaces if and only if they are isomorphic as $K \otimes_{\mathbb{Q}} \mathbb{C}$-modules. The natural complex structure on $H_{\mathbb{R}} \cong \mathrm{Lie}(A)$ gives a decomposition of $H_{\mathbb{C}}$ as $H_1 \oplus H_2$, and the determinant condition implies that H_1 is isomorphic to V_1 as $K \otimes_{\mathbb{Q}} \mathbb{C}$-modules. Thus, in order to deduce that $H_{\mathbb{R}}$ and V are isomorphic as $K \otimes_{\mathbb{Q}} \mathbb{C}$-modules it suffices to prove that V_2 and H_2 are isomorphic as $K \otimes_{\mathbb{Q}} \mathbb{C}$-modules. Note that if we denote by W_τ the $\mathbb{C}$-subspace where K acts via τ, for $\tau : K \hookrightarrow \mathbb{C}$ a complex embedding ($\tau \in \Sigma_K$) and W any $K \otimes_{\mathbb{Q}} \mathbb{C}$-module, then two $K \otimes_{\mathbb{Q}} \mathbb{C}$-modules W, W' are isomorphic if and only if $\dim_{\mathbb{C}} W_\tau = \dim_{\mathbb{C}} W'_\tau$, for all $\tau \in \Sigma_K$. Let $\tau \in \Sigma_K$. The determinant condition implies that $\dim_{\mathbb{C}} V_{1,\tau} = \dim_{\mathbb{C}} H_{1,\tau}$, the decompositions $V_{\mathbb{C}} = V_1 \oplus V_2$ and $H_{\mathbb{C}} = H_1 \oplus H_2$ imply $\dim_{\mathbb{C}} V_\tau + \dim_{\mathbb{C}} V_{\tau^c} = \frac{1}{2}(\dim_{\mathbb{C}} V_{1,\tau} + \dim_{\mathbb{C}} V_{2,\tau^c} + \dim_{\mathbb{C}} V_{1,\tau^c} + \dim_{\mathbb{C}} V_{2,\tau}) = \dim_{\mathbb{C}} V_{1,\tau} + \dim_{\mathbb{C}} V_{2,\tau}$, and $\dim_{\mathbb{C}} H_\tau + \dim_{\mathbb{C}} H_{\tau^c} = \dim_{\mathbb{C}} H_{1,\tau} + \dim_{\mathbb{C}} H_{2,\tau}$, and the equality $\dim_K L_{\mathbb{Q}} = \dim_K H$ implies $\dim_{\mathbb{C}} V_\tau + \dim_{\mathbb{C}} V_{\tau^c} = \dim_{\mathbb{C}} H_\tau + \dim_{\mathbb{C}} H_{\tau^c}$. We deduce that $\dim_{\mathbb{C}} V_{2,\tau} = \dim_{\mathbb{C}} H_{2,\tau}$ for all $\tau \in \Sigma_K$, i.e., that H_2 and V_2 are also isomorphic as $K \otimes_{\mathbb{Q}} \mathbb{C}$-modules. We conclude that the skew-Hermitian K-vector spaces H and $L_{\mathbb{Q}}$ are isomorphic over any place v of $\mathbb{Q}$.

When the Hasse principle holds, this implies the existence of an isomorphism of skew-Hermitian K-vector spaces between H and $L_{\mathbb{Q}}$. In general, there are

$$\left|\ker^1(\mathbb{Q}, GU) := \ker\left(H^1(\mathbb{Q}, GU) \to \prod_{v \text{ place of } \mathbb{Q}} H^1(\mathbb{Q}_v, GU)\right)\right|$$

isomorphism classes $L^{(i)}$ of skew-Hermitian K-vector spaces isomorphic to $L_{\mathbb{Q}}$ at every place of $\mathbb{Q}$ (and let $L^{(1)} = L_{\mathbb{Q}}$).

Let $1 \leq i \leq |\ker^1(\mathbb{Q}, GU)|$. We define $GU^{(i)}$ to be the unitary similitude group over $\mathbb{Q}$ defined by $L^{(i)}$ (so, in particular, $GU^{(1)} = GU$). Choose local isomorphisms $L_{\mathbb{Q}_v} \cong L^{(i)}_{\mathbb{Q}_v}$ for all places v of $\mathbb{Q}$, and let $GU^{(i)}(\mathbb{Q})$ act on $GU(\mathbb{A}^\infty)/\mathcal{U} \times \mathfrak{h}$ via the induced isomorphisms $GU^{(i)}_{\mathbb{Q}_v} \cong GU_{\mathbb{Q}_v}$. We define $X^{(i)}_{\mathcal{U}} = GU^{(i)}(\mathbb{Q})\backslash(GU(\mathbb{A}^\infty)/\mathcal{U} \times \mathfrak{h})$, and we define $M^{(i)}_{\mathcal{U}}(\mathbb{C})$ to be the subset of $M_{\mathcal{U}}(\mathbb{C})$ parameterizing tuples such that H is isomorphic to $L^{(i)}$. Thus, $M_{\mathcal{U}}(\mathbb{C}) = \coprod_i M^{(i)}_{\mathcal{U}}(\mathbb{C})$. We show that $M^{(i)}_{\mathcal{U}}(\mathbb{C})$ naturally identifies with $X^{(i)}_{\mathcal{U}}$.

Let i, where $1 \leq i \leq |\ker^1(\mathbb{Q}, GU)|$, be such that H and $L^{(i)}$ are isomorphic skew-Hermitian K-vector spaces, and choose an automorphism $\alpha_{\mathbb{Q}} : H \xrightarrow{\sim} L^{(i)}$. Then, the automorphism $(\alpha_{\mathbb{Q}} \otimes \mathbb{I}_{\mathbb{A}^\infty}) \circ \alpha$ of $L^{(i)} \otimes_{\mathbb{Q}} \mathbb{A}^\infty$ defines an element of $GU(\mathbb{A}^\infty)$, but since α is only well-defined up to its orbit in $\mathcal{U}$, such an isomorphism determines an element of $GU(\mathbb{A}^\infty)/\mathcal{U}$. Under $\alpha_{\mathbb{Q}}$, the complex structure on $H_1(A, \mathbb{R})$ defines a complex structure on V, which is conjugate to h by an element in $GU(\mathbb{R})$, i.e., an element in $\mathfrak{h}$.

Therefore, each class of tuples (A, i, λ, α) along with a choice of isomorphism $\alpha_{\mathbb{Q}}$ determines an element of $GU(\mathbb{A}^\infty)/\mathcal{U} \times \mathfrak{h}$. Forgetting the isomorphism $\alpha_{\mathbb{Q}}$ is equivalent to taking the quotient by the left action of $GU^{(i)}(\mathbb{Q})$. Thus, to each point of $M^{(i)}_{\mathcal{U}}(\mathbb{C})$ we associated a point on $X^{(i)}_{\mathcal{U}}$, and this map is in fact a bijection.

Note that in [22, Sects. 7 and 8] Kottwitz shows that under our assumptions (case A in loc. cit.) if n is even the Hasse principle holds, and if n is odd the natural map $\ker^1(\mathbb{Q}, Z) \to \ker^1(\mathbb{Q}, GU)$, for Z the center of GU, is a bijection, and furthermore that the subvarieties $M^{(i)}_{\mathcal{U}}(\mathbb{C})$ are all isomorphic to $M^{(1)}_{\mathcal{U}}(\mathbb{C}) = X_{\mathcal{U}}$.

For the later sections, we will denote a connected component of $M_{\mathcal{U}}(\mathbb{C})$ (or equivalently, of $X_{\mathcal{U}}(\mathbb{C})$) as $S_{\mathcal{U}}(\mathbb{C})$. Note that any two connected components are isomorphic as complex manifolds.

3 Classical Automorphic Forms

In this section we will first recall the classical definition of automorphic forms on unitary groups over $\mathbb{C}$ following [31], and then describe equivalent viewpoints that let us generalize to work over base rings other than $\mathbb{C}$.

3.1 *Classical Definition of Complex Automorphic Forms on Unitary Groups*

For the moment, suppose $a_{+\tilde{\tau}}a_{-\tilde{\tau}} \neq 0$ for all $\tau \in \Sigma$. Consider the domain $\mathcal{H}$ for $GU_+(\mathbb{R})$:

$$\mathcal{H} = \prod_{\tau\in\Sigma} \mathcal{H}_{a_{+\tilde{\tau}}\times a_{-\tilde{\tau}}} \text{ with } \mathcal{H}_{a_{+\tilde{\tau}}\times a_{-\tilde{\tau}}}$$
$$= \{z \in \mathrm{Mat}_{a_{-\tilde{\tau}}\times a_{+\tilde{\tau}}}(\mathbb{C}) \mid 1 - {}^t z^c z \text{ is positive definite}\}.$$

Note that $\mathrm{Isom}_{\mathcal{O}_K\otimes_{\mathbb{Z}}\mathbb{R}}(L\otimes_{\mathbb{Z}}\mathbb{R}) \simeq \mathrm{GL}_n(\mathcal{O}_K\otimes\mathbb{R}) \simeq \mathrm{GL}_n(\prod_{\tau\in\Sigma}\mathbb{C}) \simeq \prod_{\tau\in\Sigma}\mathrm{GL}_n(\mathbb{C})$. We use this identification to write $g \in GU_+(\mathbb{R})$ as $\begin{pmatrix} a_{g,\tau} & b_{g,\tau} \\ c_{g,\tau} & d_{g,\tau}\end{pmatrix}_{\tau\in\Sigma} \in GU_+(\mathbb{R})$, where $a_{g,\tau} \in \mathrm{GL}_{a_{+\tilde{\tau}}}(\mathbb{C})$ and $d_{g,\tau} \in \mathrm{GL}_{a_{-\tilde{\tau}}}(\mathbb{C})$. Then the action of g on $\mathcal{H}$ is given by

$$gz = ((a_{g,\tau}z_\tau + b_{g,\tau})(c_{g,\tau}z_\tau + d_{g,\tau})^{-1})_{\tau\in\Sigma} \text{ for } z = (z_\tau)_{\tau\in\Sigma} \in \prod_{\tau\in\Sigma}\mathcal{H}_{a_{+\tilde{\tau}}\times a_{-\tilde{\tau}}}.$$

By [31, 12.1], $\mathcal{H}$ is the irreducible (Hermitian) symmetric domain for $SU(\mathbb{R})$. By the classification of Hermitian symmetric domains and [26, Corollary 5.8], $\mathcal{H}$ is uniquely determined by the adjoint group of a connected component of $GU(\mathbb{R})$. Recall from Sect. 2.3.2 that we can identify $\mathfrak{h}$ with a finite union of copies of the symmetric domains for the identity component of $GU(\mathbb{R})$. Hence $\mathfrak{h}$ can be identified with a finite (disjoint) union of copies of $\mathcal{H}$.

In order to define the desired transformation properties that automorphic forms should satisfy, we need to introduce a few more definitions. Using the above notation, for $g \in GU_+(\mathbb{R})$ and $z = (z_\tau)_{\tau\in\Sigma} \in \mathcal{H}$ the *factors of automorphy* for each $\{\tilde{\tau}, \tilde{\tau}^c\} \subset \Sigma_K$ above $\tau \in \Sigma$ are defined by

$$\mu_{\tilde{\tau}}(g, z) := c_{g,\tau}z_\tau + d_{g,\tau} \text{ and } \mu_{\tilde{\tau}^c}(g, z) := \overline{b}_{g,\tau}{}^t z_\tau + \overline{a}_{g,\tau},$$

and the *scalar factors of automorphy* are

$$j_\tau(g, z) := \det(\mu_\tau(g, z)) \text{ for } \tau \in \Sigma_K.$$

So far, we have considered the case in which $a_{+\tilde{\tau}}a_{-\tilde{\tau}} \neq 0$. Now, suppose $a_{+\tilde{\tau}}a_{-\tilde{\tau}} = 0$. In this case, $\mathcal{H}_{a_{+\tilde{\tau}} \times a_{-\tilde{\tau}}}$ is defined to be the element 0, with the group acting trivially on it. Following [31, Sect. 3.3], if $a_{-\tilde{\tau}} = 0$, we define

$$\begin{aligned} \mu_{\tilde{\tau}^c}(g,z) &= \overline{g} \\ \mu_{\tilde{\tau}}(g,z) &= 1 \\ j_{\tau}(g,z) &= 1, \end{aligned}$$

and if $a_{+\tilde{\tau}} = 0$, we define

$$\begin{aligned} \mu_{\tilde{\tau}}(g,z) &= g \\ \mu_{\tilde{\tau}^c}(g,z) &= 1 \\ j_{\tau}(g,z) &= \det(g). \end{aligned}$$

Remark 3.1. By [30, Eq. (1.19)], for all $\tau \in \Sigma$ and $g \in GU_+(\mathbb{R})$,

$$\det(\mu_{\tilde{\tau}^c}(g,z)) = \det(g)^{-1}\, \nu(g)^{a_{+,\tilde{\tau}}} \det(\mu_{\tilde{\tau}}(g,z)).$$

Define

$$M_g(z) := (\mu_{\tilde{\tau}^c}(g,z), \mu_{\tilde{\tau}}(g,z))_{\tau\in\Sigma} \quad \in \prod_{\tau\in\Sigma} \mathrm{Mat}_{a_{+\tilde{\tau}} \times a_{+\tilde{\tau}}} \times \mathrm{Mat}_{a_{-\tilde{\tau}} \times a_{-\tilde{\tau}}}$$

If $\rho : H(\mathbb{C}) = \prod_{\tau\in\Sigma} \mathrm{GL}_{a_{+\tilde{\tau}}}(\mathbb{C}) \times \mathrm{GL}_{a_{-\tilde{\tau}}}(\mathbb{C}) \to \mathrm{GL}(X)$ is a rational representation into a finite-dimensional complex vector space X, $f : \mathcal{H} \to X$ a map and $g \in GU_+(\mathbb{R})$, then denote by $f||_\rho g : \mathcal{H} \to X$ and $f|_\rho g : \mathcal{H} \to X$ the maps given by

$$\begin{aligned} (f||_\rho g)(z) &:= \rho(M_g(z))^{-1} f(gz) \\ f|_\rho g &:= f||_\rho \left(\nu(g)^{-1/2} g\right) \end{aligned}$$

for all $z \in \mathcal{H}$. Note that for all $g \in U(\mathbb{R})$,

$$f|_\rho g = f||_\rho g.$$

Associated with an $\mathcal{O}_F$-lattice L_F in $L \otimes_{\mathbb{Z}} \mathbb{Q}$ and an integral $\mathcal{O}_F$-ideal $\mathfrak{c}$, we define the subgroup

$$\Gamma(L_F, \mathfrak{c}) := \{g \in GU_+(\mathbb{Q}) \mid {}^tL_F g = {}^tL_F \text{ and } {}^tL_F(1-g) \subset \mathfrak{c}\, {}^tL_F\}.$$

Then a *congruence subgroup* Γ of $GU_+(\mathbb{Q})$ is a subgroup of $GU_+(\mathbb{Q})$ that contains $\Gamma(L_F, \mathfrak{c})$ as subgroup of finite index for some choice of $(L_F, \mathfrak{c})$ as above.

Definition 3.2. Let Γ be a congruence subgroup of $GU_+(\mathbb{Q})$, X a finite-dimensional complex vector space, and $\rho : H(\mathbb{C}) \to \mathrm{GL}(X)$ a rational representation. A function $f : \mathcal{H} \to X$ is called a *(holomorphic) automorphic form* of weight ρ with respect to Γ if it satisfies the following properties:

(1) f is holomorphic,
(2) $f||_\rho\gamma = f$ for every γ in Γ,
(3) if Σ consists of only one place τ and $(a_{+\tilde{\tau}}, a_{-\tilde{\tau}}) = (1, 1)$, then f is holomorphic at every cusp.

We will call a function $f : \mathcal{H} \to X$ that satisfies property (2), but not necessarily (1) and (3), an *automorphic function*.

Remark 3.3. Note that if we are not in the case in which both Σ consists of only one place τ and $(a_{+\tilde{\tau}}, a_{-\tilde{\tau}}) = (1, 1)$, then Koecher's principle implies that an automorphic form is automatically holomorphic at the boundary. (See [25, Theorem 2.5] for a very general version.)

Remark 3.4. Sometimes, in the definition of an automorphic form, the second condition of Definition 3.2 is replaced by $f|_\rho\gamma = f$ for every γ in Γ. The condition that arises from geometry, though, is $f||_\rho\gamma = f$. Since our main results and proofs are geometric (and since we want this definition of automorphic forms to agree with the geometric definitions we give later), we require $f||_\rho\gamma = f$ instead of $f|_\rho\gamma = f$ in this paper.

3.1.1 Weights of an Automorphic Form

The irreducible algebraic representations of $H = \prod_{\tau\in\Sigma} \mathrm{GL}_{a_{+\tilde{\tau}}} \times \mathrm{GL}_{a_{-\tilde{\tau}}}$ over $\mathbb{C}$ are in one-to-one correspondence with dominant weights of a maximal torus T (over $\mathbb{C}$). More precisely, let T be the product of the diagonal tori $T_{a_{+\tilde{\tau}}} \times T_{a_{-\tilde{\tau}}}$ for $\tau \in \Sigma$. For $1 \le i \le a_{+\tilde{\tau}} + a_{-\tilde{\tau}}$, let ε_i^τ in $X(T) := \mathrm{Hom}_\mathbb{C}(T, \mathbb{G}_m)$ be the character defined by

$$T(\mathbb{C}) = \prod_{\sigma\in\Sigma} T_{a_{+\tilde{\sigma}}}(\mathbb{C}) \times T_{a_{-\tilde{\sigma}}}(\mathbb{C}) \ni \mathrm{diag}(x_1^\sigma, \dots, x_{a_{+\tilde{\sigma}}+a_{-\tilde{\sigma}}}^\sigma)_{\sigma\in\Sigma} \mapsto x_i^\tau \in \mathbb{G}_m(\mathbb{C}).$$

These characters form a basis of the free $\mathbb{Z}$-module $X(T)$, and we choose $\Delta = \{\alpha_i^\tau := \varepsilon_i^\tau - \varepsilon_{i+1}^\tau\}_{\tau\in\Sigma, 1\le i<a_{+\tilde{\tau}}+a_{-\tilde{\tau}}, i\ne a_{+\tilde{\tau}}}$ as a basis for the root system of H. Then the *dominant weights* of T with respect to Δ are $X(T)_+ = \{\kappa \in X(T) \mid \langle\kappa, \check{\alpha}\rangle \ge 0\ \forall\alpha \in \Delta\}$, and using the above basis of $X(T)$ they can be identified as follows:

$$X(T)_+ \cong \{(n_1^\tau, \dots, n_{a_{+\tilde{\tau}}+a_{-\tilde{\tau}}}^\tau)_{\tau\in\Sigma} \in \prod_{\tau\in\Sigma} \mathbb{Z}^{a_{+\tilde{\tau}}+a_{-\tilde{\tau}}} : n_i^\tau \ge n_{i+1}^\tau \forall i \ne a_{+\tilde{\tau}}\}.$$

For such a dominant weight κ, let $\rho_\kappa : H_\mathbb{C} \to M_\rho$ denote the irreducible algebraic representation of the highest weight κ. (See, for example, [18, Part II. Chap. 2].) We call f an automorphic form of weight κ, where $\kappa \in X(T)_+$, if f is an automorphic form of weight ρ_κ.

Remark 3.5. In view of later generalizations, note that ρ_κ can be defined as an algebraic (i.e., schematic) representation of the algebraic group H over the integers $\mathbb{Z}$. More precisely, let T be the maximal split torus of H extending the diagonal torus over $\mathbb{C}$ constructed above and note that $\mathrm{Hom}_{\mathbb{Z}}(T, \mathbb{G}_m) = \mathrm{Hom}_{\mathbb{C}}(T, \mathbb{G}_m)$, i.e., we can view $\kappa \in X(T)_+$ as a character of T defined over $\mathbb{Z}$. Let B be a Borel subgroup containing T (corresponding to upper triangular matrices in H) and B^- the opposite Borel, and denote by N and N^- the unipotent radicals of B and B^-, respectively. Then κ can be viewed as a character of B^- acting trivially on N^- via the quotient $B^- \twoheadrightarrow T$, and we define the representation ρ_κ of highest weight κ by

$$\mathrm{Ind}_{B^-}^{H}(-\kappa) = \{f : H/N^- \to \mathbb{A}^1 \,|\, f(ht) = \kappa(t)^{-1} f(h), t \in T\},$$

where $\mathbb{A}^1$ is the affine line and on which H acts via

$$H \ni h : f(x) \mapsto \rho_\kappa(h) f(x) = f(h^{-1}x),$$

see [18]. This representation is irreducible of the highest weight κ over any field of characteristic zero.

3.1.2 Unitary Domains and Moduli Problems

We will now reformulate Definition 3.2 in order to generalize it in Sect. 3.2. We quickly recall the correspondence between quadruples described in Sect. 2.3.2 and points of $\mathcal{H}$.

Fix a PEL datum $(K, c, L, \langle , \rangle, h)$ and a neat open compact subgroup $\mathcal{U} \subset GU(\mathbb{A}^\infty)$; let $M_{\mathcal{U}}(\mathbb{C})$ denote the complex Shimura variety of level $\mathcal{U}$ associated with the PEL datum discussed in Sect. 2.3.2, and let $S_{\mathcal{U}}(\mathbb{C})$ be a connected component in $M_{\mathcal{U}}^{(1)}(\mathbb{C})$. Recall that $V = L \otimes_{\mathbb{Z}} \mathbb{R}$. We now describe the identification between elements $z \in \Gamma \backslash \mathcal{H}$ and abelian varieties $\underline{A} \in S_{\mathcal{U}}(\mathbb{C})$ given by [31, Chap. I]. Shimura defines for $z \in \mathcal{H}$ an $\mathbb{R}$-linear isomorphism $p_z : V \to \mathbb{C}^g$ which induces a Riemann form on $\mathbb{C}^g$:

$$E_z(p_z(x), p_z(y)) = \langle x, y \rangle \quad x, y \in V.$$

This implies that $A_z = \mathbb{C}^g / p_z(L)$ is an abelian variety together with a polarization λ_z corresponding to E_z. For $k \in K$, define $i_z(k)$ to be the element of $\mathrm{End}_{\mathbb{Q}}(A_z)$ induced by the action of $h(k_\tau)$ acting on V_τ and let α_z denote the $\mathcal{U}$-orbit of isomorphisms $V \otimes \mathbb{A}^\infty \cong H_1(A_z, \mathbb{A}^\infty)$ induced by p_z. Altogether, p_z gives rise to the Riemann form E_z, the following commutative diagram:

$$\begin{array}{ccccccccc} 0 & \longrightarrow & L & \longrightarrow & V & \longrightarrow & V/L & \longrightarrow & 0 \\ & & \downarrow & & \downarrow & & \downarrow & & \\ 0 & \longrightarrow & \Lambda & \longrightarrow & \mathbb{C}^g & \longrightarrow & A & \longrightarrow & 0, \end{array} \qquad (2)$$

and $\underline{A}_z := (A_z, i_z, \lambda_z, \alpha_z)$. Shimura [31, Theorem 4.8] (together with [27, Proposition 4.1]) further proves that

Proposition 3.6. *For each $z \in \mathcal{H}$, $\underline{A}_z \in S_{\mathcal{U}}(\mathbb{C})$ for some neat open $\mathcal{U}$. Conversely, if $\underline{A} \in S_{\mathcal{U}}(\mathbb{C})$ for some $\mathcal{U}$, then there is a $z \in \mathcal{H}$ such that $\underline{A}$ is equivalent to $\underline{A}_z$. Furthermore $\underline{A}_z$ and $\underline{A}_w$ for $z, w \in \mathcal{H}$ are equivalent if and only if $w = \gamma z$ for some $\gamma \in \Gamma_{\mathcal{U}} := \mathcal{U} \cap GU_+(\mathbb{Q})$.*

3.1.3 Second Definition of Classical Automorphic Functions

For $\underline{A} \in S_{\mathcal{U}}(\mathbb{C})(\mathbb{C})$, let $\underline{\Omega} = H^1(A, \mathbb{Z}) \otimes \mathbb{C}$ which has a decomposition into $\underline{\Omega}^{\pm}$ depending on the (induced) action of $h(i)$ on $\underline{\Omega}$. Set

$$\mathcal{E}^{\pm}_{\underline{A}\ \tau} = \mathrm{Isom}_{\mathbb{C}}(\mathbb{C}^{a_{\pm\bar{\tau}}}, \underline{\Omega}^{\pm}_{\tau}) \quad \text{and} \quad \mathcal{E}_{\underline{A}} = \bigoplus_{\tau \in \Sigma} (\mathcal{E}^{+}_{\underline{A}\ \tau} \oplus \mathcal{E}^{-}_{\underline{A}\ \tau}).$$

The group $\mathrm{GL}_{a_{\pm\bar{\tau}}}(\mathbb{C})$ acts on $\mathcal{E}^{\pm}_{\underline{A}\ \tau}$ by

$$(g.f)(x) = f({}^t g x) \quad \text{for } g \in \mathrm{GL}_{a_{\pm\bar{\tau}}}(\mathbb{C}), f \in \mathcal{E}^{\pm}_{\underline{A}\ \tau}, x \in \mathbb{C}^{a_{\pm\bar{\tau}}},$$

which induces a (diagonal) action of $H(\mathbb{C})$ on $\mathcal{E}_{\underline{A}}$.

For $z \in \mathcal{H}$, the map p_z induces a choice of basis on $H_1(A_z, \mathbb{Z})$ via the identification with $p_z(L)$, which by duality induces a choice of basis $\mathcal{B}_z$ of $\underline{\Omega}$. This is equivalent to giving an element l_z of $\mathcal{E}_{\underline{A}_z}$.

Lemma 3.7. *Let $\rho : H(\mathbb{C}) \to \mathrm{GL}(X)$ be a rational representation. Then there exists a one-to-one correspondence between automorphic functions of weight ρ with respect to $\Gamma_{\mathcal{U}}$ and the set of functions F from pairs $(\underline{A}, l)$, where $\underline{A} \in S_{\mathcal{U}}(\mathbb{C})$ and $l \in \mathcal{E}_{\underline{A}}$, to X satisfying*

$$F(\underline{A}, gl) = \rho({}^t g)^{-1} F(\underline{A}, l) \text{ for all } g \in H(\mathbb{C}). \qquad (3)$$

The bijection is given by sending a function F to the automorphic function $f_F : z \mapsto F(\underline{A}_z, l_z)$.

Proof (sketch). We will first show that the map $F \mapsto (f_F : z \mapsto F(\underline{A}_z, l_z))$ is well-defined and afterwards construct an inverse for it. To do the former, let $z \in \mathcal{H}$, $\gamma \in \Gamma_{\mathcal{U}}$. We need to show that $F(\underline{A}_z, l_z) = \rho(M_\gamma(z))^{-1} F(\underline{A}_{\gamma z}, l_{\gamma z})$. It follows from [31, p. 27] that ${}^t M_\gamma(z)$ maps $p_{\gamma z}(L)$ to $p_z(L)$ and defines an isomorphism between

$\underline{A}_{\gamma z}$ and $\underline{A}_z$. Note that under this isomorphism, $l_{\gamma z}$ of $\underline{\Omega}$ gets mapped to ${}^t M_\gamma(z)^{-1} l_z$. Hence using property (3), we obtain

$$F(\underline{A}_{\gamma z}, l_{\gamma z}) = F(\underline{A}_z, {}^t M_\gamma(z)^{-1} l_z) = \rho(M_\gamma(z)) F(\underline{A}_z, l_z)$$

as desired.

We define the inverse of $F \mapsto f_F$ as follows: Let f be an automorphic function of weight ρ with respect to $\Gamma_{\mathcal{U}}$, and $(\underline{A}, l)$ where $\underline{A} \in S_{\mathcal{U}}(\mathbb{C})$ and $l \in \mathcal{E}_{\underline{A}}$. Then, by Proposition 3.6, there exists $z \in \mathcal{H}$ such that $\underline{A}_z$ is isomorphic to $\underline{A}$, and there exists a unique $g \in H(\mathbb{C})$ such that $l = g l_z$. We define $F_f(\underline{A}, l) = \rho({}^t g)^{-1} f(z)$. By the transformation property of automorphic function we obtain analogously to above that $f \mapsto F_f$ is well-defined and F_f satisfies (3). Moreover, $f \mapsto F_f$ is obviously an inverse of $F \mapsto f_F$. □

Automorphic functions of weight ρ with respect to congruence subgroups Γ that strictly contain $\Gamma_{\mathcal{U}}$ can be characterized in the same style as follows:

Lemma 3.8. *Let $\rho : H(\mathbb{C}) \to \mathrm{GL}(X)$ be a rational representation, and Γ a congruence subgroup of $GU_+(\mathbb{Q})$ containing $\Gamma_{\mathcal{U}} = \mathcal{U} \cap GU_+(\mathbb{Q})$. Then there exists a one-to-one correspondence between automorphic functions of weight ρ with respect to Γ and the set of functions F from pairs $(\underline{A}, l)$, where $\underline{A} = (A, i, \lambda, \alpha) \in S_{\mathcal{U}}(\mathbb{C})$ and $l \in \mathcal{E}_{\underline{A}}$, to X satisfying*

$$F(\underline{A}, gl) = \rho({}^t g)^{-1} F(\underline{A}, l) \text{ for all } g \in H(\mathbb{C}),$$

and such that for all $\gamma \in \Gamma$ and $z \in \mathcal{H}$, we have

$$F(\underline{A}_z, l_z) = \rho(M_\gamma(z))^{-1} F(\underline{A}_{\gamma z}, l_{\gamma z}). \tag{4}$$

It suffices to check condition (4) for a set of representatives of $\Gamma / \Gamma_{\mathcal{U}}$.

This lemma follows easily from Lemma 3.7 and Definition 3.2.

3.1.4 Algebraic Definition of Classical Automorphic Functions

Finally, we would like to view automorphic forms as functions on abelian varieties on $S_{\mathcal{U}}(\mathbb{C})$. For this, we define the *contracted product* $\mathcal{E}_{\underline{A},\rho} = \mathcal{E}_{\underline{A}} \times^\rho X$ of $\mathcal{E}_{\underline{A}}$ and X to be the product $\mathcal{E}_{\underline{A}} \times X$ modulo the equivalence relation given by $(l, v) \sim (gl, \rho({}^t g)^{-1} v)$ for $g \in H(\mathbb{C})$. Note that for $g \in GU_+(\mathbb{Q})$, the identification $H_1(A_z, \mathbb{Z}) \otimes \mathbb{C} = \mathbb{C}^g = H_1(A_{gz}, \mathbb{Z}) \otimes \mathbb{C}$ induces an identification $\iota_g : \mathcal{E}_{\underline{A}_z} \to \mathcal{E}_{\underline{A}_{gz}}$, and we can define the isomorphism $i_g : \mathcal{E}_{\underline{A}_z,\rho} \to \mathcal{E}_{\underline{A}_{gz},\rho}$ by $(l, v) \to (\iota_g(l), \rho(M_g(z))v)$. Note that i_g is the identity for $g \in \Gamma_{\mathcal{U}}$. It is an easy exercise to see that Lemma 3.8 can be reformulated as follows.

Lemma 3.9. *Let $\rho : H(\mathbb{C}) \to \mathrm{GL}(X)$ be a rational representation, and Γ a congruence subgroup of $GU_+(\mathbb{Q})$ containing $\Gamma_{\mathcal{U}}$. Then there exists a one-to-one correspondence between automorphic functions of weight ρ and level Γ and the set of functions $\widetilde{F}$ from $\underline{A} \in S_{\mathcal{U}}(\mathbb{C})$ to $\mathcal{E}_{\underline{A},\rho}$ satisfying*

$$i_\gamma\left(\widetilde{F}(\underline{A}_z)\right) = \widetilde{F}(\underline{A}_{\gamma z}) \text{ for all } z \in \mathcal{H}, \gamma \in \Gamma. \tag{5}$$

Remark 3.10. Note that by Proposition 3.6 giving a function $\widetilde{F}$ from $S_{\mathcal{U}}(\mathbb{C})$ to $\mathcal{E}_{\underline{A},\rho}$ satisfying (5) is the same as giving a global (point-theoretic) section of the vector bundle $\mathcal{E}_{\underline{A}_z,\rho}$ over $\Gamma\backslash\mathcal{H}$.

3.2 (Classical) Algebraic Automorphic Forms

In Sect. 3.1, we considered automorphic forms over $\mathbb{C}$. Building on the discussion from Sect. 3.1, we now consider automorphic forms over other base rings. The approach in this section is similar to the approach in [8, Sect. 2.5] and [20, Sect. 1.2]. Note that [8] only considers the case in which the signature is $(a_{+\tau}, a_{-\tau})_{\tau\in\Sigma_K}$ with $a_{+\tau} = a_{-\tau}$ for every $\tau \in \Sigma_K$, but the definitions from [8, Sect. 2.5] carry over to the general case with only trivial modifications. In order not to worry about the holomorphy conditions at cusps, we exclude the case of $\Sigma = \{\tau\}$ with $(a_{+\tilde{\tau}}, a_{-\tilde{\tau}}) = (1, 1)$ from the discussion in this section, compare Remark 3.3.

For any neat open compact subgroup $\mathcal{U}$, consider the integral model $\mathcal{M}_{\mathcal{U}}/\mathcal{O}_{E,(p)}$ introduced in Sect. 2.2. For any scheme S over $\mathrm{Spec}(\mathcal{O}_{E,(p)})$, we put

$$\mathcal{M}_{\mathcal{U},S} := \mathcal{M}_{\mathcal{U}} \times_{\mathcal{O}_{E,(p)}} S.$$

When $S = \mathrm{Spec}(R)$ for a ring R, we will often write $\mathcal{M}_{\mathcal{U},R}$ instead of $\mathcal{M}_{\mathcal{U},\mathrm{Spec}(R)}$. If $\mathbb{W}$ denotes the ring of Witt vectors associated with $\overline{\mathbb{F}}_p$, then consider $\mathcal{M}_{\mathcal{U},\mathbb{W}}$ (note that since p splits completely, we can base change to $\mathbb{W}$). In the sequel, we consider (locally noetherian) schemes S over $\mathbb{W}$. Note that instead of working over $\mathbb{W}$, we could also work over $\mathcal{O}_{E',(p)}$ in this section, where E' is a finite extension of E that contains $\tau(K)$ for all $\tau \in \Sigma_K$.

For any S-point $\underline{A} = (A, i, \lambda, \alpha^p)$ of $\mathcal{M}_{\mathcal{U},\mathbb{W}}$, let $\underline{\Omega}_{A/S}$ denote the locally free $\mathbb{W} \otimes \mathcal{O}_K$-module defined as the pullback via the identity section of the relative differentials. We have a natural decomposition $\underline{\Omega}_{A/S} = \bigoplus_{\tau\in\Sigma}(\underline{\Omega}^+_{A/S,\tilde{\tau}} \bigoplus \underline{\Omega}^-_{A/S,\tilde{\tau}})$ where $\underline{\Omega}^\pm_{A/S,\tilde{\tau}}$ is rank $a_{+\tilde{\tau}}$ and $a_{-\tilde{\tau}}$, respectively. Note that the element $x \in \mathcal{O}_K$ acts on $\underline{\Omega}^+_{A/S,\tilde{\tau}}$ (resp., $\underline{\Omega}^-_{A/S,\tilde{\tau}}$) via $\tilde{\tau}(x)$ (resp., $\tilde{\tau}^c(x)$). (Here, we view $\tilde{\tau}$ as an embedding of K which factors through $\mathrm{Frac}(\mathbb{W})$). Define

$$\mathcal{E}^\pm_{\underline{A}/S} := \bigoplus_{\tau\in\Sigma} \mathrm{Isom}_{\mathcal{O}_S}\left(\mathcal{O}_S^{a_{\pm\tilde{\tau}}}, \underline{\Omega}^\pm_{A/S,\tau}\right) \qquad \text{and} \qquad \mathcal{E}_{\underline{A}/S} := \mathcal{E}^+_{\underline{A}/S} \oplus \mathcal{E}^-_{\underline{A}/S},$$

respectively. Let R be a $\mathbb{W}$-algebra, and consider an algebraic representation ρ of H_R into a finite free R-module M_ρ.

Definition 3.11 (First Equivalent Definition of Algebraic Automorphic Forms). An automorphic form of weight ρ and level $\mathcal{U}$ defined over R is a function f

$$(\underline{A}, \ell) \mapsto f(\underline{A}, \ell) \in (M_\rho)_{R'}$$

defined for all R-algebras R', $\underline{A} \in \mathcal{M}_{\mathcal{U}}(R')$, and $\ell \in \mathcal{E}_{\underline{A}/R'}$, such that all of the following hold:

(1) $f(\underline{A}, \alpha\ell) = \rho\left(({}^t\alpha)^{-1}\right) f(\underline{A}, \ell)$ for all $\alpha \in H(R')$ and all $\ell \in \mathcal{E}_{\underline{A}/R'}$
(2) The formation of $f(\underline{A}, \ell)$ commutes with extension of scalars $R_2 \to R_1$ for any R-algebras R_1 and R_2. More precisely, if $R_2 \to R_1$ is a ring homomorphism of R-algebras, then

$$f(\underline{A} \times_{R_1} R_2, \ell \otimes_{R_1} 1) = f(\underline{A}, \ell) \otimes_{R_1} 1_{R_2} \in (M_\rho)_{R_2}$$

In order to give a different equivalent definition (Definition 3.12 below), we define for any algebraic representation ρ of H_R into a finite free R-module M_ρ and R-algebra R'

$$\mathcal{E}_{(\underline{A}/R',\rho)} = \mathcal{E}_{\underline{A}/R'} \times^\rho (M_\rho)_{R'} := \left(\mathcal{E}_{\underline{A}/R'} \times (M_\rho)_{R'}\right) / (\ell, m) \sim \left(g\ell, \rho({}^t g^{-1})m\right),$$

where $g \in H(R')$ acts on $\mathcal{E}_{\underline{A}/R'}$ by precomposing with ${}^t g$.

Definition 3.12 (Second Equivalent Definition of Algebraic Automorphic Forms). An automorphic form of weight ρ and level $\mathcal{U}$ defined over a $\mathbb{W}$-algebra R is a function $\tilde{f}$

$$\underline{A} \mapsto \tilde{f}(\underline{A}) \in \mathcal{E}_{(\underline{A}/R',\rho)}$$

defined for all R-algebras R' and $\underline{A} \in \mathcal{M}_{\mathcal{U}}(R')$ such that the formation of $\tilde{f}(\underline{A})$ commutes with extension of scalars $R_2 \to R_1$ for any R-algebras R_1 and R_2. More precisely, if $R_2 \to R_1$ is a ring homomorphism of R-algebras, then

$$\tilde{f}(\underline{A} \times_{R_1} R_2) = \tilde{f}(\underline{A}) \otimes_{R_1} 1_{R_2}.$$

Remark 3.13. The equivalence between Definitions 3.11 and 3.12 is given by

$$\tilde{f}(\underline{A}) = (\ell, f(\underline{A}, \ell))$$

for all abelian varieties $\underline{A}/R$ (corresponding to $\mathcal{M}(R)$) and $\ell \in \mathcal{E}_{\underline{A}/R}$.

Finally, we want to view automorphic forms as global sections of a certain sheaf. In order to do so, let $\mathcal{A} = (A, i, \lambda, \alpha^p)^{\text{univ}}$ denote the universal abelian variety over $\mathcal{M}_{\mathcal{U},\mathbb{W}}$, and define the sheaf

$$\mathcal{E} = \mathcal{E}_{\mathcal{U}} := \bigoplus_{\tau\in\Sigma} \underline{\mathrm{Isom}}_{\mathcal{O}_{\mathcal{M}_{\mathcal{U},\mathbb{W}}}}\left(\mathcal{O}^{a_{+\tau}}_{\mathcal{M}_{\mathcal{U},\mathbb{W}}}, \underline{\Omega}^{+}_{\mathcal{A}/\mathcal{M}_{\mathcal{U}},\tau}\right)$$

$$\oplus \bigoplus_{\tau\in\Sigma} \underline{\mathrm{Isom}}_{\mathcal{O}_{\mathcal{M}_{\mathcal{U},\mathbb{W}}}}\left(\mathcal{O}^{a_{-\tau}}_{\mathcal{M}_{\mathcal{U},\mathbb{W}}}, \underline{\Omega}^{-}_{\mathcal{A}/\mathcal{M}_{\mathcal{U},\mathbb{W}},\tau}\right),$$

i.e., for every open immersion $S \hookrightarrow \mathcal{M}_{\mathcal{U},\mathbb{W}}$, we set $\mathcal{E}_{\mathcal{U}}(S) = \mathcal{E}_{\mathcal{A}_S/S}$. Moreover, for any algebraic representation ρ of H_R over a free finite R-module M_ρ, we define the sheaf $\mathcal{E}_\rho = \mathcal{E}_{\mathcal{U},\rho} := \mathcal{E} \times^\rho M_\rho$, i.e., for each open immersion $\operatorname{Spec} R' \hookrightarrow \mathcal{M}_{U,\mathbb{W}}$, set $\mathcal{E}_{\mathcal{U},\rho}(R') = \mathcal{E}_{(\mathcal{A}_{R'}/R',\rho)}$.

Definition 3.14 (Third Equivalent Definition of Algebraic Automorphic Forms). An automorphic form of weight ρ and level $\mathcal{U}$ defined over R is a global section of the sheaf $\mathcal{E}_{\mathcal{U},\rho}$ on $\mathcal{M}_{\mathcal{U},R}$.

Remark 3.15. When we are working with a representation ρ which are uniquely determined by its highest weight κ, we shall sometimes write $\mathcal{E}_{\mathcal{U},\kappa}$ or $\mathcal{E}_\kappa$, in place of $\mathcal{E}_{\mathcal{U},\rho}$.

Remark 3.16. Usually automorphic forms are defined over a compactification of $\mathcal{M}_{\mathcal{U},R}$, but in our case, i.e., excluding the case of Σ consisting only of one place τ and $(a_{+\tau}, a_{-\tau}) = (1, 1)$, both definitions are equivalent by Koecher's principle.

3.2.1 Comparison with Classical Definition of Complex Automorphic Forms

Having defined algebraic automorphic forms over general base rings, we will show that in the special case of the base ring being $\mathbb{C}$ the definition coincides with the classical definition of complex automorphic forms given in Sect. 3.1.

For an integer N, we define $\mathcal{U}_N$ to be a compact open subgroup of $GU(\mathbb{A}^\infty)$ such that

$$GU_+(\mathbb{Q}) \cap \mathcal{U}_N = \Gamma(N) := \{(g, \nu) \in GU_+(\mathbb{Q}) : g \equiv 1 \bmod N\}.$$

Proposition 3.17. *Let N be a large enough integer so that $\mathcal{U}_N$ is neat, and let ρ be an algebraic representation of H over $\mathbb{C}$. Then there is a bijection between the (algebraic) automorphic forms of weight ρ defined in Definition 3.14 as global sections of $\mathcal{E}_\rho$ on $M_{\mathcal{U}_N}(\mathbb{C})$ and a finite set of holomorphic automorphic forms of weight ρ with respect to $\Gamma(N)$ as defined in Definition 3.2 in Sect. 3.1.*

Proof (sketch). From the classification of Hermitian symmetric spaces mentioned in Sect. 3.1, one can deduce that $M^{(i)}_{\mathcal{U}_N}(\mathbb{C})$ (as defined in Sect. 2.3.2) is isomorphic to a finite union of copies of $\Gamma\backslash\mathcal{H}$. By GAGA and Lemma 3.9 together with Remark 3.10, we conclude that the global sections of $\mathcal{E}_\rho = \mathcal{E}_{\mathcal{U},\rho}$ on $M_{\mathcal{U}_N}(\mathbb{C})$ are in one-to-one correspondence with a finite set of holomorphic automorphic forms of weight ρ with respect to $\Gamma(N)$, one for each connected component of $M^{(i)}_{\mathcal{U}_N}(\mathbb{C})$ for all i. □

4 p-Adic Theory

Section 4.1 introduces the Igusa tower, a tower of finite étale Galois coverings of the ordinary locus of $\mathcal{M}_{\mathcal{U}}$, which we denote by $\mathcal{M}$ from now on, since we fix the neat level $\mathcal{U}$ throughout the rest of the paper. Section 4.2 introduces p-adic automorphic forms, which arise as global sections of the structure sheaf of the Igusa tower.

We are mainly following [14, Sect. 8] and [15].

4.1 The Igusa Tower Over the Ordinary Locus

Recall that our Shimura varieties with hyperspecial level at p admit integral models $\mathcal{M}$ over $\mathcal{O}_{E,(p)}$. These Shimura varieties have a neat level away from p, but we suppress it from the notation since the tame level won't affect the geometry of our integral models. As we will see below, the geometry of the integral models $\mathcal{M}$ is governed by the p-divisible group of the universal abelian variety over $\mathcal{M}$.

In order to guarantee that the ordinary locus (defined below) over the special fiber of $\mathcal{M}$ is non-empty, we make the following assumption: the prime p splits completely in the reflex field E (in [33] Wedhorn proves that such an assumption is both necessary and sufficient). In this case, the ordinary locus is open and dense in the special fiber of $\mathcal{M}$. Choose a place P of E above p and let E_P be the corresponding completion of E, with ring of integers $\mathcal{O}_{E_P}$ and residue field k. By abuse of notation, we will still denote the base change of $\mathcal{M}$ to $\mathcal{O}_{E_P}$ by $\mathcal{M}$. Let S be a scheme of characteristic p.

Definition 4.1. An abelian variety A/S of dimension g is ordinary if for all geometric points s of S, the set $A[p](s)$ has p^g elements.

For every abelian variety A/S, the Hasse invariant $\mathrm{Ha}_{p-1}(A/S)$ is a global section of $\omega_{A/S}^{\otimes(p-1)}$, where $\omega_{A/S}$ is the top exterior power of the pushforward to S of the sheaf of invariant differentials on A. It is easy to show that an abelian variety A is ordinary if and only if $\mathrm{Ha}(A/S)$ is invertible. (We now sketch an argument for this, as in [29, Lemma III.2.5]: the Hasse invariant, which corresponds to pullback along the Verschiebung isogeny, is invertible if and only if Verschiebung is an isomorphism on tangent spaces, which happens if and only if Verschiebung is finite etale. A degree computation shows that this is equivalent to the condition that A be ordinary.) We define the ordinary locus

$$\overline{\mathcal{M}}^{\mathrm{ord}} \subset \overline{\mathcal{M}} := \mathcal{M} \times_{\mathcal{O}_{E_P}} k$$

to be the complement of the zero set of the Hasse invariant. Since p splits completely in E, the nonemptiness of $\overline{\mathcal{M}}^{\mathrm{ord}}$ follows from [33]. In fact, Wedhorn proves something stronger, namely that $\overline{\mathcal{M}}^{\mathrm{ord}}$ is dense in the special fiber $\overline{\mathcal{M}}$. We also define the ordinary locus $\mathcal{M}^{\mathrm{ord}}$ over $\mathcal{O}_{E_P}$ to be the complement of the zero set of a lift of some power of the Hasse invariant.

In addition, we define $\mathcal{S}^{\mathrm{ord}}$ over $\mathbb{W}$ to be a connected component of $\mathcal{M}^{\mathrm{ord}}_{\mathbb{W}} := \mathcal{M}^{\mathrm{ord}} \times_{\mathcal{O}_{E_p}} \mathbb{W}$. (Recall that p splits completely in E, so the base change to $\mathbb{W}$ makes sense.) In other words, $\mathcal{S}^{\mathrm{ord}}$ is the ordinary locus of one connected component $\mathcal{S}$ of $\mathcal{M}_{\mathbb{W}}$.

Remark 4.2. We note that we could define the ordinary locus in an alternate way, using the stratification of $\overline{\mathcal{M}}$ in terms of the isogeny class of the p-divisible group $\mathcal{G} := \mathcal{A}_{\overline{\mathcal{M}}}[p^\infty]$, where $\mathcal{A}_{\overline{\mathcal{M}}}$ denotes the universal abelian variety over $\overline{\mathcal{M}}$. The isogeny class of the p-divisible group $\mathcal{G}$ (equipped with all its extra structures) defines a stratification of $\overline{\mathcal{M}}$ with locally closed strata, which is called the *Newton stratification*. The ordinary locus, corresponding to the constant isogeny class $(\mu_{p^\infty})^g \times (\mathbb{Q}_p/\mathbb{Z}_p)^g$, is the unique open stratum.

Let $\mathcal{A} := \mathcal{A}_{\mathcal{M}^{\mathrm{ord}}}$ be the universal ordinary abelian variety over $\mathcal{M}^{\mathrm{ord}}$. Pick a $\mathbb{W}$-point x of $\mathcal{S}^{\mathrm{ord}}$, with an $\overline{\mathbb{F}}_p$-point $\bar{x} \in \mathcal{S}^{\mathrm{ord}}$ below it. We can identify L_p (defined as in Sect. 2.1) with the p-adic Tate module of the p-divisible group $\mathcal{G}_x$, i.e., the p-adic Tate module of $\mathcal{A}_x$. Choose such an identification $L_p \simeq T_p\mathcal{A}_x[p^\infty]$, compatible with the $\mathcal{O}_K$-action and with the Hermitian pairings. The kernel of the reduction map

$$T_p\mathcal{A}_x[p^\infty] \to T_p\mathcal{A}_{\bar{x}}[p^\infty]^{\acute{e}t}$$

corresponds to an $\mathcal{O}_K$-direct summand of $\mathcal{L} \subset L_p$. Note that the lattice $\mathcal{L}$ is independent of the choice of x inside the connected component $\mathcal{S}^{\mathrm{ord}}$, and hence the different connected components of $\mathcal{M}^{\mathrm{ord}}$ can be labeled by lattices $\mathcal{L}$. Moreover, using the self-duality of L_p under the Hermitian pairing $\langle\cdot,\cdot\rangle$ and the compatibility with the Weil pairing, we can identify the dual $\mathcal{L}^\vee$ of $\mathcal{L}$ with the orthogonal complement of $\mathcal{L}$ inside L_p.

Remark 4.3. By considering the primes in K^+ above p individually, we can write down an explicit formula for $\mathcal{L}$, using the fact that each such prime splits from K^+ to K. The exact formula for $\mathcal{L}$ as an $\mathcal{O}_K$-module will depend on the set of signatures of the unitary similitude group $GU(\mathbb{R})$. More precisely, recall that for each embedding $\tau : K \hookrightarrow \mathbb{C}$, $(a_{+\tau}, a_{-\tau})$ is the signature of GU at the infinite place τ. Choose an isomorphism $\iota_p : \mathbb{C} \xrightarrow{\sim} \bar{\mathbb{Q}}_p$. By composing with ι_p, each τ determines a place of K above p. Let $p = \prod_{i=1}^r \mathfrak{p}_i$ be the decomposition of p into prime ideals of K^+. Each $\mathfrak{p}_i$ splits in K as $\mathfrak{p}_i = \mathfrak{P}_i\mathfrak{P}_i^c$, where $\mathfrak{P}_i$ lies above the prime w of F. The i-term of L_p (obtained from the decomposition $\mathcal{O}_K \otimes_{\mathbb{Z}} \mathbb{Z}_p = \bigoplus_{i=1}^r (\mathcal{O}_{K_{\mathfrak{P}_i}} \oplus \mathcal{O}_{K_{\mathfrak{P}_i^c}})$) can be identified with

$$\mathcal{O}^n_{K_{\mathfrak{P}_i}} \oplus \mathcal{O}^n_{K_{\mathfrak{P}_i^c}}.$$

Then the determinant condition implies that

$$\mathcal{L} \simeq \bigoplus_{i=1}^r (\mathcal{O}^{a_{+\tau_i}}_{K_{\mathfrak{P}_i}} \oplus \mathcal{O}^{a_{-\tau_i}}_{K_{\mathfrak{P}_i^c}}),$$

where τ_i is a place inducing $\mathfrak{P}_i$. Note that there is a natural decomposition $\mathcal{L} = \mathcal{L}^+ \oplus \mathcal{L}^-$, coming from the splitting $p = w \cdot w^c$. Also note that $H(\mathbb{Z}_p) \cong \prod_{i=1}^r \left(GL_{a_{+\tau_i}}(\mathcal{O}_{K_{\mathfrak{P}_i}}) \times GL_{a_{-\tau_i}}(\mathcal{O}_{K_{\mathfrak{P}_i^c}}) \right)$ can be identified with the $\mathcal{O}_K$-linear automorphism group of $\mathcal{L}$, which induces a natural action of H on the dual $\mathcal{L}^\vee$ of $\mathcal{L}$ (by precomposing with the inverse) and on all other spaces defined in terms of $\mathcal{L}$.

Now, we introduce the *Igusa tower* over the component $\mathcal{S}^{\mathrm{ord}}$. For $n \in \mathbb{N}$, consider the functor

$$\mathrm{Ig}_n^{\mathrm{ord}} : \{\mathrm{Schemes}/\mathcal{S}^{\mathrm{ord}}\} \to \{\mathrm{Sets}\}$$

that takes an $\mathcal{S}^{\mathrm{ord}}$-scheme S to the set of $\mathcal{O}_K$-linear closed immersions

$$\mathcal{L} \otimes_{\mathbb{Z}} \mu_{p^n} \hookrightarrow \mathcal{A}_S[p^n], \tag{6}$$

where $\mathcal{A}_S := \mathcal{A}_{\mathcal{S}^{\mathrm{ord}}} \times_{\mathcal{S}^{\mathrm{ord}}} S$. This functor is representable by an $\mathcal{S}^{\mathrm{ord}}$-scheme, which we also denote by $\mathrm{Ig}_n^{\mathrm{ord}}$. Let $\mathbb{W}_m := \mathbb{W}/p^m\mathbb{W}$, and for each $m \in \mathbb{Z}_{\geq 1}$, define $\mathcal{S}_m^{\mathrm{ord}} := \mathcal{S}^{\mathrm{ord}} \times_{\mathbb{W}} \mathbb{W}_m$. Then $\mathrm{Ig}_{n,m}^{\mathrm{ord}} := \mathrm{Ig}_n^{\mathrm{ord}} \times_{\mathbb{W}} \mathbb{W}_m$ is a scheme over $\mathbb{W}_m$, whose functor takes an $\mathcal{S}_m^{\mathrm{ord}}$-scheme S to the set of $\mathcal{O}_K$-linear closed immersions

$$\mathcal{L} \otimes_{\mathbb{Z}} \mu_{p^n} \hookrightarrow \mathcal{A}_S[p^n].$$

For each $n \geq 1$, $\mathrm{Ig}_{n,m}^{\mathrm{ord}}$ is a finite étale and Galois covering of $\mathcal{S}_m^{\mathrm{ord}}$ whose Galois group is the group of $\mathcal{O}_K$-linear automorphisms of $\mathcal{L}^\vee/p^n\mathcal{L}^\vee$. (See Section 8.1.1 of [14] for a discussion of representability and of the fact that these Igusa varieties are finite étale covers of $\mathcal{S}_m^{\mathrm{ord}}$: the key point is that $\mathcal{L}^\vee/p^n\mathcal{L}^\vee$ is an étale sheaf.)

We also define the formal scheme $\mathfrak{Ig}_n^{\mathrm{ord}}$ to be the formal completion of $\mathrm{Ig}_n^{\mathrm{ord}}$ along the special fiber $\mathcal{S}_{\overline{\mathbb{F}}_p}^{\mathrm{ord}}$, i.e., as a functor from $\{\mathcal{S}^{\mathrm{ord}}\text{-schemes on which } p \text{ is nilpotent}\}$ to $\{\mathrm{Sets}\}$, we have $\mathfrak{Ig}_n^{\mathrm{ord}} = \varinjlim_m \mathrm{Ig}_{n,m}^{\mathrm{ord}}$.

This formal scheme is a finite étale and Galois cover of $\mathfrak{S}^{\mathrm{ord}}$, the formal completion of $\mathcal{S}^{\mathrm{ord}}$ along its special fiber. As we let n vary, we obtain a tower of finite étale coverings of $\mathfrak{S}^{\mathrm{ord}}$, called the *Igusa tower*. The Galois group of the whole Igusa tower over $\mathfrak{S}^{\mathrm{ord}}$ can be identified with $H(\mathbb{Z}_p)$.

The inverse limit of formal schemes $\mathfrak{Ig}_n^{\mathrm{ord}}$ also exists as a formal scheme, which we denote by $\mathfrak{Ig}^{\mathrm{ord}}$. The point is that $(\mathfrak{Ig}_n^{\mathrm{ord}})_{n \in \mathbb{Z}_{\geq 1}}$ is a projective system of formal schemes, with affine transition maps, so the inverse limit exists in the category of formal schemes. (See, for example, [11, Proposition D.4.1].) This is a pro-finite étale cover of $\mathfrak{S}^{\mathrm{ord}}$, with Galois group $H(\mathbb{Z}_p)$.

We now give a different way of thinking about the Igusa tower. Let $\mathfrak{A}_{\mathfrak{S}^{\mathrm{ord}}}$ be the universal abelian variety over $\mathfrak{S}^{\mathrm{ord}}$. For p-divisible groups over $\mathfrak{S}^{\mathrm{ord}}$ there is a connected-étale exact sequence, so it makes sense to define the connected part $\mathfrak{A}_{\mathfrak{S}^{\mathrm{ord}}}[p^\infty]^\circ$ of $\mathfrak{A}_{\mathfrak{S}^{\mathrm{ord}}}[p^\infty]$. Then the formal completion $\mathfrak{Ig}_n^{\mathrm{ord}}$ can be identified with the formal scheme $\mathrm{Isom}_{\mathfrak{S}^{\mathrm{ord}}}(\mathcal{L} \otimes_{\mathbb{Z}} \mu_{p^n}, \mathfrak{A}_{\mathfrak{S}^{\mathrm{ord}}}[p^n]^\circ)$. Using the duality induced by the

Hermitian pairing on L_p and by λ on $\mathfrak{A}_{\mathfrak{S}^{\mathrm{ord}}}[p^\infty]$ (and noting that duality interchanges the connected and étale parts), we can further identify $\mathfrak{Ig}_n^{\mathrm{ord}}$ with the formal scheme $\mathrm{Isom}_{\mathfrak{S}^{\mathrm{ord}}}\left(\mathcal{L}^\vee/p^n\mathcal{L}^\vee, \mathfrak{A}_{\mathfrak{S}^{\mathrm{ord}}}[p^n]^{\acute{e}t}\right)$. This is finite étale over $\mathfrak{S}^{\mathrm{ord}}$.

4.1.1 Irreducibility

In this section, we show that the Igusa tower $\{\mathfrak{Ig}_n^{\mathrm{ord}}\}_{n\in\mathbb{N}}$, or equivalently $\{\mathrm{Ig}_n^{\mathrm{ord}}\}_{n\in\mathbb{N}}$, is not irreducible, but we also sketch how one can pass to a partial SU-tower that is irreducible.

As explained above, $\mathfrak{Ig}_n^{\mathrm{ord}}$ can be identified with

$$\mathrm{Isom}_{\mathfrak{S}^{\mathrm{ord}}}\left(\mathcal{L}^\vee/p^n\mathcal{L}^\vee, \mathfrak{A}_{\mathfrak{S}^{\mathrm{ord}}}[p^n]^{\acute{e}t}\right). \tag{7}$$

Such an isomorphism of sheaves on $\mathfrak{S}^{\mathrm{ord}}$ induces an isomorphism of the top exterior powers of these sheaves, so there is a morphism

$$\begin{aligned} \det : &\ \mathrm{Isom}_{\mathfrak{S}^{\mathrm{ord}}}\left(\mathcal{L}^\vee/p^n\mathcal{L}^\vee, \mathfrak{A}_{\mathfrak{S}^{\mathrm{ord}}}[p^n]^{\acute{e}t}\right) \\ &\to \mathrm{Isom}_{\mathfrak{S}^{\mathrm{ord}}}\left(\wedge^{\mathrm{top}}(\mathcal{L}^\vee/p^n\mathcal{L}^\vee), \wedge^{\mathrm{top}}(\mathfrak{A}_{\mathfrak{S}^{\mathrm{ord}}}[p^n]^{\acute{e}t})\right). \end{aligned}$$

Hida [15] shows that the sheaf $\wedge^{\mathrm{top}}(\mathfrak{A}_{\mathfrak{S}^{\mathrm{ord}}}[p^n]^{\acute{e}t})$ on $\mathfrak{S}^{\mathrm{ord}}$ is isomorphic to the constant sheaf $\mathcal{O}_K/p^n\mathcal{O}_K$. This gives an isomorphism

$$\mathrm{Isom}_{\mathfrak{S}^{\mathrm{ord}}}\left(\wedge^{\mathrm{top}}(\mathcal{L}^\vee/p^n\mathcal{L}^\vee), \wedge^{\mathrm{top}}(\mathfrak{A}_{\mathfrak{S}^{\mathrm{ord}}}[p^n]^{\acute{e}t})\right) \xrightarrow{\sim} (\mathcal{O}_K/p^n\mathcal{O}_K)^\times$$

and shows that the full Igusa tower $\{\mathfrak{Ig}_n^{\mathrm{ord}}\}_{n\in\mathbb{N}}$ is not irreducible.

As n varies, the determinant morphisms above are compatible. Let $\mathfrak{Ig}^{\mathrm{ord},SU}$ be the inverse image of $(1)_{n\in\mathbb{N}} \in ((\mathcal{O}_K/p^n\mathcal{O}_K)^\times)_{n\in\mathbb{N}}$ under det.

Theorem 4.4. *(Hida) $\mathfrak{Ig}^{\mathrm{ord},SU}$ is a geometrically irreducible component of $\mathfrak{Ig}^{\mathrm{ord}}$.*

Proof (sketch). One proof of this statement can be found in [15, Sects. 3.4–3.5]; we merely sketch the argument here.

It is sufficient to show that $\mathfrak{Ig}_n^{\mathrm{ord},SU}$ is irreducible for each n. This is an étale cover of $\mathfrak{S}^{\mathrm{ord}}$, the formal completion of the smooth irreducible variety $\mathcal{S}^{\mathrm{ord}}$ over $\mathbb{W}$ along its special fiber. One of Hida's strategies (in, for example, [15, Sect. 3.4]) for proving the irreducibility of the étale cover in this situation is to consider a compatible group action of a product $\mathcal{G}_1 \times \mathcal{G}_2$ on $\mathfrak{Ig}_n^{\mathrm{ord},SU}$ and $\mathfrak{S}^{\mathrm{ord}}$, in such a way that $\mathcal{G}_1 \subset \mathrm{Aut}(\mathfrak{S}^{\mathrm{ord}})$ fixes and $\mathcal{G}_2 \subset \mathrm{Aut}(\mathfrak{Ig}_n^{\mathrm{ord},SU}/\mathfrak{S}^{\mathrm{ord}})$ acts transitively on the connected components of $\mathfrak{Ig}_n^{\mathrm{ord},SU}$. The group $\mathcal{G}_1$ can be identified with the finite adelic points $SU(\mathbb{A}^\Sigma)$ away from certain bad places Σ containing p. This group will not have any finite quotient and therefore will preserve the connected components of the Igusa tower. The group

$\mathcal{G}_2$ can be identified with the Levi subgroup $\mathrm{Levi}_1(\mathbb{Z}_p) := H(\mathbb{Z}_p) \cap SU(\mathbb{Z}_p)$, where $H(\mathbb{Z}_p)$ is as in Remark 4.3. The action of $\mathrm{Levi}_1(\mathbb{Z}_p)$ on the connected components is transitive, since $\mathfrak{Ig}_n^{\mathrm{ord},SU}/\mathfrak{S}^{\mathrm{ord}}$ is a $\mathrm{Levi}_1(\mathbb{Z}_p/p^n\mathbb{Z}_p)$-torsor via the action on the Igusa level structure.

Following Hida's argument, we choose a base point x_0 on the Igusa tower. Hida considers the group $\mathcal{T}_{x_0}$ generated by $\mathcal{G}_1$ and the stabilizer of x_0 in $\mathcal{G}_1 \times \mathcal{G}_2$. In [15, Sect. 3.5], he shows one can choose the point x_0 such that $\mathcal{T}_{x_0}$ is dense in $\mathcal{G}_1 \times \mathcal{G}_2$, which amounts to choosing a point whose stabilizer has p-adically dense image in $\mathrm{Levi}_1(\mathbb{Z}_p)$. This density is obtained as a by-product of the fact that the abelian variety with structures corresponding to the point x_0 has many extra endomorphisms over $\mathbb{Q}$ and the fact that $\mathrm{Levi}_1(\mathbb{Z}_p) \cap SU(\mathbb{Q})$ is dense in $\mathrm{Levi}_1(\mathbb{Z}_p)$. On one hand, $\mathcal{T}_{x_0}$ is dense in a group acting transitively on the connected components of $\mathfrak{Ig}_n^{\mathrm{ord},SU}$; on the other hand, $\mathcal{T}_{x_0}$ fixes the connected components by definition. This shows that $\mathfrak{Ig}_n^{\mathrm{ord},SU}$ has only one connected component to start with and, therefore, that it is irreducible. In fact, Hida shows that one can take x_0 to be a CM point; this is the entire subject of [15, Sect. 3.5]. □

4.2 p-Adic Automorphic Forms

In order to define p-adic automorphic forms, we define the global sections

$$V_{n,m} = H^0(\mathrm{Ig}_{n,m}^{\mathrm{ord}}, \mathcal{O}_{\mathrm{Ig}_{n,m}^{\mathrm{ord}}}),$$

and let $V_{\infty,m} := \varinjlim_n V_{n,m}$ and $V := V_{\infty,\infty} := \varprojlim_m V_{\infty,m}$.

The space V is endowed with a left action of $H(\mathbb{Z}_p)$, $f \mapsto g \cdot f$, induced by the natural right action of $g \in H(\mathbb{Z}_p)$ on the Igusa tower by $g \cdot f := g^*(f) = f \circ g$. For any point $x = (x_n) \in (\mathrm{Ig}_n^{\mathrm{ord}}(\mathbb{W}))_{n\in\mathbb{N}}$, where x_{n+1} maps to x_n under the projection, if $\underline{A} = \underline{A}_x$ denotes the associated abelian scheme over $\mathbb{W}$ endowed with additional structures, and $\iota_x : \mu_{p^\infty} \otimes \mathcal{L} \hookrightarrow A[p^\infty]$ denotes the Igusa structure of infinite level on A, then for any $g \in H(\mathbb{Z}_p)$ the image of x under the morphism g is the point $x^g = (x_n^g) \in (\mathrm{Ig}_n^{\mathrm{ord}}(\mathbb{W}))_{n\in\mathbb{N}}$ corresponding to the data of the abelian scheme $\underline{A}$ (with the associated additional structures), together with the Igusa structure of infinite level $\iota_{x^g} = \iota_x \circ (1 \otimes g)$.

Definition 4.5. We call $V^N := V_{\infty,\infty}^{N(\mathbb{Z}_p)}$ the *space of p-adic automorphic forms*.

It is worth noting that taking invariants by $N(\mathbb{Z}_p)$ commutes with both direct and inverse limits, thus we could define V^N also as $\varprojlim_m \varinjlim_n V_{n,m}^N$ with $V_{n,m}^N := V_{n,m}^{N(\mathbb{Z}_p)}$.

Remark 4.6. In Definition 4.5, we have defined p-adic automorphic forms over the non-compactified Shimura variety. Typically, p-adic automorphic forms are defined over the compactified Shimura variety by constructing sheaves on any toroidal compactification of $\mathcal{M}$ which descend to the minimal compactification of $\mathcal{M}$ and are canonically identified with $\mathcal{E}_{\mathcal{U},\rho}$ when restricted to $\mathcal{M}$. (See §8.3.5 of [24].) For the present paper, we are interested in local properties of automorphic forms, namely their local behavior at ordinary CM points. Thus, compactifications have no bearing on the main results of this paper. Furthermore, by Koecher's principle (see Remark 3.3), so long as we are not in the case in which both Σ consists of only one place and the signature is $(1, 1)$, both definitions agree.

4.2.1 Comparison of Automorphic Forms and p-Adic Automorphic Forms

We now construct an embedding of the global sections of automorphic vector bundles on the connected component $\mathcal{S}^{\mathrm{ord}}$ of the ordinary locus into V^N, our newly constructed space of p-adic automorphic forms.

Let $n \geq m$. Recall that each element $f \in V^N_{n,m}$ can be viewed as a function

$$(\underline{A}, j) \mapsto f(\underline{A}, j) \in \mathbb{W}_m, \tag{8}$$

where $\underline{A}$ consists of an abelian variety $A/\mathbb{W}_m$ together with a polarization, an endomorphism, and a level structure, and an $\mathcal{O}_K$-linear isomorphism $j : A[p^n]^{\acute{e}t} \xrightarrow{\sim} (\mathbb{Z}/p^n\mathbb{Z}) \otimes \mathcal{L}^\vee$.

Recall that each element g in $H(\mathbb{Z}/p^n\mathbb{Z})$ acts on elements f in $V_{n,m}$ via

$$(g \cdot f)(\underline{A}, j) := f(\underline{A}, gj).$$

We may view each element $f \in H^0\left(\mathcal{S}^{\mathrm{ord}}_m, \mathcal{E}_\kappa\right)$ as a function

$$(\underline{A}, \ell) \mapsto f(\underline{A}, \ell) \in \mathrm{Ind}_{B^-}^H(-\kappa)_{\mathbb{W}_m},$$

where $\underline{A}$ is as in Eq. (8), $\ell \in \mathrm{Isom}_{\mathcal{O}_{\mathcal{S}^{\mathrm{ord}}_m}}(\mathbb{Z}/p^n\mathbb{Z} \otimes \mathcal{L}^\vee, \underline{\Omega}_{\mathcal{A}/\mathcal{S}^{\mathrm{ord}}_m})$, and $\mathbb{A}^1$ is the affine line, satisfying $f(\underline{A}, g\ell) = \rho_\kappa\left(({}^t g)^{-1}\right) f(\underline{A}, \ell)$ for all $g \in H(\mathbb{W}_m)$. (Compare with Definition 3.11.) Equivalently, we may view f as a function

$$(\underline{A}, \ell) \mapsto f(\underline{A}, \ell) \in \mathrm{Ind}_B^H(\kappa)_{\mathbb{W}_m} = \{f : H_{\mathbb{W}_m}/N_{\mathbb{W}_m} \to \mathbb{A}^1_{\mathbb{W}_m} : f(ht) = \kappa(t)f(h), t \in T\}$$

that satisfies $f(\underline{A}, g\ell) = \rho(g) f(\underline{A}, \ell)$ for all $g \in H(\mathbb{W}_m)$, where the action of H on $\mathrm{Ind}_B^H(\kappa)_{\mathbb{W}_m}$ via ρ is given by

$$H \ni h : f(x) \mapsto \rho(h)f(x) = f(h^{-1}x).$$

For $n \geq m$ and A ordinary over $\mathbb{W}_m$, we have

$$\mathrm{Lie}\, A = \mathrm{Lie}\, A[p^n]^\circ.$$

Thus, for the universal abelian variety $\mathcal{A}$, we have (compare [21, Sect. 3.3–3.4])

$$\underline{\Omega}_{\mathcal{A}/\mathcal{S}_m^{\mathrm{ord}}} = (\mathrm{Lie}\,\mathcal{A})^\vee = (\mathrm{Lie}\,\mathcal{A}[p^n]^\circ)^\vee \cong (T_p(\mathcal{A}[p^n]^{\acute{e}t})^\vee)^\vee \cong \mathcal{A}[p^n]^{\acute{e}t} \otimes \mathcal{O}_{\mathcal{S}_m^{\mathrm{ord}}}. \tag{9}$$

By Eq. (9), there is a canonical isomorphism

$$\underline{\Omega}_{\mathcal{A}/\mathcal{S}_m^{\mathrm{ord}}} \xrightarrow{\sim} \mathcal{A}[p^n]^{\acute{e}t} \otimes \mathcal{O}_{\mathcal{S}_m^{\mathrm{ord}}}.$$

Thus, we may view $j : A[p^n]^{\acute{e}t} \xrightarrow{\sim} (\mathbb{Z}/p^n\mathbb{Z}) \otimes \mathcal{L}^\vee$ as an element of $\mathrm{Isom}_{\mathcal{O}_{\mathcal{S}_m^{\mathrm{ord}}}}(\mathbb{Z}/p^n\mathbb{Z} \otimes \mathcal{L}^\vee, \underline{\Omega}_{\mathcal{A}/\mathcal{S}_m^{\mathrm{ord}}})$ and each element $f \in H^0\left(\mathcal{S}_m^{\mathrm{ord}}, \mathcal{E}_\kappa\right)$ yields a function

$$(\underline{A}, j) \mapsto f(\underline{A}, j) \in \mathrm{Ind}_B^H(\kappa)_{\mathbb{W}_m},$$

where $\underline{A}$ and j are as in Eq. (8), satisfying $f(\underline{A}, gj) = \rho\left(({}^tg)^{-1}\right) f(\underline{A}, j)$ for $g \in H(\mathbb{Z}/p^n\mathbb{Z})$ (due to the action of g on ℓ by precomposition with ${}^tg$, as described in Sect. 3.2).

As explained in [14, Sect. 8.1.2], there is a unique (up to $\mathbb{W}$-unit multiple) N-invariant element $\ell_{\mathrm{can}} \in \left(\mathrm{Ind}_B^H(\kappa)_{\mathbb{W}}\right)^\vee = \mathrm{Hom}_{\mathbb{W}}(\mathrm{Ind}_B^H(\kappa)_{\mathbb{W}}, \mathbb{W})$. The element ℓ_{can} generates $\left(\left(\mathrm{Ind}_B^H(\kappa)_{\mathbb{W}}\right)^\vee\right)^N$. We may normalize ℓ_{can} so that it is evaluation at the identity in H.

Now, we define a map of functions that turns out to be a map of global sections

$$\begin{aligned} \Psi_{n,m} : H^0\left(\mathcal{S}_m^{\mathrm{ord}}, \mathcal{E}_\kappa\right) &\to V_{n,m}^N[\kappa] \\ f &\mapsto \tilde{f} : (\underline{A}, j) \mapsto \ell_{\mathrm{can}}(f(\underline{A}, j)), \end{aligned}$$

where $V_{n,m}^N[\kappa]$ denotes the κ-eigenspace of the torus. Note that for each $b \in B(\mathbb{Z}/p^n\mathbb{Z})$,

$$\begin{aligned} (b\tilde{f})(\underline{A}, j) &= \tilde{f}(\underline{A}, bj) \\ &= \ell_{\mathrm{can}}(f(\underline{A}, bj)) \\ &= f(\underline{A}, bj)(1) \\ &= \rho\left(({}^tg)^{-1}\right) f(\underline{A}, j)(1) \\ &= f(\underline{A}, j)({}^tb) \\ &= \kappa(b) f(\underline{A}, j)(1) = \kappa(b)\ell_{\mathrm{can}}(f(\underline{A}, j)), \end{aligned}$$

for all $\underline{A}$ and j as in Eq. (8). So $\tilde{f}$ is indeed in $V_{n,m}^N[\kappa] \subset V_{n,m}$.

We therefore have a map

$$\Psi_\kappa : H^0\left(\mathcal{S}^{\mathrm{ord}}, \mathcal{E}_\kappa\right) \to V^N[\kappa],$$

where $V^N[\kappa]$ denotes the κ-eigenspace of the torus, and we define

$$\Psi = \oplus_\kappa \Psi_\kappa : \oplus_\kappa H^0(\mathcal{S}^{\mathrm{ord}}, \mathcal{E}_\kappa) \to V^N.$$

This map yields an embedding.

Proposition 4.7 ([14, Sect. 8.1.3, p. 335]). *The map Ψ is injective.*

Remark 4.8. While it is not the subject of this paper, it is natural to ask about results on the density of classical automorphic forms within the space of p-adic automorphic forms. Because our main theorems do not use such density results in our arguments, we refer the reader to [14, Chap. 8] and [13, Lemma 6.1]. Additionally, it is also true by [29, Theorem IV.3.1] that classes in the completed cohomology of Shimura varieties are also p-adically interpolated from classical automorphic forms. This statement is stronger than all previous results, since it also applies to torsion classes which contribute to completed cohomology.

5 Serre–Tate Expansions

The goal of this section is to establish a p-adic analogue of the q-expansion principle for automorphic forms as a consequence of Hida's irreducibility result for the Igusa tower.

Classically, q-expansions arise by localization at a cusp, i.e., the q-expansion of a scalar-valued form f is the image of f in the complete local ring at the cusp, regarded as a power series in q, for q a canonical choice of the local parameter at the cusp. In these terms, the q-expansion principle states that localization is injective, and it is an immediate consequence of the fact that the space is connected. Alternatively, when working over the whole Shimura variety, it becomes necessary to choose a cusp on each connected component, and consider all localizations at once.

In this paper, we work over a connected component of $\mathcal{M}$, and we replace cusps with integral ordinary CM points (i.e., points of $\mathcal{S}^{\mathrm{ord}}$ defined over $\mathbb{W}$ corresponding to abelian varieties with complex multiplication). The crucial observation is that given an integral ordinary CM point x_0, the choice of a lift x of x_0 to the Igusa tower uniquely determines a choice of Serre–Tate local parameters at x_0, i.e., x defines an isomorphism of the p-adic completion of the complete local ring at x_0 with a power series ring over $\mathbb{W}$. We call the power series corresponding to the localization at x of an automorphic form its t-expansion, for t denoting the Serre–Tate local parameters.

By abuse of notation, we denote by $g : \mathrm{Ig}^{\mathrm{ord}} \to \mathrm{Ig}^{\mathrm{ord}}$ the action of $g \in H(\mathbb{Z}_p)$ on the Igusa tower described in Sect. 4.2, and we write $\otimes$ in place of $\otimes_{\mathbb{Z}_p}$ for the tensor product over $\mathbb{Z}_p$.

5.1 Localization

Let $\bar{x}_0 \in \mathcal{S}^{\mathrm{ord}}(\overline{\mathbb{F}}_p)$ be a geometric point, and let $x_0 \in \mathcal{S}^{\mathrm{ord}}(\mathbb{W})$ be any integral lift of $\bar{x}_0$. (Without loss of generality, we may chose x_0 to be a CM point, or even the canonical CM lift of $\bar{x}_0$; see Remark 5.9.) We write $\mathcal{S}^{\mathrm{ord}\wedge}_{x_0}$ for the formal completion of $\mathcal{S}^{\mathrm{ord}}$ at x_0. Then

$$\mathcal{S}^{\mathrm{ord}\wedge}_{x_0} = \mathrm{Spf}(\mathcal{R}_{\mathcal{S}^{\mathrm{ord}},x_0})$$

where $\mathcal{R}_{\mathcal{S}^{\mathrm{ord}},x_0}$ is a p-adically complete local ring.

More explicitly, $\mathcal{R}_{\mathcal{S}^{\mathrm{ord}},x_0}$ can be constructed as follows. For each $m \geq 1$, we write $x_{0,m}$ for the reduction of x_0 modulo p^m, regarded as a point of $\mathcal{S}^{\mathrm{ord}}_m := \mathcal{S}^{\mathrm{ord}} \times_{\mathbb{W}} \mathbb{W}/p^m\mathbb{W}$ (in particular, $\bar{x}_0 = x_{0,1}$). Let $\mathcal{O}^{\wedge}_{\mathcal{S}^{\mathrm{ord}}_m,x_0}$ denote the completed local ring of $\mathcal{S}^{\mathrm{ord}}_m$ at $x_{0,m}$. Then, the local ring $\mathcal{R}_{\mathcal{S}^{\mathrm{ord}},x_0}$ can be identified with $\varprojlim_m \mathcal{O}^{\wedge}_{\mathcal{S}^{\mathrm{ord}}_m,x_0}$. Alternatively, $\mathcal{R}_{\mathcal{S}^{\mathrm{ord}},x_0}$ can also be identified with $\mathcal{O}^{\wedge}_{\mathcal{S}^{\mathrm{ord}},\bar{x}_0}$, the completed local ring of $\mathcal{S}^{\mathrm{ord}}$ at $\bar{x}_0$.

Let $x \in \mathrm{Ig}^{\mathrm{ord}}(\mathbb{W})$ denote a compatible system of integral points $x = (x_n)_{n\geq 0}$ on the Igusa tower $\{\mathrm{Ig}^{\mathrm{ord}}_n\}_{n\geq 0}$ above x_0; i.e., for each $n \geq 0$, x_n is an integral point in $\mathrm{Ig}^{\mathrm{ord}}_n$, $\mathrm{Ig}^{\mathrm{ord}}_0 = \mathcal{S}^{\mathrm{ord}}$, with x_n mapping to x_{n-1} under the natural projections. Given the point x_0 of $\mathcal{S}^{\mathrm{ord}}$, the choice of a point x of $\mathrm{Ig}^{\mathrm{ord}}$ lying above x_0 is equivalent to the choice of an Igusa structure of infinite level (i.e., of compatible closed immersions as in Eq. (6) on the corresponding ordinary abelian variety.

For all m, as n varies, the natural finite étale projections $j = j_{n,m} : \mathrm{Ig}^{\mathrm{ord}}_{n,m} \to \mathcal{S}^{\mathrm{ord}}_m$ induce a compatible system of isomorphisms

$$j^*_x : \mathcal{O}^{\wedge}_{\mathcal{S}^{\mathrm{ord}}_m,x_0} \xrightarrow{\sim} \mathcal{O}^{\wedge}_{\mathrm{Ig}^{\mathrm{ord}}_{n,m},x_n}$$

which allow us to canonically identify $\mathcal{O}^{\wedge}_{\mathrm{Ig}^{\mathrm{ord}},x} := \varprojlim_m \varinjlim_n \mathcal{O}^{\wedge}_{\mathrm{Ig}^{\mathrm{ord}}_{n,m},x_n}$ with $\mathcal{R}_{\mathcal{S}^{\mathrm{ord}},x_0}$.

Remark 5.1. Given $x_0 \in \mathcal{S}^{\mathrm{ord}}(\mathbb{W})$, let $\bar{x}_0 \in \mathcal{S}^{\mathrm{ord}}(\overline{\mathbb{F}}_p)$ denote its reduction modulo p. We observe that, as a consequence of the fact that the morphisms in the Igusa tower are étale, the reduction modulo p gives a canonical bijection between the points on the Igusa tower above x_0 and those above $\bar{x}_0$. Moreover, if we denote by $\bar{x} \in \mathrm{Ig}^{\mathrm{ord}}(\overline{\mathbb{F}}_p)$, $\bar{x} = (\bar{x}_n)_{n\geq 0}$, the reduction of x modulo p, then the previous isomorphisms agree with the compatible system of isomorphisms $j^*_{n,\bar{x}_n} : \mathcal{O}^{\wedge}_{\mathcal{S}^{\mathrm{ord}},\bar{x}_0} \to \mathcal{O}^{\wedge}_{\mathrm{Ig}^{\mathrm{ord}}_n,\bar{x}_n}$.

Definition 5.2. Let $x \in \mathrm{Ig}^{\mathrm{ord}}(\mathbb{W})$, lying above $x_0 \in \mathcal{S}^{\mathrm{ord}}(\mathbb{W})$. We define

$$\mathrm{loc}_x : V \to \mathcal{R}_{\mathcal{S}^{\mathrm{ord}},x_0}$$

as the localization at x composed with j^{*-1}_x.

By abuse of language, we will still refer to $\mathrm{loc}_x(f) \in \mathcal{R}_{\mathcal{S}^{\mathrm{ord}},x_0}$ as the localization of f at x, for all $f \in V$. Furthermore, with abuse of notation, we will still denote by loc_x the restriction of loc_x to V^N (resp., $V^N[\kappa]$, for any weight κ).

We compare localizations at different points of the Igusa tower. Recall that given a point $x \in \mathrm{Ig}^{\mathrm{ord}}(\mathbb{W})$, the set of all points of the Igusa tower above $x_0 = j(x) \in \mathcal{S}^{\mathrm{ord}}(\mathbb{W})$ is a principle homogeneous space for the action of $H(\mathbb{Z}_p)$.

Lemma 5.3. *Let* $x \in \mathrm{Ig}^{\mathrm{ord}}(\mathbb{W})$. *For any* $g \in H(\mathbb{Z}_p)$ *and* $f \in V$, *we have*

$$\mathrm{loc}_{xg}(f) = \mathrm{loc}_x(g \cdot f).$$

In particular, if $g \in T(\mathbb{Z}_p)$ *and* $f \in V^N[\kappa]$, *for a weight* κ, *we have*

$$\mathrm{loc}_{xg}(f) = \kappa(g)\mathrm{loc}_x(f).$$

Proof. For all $g \in H(\mathbb{Z}_p)$ and $x \in \mathrm{Ig}^{\mathrm{ord}}(\mathbb{W})$, the morphism $g : \mathrm{Ig}^{\mathrm{ord}} \to \mathrm{Ig}^{\mathrm{ord}}$ induces an isomorphism of complete local rings $g^* : \mathcal{O}^{\wedge}_{\mathrm{Ig}^{\mathrm{ord}},xg} \to \mathcal{O}^{\wedge}_{\mathrm{Ig}^{\mathrm{ord}},x}$, $\phi \mapsto \phi \circ g$. Then, the first statement follows from the definition given the equality $j_{xg}^{*\,-1} = j_x^{*-1} \circ g^*$. The second statement is an immediate consequence of the first one, given that for all $f \in V^N[\kappa]$ and $g \in T(\mathbb{Z}_p)$ we have $g \cdot f = \kappa(g)f$. □

Proposition 5.4. *Let* $x \in \mathrm{Ig}^{\mathrm{ord}}(\mathbb{W})$. *For any weight* κ, *the map* $\mathrm{loc}_x : V^N[\kappa] \to \mathcal{R}_{\mathcal{S}^{\mathrm{ord}},x_0}$ *is injective.*

Proof. Were the Igusa covers $\mathrm{Ig}_n^{\mathrm{ord}}$ irreducible over $\mathcal{S}^{\mathrm{ord}}$, the statement would immediately follow. As it happens $\mathrm{Ig}_n^{\mathrm{ord}}$ is not irreducible, thus a priori the vanishing under the localization map loc_x only implies the vanishing on the connected component containing x. Yet, as the torus $T(\mathbb{Z}_p) \subset H(\mathbb{Z}_p)$ acts transitively on the connected components of $\mathrm{Ig}^{\mathrm{ord}}$ over $\mathcal{S}^{\mathrm{ord}}$, it suffices to prove that for all $f \in V^N[\kappa]$, the identity $\mathrm{loc}_x(f) = 0$ implies $\mathrm{loc}_{xg}(f) = 0$ for all $g \in T(\mathbb{Z}_p)$. The last statement follows immediately from the second part of Lemma 5.3. □

We note that for a general function $f \in V$ the above statement is false. Yet, by the same argument, Lemma 5.3 implies the following weaker statement for all $f \in V$.

Proposition 5.5. *Let* $x \in \mathrm{Ig}^{\mathrm{ord}}(\mathbb{W})$, *lying above* $x_0 \in \mathcal{S}^{\mathrm{ord}}(\mathbb{W})$. *For any* $f \in V$, *if* $\mathrm{loc}_x(g \cdot f) \in \mathcal{R}_{\mathcal{S}^{\mathrm{ord}},x_0}$ *vanishes for all* $g \in T(\mathbb{Z}_p)$, *then* $f = 0$.

5.2 Serre–Tate Coordinates

It follows from the smoothness of $\mathcal{S}^{\mathrm{ord}}$ that for any $\mathbb{W}$-point x_0 the ring $\mathcal{R}_{\mathcal{S}^{\mathrm{ord}},x_0}$ is (non-canonically) isomorphic to a power series ring over $\mathbb{W}$. The goal of this section is to explain how Serre–Tate theory implies that for x a lift of x_0 to $\mathrm{Ig}^{\mathrm{ord}}$, the ring $\mathcal{O}^{\wedge}_{\mathrm{Ig}^{\mathrm{ord}},x}$ is canonically isomorphic to a ring of power series over $\mathbb{W}$.

We recall Serre–Tate theory following [21]. The first theorem describes the deformation space of an ordinary abelian variety, and the second explains how to address the lifting of additional structures (such as a polarization and extra endomorphisms). As an application we deduce a description of the ring $\mathcal{R}_{\mathcal{S}^{\mathrm{ord}},x_0}$, for any $x_0 \in \mathcal{S}^{\mathrm{ord}}(\mathbb{W})$.

We introduce some notation. Let A be an ordinary abelian variety over $\overline{\mathbb{F}}_p$, of dimension g. The *physical Tate module* of A, $T_pA(\overline{\mathbb{F}}_p)$, is the Tate module of the maximal étale quotient of $A[p^\infty]$, i.e.,

$$T_pA(\overline{\mathbb{F}}_p) = \varprojlim_n A[p^n](\overline{\mathbb{F}}_p) = \varprojlim_n A[p^n]^{ét}(\overline{\mathbb{F}}_p).$$

As A is ordinary, $T_pA(\overline{\mathbb{F}}_p)$ is a free $\mathbb{Z}_p$-module of rank g. Like above, we denote by $A^\vee$ the dual abelian variety, and we denote by $T_pA^\vee(\overline{\mathbb{F}}_p)$ the physical Tate module of $A^\vee$.

Let R be an Artinian local ring, with residue field $\overline{\mathbb{F}}_p$. We denote by $\mathfrak{m}_R$ the maximal ideal of R. A *lifting* (or *deformation*) of A over R is a pair $(\mathcal{A}/R, j)$, consisting of an abelian scheme $\mathcal{A}$ over R, together with an isomorphism j : $\mathcal{A} \otimes_R \overline{\mathbb{F}}_p \to A$. By abuse of notation we sometimes simply write $\mathcal{A}/R$ for the pair $(\mathcal{A}/R, j)$. To each lifting $(\mathcal{A}/R, j)$, as explained in [21, Sect. 2.0, p. 148], Serre and Tate associated a $\mathbb{Z}_p$-bilinear form

$$q_{\mathcal{A}/R} : T_pA(\overline{\mathbb{F}}_p) \times T_pA^\vee(\overline{\mathbb{F}}_p) \to \hat{\mathbb{G}}_m(R) = 1 + \mathfrak{m}_R.$$

Theorem 5.6 ([21] Theorem 2.1, p. 148). *Let the notation be as above.*

(1) *The map* $(\mathcal{A}/R, j) \mapsto q_{\mathcal{A}/R}$ *is a bijection from the set of isomorphism classes of liftings of* A *over* R *to the group* $\mathrm{Hom}_{\mathbb{Z}_p}(T_pA(\overline{\mathbb{F}}_p) \otimes T_pA^\vee(\overline{\mathbb{F}}_p), \hat{\mathbb{G}}_m(R))$.
(2) *The above construction defines an isomorphism of functors between the deformation space* $\mathcal{M}_{A/\overline{\mathbb{F}}_p}$ *and* $\mathrm{Hom}_{\mathbb{Z}_p}(T_pA(\overline{\mathbb{F}}_p) \otimes T_pA^\vee(\overline{\mathbb{F}}_p), \hat{\mathbb{G}}_m)$.

In the following we refer to the above isomorphism as the *Serre–Tate isomorphism*.

Let A and B be ordinary abelian varieties over $\overline{\mathbb{F}}_p$, and let $f : A \to B$ be an $\overline{\mathbb{F}}_p$-isogeny. We write $f^\vee : B^\vee \to A^\vee$ for the dual isogeny. A theorem of Drinfield ([21, Lemma 1.1.3, p.141]) proves that for any Artinian local ring R, and pair of liftings $(\mathcal{A}/R, j_\mathcal{A}), (\mathcal{B}/R, j_\mathcal{B})$ of A, B, respectively, if there exists an isogeny $\phi : \mathcal{A} \to \mathcal{B}$ lifting f (i.e., satisfying $f = j_\mathcal{B} \circ (\phi \otimes 1_{\overline{\mathbb{F}}_p}) \circ j_\mathcal{A}^{-1}$), then ϕ is unique. Yet, in general such a lifting of f will not exist. Theorem 5.7 gives a necessary and sufficient condition for the existence of ϕ in terms of the Serre–Tate isomorphism.

Theorem 5.7 ([21] Theorem 2.1, Part 4, p.149). *Let the notation be as above. Given* $\mathcal{A}$ *and* $\mathcal{B}$ *lifting* A *and* B*, respectively, over an Artinian local ring* R*. A morphism* $f : A \to B$ *lifts to a morphism* $\mathcal{A} \to \mathcal{B}$ *if and only if* $q_{\mathcal{A}/R} \circ (1 \times f^\vee) = q_{\mathcal{B}/R} \circ (f \times 1)$.

We apply the above results in our setting, adapting to our context the arguments in [14, Sects. 8.2.4 and 8.2.5]. (In loc. cit., Hida deals, respectively, with the cases of Siegel varieties and of the unitary Shimura varieties over a quadratic imaginary field in which p splits.)

Let $\mathcal{O}_{K,p} := \mathcal{O}_K \otimes_{\mathbb{Z}} \mathbb{Z}_p$. Recall that under our assumptions p splits completely in K. That is, we have $\mathcal{O}_{K,p} = \prod_{i=1}^r \mathcal{O}_{K_{\mathfrak{P}_i}} \times \prod_{i=1}^r \mathcal{O}_{K_{\mathfrak{P}_i^c}}$, where $\mathcal{O}_{K_{\mathfrak{P}_i}} \simeq \mathbb{Z}_p$ and $\mathcal{O}_{K_{\mathfrak{P}_i^c}} \simeq \mathbb{Z}_p$ for all i.

Let $\bar{x}_0 \in \mathcal{S}^{\mathrm{ord}}(\overline{\mathbb{F}}_p)$ be an ordinary point, and $\underline{A} := \underline{\mathcal{A}}_{\bar{x}_0}$ be the associated abelian variety over $\overline{\mathbb{F}}_p$, together with its additional structures. Then, the physical Tate module $T_pA(\overline{\mathbb{F}}_p)$ of A is a free $\mathcal{O}_{K,p}$-module, and the prime-to-p polarization λ of A induces a conjugate-linear isomorphism $T_p(\lambda) : T_p(A)(\overline{\mathbb{F}}_p) \xrightarrow{\sim} T_pA^\vee(\overline{\mathbb{F}}_p)$. We deduce that

$$T_pA(\overline{\mathbb{F}}_p) = \left(\oplus_{i=1}^r T_{\mathfrak{P}_i}A(\overline{\mathbb{F}}_p)\right) \oplus \left(\oplus_{i=1}^r T_{\mathfrak{P}_i^c}A(\overline{\mathbb{F}}_p)\right)$$

and that $T_p(\lambda)$ induces a $\mathbb{Z}_p$-linear isomorphism between $T_{\mathfrak{P}_i^c}A(\overline{\mathbb{F}}_p)$ and $T_{\mathfrak{P}_i}A^\vee(\overline{\mathbb{F}}_p)$.

Finally, we recall that the formal completion $\mathcal{S}^{\mathrm{ord}\wedge}_{\bar{x}_0}$ of $\mathcal{S}^{\mathrm{ord}}$ at $\bar{x}_0$ represents the deformation problem naturally associated with the moduli problem. This observation allows us to canonically identify $\mathcal{S}^{\mathrm{ord}\wedge}_{\bar{x}_0}$ with the closed subspace of the deformation space $\mathcal{M}_{A/\overline{\mathbb{F}}_p}$ consisting of all deformations of A which are (can be) endowed with additional structures (a polarization and an $\mathcal{O}_K$-action) lifting those of A. We deduce the following description of $\mathcal{S}^{\mathrm{ord}\wedge}_{\bar{x}_0}$.

Proposition 5.8. *Let the notation be as above.*

The map $x \mapsto q_x := (1 \times T_p(\lambda)) \circ q_{\mathcal{A}_x}$, from $\mathcal{S}^{\mathrm{ord}\wedge}_{\bar{x}_0}$ to $\mathrm{Hom}_{\mathbb{Z}_p}(T_pA(\overline{\mathbb{F}}_p) \otimes T_pA(\overline{\mathbb{F}}_p), \hat{\mathbb{G}}_m)$, induces an isomorphism between $\mathcal{S}^{\mathrm{ord}\wedge}_{\bar{x}_0}$ and $\oplus_{i=1}^r \mathrm{Hom}_{\mathbb{Z}_p}(T_{\mathfrak{P}_i}A(\overline{\mathbb{F}}_p) \otimes T_{\mathfrak{P}_i^c}A(\overline{\mathbb{F}}_p), \hat{\mathbb{G}}_m)$.

Proof. For any local Artinian ring R and $x \in \mathcal{S}^{\mathrm{ord}}(R)$, lifting $\bar{x}_0$, let q_x denote the $\mathbb{Z}_p$-bilinear form defined in the statement

$$q_x = (1 \times T_p(\lambda)) \circ q_{\mathcal{A}_x} : T_pA(\overline{\mathbb{F}}_p) \times T_pA(\overline{\mathbb{F}}_p) \to \hat{\mathbb{G}}_m(R).$$

By Theorem 5.7 the additional structures on A (namely, the polarization and the $\mathcal{O}_K$-action, respectively) lift to $\mathcal{A}_x$ if and only if q_x is symmetric and c-Hermitian.

Indeed, let $\mathrm{sw} : T_pA(\overline{\mathbb{F}}_p) \times T_pA(\overline{\mathbb{F}}_p) \to T_pA(\overline{\mathbb{F}}_p) \times T_pA(\overline{\mathbb{F}}_p)$ denote the map $\mathrm{sw}(v, w) = (w, v)$, for all $v, w \in T_pA(\overline{\mathbb{F}}_p)$. Then, given any abelian variety $\mathcal{A}$ lifting A, $q_{\mathcal{A}^\vee} = q_{\mathcal{A}} \circ \mathrm{sw}$. Also, note that since the polarization λ has degree prime-to-p then the induced maps on the physical Tate modules satisfy the condition $T_p(\lambda^\vee) = T_p(\lambda)^{-1}$. We deduce that the polarization λ of A lifts to $\mathcal{A}_x$ if and only if $q_{\mathcal{A}_x} \circ (1 \times T_p(\lambda^\vee)) = q_{\mathcal{A}_x^\vee} \circ (T_p(\lambda) \times 1)$, or equivalently if and only if $q_x = q_x \circ \mathrm{sw}$. Similarly, for all $b \in \mathcal{O}_K$, let $i(b)$ denote the action on A. We deduce that the action of $\mathcal{O}_K$ on A lifts to $\mathcal{A}$ if and only if $q_{\mathcal{A}_x} \circ (1 \times i(b)^\vee) = q_{\mathcal{A}_x} \circ (i(b) \times 1)$, for all $b \in \mathcal{O}_K$, or equivalently if and only if $q_x(1 \times i(b^c)) = q_x(i(b) \times 1)$, because $\lambda \circ i(b^c) = i(b)^\vee \circ \lambda$ by definition. (Recall Sect. 2.2.)

For all $i, j = 1, \cdots, r$, let the forms $q_{x,i,j} : T_{\mathfrak{P}_i}A(\overline{\mathbb{F}}_p) \times T_{\mathfrak{P}_j^c}A(\overline{\mathbb{F}}_p) \to \hat{\mathbb{G}}_m(R)$ and $q_{x,i,j,c} : T_{\mathfrak{P}_i^c}A(\overline{\mathbb{F}}_p) \times T_{\mathfrak{P}_j}A(\overline{\mathbb{F}}_p) \to \hat{\mathbb{G}}_m(R)$ denote the restrictions of q_x. Then, since q_x is symmetric, for all $i, j = 1, \cdots, r$, we have $q_{x,i,j} = q_{x,j,i,c} \circ \mathrm{sw}$, where with abuse of notation we still write sw for its restriction $T_{\mathfrak{P}_i}A(\overline{\mathbb{F}}_p) \times T_{\mathfrak{P}_j^c}A(\overline{\mathbb{F}}_p) \to T_{\mathfrak{P}_j^c}A(\overline{\mathbb{F}}_p) \times T_{\mathfrak{P}_i}A(\overline{\mathbb{F}}_p)$. Furthermore, since q_x is c-Hermitian, we deduce that for all $i \neq j$ the forms $q_{x,i,j}$ and $q_{x,i,j,c}$ necessarily vanish. Thus,

$$q_x = (\oplus_{i=1}^r q_{x,i,i}) \oplus (\oplus_{i=1}^r q_{x,i,i} \circ \mathrm{sw}).$$

In particular, the form q_x is uniquely determined by its restrictions

$$q_{x,i,i} : T_{\mathfrak{P}_i}A(\overline{\mathbb{F}}_p) \times T_{\mathfrak{P}_i^c}A(\overline{\mathbb{F}}_p) \to \hat{\mathbb{G}}_m(R),$$

for $i = 1, \ldots r$. To conclude we observe that any collection $(q_i)_{i=1,\ldots,r}$ of $\mathbb{Z}_p$-bilinear morphisms, $q_i : T_{\mathfrak{P}_i}A(\overline{\mathbb{F}}_p) \times T_{\mathfrak{P}_i^c}A(\overline{\mathbb{F}}_p) \to \hat{\mathbb{G}}_m(R)$, extends uniquely to a symmetric c-Hermitian bilinear form $q : T_pA(\overline{\mathbb{F}}_p) \otimes T_pA(\overline{\mathbb{F}}_p) \to \hat{\mathbb{G}}_m(R)$, namely $q = (\oplus_{i=1}^r q_i) \oplus (\oplus_{i=1}^r q_i \circ \mathrm{sw})$. □

Remark 5.9. As a consequence of the Serre–Tate theory, we see that any point $\bar{x}_0 \in \mathcal{S}^{\mathrm{ord}}(\overline{\mathbb{F}}_p)$ always lifts to a point $x_0 \in \mathcal{S}^{\mathrm{ord}}(\mathbb{W})$. In fact, it even admits a canonical lift y_0, namely the one corresponding to the form $q = 0$. The abelian variety $\mathcal{A}_{y_0}$ is the unique deformation of $\mathcal{A}_{\bar{x}_0}$ to which all endomorphisms lift. Thus, in particular, $\mathcal{A}_{y_0}$ is a CM abelian variety with ordinary reduction. The point y_0 is called the canonical CM lift of $\bar{x}_0$.

We finally describe how the choice of an Igusa structure (of infinite level) on $\underline{A} = \underline{\mathcal{A}}_{x_0}$ determines a choice of local parameters.

We consider the $\mathcal{O}_{K,p}$-modules introduced in Sect. 4.1:

$$\mathcal{L} = \mathcal{L}^+ \oplus \mathcal{L}^- \simeq \oplus_{i=1}^r (\mathcal{O}_{K_{\mathfrak{P}_i}}^{a_{\tau_i+}} \oplus \mathcal{O}_{K_{\mathfrak{P}_i^c}}^{a_{\tau_i-}}),$$

where the decomposition comes from the splitting $p = w \cdot w^c$, and for each $i = 1, \ldots, r$, τ_i is a place inducing $\mathfrak{P}_i$. For each i, we write $\mathcal{L}_i^+ \subset \mathcal{L}^+$ for the submodule corresponding to the place τ_i, $\mathcal{L}_i^+ \simeq \mathcal{O}_{K_{\mathfrak{P}_i}}^{a_{\tau_i+}}$. Similarly, we define $\mathcal{L}_i^- \subset \mathcal{L}^-$, $\mathcal{L}_i^- \simeq \mathcal{O}_{K_{\mathfrak{P}_i}}^{a_{\tau_i-}}$. Then, $\mathcal{L}^+ = \oplus_{i=1}^r \mathcal{L}_i^+$ and $\mathcal{L}_i^- = \oplus_{i=1}^r \mathcal{L}_i^-$.

We define the $\mathcal{O}_{K,p}$-module

$$\mathcal{L}^2 \simeq \oplus_{i=1}^r \mathcal{L}_i^+ \otimes \mathcal{L}_i^-$$

naturally regarded as a submodule of $\mathcal{L} \otimes \mathcal{L}$.

Proposition 5.10. *Let $x_0 \in \mathcal{S}^{\mathrm{ord}}(\mathbb{W})$. Each point $x \in \mathrm{Ig}^{\mathrm{ord}}(\mathbb{W})$ above x_0 determines a unique isomorphism*

$$\beta_x : \mathcal{S}^{\mathrm{ord}\wedge}_{\bar{x}_0} \to \hat{\mathbb{G}}_m \otimes \mathcal{L}^2.$$

In particular, for each x we have an isomorphism of local rings

$$\beta_x^* : \mathbb{W}[[t]] \otimes (\mathcal{L}^2)^\vee \xrightarrow{\sim} \mathcal{R}_{\mathcal{S}^{\mathrm{ord}}, x_0},$$

where $\mathbb{W}[[t]] \otimes (\mathcal{L}^2)^\vee$ *denotes the complete ring corresponding to the formal scheme* $\hat{\mathbb{G}}_m \otimes \mathcal{L}^2$*, i.e., a choice of basis of* $(\mathcal{L}^2)^\vee$ *yields an isomorphism* $\mathbb{W}[[t]] \otimes (\mathcal{L}^2)^\vee \simeq \mathbb{W}[[t_j \mid 1 \leq j \leq \sum_{i=1}^r a_{\tau_i+} a_{\tau_i-}]]$.

Proof. The choice of an Igusa structure of infinite level $\iota = \iota_x : \mu_{p^\infty} \otimes \mathcal{L} \hookrightarrow A[p^\infty]$ is equivalent to the choice of an $\mathcal{O}_{K,p}$-linear isomorphism $T_p(\iota^\vee) : T_pA(\overline{\mathbb{F}}_p) \xrightarrow{\sim} \mathcal{L}^\vee$. We observe that by linearity

$$T_p(\iota^\vee)(T_wA(\overline{\mathbb{F}}_p)) = (\mathcal{L}^+)^\vee \text{ and } T_p(\iota^\vee)(T_{w^c}A(\overline{\mathbb{F}}_p)) = (\mathcal{L}^-)^\vee.$$

More precisely, $T_p(\iota^\vee)$ induces isomorphisms $T_{\mathfrak{P}_i}A(\overline{\mathbb{F}}_p) \simeq (\mathcal{L}_i^+)^\vee$ and $T_{\mathfrak{P}_i^c}A(\overline{\mathbb{F}}_p) \simeq (\mathcal{L}_i^-)^\vee$, for all $i = 1, \ldots, r$. The isomorphism β_x is defined as the composition of the Serre–Tate isomorphism in Proposition 5.8 with the inverses of the trivializations induced by $T_p(\iota^\vee)$, while β_x^* is the corresponding ring homomorphism, for t the canonical parameter on $\hat{\mathbb{G}}_m$. □

Let $\mathbb{I}$ denote the identity on $\mathbb{W}[[t]]$. For all $g \in H(\mathbb{Z}_p)$, we write $\mathbb{I} \otimes g$ for the natural left $\mathbb{W}$-linear action on $\mathbb{W}[[t]] \otimes (\mathcal{L}^2)^\vee$.

As in Lemma 5.3, the action of $H(\mathbb{Z}_p)$ on the points of the Igusa tower lying above x_0 allows us to compare the above construction for different points x.

Lemma 5.11. *For all* $x \in \mathrm{Ig}^{\mathrm{ord}}(\mathbb{W})$ *and* $g \in H(\mathbb{Z}_p)$*, we have*

$$\beta_{x^g}^* = \beta_x^* \circ (\mathbb{I} \otimes g).$$

Proof. The statement follows immediately from the definition since $\iota_{x^g} = \iota_x \circ (1 \otimes g)$. □

Definition 5.12. Given a point $x \in \mathrm{Ig}^{\mathrm{ord}}(\mathbb{W})$, for any $f \in V$ we define the t-expansion of f at the point x as

$$f_x(t) := \beta_x^{*-1}(\mathrm{loc}_x(f)).$$

Proposition 5.13. *Let* $x \in \mathrm{Ig}^{\mathrm{ord}}(\mathbb{W})$*. For all* $g \in H(\mathbb{Z}_p)$ *and* $f \in V$ *we have*

$$f_{x^g}(t) = (\mathbb{I} \otimes g^{-1})(g \cdot f)_x(t).$$

Proof. The statement follows from Lemmas 5.3, 5.11 combined. □

Finally, we can state an appropriate analogue of the q-expansion principle for p-adic automorphic forms.

Theorem 5.14 ("t-Expansion Principle" or "Serre–Tate Expansion Principle"). *Let $x \in \mathrm{Ig}^{\mathrm{ord}}(\mathbb{W})$. For any weight κ and any $f \in V^N[\kappa]$, the t-expansion $f_x(t)$ vanishes if and only if f vanishes.*

Proof. The statement follows from Propositions 5.4 and 5.10 combined. □

Furthermore, by combining Propositions 5.5 and 5.10 together we deduce the following weaker statement for all $f \in V$.

Proposition 5.15. *Let $x \in \mathrm{Ig}^{\mathrm{ord}}(\mathbb{W})$. For any $f \in V$, if $(g \cdot f)_x(t) \in \mathbb{W}[[t]] \otimes (\mathcal{L}^2)^\vee$ vanish for all $g \in T(\mathbb{Z}_p)$, then $f = 0$.*

As an immediate consequence of Proposition 5.15 we obtain the following corollary.

Corollary 5.16. *For each $m \in \mathbb{N}$, given two p-adic automorphic forms f, f' of any weight κ, κ', respectively, we have that $f \equiv f' (\mathrm{mod}\, p^m)$ if and only if*

$$\kappa(g) f_x(t) \equiv \kappa'(g) f'_x(t)\ (\mathrm{mod}\, p^m)$$

for all $g \in T(\mathbb{Z}_p)$.

Remark 5.17. A crucial advantage of Hida's realization of (vector-valued) automorphic forms as function on the infinite Igusa tower is that one can define congruences between forms without any restriction on their weights. The t-expansion principle as stated above implies that those congruence relations can be detected by the coefficients of the associated power series, as in the classical setting. Indeed, for f a vector-valued automorphic form of weight κ, one may also consider the localization of f at a point x_0 of $\mathcal{S}^{\mathrm{ord}}$. Such a localization is an element in a free module M_κ over $\mathcal{R}_{\mathcal{S}^{\mathrm{ord}}, x_0}$, with rank depending on the weight κ. Even with canonical choices of trivializations of the modules M_κ's, such an approach only allows to detect congruences when the two ranks agree, e.g., when the difference among the weights is parallel, i.e., equals $(n^\tau, \dots, n^\tau)_{\tau \in \Sigma}$ in the notation of Sect. 3.1.1. This restriction is not necessary for the above corollary.

6 Restriction of t-Expansions

As an example of how Serre–Tate coordinates may be used to understand certain operators on p-adic automorphic forms, we consider the case of the restriction map from our unitary Shimura variety to a lower dimensional unitary Shimura subvariety, i.e., the pullback map on global sections of the automorphic sheaves.

6.1 Description of the Geometry

We start by introducing the PEL data defining this setting. We maintain the notation introduced in Sect. 2.1.

Let $L = \oplus_{i=1}^{s} W_i$ be a self-dual decomposition of the lattice L. We denote by $\langle , \rangle_i$ the pairing on W_i induced by $\langle , \rangle$ on L, and define $GU_i = GU(W_i, \langle , \rangle_i)$, a unitary group of signature $(a_{+\tau,i}, a_{-\tau,i})_{\tau \in \Sigma_K}$. Then, the signatures $(a_{+\tau,i}, a_{-\tau,i})_{i=1,\dots,s}$ form a partition of the signature $(a_{+\tau}, a_{-\tau})$.

We define $G' \subset \prod_i GU_i$ to be the subgroup of elements with the same similitude factor; i.e., if $\nu_i : GU_i \to \mathbb{G}_m$ denote the similitude factors, then $G' = \nu^{-1}(\mathbb{G}_m)$, for $\nu = \prod_i \nu_i$ and $\mathbb{G}_m \subset \mathbb{G}_m^s$ embedded diagonally. Then, there is a natural closed immersion $G' \to GU$ of algebraic groups which is compatible with the partition of the signature. That is, the above data defines a morphism of Shimura data $(G', X') \to (GU, X)$, for X (resp., X') the $GU(\mathbb{R})$- (resp., $G'(\mathbb{R})$-) conjugacy class of the homomorphism $h : \mathbb{C} \to \mathrm{End}_{K \otimes_{\mathbb{Z}} \mathbb{R}}(L \otimes_{\mathbb{Z}} \mathbb{R})$, where the map $G' \to GU$ is a closed immersion.

As in Sect. 2.1, let H be $\prod_{\tau \in \Sigma} \mathrm{GL}_{a_{+\tilde{\tau}}} \times \mathrm{GL}_{a_{-\tilde{\tau}}}$, which can be identified with a Levi subgroup of U over $\mathbb{Z}_p$. Similarly, we define H' to be $\prod_{\tau \in \Sigma, 1 \le i \le s} \mathrm{GL}_{a_{+\tilde{\tau},i}} \times \mathrm{GL}_{a_{-\tilde{\tau},i}}$ corresponding to the partition $(a_{+\tau,i}, a_{-\tau,i})_{i=1,\dots,s}$ of the signature $(a_{+\tau}, a_{-\tau})$, which can be identified over $\mathbb{Z}_p$ with a Levi subgroup of $G' \cap U$. Note that we have a closed immersion $H' \to H$ that comes from the natural inclusion of G' in GU over $\mathbb{Z}_p$. We let B' be the intersection of the Borel $B \subset H$ with H', and (by abuse of notation) we denote by N and N' the $\mathbb{Z}_p$-points of the unipotent radicals of the Borel subgroups B and B', respectively. Then $N' = N \cap H'(\mathbb{Z}_p)$. Furthermore, we may identify the maximal torus T of H with a maximal torus T' in H'. Note that for any character κ of $T = T'$, if κ is dominant in $X^*(T)$, then it is also dominant in $X^*(T')$, but the converse is false in general.

The morphism of Shimura data $(G', X') \to (GU, X)$ described above defines a morphism of Shimura varieties, $\theta : M' \hookrightarrow M$, from the Shimura variety M' associated with G' into the Shimura variety M associated with GU, which is a closed immersion since $G' \to GU$ is a closed immersion ([7, Theorem 1.15]). The morphism θ extends to a map between the canonical integral models (which with abuse of notation we still denote by θ)

$$\theta : \mathcal{M}' \hookrightarrow \mathcal{M}.$$

A point x in $\mathcal{M}$ is in the image of θ if and only if the corresponding abelian variety $\underline{\mathcal{A}}_x$ decomposes as a Cartesian product of abelian varieties, with dimensions and additional structures prescribed by the above data.

Remark 6.1. We describe an example. Let $K = F$ be a quadratic imaginary field, V_n be an n-dimensional $\mathcal{O}_F$-lattice equipped with a Hermitian pairing $\langle , \rangle$, and assume that the associated group $GU_n = GU(V_n, \langle , \rangle)$ has real signature $(1, n-1)$. We fix the partition $\{(1, n-2), (0, 1)\}$ of the signature $(1, n-1)$, and we realize V_{n-1} as a direct summand of V_n. We choose a neat level Γ hyperspecial at p and denote by Sh_n the simple Shimura variety of level Γ associated with GU_n. Sh_n is a classifying space for polarized abelian varieties of dimension n, equipped with a compatible action of $\mathcal{O}_F$. We write $\mathcal{A}_n$ for the universal abelian scheme on Sh_n. Then, for each elliptic curve $E_0/\mathbb{Z}_p$ with complex multiplication by $\mathcal{O}_F$ (corresponding to the choice of a connected component of the 0-dimensional Shimura variety associated with $GU(0,1)$), the morphism θ is defined by $\theta^*\mathcal{A}_n = \mathcal{A}_{n-1} \times E_0$, and its image $\theta(\mathrm{Sh}_{n-1})$ is a divisor in Sh_n.

We now assume that p splits completely in all the reflex fields E_i (and thus also in E), where E_i is the reflex field for the integral model $\mathcal{M}_i$ associated with the group GU_i. Then the ordinary loci $\mathcal{M}_i^{\mathrm{ord}}$, $\mathcal{M}'^{\mathrm{ord}}$, and $\mathcal{M}^{\mathrm{ord}}$ are non-empty. In particular, a split abelian variety $A = \prod_i A_i$ is ordinary if and only if each of its constituents A_i are ordinary.

Each connected component $\mathcal{S}'$ of $\mathcal{M}'$ can be identified with a product of connected components $\mathcal{S}_i$ of $\mathcal{M}_i$. We choose a connected component $\mathcal{S}'$ of $\mathcal{M}'$, and identify $\mathcal{S}' = \prod_{i=1}^s \mathcal{S}_i$. Then, there is a unique connected component $\mathcal{S}$ of $\mathcal{M}$ such that $\theta(\mathcal{S}') \subset \mathcal{S}$. We write $\mathcal{S}_i^{\mathrm{ord}}$ (resp., $\mathcal{S}'^{\mathrm{ord}}$ and $\mathcal{S}^{\mathrm{ord}}$) for the ordinary locus of $\mathcal{S}_i$ (resp., $\mathcal{S}'$ and $\mathcal{S}$). Thus, $\theta(\mathcal{S}'^{\mathrm{ord}}) \subset \mathcal{S}^{\mathrm{ord}}$, and we may identify $\mathcal{S}'^{\mathrm{ord}} = \prod_i \mathcal{S}_i^{\mathrm{ord}}$. Corresponding to our choice of connected components, there are two $\mathcal{O}_{F,p}$-linear decompositions $\mathcal{L}^+ = \oplus_{i=1}^s \mathfrak{L}_i^+$ and $\mathcal{L}^- = \oplus_{i=1}^s \mathfrak{L}_i^-$, the ranks of the summands determined by the partition $(a_{+i} = \sum_{\tau\in\Sigma} a_{+\tilde{\tau},i}, a_{-i} = \sum_{\tau\in\Sigma} a_{-\tilde{\tau},i})_{i=1,\dots,s}$ of the signature $(a_+ = \sum_{\tau\in\Sigma} a_{+\tilde{\tau}}, a_- = \sum_{\tau\in\Sigma} a_{-\tilde{\tau}})$. For each $i = 1,\dots,s$, we write $\mathfrak{L}_i = \mathfrak{L}_i^+ \oplus \mathfrak{L}_i^-$. Thus, $\mathcal{L} = \oplus_{i=1}^s \mathfrak{L}_i$.

Proposition 6.2. *For every level $n \geq 1$, the homomorphism $\theta : \mathcal{S}'^{\mathrm{ord}} = \prod_i \mathcal{S}_i^{\mathrm{ord}} \to \mathcal{S}^{\mathrm{ord}}$ lifts canonically to a compatible system of homomorphisms $\Theta = (\Theta_n)_n$, among the Igusa towers,*

$$\Theta_n : \mathrm{Ig}_n'^{\mathrm{ord}} := \prod_i \mathrm{Ig}_{n,i}^{\mathrm{ord}} \to \mathrm{Ig}_n^{\mathrm{ord}},$$

where $\mathrm{Ig}_{n,i}^{\mathrm{ord}}$ denotes the nth level of the Igusa tower over $\mathcal{S}_i^{\mathrm{ord}}$, for each $i = 1,\dots,s$.

Proof. For each $i = 1,\dots,s$, let $\underline{A}_i$ denote the universal abelian scheme over $\mathcal{S}_i^{\mathrm{ord}}$, and $\iota_i : \mu_{p^n} \otimes \mathfrak{L}_i \hookrightarrow A_i[p^n]$ denote the universal Igusa structure of level n on A_i. By the universal property of $\mathrm{Ig}_n^{\mathrm{ord}}$ (i.e., using the fact that the Igusa tower represents the functor classifying ordinary points with additional structure), constructing a morphism Θ_n lifting θ is equivalent to defining an Igusa structure of level n on the abelian scheme $\theta^*\underline{A} = \prod_i \underline{A}_i$ over $\mathcal{S}'^{\mathrm{ord}}$. We define $\iota : \mu_{p^n} \otimes \mathcal{L} \hookrightarrow \theta^*\mathcal{A}[p^n]$ as $\iota = \oplus_{i=1}^s \iota_i$. □

Remark 6.3. For each integer $n \geq 0$, the morphism Θ_n defines a closed embedding of the Igusa covers over $\mathcal{S}'^{\mathrm{ord}}$,

$$\mathrm{Ig}_n'^{\mathrm{ord}} \hookrightarrow \mathcal{S}'^{\mathrm{ord}} \times_{\mathcal{S}^{\mathrm{ord}}} \mathrm{Ig}_n^{\mathrm{ord}}.$$

Indeed, since both projections $\mathrm{Ig}_n'^{\mathrm{ord}} \to \mathcal{S}'^{\mathrm{ord}}$ and $\mathcal{S}'^{\mathrm{ord}} \times_{\mathcal{S}^{\mathrm{ord}}} \mathrm{Ig}_n^{\mathrm{ord}} \to \mathcal{S}'^{\mathrm{ord}}$ are finite and étale, it suffices to check that the map is one-to-one on points. Given a point x_0 of $\mathcal{S}'^{\mathrm{ord}}$, corresponding to a split ordinary abelian variety $\underline{A} = \prod_i \underline{A}_i$. A point x of

$\mathrm{Ig}_n^{\mathrm{ord}}$, lying above x_0, is in the image of Θ_n if and only if the corresponding Igusa structure $\iota_x : \mu_{p^n} \otimes \mathcal{L} \hookrightarrow A[p^n]$ satisfies the conditions $\iota_x(\mu_{p^n} \otimes \mathfrak{L}_i) \subset A_i[p^n]$, for all $i = 1, \ldots, s$. Then, $\iota_x = \oplus_i \iota_{x,i}$, for $\iota_{x,i} : \mu_{p^n} \otimes \mathfrak{L}_i \to A_i[p^n]$ the restrictions of ι_x, i.e., there is a unique point $y \in \mathrm{Ig}_n^{'\mathrm{ord}}$ such that $\Theta_n(y) = x$.

We observe that, for non-trivial partitions of the signature $(a_{+\tau}, a_{-\tau})_{\tau \in \Sigma_K}$, such a closed immersion is not an isomorphism. In fact, given a point x in $\Theta_n(\mathrm{Ig}_n^{'\mathrm{ord}}) \subset \mathcal{S}'^{\mathrm{ord}} \times_{\mathcal{S}^{\mathrm{ord}}} \mathrm{Ig}_n^{\mathrm{ord}}$, for any $g \in H(\mathbb{Z}_p)$, the point x^g is in the image of Θ_n if and only if $g \in H'(\mathbb{Z}_p)$.

6.2 Restriction of Automorphic Forms

For κ a dominant character of T, let ρ_κ denote the irreducible representation of $H_{\mathbb{Z}_p}$ with the highest weight κ described in Sect. 3.1.1. Similarly, for κ' a dominant character of T', let $\rho'_{\kappa'}$ denote the irreducible representation of $H'_{\mathbb{Z}_p}$ with the highest weight κ'.

Assume that the restriction of ρ_κ from $H_{\mathbb{Z}_p}$ to $H'_{\mathbb{Z}_p}$ has an irreducible quotient isomorphic to $\rho'_{\kappa'}$ (e.g., when $\kappa' = \kappa$), and fix a projection $\pi_{\kappa,\kappa'} : \rho_\kappa \to \rho'_{\kappa'}$. If $\kappa' = \kappa^\sigma$ for some σ in the Weyl group $W_H(T)$, then we choose $\pi_{\kappa,\kappa'}$ to satisfy the equality $\ell^\kappa_{\mathrm{can}} = \ell^{\kappa'}_{\mathrm{can}} \circ \pi_{\kappa,\kappa'} \circ g_\sigma$, where $g_\sigma \in N_H(T)(\mathbb{Z}_p)$ is a lifting of σ, and where we view $\ell^\kappa_{\mathrm{can}}$ as a functional on the space of ρ_κ by identifying the space of ρ_κ with the space of $\mathrm{Ind}_B^H(\kappa)$ on which H acts by the usual left translation action (described in Sect. 3.1.1) precomposed with transpose-inverse. In other words, $\ell^\kappa_{\mathrm{can}}$ is the unique (up to multiple) N^--invariant functional on ρ_κ.

Recall that $\mathcal{E}_\kappa = \mathcal{E}_{\mathcal{U},\kappa}$ denotes the automorphic sheaf of weight κ over $\mathcal{M}$, as defined in Sect. 3.2, and denote by $\mathcal{E}'_{\kappa'}$ the automorphic sheaf of weight κ' over $\mathcal{M}'$. Then, on $\mathcal{M}'$ we have a canonical morphism of sheaves $r_{\kappa,\kappa'} : \theta^*\mathcal{E}_\kappa \to \mathcal{E}'_{\kappa'}$.

Definition 6.4. We define

$$\mathrm{res}_{\kappa,\kappa'} := r_{\kappa,\kappa'} \circ \theta^* : H^0(\mathcal{M}, \mathcal{E}_\kappa) \to H^0(\mathcal{M}', \theta^*\mathcal{E}_\kappa) \to H^0(\mathcal{M}', \mathcal{E}'_{\kappa'}).$$

We call $\mathrm{res}_{\kappa,\kappa'}$ the *weight* (κ, κ')*-restriction.*

In the following, for $\kappa' = \kappa$, we write $\mathrm{res}_\kappa = \mathrm{res}_{\kappa,\kappa'}$ and call it the *weight* κ*-restriction.*

By abuse of notation we still denote by $\mathrm{res}_{\kappa,\kappa'}$ (and res_κ) the restriction of this map to the space of sections over the ordinary loci; i.e.,

$$\mathrm{res}_{\kappa,\kappa'} := r_{\kappa,\kappa'} \circ \theta^* : H^0(\mathcal{S}^{\mathrm{ord}}, \mathcal{E}_\kappa) \to H^0(\mathcal{S}'^{\mathrm{ord}}, \theta^*\mathcal{E}_\kappa) \to H^0(\mathcal{S}'^{\mathrm{ord}}, \mathcal{E}'_{\kappa'}).$$

Let V, V' denote the spaces of global functions of the Igusa towers $\mathrm{Ig}^{\mathrm{ord}}/\mathcal{S}^{\mathrm{ord}}$ and $\mathrm{Ig}^{\mathrm{ord}'}/\mathcal{S}'^{\mathrm{ord}}$, respectively, as introduced in Sect. 4.2. We write Θ^* for the pullback on global functions of the Igusa towers,

$$\Theta^* : V \to V'.$$

In the following, we refer to Θ^* as *restriction*. Note that Θ maps V^N and $V^N[\kappa]$, for any weight κ, to $V'^{N'}$ and $V'^{N'}[\kappa]$, respectively. With abuse of notation, we will still denote by Θ^* the restrictions of Θ^* to V^N and $V^N[\kappa]$.

Finally, for any κ, κ', we write $\Psi_\kappa : H^0(\mathcal{S}^{\mathrm{ord}}, \mathcal{E}_\kappa) \to V^N$ and $\Psi'_{\kappa'} : H^0(\mathcal{S}'^{\mathrm{ord}}, \mathcal{E}'_\kappa) \to V'^{N'}$ for the inclusions defined in Sect. 4.2.1.

Proposition 6.5. *For κ a dominant character of T, and any $f \in H^0(\mathcal{S}^{\mathrm{ord}}, \mathcal{E}_\kappa)$:*

$$\Theta^*(\Psi_\kappa(f)) = \Psi'_\kappa(\mathrm{res}_\kappa(f)).$$

Proof. The statement follows from the equality $\Theta^* \circ j^* = j^* \circ \theta^*$, together with the observation that, for any dominant weight κ of T, the functional ℓ_{can} appearing in the definitions of Ψ_κ (in Sect. 4.2.1) factors by our choice via the projection $\pi_{\kappa,\kappa} : \rho_\kappa \to \rho'_\kappa$. □

Note that if $\kappa' \neq \kappa$, then the maps $\Theta^* \circ \Psi_\kappa$ and $\Psi'_{\kappa'} \circ \mathrm{res}_{\kappa,\kappa'}$ do not agree, as a consequence of Proposition 6.5 and the injectivity of the Ψ (see Proposition 4.7). Instead, we have the following result.

Proposition 6.6. *The notation is as above. Assume the weight κ' is conjugate to κ under the action of the Weil group $W_H(T)$, i.e., $\kappa' = \kappa^\sigma$ for some $\sigma \in W_H(T)$, and choose $g_\sigma \in N_H(T)(\mathbb{Z}_p)$ lifting σ.*

Then, for all $f \in H^0(\mathcal{S}^{\mathrm{ord}}, \mathcal{E}_\kappa)$, we have

$$\Theta^*(g_\sigma \cdot \Psi_\kappa(f)) = \Psi'_{\kappa'}(\mathrm{res}_{\kappa,\kappa'}(f)).$$

Proof. The same argument as in the proof of Proposition 6.5 applies here, because we chose $\pi_{\kappa,\kappa'}$ such that the functional ℓ_{can} appearing in the definition of Ψ_κ factors via the map $\pi_{\kappa,\kappa'} \circ g_\sigma : \rho_\kappa \to \rho_\kappa \to \rho'_{\kappa'}$. □

Our goal is to give a simple description of Θ^* in Serre–Tate coordinates, and deduce an explicit criterion for the vanishing of the restriction of a p-adic automorphic form in terms of vanishing of some of the coefficients in its t-expansion.

Let $x_0 \in \mathcal{S}'^{\mathrm{ord}}(\mathbb{W})$, $x_0 = (x_0^i)_{i=1,\dots,s}$ where $x_0^i \in \mathcal{S}_i^{\mathrm{ord}}(\mathbb{W})$, for each $i = 1, \dots, s$. We write $\bar{x}_0$ and $\theta(\bar{x}_0)$ for the reductions modulo p of x_0 and of $\theta(x_0) \in \mathcal{S}^{\mathrm{ord}}(\mathbb{W})$, respectively. Let $\underline{A} = \underline{\mathcal{A}}_{\theta(\bar{x}_0)} = \underline{\mathcal{A}}_{\bar{x}_0} = \prod_i \underline{A}_i$ be the corresponding split ordinary abelian variety over $\overline{\mathbb{F}}_p$. We deduce that the physical Tate module of A decomposes as

$$T_pA(\overline{\mathbb{F}}_p) = \oplus_i T_pA_i(\overline{\mathbb{F}}_p),$$

and by linearity we also have $T_{\mathfrak{P}_j}A(\overline{\mathbb{F}}_p) = \oplus_i T_{\mathfrak{P}_j}A_i(\overline{\mathbb{F}}_p)$, for each $j = 1, \dots, r$.

Proposition 6.7. *The notation is the same as above. Under the isomorphism in Proposition 5.8, $x \mapsto q_x$, the map $\theta_{\bar{x}_0} : \mathcal{S}'^{\mathrm{ord}\wedge}_{\bar{x}_0} \to \mathcal{S}^{\mathrm{ord}\wedge}_{\theta(\bar{x}_0)}$ is the closed immersion corresponding to the collection of the natural inclusions*

$$\oplus_{i=1}^{s}\mathrm{Hom}_{\mathbb{Z}_p}(T_{\mathfrak{P}_j}A_i(\overline{\mathbb{F}}_p) \otimes T_{\mathfrak{P}_j^c}A_i(\overline{\mathbb{F}}_p), \hat{\mathbb{G}}_m) \subset \mathrm{Hom}_{\mathbb{Z}_p}(T_{\mathfrak{P}_j}A(\overline{\mathbb{F}}_p) \otimes T_{\mathfrak{P}_j^c}A(\overline{\mathbb{F}}_p), \hat{\mathbb{G}}_m),$$

for $j = 1, \ldots, r$.

Proof. By the definition of θ, a point $x \in \mathcal{S}^{\mathrm{ord}\wedge}_{\theta(\bar{x}_0)}$ is in the image of $\theta_{\bar{x}_0}$ if and only if the corresponding abelian variety $\underline{\mathcal{A}}_x$ decomposes as a Cartesian product of abelian varieties with additional structures, compatibly with the decomposition $\underline{A} = \prod_i \underline{A}_i$. We argue that such a decomposition exists if and only if the endomorphisms $e_i : A \to A_i \hookrightarrow A$ lift to $\underline{\mathcal{A}}_x$.

Clearly, if the decomposition lifts to $\underline{\mathcal{A}}_x$ so do the endomorphisms e_i for all $i = 1, \ldots, s$. Vice versa, let us assume there exists an endomorphism $\tilde{e}_i$ of $\mathcal{A}_x$ lifting the e_i. By Theorem 5.7 the endomorphisms $\tilde{e}_i$ are unique, thus in particular they are orthogonal idempotents (since the e_i are) and the identity of $\mathcal{A}_x$ decomposes as $1_{\mathcal{A}_x} = \sum_i \tilde{e}_i$ (lifting the equality $1_A = \sum_i e_i$). We deduce that for each i, the image $\mathcal{A}_i = \tilde{e}_i(\mathcal{A}_x)$ is an abelian subvariety of $\mathcal{A}_x$ lifting A_i, and $\mathcal{A}_x = \prod \mathcal{A}_i$. Furthermore, for each i, the additional structures on $\mathcal{A}_x$ define unique additional structures on $\mathcal{A}_i$ (by the properties of the Cartesian product) which lift those on A_i. To conclude, we observe that since such lifts are unique, the decomposition of $\mathcal{A}_x$ is compatible with the additional structures, i.e., $\underline{\mathcal{A}}_x = \prod_i \underline{\mathcal{A}_i}$ lifting the decomposition of $\underline{A}$. Furthermore, such lifting is unique.

Finally, by Theorem 5.7, for each $i = 1, \ldots, s$, the endomorphism e_i of A lifts to $\mathcal{A}_x$ if and only if $q_{\mathcal{A}_x} \circ (1 \times e_i^\vee) = q_{\mathcal{A}_x} \circ (e_i \times 1)$. Equivalently, if and only if

$$q_x \circ (1 \times e_i) = q_x \circ (e_i \times 1)$$

(recall $q_x = q_{\mathcal{A}_x} \circ (1 \times T_p(\lambda))$ and under our assumption $e_i^\vee \circ \lambda = \lambda \circ e_i$). We deduce that this is the case if and only if for all $j = 1, \ldots, r$ and any $i, k = 1, \ldots, s$, the restriction of the bilinear form q_x to the subspaces $T_{\mathfrak{P}_j}A_i(\overline{\mathbb{F}}_p) \otimes T_{\mathfrak{P}_j^c}A_k(\overline{\mathbb{F}}_p)$ of $T_{\mathfrak{P}_j}A(\overline{\mathbb{F}}_p) \otimes T_{\mathfrak{P}_j^c}A(\overline{\mathbb{F}}_p)$ vanishes unless $i = k$. □

For each $i = 1, \ldots, s$, we define $\mathcal{O}_{K,p}$-modules $\mathfrak{L}_i^2 \subset \mathfrak{L}_i^{\otimes 2}$ similarly to $\mathcal{L}^2 \subset \mathcal{L}^{\otimes 2}$ in Sect. 5.2. We consider the $\mathcal{O}_{K,p}$-module $\mathfrak{L}^2 = \oplus_{i=1}^{s}\mathfrak{L}_i^2$. By the definition $\mathfrak{L}^2$ is a direct summand of $\mathcal{L}^2$, we write $\epsilon : \mathfrak{L}^2 \to \mathcal{L}^2$ for the natural inclusion.

We choose a point $x \in \mathrm{Ig}^{\mathrm{ord}'}(\mathbb{W})$, lying above x_0, $(x^i)_i = x$. By Proposition 5.10, associated with the point x on the Igusa tower, we have isomorphisms

$$\beta_x : \mathcal{S}'^{\mathrm{ord}\wedge}_{\bar{x}} \xrightarrow{\sim} \hat{\mathbb{G}}_m \otimes \mathfrak{L}^2 \text{ and } \beta_{\Theta(x)} : \mathcal{S}^{\mathrm{ord}\wedge}_{\theta(\bar{x}_0)} \xrightarrow{\sim} \hat{\mathbb{G}}_m \otimes \mathcal{L}^2.$$

For all $i = 1, \dots, s$, we may also consider the isomorphism $\beta_{x^i} : (\mathcal{S}_i^{\mathrm{ord}})^{\wedge}_{\bar{x}_0^i} \xrightarrow{\sim} \hat{\mathbb{G}}_m \otimes \mathfrak{L}_i^2$. Then, $\beta_x = \oplus_{i=1}^{s} (\beta_{x^i})_{i=1,\dots,s}$. As before, we denote, respectively, by β_x^* and $\beta_{\Theta(x)}^*$ the corresponding ring homomorphisms.

Proposition 6.8. *The notation is the same as above. The map $\beta_{\Theta(x)} \circ \theta_{\bar{x}_0} \circ \beta_x^{-1}$ agrees with the inclusion $\mathbb{I} \otimes \epsilon : \hat{\mathbb{G}}_m \otimes \mathfrak{L}^2 \to \hat{\mathbb{G}}_m \otimes \mathcal{L}^2$.*

Proof. The statement follows from Propositions 6.7 and 5.10 combined. □

Equivalently, in terms of the local Serre–Tate coordinates, Proposition 6.8 states that the two homomorphisms

$$\theta_{x_0}^* : \mathcal{R}^{\wedge}_{\mathcal{S}^{\mathrm{ord}},\theta(x_0)} \to \mathcal{R}^{\wedge}_{\mathcal{S}'^{\mathrm{ord}},x_0} \text{ and } \mathbb{I} \otimes \epsilon^{\vee} : \mathbb{W}[[t]] \otimes (\mathcal{L}^2)^{\vee} \twoheadrightarrow \mathbb{W}[[t]] \otimes (\mathfrak{L}^2)^{\vee}$$

satisfy the equality $\mathbb{I} \otimes \epsilon^{\vee} = \beta_x^{*-1} \circ \theta_{x_0}^* \circ \beta_{\Theta(x)}^*$.

Corollary 6.9. *The notation is the same as above. For any $f \in V$, we have*

$$(\Theta^* f)_x(t) = (\mathbb{I} \otimes \epsilon^{\vee})(f_{\Theta(x)}(t)).$$

Proof. By definition $(\Theta^* f)_x(t) = \beta_x^{*-1} \circ j_x^{*-1}((\Theta^* f)_x)$, and $f_{\Theta(x)}(t) = {\beta_{\Theta(x)}^*}^{-1} \circ {j_{\Theta(x)}^*}^{-1}(f_{\Theta(x)})$. Also by definition, $j \circ \Theta = \theta \circ j$ and $x_0 = j(x)$. Then, for all $f \in V$,

$$(\Theta^* f)_x(t) = \beta_x^{*-1} \circ j_x^{*-1}((\Theta^* f)_x) = \beta_x^{*-1} \circ j_x^{*-1} \circ \Theta_x^* (f_{\Theta(x)}) = \beta_x^{*-1} \circ \theta_{x_0}^* \circ {j_{\Theta(x)}^*}^{-1}(f_{\Theta(x)}),$$

and

$$(\mathbb{I} \otimes \epsilon^{\vee})(f_{\Theta(x)}(t)) = (1 \otimes \epsilon^{\vee}) \circ {\beta_{\Theta(x)}^*}^{-1} \circ {j_{\Theta(x)}^*}^{-1}(f_{\Theta(x)}).$$

Thus, the equality $\mathbb{I} \otimes \epsilon^{\vee} \circ {\beta_{\Theta(x)}^*}^{-1} = \beta_x^{*-1} \circ \theta_{x_0}^*$ (following Proposition 6.8) suffices to conclude. □

We observe that the statement in Corollary 6.9 is equivariant for the action of $H'(\mathbb{Z}_p)$. More precisely, the following equalities hold.

Lemma 6.10. *For any $g \in H'(\mathbb{Z}_p) \subset H(\mathbb{Z}_p)$, $x \in \mathrm{Ig}^{\mathrm{ord}'}(\mathbb{W})$, and $f \in V$, we have*

$$(\Theta^* f)_{x^g}(t) = (\mathbb{I} \otimes g^{-1})(\Theta^*(g \cdot f))_x(t) \textit{ and } (\mathbb{I} \otimes \epsilon^{\vee})(f_{\Theta(x^g)}(t))$$
$$= (\mathbb{I} \otimes \epsilon^{\vee} \circ g^{-1})((g \cdot f)_{\Theta(x)}(t)).$$

Proof. Recall that $\epsilon \circ g = g \circ \epsilon$ and $\Theta \circ g = g \circ \Theta$, for all $g \in H'(\mathbb{Z}_p) \subset H(\mathbb{Z}_p)$. Thus, for any $x \in \mathrm{Ig}^{\mathrm{ord}'}(\mathbb{W})$, the point x^g is another point of $\mathrm{Ig}^{\mathrm{ord}'}$ satisfying $\Theta(x)^g = \Theta(x^g)$, and for all $f \in V$, we have $\Theta^*(g \cdot f) = g \cdot \Theta^* f$. The statement then follows immediately from Proposition 5.13. □

By the t-expansion principle, we deduce the following vanishing criteria.

Corollary 6.11. *Let* $x \in \mathrm{Ig}^{\mathrm{ord}'}(\mathbb{W})$, $f \in H^0(\mathcal{S}^{\mathrm{ord}}, \mathcal{E}_\kappa)$, *for any* κ.
The κ*-restriction* $\mathrm{res}_\kappa(f)$ *of* f *to the subgroup* G' *vanishes if and only if*

$$(\mathbb{I} \otimes \epsilon^\vee)(\Psi_\kappa(f)_{\Theta(x)}(t)) = 0.$$

Proof. By Theorem 5.14, $\mathrm{res}_\kappa(f)$ vanishes if and only if $\Psi'_\kappa(\mathrm{res}_\kappa(f))_x(t)$ vanishes. On the other hand, Proposition 6.5 and Corollary 6.9 combined imply

$$\Psi'_\kappa(\mathrm{res}_\kappa(f))_x(t) = \Theta^*(\Psi_\kappa(f))_x(t) = (\mathbb{I} \otimes \epsilon^\vee)(\Psi_\kappa(f)_{\Theta(x)}(t)).$$

□

Corollary 6.12. *Let* $x \in \mathrm{Ig}^{\mathrm{ord}'}(\mathbb{W})$, $f \in H^0(\mathcal{S}^{\mathrm{ord}}, \mathcal{E}_\kappa)$, *for any* κ. *Let* $\kappa' \neq \kappa$ *be a dominant weight of* T'.
Assume $\kappa' = \kappa^\sigma$, *for some* $\sigma \in W_H(T)$, *and choose* $g_\sigma \in N_H(T)(\mathbb{Z}_p)$ *lifting* σ.
The (κ, κ')*-restriction* $\mathrm{res}_{\kappa,\kappa'}(f)$ *of* f *to the subgroup* G' *vanishes if and only if*

$$(\mathbb{I} \otimes (\epsilon^\vee \circ g_\sigma))(\Psi_\kappa(f)_{\Theta(x)^{g_\sigma}}(t)) = 0.$$

Proof. Theorem 5.14 implies that $\mathrm{res}_{\kappa,\kappa'}(f)$ vanishes if and only if $\Psi'_{\kappa'}(\mathrm{res}_{\kappa,\kappa'}(f))_x(t)$ vanishes. By combining Proposition 6.6 and Corollary 6.9, we have

$$\Psi'_{\kappa'}(\mathrm{res}_{\kappa,\kappa'}(f))_x(t) = \Theta^*(g_\sigma \cdot \Psi_\kappa(f))_x(t) = (\mathbb{I} \times \epsilon^\vee)((g_\sigma \cdot \Psi_\kappa(f))_{\Theta(x)}(t)).$$

Finally, Proposition 5.13 implies

$$(g_\sigma \cdot \Psi_\kappa(f))_{\Theta(x)}(t) = (\mathbb{I} \otimes g_\sigma)(\Psi_\kappa(f)_{\Theta(x)^{g_\sigma}}(t)).$$

□

Note that, in the above corollary, since $\kappa' \neq \kappa$, $g_\sigma \notin H'(\mathbb{W})$ and the point $\Theta(x)^{g_\sigma}$ is not in the image of Θ, for any $x \in \mathrm{Ig}^{\mathrm{ord}'}(\mathbb{W})$.

Corollary 6.13. *Let* $x \in \mathrm{Ig}^{\mathrm{ord}'}(\mathbb{W})$. *Let* f *be a global function on the Igusa tower* $\{\mathrm{Ig}_n^{\mathrm{ord}}\}_{n\geq 0}$, *i.e.,* $f \in V$. *The restriction* $\Theta^* f$ *of* f *to the subgroup* G' *vanishes if and only if* $(\mathbb{I} \otimes \epsilon^\vee)((g \cdot f)_{\Theta(x)}(t)) = 0$, *or equivalently* $(\mathbb{I} \otimes \epsilon^\vee)(f_{\Theta(x)^g}(t)) = 0$, *for all* $g \in T(\mathbb{Z}_p) = T'(\mathbb{Z}_p)$.

Proof. By Proposition 5.15, $\Theta^* f$ vanishes if and only if $(g \cdot \Theta^* f)_x(t)$ vanish for all $g \in T(\mathbb{Z}_p)$. From Lemma 6.10 and Corollary 6.9, for all $g \in T(\mathbb{Z}_p)$ (recall $T(\mathbb{Z}_p) \subset H'(\mathbb{Z}_p)$), we deduce

$$(g\cdot\Theta^* f)_x(t) = (\mathbb{I}\otimes g)(\Theta^* f)_{x^g}(t) = (\mathbb{I}\otimes(g\circ\epsilon^\vee))(f_{\Theta(x^g)}(t)) = (\mathbb{I}\otimes(g\circ\epsilon^\vee))(f_{\Theta(x)^g}(t))$$

On the other hand, we also have

$$(g \cdot \Theta^* f)_x(t) = (\Theta^*(g \cdot f))_x(t) = (\mathbb{I} \otimes \epsilon^\vee)(g \cdot f)_{\Theta(x)}(t).$$

□

Corollary 6.14. *Let* $x \in \mathrm{Ig}^{\mathrm{ord}'}(\mathbb{W})$. *Let* $r \in \mathbb{N}$, *and let* f, f' *be two* p*-adic automorphic forms on* GU *of weights* κ, κ', *for any* κ, κ', *i.e.*, $f \in V^N[\kappa], f' \in V^N[\kappa']$. *We denote by* $\mathrm{res}_\kappa(f)$ *and* $\mathrm{res}_{\kappa'}(f')$ *their restrictions to the subgroup* G'. *Then* $\mathrm{res}_\kappa(f) \equiv \mathrm{res}_{\kappa'}(f') \bmod p^r$ *if and only if*

$$\kappa(g)(\mathbb{I} \otimes \epsilon^\vee)(f_{\Theta(x)}(t)) \equiv \kappa'(g)(\mathbb{I} \otimes \epsilon^\vee)(f'_{\Theta(x)}(t)) \bmod p^r,$$

for all $g \in T(\mathbb{Z}_p)$.

Proof. The statement is an immediate consequence of the Corollary 6.13. □

Acknowledgements We are grateful to L. Long, R. Pries, and K. Stange for organizing the Women in Numbers 3 workshop and facilitating this collaboration. We would like to thank the referee for carefully reading the paper and providing many helpful comments, including suggestions for how to improve the introduction. We would also like to thank M. Harris, H. Hida, and K.-W. Lan for answering questions about q-expansion principles. We are grateful to the Banff International Research Station for creating an ideal working environment.

References

1. Ando, M., Hopkins, M., Rezk, C.: Multiplicative orientations of *KO*-theory and of the spectrum of topological modular forms. http://www.math.uiuc.edu/~mando/papers/koandtmf.pdf (2007)
2. Behrens, M.: Eisenstein orientation. Typed notes available at http://www-math.mit.edu/~mbehrens/other/coredump.pdf (2009)
3. Brooks, E.H.: Generalized Heegner cycles. Shimura curves, and special values of p-adic L-functions. Ph.D. Thesis, University of Michigan (2013)
4. Burungale, A., Hida, H.: $\mathfrak{p}$-rigidity and Iwasawa μ-invariants. http://www.math.ucla.edu/~ashay/rigidity.pdf (2014)
5. Buzzard, K., Taylor, R.: Companion forms and weight one forms. Ann. Math. (2) **149**(3), 905–919 (1999)
6. Chai, C.-L., Faltings, G.: Degeneration of abelian varieties. In: Ergebnisse der Mathematik und ihrer Grenzgebiete (3) [Results in Mathematics and Related Areas (3)], vol. 22. Springer, Berlin (1990). With an appendix by David Mumford
7. Deligne, P.: Travaux de Shimura, Séminaire Bourbaki, 23ème année (1970/71), Exp. No. 389. Lecture Notes in Mathematics, vol. 244, pp. 123–165. Springer, Berlin (1971)
8. Eischen, E.E.: p-adic differential operators on automorphic forms on unitary groups. Ann. Inst. Fourier (Grenoble) **62**(1), 177–243 (2012)
9. Eischen, E.E.: A p-adic Eisenstein measure for unitary groups. J. Reine Angew. Math. **699**, 111–142 (2015)
10. Eischen, E.: Differential operators, pullbacks, and families of automorphic forms. Ann. Math. Québec 1–8 (2016). Accepted for publication. doi:10.1007/s40316-015-0049-z

11. Fargues, L.: L'isomorphisme entre les tours de Lubin-Tate et de Drinfeld et applications cohomologiques. L'isomorphisme entre les tours de Lubin-Tate et de Drinfeld. Progress in Mathematics, vol. 262, pp. 1–325. Birkhäuser, Basel (2008)
12. Harris, M., Li, J.-S., Skinner, C.M.: p-adic L-functions for unitary Shimura varieties. I. In: Construction of the Eisenstein measure. Documenta Mathematica, Extra volume, pp. 393–464 (2006) (electronic)
13. Harris, M., Lan, K.-W., Taylor, R., Thorne, J.: On the rigid cohomology of certain Shimura varieties. http://www.math.ias.edu/~rtaylor/rigcoh.pdf (2013)
14. Hida, H.: Springer Monographs in Mathematics. Springer, New York (2004)
15. Hida, H.: Irreducibility of the Igusa tower over unitary Shimura varieties. In: On Certain L-Functions. Clay Mathematics Proceedings, vol. 13, pp. 187–203. American Mathematical Society, Providence, RI (2011)
16. Hopkins, M.J.: Topological modular forms, the Witten genus, and the theorem of the cube. In: Proceedings of the International Congress of Mathematicians, Zürich, 1994, vol. 1, 2, pp. 554–565. Birkhäuser, Basel (1995)
17. Hopkins, M.J.: Algebraic topology and modular forms. In: Proceedings of the International Congress of Mathematicians, Beijing, 2002, vol. I, pp. 291–317. Higher Education Press, Beijing (2002)
18. Jantzen, J.C.: Representations of Algebraic Groups, 2nd edn. Mathematical Surveys and Monographs, vol. 107. American Mathematical Society, Providence, RI (2003)
19. Katz, N.M.: p-adic properties of modular schemes and modular forms. In: Modular functions of one variable, III: Proceedings International Summer School. University of Antwerp, Antwerp, 1972. Lecture Notes in Mathematics, vol. 350, pp. 69–190. Springer, Berlin (1973)
20. Katz, N.M.: p-adic L-functions for CM fields. Invent. Math. **49**(3), 199–297 (1978)
21. Katz, N.M.: Serre-Tate local moduli. In: Algebraic Surfaces (Orsay, 1976–78). Lecture Notes in Mathematics, vol. 868, pp. 138–202. Springer, Berlin (1981)
22. Kottwitz, R.E.: Points on some Shimura varieties over finite fields. J. Am. Math. Soc. **5**(2), 373–444 (1992)
23. Lan, K.-W.: Arithmetic compactifications of PEL-type Shimura varieties. London Mathematical Society Monographs, vol. 36. Princeton University Press, Princeton, NJ (2013)
24. Lan, K.-W.: Compactifications of PEL-type Shimura varieties and Kuga families with ordinary loci. Preprint available at http://www.math.umn.edu/~kwlan/articles/cpt-ram-ord.pdf (2014)
25. Lan, K.-W.: Higher Koecher's principle. Preprint available at http://www-users.math.umn.edu/~kwlan/articles/Koecher.pdf
26. Milne, J.: Introduction to Shimura varieties. Notes available online at http://www.jmilne.org/math/xnotes/svi.pdf
27. Milne, J.S.: Introduction to Shimura varieties. In: Harmonic Analysis, The Trace Formula, and Shimura Varieties. Clay Mathematics Proceedings, vol. 4, pp. 265–378. American Mathematical Society, Providence, RI (2005)
28. Mumford, D.: Ergebnisse der Mathematik und ihrer Grenzgebiete, Neue Folge, Band 34. Springer, Berlin, New York (1965)
29. Scholze, P.: On torsion in the cohomology of locally symmetric varieties. Preprint. Available at http://arxiv.org/pdf/1306.2070v1.pdf
30. Shimura, G.: The arithmetic of automorphic forms with respect to a unitary group. Ann. Math. (2) **107**(3), 569–605 (1978)
31. Shimura, G.: Arithmeticity in the theory of automorphic forms. Mathematical Surveys and Monographs, vol. 82. American Mathematical Society, Providence, RI (2000)
32. Shin, S.W.: Galois representations arising from some compact Shimura varieties. Ann. Math. (2) **173**(3), 1645–1741 (2011)
33. Wedhorn, T.: Ordinariness in good reductions of Shimura varieties of PEL-type. Ann. Sci. École Norm. Sup. (4) **32**(5), 575–618 (1999)

Kneser–Hecke-Operators for Codes over Finite Chain Rings

Amy Feaver, Anna Haensch, Jingbo Liu, and Gabriele Nebe

Abstract In this paper we extend results on Kneser–Hecke-operators for codes over finite fields, to the setting of codes over finite chain rings. In particular, we consider chain rings of the form $\mathbb{Z}/p^2\mathbb{Z}$ for p prime. On the set of self-dual codes of length N, we define a linear operator, T, and characterize its associated eigenspaces.

Keywords Coding theory • Hecke operators

2010 *Mathematics Subject Classification.* Primary 11T71, 94B05

This work was started at the WIN3 workshop at the Banff International Research Station in Banff, Canada. The authors wish to thank the organizers for their tremendous efforts, and for providing the opportunity to begin this collaboration. They also wish to thank the referees for their helpful comments and suggestions.

A. Feaver
Department of Mathematics and Computing Science, The King's University, 9125 50 St NW, Edmonton, AB, T6B 2H3, Canada
e-mail: amy.feaver@kingsu.ca

A. Haensch
Department of Mathematics and Computer Science, Duquesne University, Pittsburgh, PA 15282, USA
e-mail: haenscha@duq.edu

J. Liu
Department of Mathematics and Computer Science, Wesleyan University, Middletown, CT 06459, USA
e-mail: jliu02@wesleyan.edu

G. Nebe (✉)
Lehrstuhl D für Mathematik, RWTH Aachen University, 52056 Aachen, Germany
e-mail: nebe@math.rwth-aachen.de

© Springer International Publishing Switzerland 2016
E.E. Eischen et al. (eds.), *Directions in Number Theory*, Association for Women in Mathematics Series 3, DOI 10.1007/978-3-319-30976-7_8

1 Introduction

Many of the concepts of lattice theory have analogues in coding theory and vice versa.

Lattices L are $\mathbb{Z}$-modules generated by a basis of Euclidean space $(\mathbb{R}^N, (,))$. So they come with a given inner product that is used to define the dual lattice

$$L^\# := \{x \in \mathbb{R}^N \mid (x, \ell) \in \mathbb{Z} \text{ for all } \ell \in L\}$$

which satisfies $\mathrm{vol}(\mathbb{R}^N/L)\,\mathrm{vol}(\mathbb{R}^N/L^\#) = 1$. The Euclidean norm enables the counting of lattice points according to their length and therewith defines a holomorphic function

$$\theta_L(z) := \sum_{\ell \in L} \exp((\ell, \ell)\pi i z), z \in \mathbb{C}, \Im(z) > 0$$

on the upper half plane, the so-called theta series of the lattice L. From the theta series, it is possible to read off important invariants of the lattice, such as the density of the associated sphere packing. Theta series have nice invariance properties, they are examples of modular forms. In particular for even unimodular lattices (i.e., $L = L^\#$ and $(\ell, \ell) \in 2\mathbb{Z}$ for all $\ell \in L$), this theta series is a modular form for the full modular group $\mathrm{SL}_2(\mathbb{Z})$ ([6, Theorem 2.1]). Good upper bounds on the sphere packing density of an even unimodular lattice can be found using the theory of modular forms.

For the purpose of this note, codes C are R-submodules of R^N, where R is a finite commutative ring. Also R^N has a standard inner product $(,)$ that is used to define the dual code

$$C^\perp = \{x \in R^N \mid (x, c) = 0 \text{ for all } c \in C\}$$

for which one has $|C||C^\perp| = |R|^N$. Important invariants of the code C are given in the complete weight enumerator of C (see Definition 10) which is a homogenous polynomial of degree N in $|R|$ variables. For self-dual codes (i.e., $C = C^\perp$) this complete weight enumerator is invariant under the associated Clifford–Weil group (as defined in [12]) which is a finite complex matrix group. Again for certain families of self-dual codes invariant theory of these groups allow to find good upper bounds on the error correcting properties of the codes.

There is a direct connection relating lattices and codes given by the well-known Construction A (cf. [4, Sect. 7.2]). This construction associates a lattice $L(C)$ to a code C over a finite prime field which inherits certain properties of the code: If C is self-dual, then so is $L(C)$, more general $L(C)^\# = L(C^\perp)$ and also the theta series of $L(C)$ is obtained from the complete weight enumerator of C by inserting certain well-defined theta functions (see [4, Theorem (7.3)]).

Also other concepts like Siegel theta series and Siegel's phi operator have their coding theory analogues: higher genus complete weight enumerators and Runge's phi operator [16]. Also theta series with harmonic coefficients have a counterpart in coding theory (see [1, 2]). One of the major tools to study modular forms are Hecke-operators. Certain Hecke operators may be expressed in terms of lattices (see, for instance, [11]). In [10], Nebe translates the notion of Hecke operators for theta series to the setting of codes over finite fields, defining the Kneser–Hecke-operator for codes when $R = \mathbb{F}_q$ is a finite field and therewith answers a question raised in 1977 in [3]. The primary goal of this paper is to extend these results to codes over finite chain rings, beginning with those chain rings of the form $R = \mathbb{Z}/p^2\mathbb{Z}$.

As in [10], we consider the family $\mathcal{F}$ of codes of a certain Type. While [10] deals with self-dual codes over finite fields, the present note starts the investigation for self-dual codes over $R = \mathbb{Z}/p^2\mathbb{Z}$ (for a complete discussion of code Types, see [12]). The general strategy of these papers is the same, namely we define some notion of equivalence for codes, define a neighboring relation, and then define a linear operator, T, on the set of equivalence classes of codes in $\mathcal{F}$. This operator maps a code C to the sum of equivalence classes containing neighboring codes to C. One new ingredient here is that self-dual codes of the same length need not be isomorphic as R-modules. This yields a natural partition of the set of equivalence classes of self-dual codes into module isomorphism classes. We describe the connected components of the restriction of the neighboring graph to each of these subsets. It turns out that odd and even primes behave quite differently very likely due to the fact that we only work with bilinear forms instead of quadratic forms. For the ring $R = \mathbb{Z}/4\mathbb{Z}$ we get a very nice description of these connected components in Theorem 38.

This note reports on research on a WIN project for which time was limited. As we are just at the starting point, this paper implicitly contains more questions than answers. Therefore, it should be considered as a motivation to continue and generalize the research on this topic.

2 Codes Over $\mathbb{Z}/p^2\mathbb{Z}$

In this note we will discuss codes over base rings R of the form $R = \mathbb{Z}/p^2\mathbb{Z}$ where $p \in \mathbb{Z}$ is prime. Our computations will be performed for $p = 2$ as there are programs available to test equivalence of quaternary codes.

Recall that a *code* C over a finite ring R is an R-submodule $C \leq R^N$ of the free R-module of rank $N \in \mathbb{N}$. The Krull–Schmidt theorem gives us valuable insight regarding structure of the R-module C. Before stating this result, it is helpful to recall some facts about the ring $R = \mathbb{Z}/p^2\mathbb{Z}$:

(a) R is a local ring with maximal ideal pR and unit group $R^* = R \setminus pR$. The ideals in R are R, pR, and $\{0\}$.
(b) There are exactly two indecomposable R-modules,

(1) the regular R-module R and
(2) the simple R-module $S \cong R/pR \cong pR$.

Applying the *Krull–Schmidt theorem*, we have that every finitely generated R-module M decomposes as $M \cong R^a \oplus S^b$ for unique $a, b \in \mathbb{N}_0 = \{0, 1, 2, \ldots\}$.

2.1 Self-Dual Codes

One class of codes of particular interest is self-dual codes. Here we present the definition:

Definition 1. The standard inner product is given by $b : R^N \times R^N \to R$ where $b(x, y) := \sum_{i=1}^{N} x_i y_i$. For a code $C \leq R^N$, the *dual code* is defined as

$$C^{\perp} := \{x \in R^N \mid b(x, c) = 0 \text{ for all } c \in C\}.$$

C is called *self-orthogonal* if $C \subseteq C^{\perp}$ and C is called *self-dual* if $C = C^{\perp}$.

Let $C \leq R^N$ be a code of length N over the ring R. We write the elements of C as rows. Let $d \in \mathbb{N}$ be the smallest integer such that there exist $r_1, \ldots, r_d \in R^N$ which generate the R-module C. Then a generator matrix of C is a matrix $G \in R^{d \times N}$ where the rows of G are d elements $r_1, \ldots, r_d$ generating C. The isomorphism type of C as an R-module may be read off from a canonical generator matrix which we will define below. There is also a more structural way to obtain this information.

Remark 2. As $|C||C^{\perp}| = |R|^N = p^{2N}$ (see, for instance, [12, Lemma 3.3.4]) any self-dual code $C = C^{\perp} \leq R^N$ is isomorphic, as an R-module, to $R^a \oplus S^b$ with $2a + b = N$.

We first define families $\mathcal{F}$ of self-dual codes. Let $\mathcal{F} := \left\{C = C^{\perp} \leq R^N\right\}$. For $N = 2a + b$ with $a, b \in \mathbb{N}_0$ we define the set

$$\mathcal{F}_{a,b} := \left\{C \leq R^N \mid C = C^{\perp}, C \cong R^a \oplus S^b\right\} \subseteq \mathcal{F}.$$

Then $\mathcal{F}$ is the disjoint union of the sets $\mathcal{F}_{a,b}$. Let $[C]$ denote the permutation equivalence class of the code $C \in \mathcal{F}$. Then any of the sets $\mathcal{F}_{a,b}$ is the disjoint union of finitely many equivalence classes

$$\mathcal{F}_{a,b} = [C_1] \,\dot{\cup} \ldots \dot{\cup} \left[C_{h(a,b)}\right].$$

We call $h(a, b)$ the *class number* of $\mathcal{F}_{a,b}$. Note that, when $a = 0$ and $b = N$ we always have $h(0, N) = 1$, as $\mathcal{F}_{0,N}$ consists of a single code:

$$\mathcal{F}_{0,N} = \left\{pR^N\right\} = \left[pR^N\right].$$

Let $C \in \mathcal{F}_{a,b}$. Then after replacing C by some equivalent code, if necessary, the code C has a generator matrix

$$G = \begin{bmatrix} I_a & X & Y \\ 0 & pI_b & pZ \end{bmatrix}$$

where I_a and I_b denote the unit matrices of size a and b, respectively, $X \in \{0,\ldots,p-1\}^{a\times b}$, $Y \in R^{a\times a}$, and $Z \in \{0,\ldots,p-1\}^{b\times a}$. The self-duality of C is equivalent to $GG^{tr} = 0$. Thus $XX^{tr} + YY^{tr} = -I_a$ and $YZ^{tr} = -X \pmod{p}$.

2.2 Torsion and Residue Codes

To any code C over R we can associate a chain of codes over $\mathbb{F}_p$ using the general constructions for codes over finite chain rings (see [5] or [13]) as follows.

Definition 3. For a vector $v \in R^N$, let $\overline{v}$ denote the canonical projection of v to the vector space $(R/pR)^N \cong \mathbb{F}_p^N$. Then the *torsion code* is given by

$$\mathrm{Tor}(C) = \{\overline{v} : pv \in C\} \leq (R/pR)^N \cong \mathbb{F}_p^N$$

and $\mathrm{Tor}(C)$ has the generator matrix

$$\begin{bmatrix} I_a & X & Y \\ 0 & I_b & Z \end{bmatrix}$$

where X, Y, and Z are given by the generator matrix G of C. The *residue code* is given by

$$\mathrm{Res}(C) = \{\overline{v} : v \in C\} \leq (R/pR)^N \cong \mathbb{F}_p^N,$$

and $\mathrm{Res}(C)$ has the generator matrix

$$\begin{bmatrix} I_a \; X \; Y \end{bmatrix}.$$

From this definition, it is clear that $\mathrm{Res}(C) \subseteq \mathrm{Tor}(C)$. It will be helpful for us to consider the following equivalent definitions for $\mathrm{Tor}(C)$ and $\mathrm{Res}(C)$. Identifying pR with $\mathbb{F}_p$ by $p \mapsto 1$, we have

$$\mathrm{Tor}(C) = C \cap pR^N \leq \mathbb{F}_p^N$$

and

$$\mathrm{Res}(C) = (C + pR^N)/pR^N \leq \mathbb{F}_p^N.$$

We also note that C is of module isomorphism type $R^a \oplus S^b$ if and only if $\dim(\mathrm{Tor}(C)) = a + b$ and $\dim(\mathrm{Res}(C)) = a$.

The lemma which follows is just [5, Lemma 5.4], but we present a proof here for the sake of exposition.

Lemma 4. *If the code $C \leq R^N$ is self-dual, then $Res(C)^\perp = Tor(C)$ with respect to the standard inner product on $\mathbb{F}_p^N$.*

Proof. Let $C = C^\perp \cong R^a \oplus S^b \leq R^N$. Then by Remark 2 we have that $2a + b = N$ and hence $\dim(\mathrm{Tor}(C)) + \dim(\mathrm{Res}(C)) = N$. Thus it is enough to show that $C \subseteq C^\perp$ implies that $\mathrm{Tor}(C) \subseteq \mathrm{Res}(C)^\perp$.

Let $v, w \in R^N$ such that $\overline{v} \in \mathrm{Tor}(C)$ and $\overline{w} \in \mathrm{Res}(C)$. Then $b(pv, w) = pb(v, w) = 0$, since $pv, w \in C$ and C is self-dual. But then $b(v, w) \in pR$, and consequently $\overline{v}$ and $\overline{w}$ have inner product 0 in $\mathbb{F}_p^N$. □

Corollary 5. *Let $C = C^\perp \leq R^N$. Then C is uniquely determined by any of its maximal free submodules.*

Proof. Assume that $C \cong R^a \oplus S^b$ and let $C_1 \leq C$ be such a maximal free submodule, so $C_1 \cong R^a$. Then $\mathrm{Res}(C) = \mathrm{Res}(C_1)$, so given such a C_1 it is possible to find $\mathrm{Res}(C)$. Combining this with Lemma 4, we have

$$\mathrm{Tor}(C) = \mathrm{Res}(C)^\perp = \mathrm{Res}(C_1)^\perp$$

and therefore we can determine $C \cap pR^N$. But since $C = C_1 + (C \cap pR^N)$, we see that C_1 uniquely determines C. □

Suppose that we have a self-orthogonal code $C_1 \subseteq C_1^\perp \leq R^N$ with $C_1 \cong R^a$. Then we also have that $R^N/C_1 \cong \mathrm{Hom}_R(C_1, R) \cong R^a$ and hence $C_1^\perp \cong R^{N-a}$. Consequently, we have $C_1 + pC_1^\perp \cong R^a \oplus S^{N-2a}$. Moreover $C_1 + pC_1^\perp$ is self-orthogonal and hence self-dual, because $|C_1 + pC_1^\perp| = p^N$. Therefore we make the following remark:

Remark 6. Given any self-orthogonal code $C_1 \subseteq C_1^\perp \leq R^N$ with $C_1 \cong R^a$ for some a, there is always a unique self-dual code $C \cong R^a \oplus S^{N-2a}$ containing C_1 as a maximal free submodule, namely $C = C_1 + pC_1^\perp$.

A code C over $\mathbb{F}_2$ is called *doubly even* if the number of nonzero entries in every codeword in C is divisible by 4. This definition will be illuminated further in the next section, but a preliminary notion is necessary for the following result.

Corollary 7. *If $C \subseteq C^\perp \leq R^N$ is a self-orthogonal code, then $Res(C)$ is a self-orthogonal code of length N over $\mathbb{F}_p$. Moreover if $p = 2$, then $Res(C)$ is doubly even.*

A special case is $b = 0$: here C is a free R-module so $\mathrm{Res}(C) = \mathrm{Tor}(C)$. If we additionally have that $C = C^\perp$, then $\mathrm{Res}(C) \leq \mathbb{F}_p^N$ is a self-dual code, which is doubly even if $p = 2$.

Doubly even self-dual binary codes exist if and only if the length N is a multiple of 8. For odd primes p self-dual codes over $\mathbb{F}_p$ exist if and only if either the length N is a multiple of 4 or N is even and $p \equiv 1 \pmod 4$.

Corollary 8. *Let $C = C^{\perp} \leq R^N$ be a self-dual code that is also a free R-module. Then $C \cong R^{N/2}$ so N is even. If $p = 2$, then the length N is a multiple of* 8 *and if* $p \equiv 3 \pmod 4$, *the length N is a multiple of* 4.

We usually get a smoother theory of orthogonal groups (such as Witt's extension theorem) if we work with quadratic forms. This is straightforward if $p \neq 2$: The function $q : R^N \to R$ given by $q(x) := \frac{1}{2}b(x,x)$ is a quadratic form with associated bilinear form b, as

$$b(x,y) = q(x+y) - q(x) - q(y).$$

Thus the orthogonal groups of b and q coincide and any self-orthogonal code C satisfies $q(C) = \{0\}$, i.e., C is isotropic.

If $p = 2$ and $R = \mathbb{Z}/4\mathbb{Z}$, the situation is more complicated: There is a well-defined quadratic form

$$q : R^N \to \mathbb{Z}/8\mathbb{Z}, q(x) := \sum_{i=1}^{N} x_i^2$$

where $x_i^2 = 1 \in \mathbb{Z}/8\mathbb{Z}$ if $x_i = 1$ or $3 \in R$, $x_i^2 = 0$ if $x_i = 0$, and $x_i^2 = 4$ if $x_i = 2 \in R$. Then $b(x,y) = \frac{1}{2}(q(x+y) - q(x) - q(y))$ for all $x, y \in R^N$. We call C *isotropic* if $q(C) = \{0\}$. Clearly isotropic codes are always self-orthogonal. Isotropic self-dual codes are also called *doubly even self-dual codes*.

Essentially the same form q may also be obtained as an R-valued quadratic form. If $C \leq R^N$ is a self-orthogonal code, then $b(c,c) = 0$ for all $c \in C$, so the number of odd entries in c is 0 mod 4. Let

$$\mathcal{X} := \{x \in R^N \mid b(x, \mathbf{1}) \equiv 0 \pmod 2\}$$

(where $\mathbf{1}$ is the all ones vector) denote the submodule of all vectors having an even number of odd entries. Then $b(x,x)$ is always even for $x \in \mathcal{X}$, so

$$\overline{q} : \mathcal{X} \to R, x \mapsto \frac{1}{2}|\{i \mid x_i \text{ is odd }\}| + 2|\{i \mid x_i = 2\}|$$

is a well-defined R-valued quadratic form with associated bilinear form b. In fact this quadratic form is obtained from the restriction of the $\mathbb{Z}/8\mathbb{Z}$-valued form to $\mathcal{X}$ and then dividing by 2. The radical of $\mathcal{X}$ is $\mathcal{X}^{\perp} = \langle 2 \cdot \mathbf{1} \rangle$ and self-dual isotropic codes correspond to the maximal isotropic subspaces of the non-degenerate quadratic module $(\mathcal{X}/\mathcal{X}^{\perp}, \overline{q})$.

Note that any self-dual code C contains $\mathcal{X}^{\perp} = \langle 2 \cdot \mathbf{1} \rangle$ as $b(x, 2 \cdot \mathbf{1}) = 2$ times the number of odd entries in x. So the doubly even self-dual codes are in natural bijection to the maximal isotropic subspaces of $\mathcal{X}/\mathcal{X}^{\perp}$.

Note that the module structure of $\mathcal{X}/\mathcal{X}^{\perp}$ is $R^{N-2} \oplus S^2$ if N is even and it is R^{N-1} if N is odd. A similar situation has been investigated by J.A. Wood [17] for the case that $R = \mathbb{F}_2$ and the quadratic form is $\mathbb{Z}/4\mathbb{Z}$ valued. The second approach to the quadratic form clarifies his "obstruction" to the extendability of isometries.

2.3 Weight Enumerators

In this section we will build to the definitions of two different types of weight enumerators of codes.

Definition 9. Let R be any ring and $N \in \mathbb{N}$.

(1) For any $c := (c_1, \ldots, c_N) \in R^N$ the *Hamming weight* of c is

$$wt(c) := |\{i, 1 \leq i \leq N : c_i \neq 0\}|.$$

(2) For any subset $C \subseteq R^N$ the *minimal Hamming weight* of C is

$$wt(C) = \min\{wt(c) | 0 \neq c \in C\}.$$

More generally, for each $c \in R^N$ we can use the notion of the *composition* of c to refine its Hamming weight. For each $r \in R$ define $a_r(c) := |\{i : c_i = r\}|$. The set $\{a_r(c) | r \in R\}$ is the *composition* of c and tells us the number of components of c which are equal to each $r \in R$. This is connected to the Hamming weight in that $\mathrm{wt}(c) = N - a_0(c)$.

For a code $C \leq R^N$ the weight enumerator of C is a polynomial attached to the code and the associated weight. These may give, for example, the number of codewords with a given weight or with a given composition.

Definition 10. Let R be a ring and $C \leq R^N$ be a code of length $N \in \mathbb{N}$.

(1) The *Hamming weight enumerator* of C is

$$hwe(C)(x, y) := \sum_{c \in C} x^{N-wt(c)} y^{wt(c)} \in \mathbb{C}[x, y]$$

(2) The *complete weight enumerator* of C is

$$cwe(C) := \sum_{c \in C} \prod_{i=1}^{N} x_{c_i} = \sum_{c \in C} \prod_{v \in V} x_v^{a_v(c)} \in \mathbb{C}[x_v : v \in V].$$

Note that in the above definition, both weight enumerators are homogeneous polynomials of degree N.

Another weight, the Lee weight, is defined on elements of rings of the form $R = \mathbb{Z}/m\mathbb{Z}$ for $m \in \mathbb{N}$ which we identify with the set $\{0, \dots, m-1\} \subset \mathbb{Z}$. The Lee weight can be thought of as the minimum "distance" of an element of $r \in R$ to $0 \in R$. More precisely:

Definition 11. Let $R = \mathbb{Z}/m\mathbb{Z}$ for some $m \in N$.

(1) The *Lee weight* of an element $r \in R$ is $\text{Lee}(r) := \min\{r, m-r\}$.
(2) For any $N \in \mathbb{N}$ and any vector $c = (c_1, \dots, c_N) \in R^N$ we define the *Lee weight* of c as

$$\text{Lee}(c) = \sum_{i=1}^{N} \text{Lee}(c_i).$$

Note that in the case where $m = 2$ we find that for any $c \in R^N$ $\text{wt}(c) = \text{Lee}(c)$. In the case of quaternary codes we have $R = \{0, 1, 2, 3\}$ and the Lee weights of these elements are, respectively, $0, 1, 2, 1$.

2.4 The Associated Clifford–Weil Groups

One main goal of the book [12] is to develop a general theory of a "Type" $\mathcal{T}$ of a self-dual code over a finite alphabet V. To such a Type one may associate in a very natural way a finite subgroup $\mathcal{C}(\mathcal{T}) \leq \text{GL}_{|V|}(\mathbb{C})$, *the associated Clifford–Weil group*, such that the complete weight enumerators of self-dual codes of Type $\mathcal{T}$ are invariant under $\mathcal{C}(\mathcal{T})$. In fact one of the main results of [12] is that for codes over finite chain rings (or more general matrix rings over finite chain rings) the complete weight enumerators of self-dual codes of Type $\mathcal{T}$ and length N span the space of degree N homogeneous invariants of $\mathcal{C}(\mathcal{T})$. This section intends to explain the recipe to compute $\mathcal{C}(\mathcal{T})$ for our situation without introducing the general, but quite heavy, machinery from [12].

In our situation the alphabet $V = R = \mathbb{Z}/p^2\mathbb{Z}$ is the ring itself. As any code $C \leq R^N$ is an R-module we have $C = rC$ for any $r \in R^*$ and hence also

$$cwe(C)(x_0, x_1, \dots, x_{p^2-1}) = cwe(C)(x_{r\cdot 0}, x_{r\cdot 1}, \dots, x_{r\cdot(p^2-1)})$$

so complete weight enumerators of codes are invariant under all variable substitutions

$$m_r : x_v \mapsto x_{rv},$$

where r is a unit in R.

There is a famous theorem, proven in the Ph.D. thesis of Jessie MacWilliams, that relates the weight enumerator of a code over a finite field to the weight enumerator of its dual code. This result is proved in greater generality in [12, Sect. 2.2]. In our situation this reads as follows: Let $\zeta := \exp(\frac{2\pi i}{p^2}) \in \mathbb{C}$ be a primitive p^2-th root of unity. For $v \in \mathbb{Z}/p^2\mathbb{Z}$ we define $h(x_v) := \sum_{w \in \mathbb{Z}/p^2\mathbb{Z}} \zeta^{vw} x_w$. Then for any code $C \leq (\mathbb{Z}/p^2\mathbb{Z})^N$ the complete weight enumerator of the dual code is

$$cwe(C^{\perp})(x_0, x_1, \ldots, x_{p^2-1}) = \frac{1}{|C|} cwe(C)(h(x_0), h(x_1), \ldots, h(x_{p^2-1})).$$

In particular if $C = C^{\perp}$, then $|C| = p^N$ and $cwe(C)$ is invariant under

$$H := \frac{1}{p} h : x_v \mapsto \frac{1}{p} \sum_{w \in \mathbb{Z}/p^2\mathbb{Z}} \zeta^{vw} x_w.$$

One additional ingredient of a Type are certain quadratic conditions. One of these conditions comes from the inner product of codewords with themselves: If $C \leq R^N$ is self-orthogonal, then $b(c, c) = \sum_{i=1}^{N} c_i^2 = 0$ for all $c \in C$ and hence $cwe(C)$ is invariant under the variable substitution d with

$$d(x_v) := \zeta^{v^2} x_v$$

Definition 12. The associated Clifford Weil group of the Type of all self-dual codes over $\mathbb{Z}/p^2\mathbb{Z}$ is

$$\mathcal{C}(p^2) := \langle m_r, H, d \mid r \in \mathbb{Z}/p^2\mathbb{Z}^* \rangle \leq \mathrm{GL}_{p^2}(\mathbb{Q}[\zeta]).$$

With this notation the main result of [12] implies that

Theorem 13. *The invariant ring of $\mathcal{C}(p^2)$ is spanned by the complete weight enumerators of all self-dual codes over $\mathbb{Z}/p^2\mathbb{Z}$.*

As an example we give explicit generators for $p = 2$ and $p = 3$:

$$\mathcal{C}(4) = \langle m_3 = \begin{pmatrix} 1 & 0 & 0 & 0 \\ 0 & 0 & 0 & 1 \\ 0 & 0 & 1 & 0 \\ 0 & 1 & 0 & 0 \end{pmatrix}, H = \frac{1}{2} \begin{pmatrix} 1 & 1 & 1 & 1 \\ 1 & \zeta & -1 & -\zeta \\ 1 & -1 & 1 & -1 \\ 1 & -\zeta & -1 & \zeta \end{pmatrix}, d = \mathrm{diag}(1, \zeta, 1, \zeta) \rangle$$

of order $2^6 = 64$ where $\zeta = i$ is a primitive fourth root of unity and

$$\mathcal{C}(9) = \langle m_2, H, d \rangle$$

where $d = \mathrm{diag}(1, \zeta, \zeta^4, 1, \zeta^7, \zeta^7, 1, \zeta^4, \zeta)$, and ζ is a primitive ninth root of unity,

$$m_2 = \begin{pmatrix} 1&0&0&0&0&0&0&0&0 \\ 0&0&1&0&0&0&0&0&0 \\ 0&0&0&0&1&0&0&0&0 \\ 0&0&0&0&0&0&1&0&0 \\ 0&0&0&0&0&0&0&0&1 \\ 0&1&0&0&0&0&0&0&0 \\ 0&0&0&1&0&0&0&0&0 \\ 0&0&0&0&0&1&0&0&0 \\ 0&0&0&0&0&0&0&1&0 \end{pmatrix}, \text{ and } H = \frac{1}{3}\begin{pmatrix} 1&1&1&1&1&1&1&1&1 \\ 1&\zeta&\zeta^2&\zeta^3&\zeta^4&\zeta^5&\zeta^6&\zeta^7&\zeta^8 \\ 1&\zeta^2&\zeta^4&\zeta^6&\zeta^8&\zeta&\zeta^3&\zeta^5&\zeta^7 \\ 1&\zeta^3&\zeta^6&1&\zeta^3&\zeta^6&1&\zeta^3&\zeta^6 \\ 1&\zeta^4&\zeta^8&\zeta^3&\zeta^7&\zeta^2&\zeta^6&\zeta&\zeta^5 \\ 1&\zeta^5&\zeta&\zeta^6&\zeta^2&\zeta^7&\zeta^3&\zeta^8&\zeta^4 \\ 1&\zeta^6&\zeta^3&1&\zeta^6&\zeta^3&1&\zeta^6&\zeta^3 \\ 1&\zeta^7&\zeta^5&\zeta^3&\zeta&\zeta^8&\zeta^6&\zeta^4&\zeta^2 \\ 1&\zeta^8&\zeta^7&\zeta^6&\zeta^5&\zeta^4&\zeta^3&\zeta^2&\zeta \end{pmatrix}.$$

We computed that $|\mathcal{C}(9)| = 2^3 3^4 = 648$.

Adding certain "quadratic" conditions we obtain overgroups of these associated Clifford Weil groups: For instance, we may consider only those self-dual codes that contain the all ones vector $\mathbf{1} = (1, \ldots, 1)$. We have that $\mathbf{1} \in C^{\perp}$ if and only if $b(c, \mathbf{1}) = \sum_{i=1}^{N} c_i = 0$ for all $c \in C$. Then the weight enumerator of C is invariant under $d_1 : x_v \mapsto \zeta^v x_v$ and we obtain the associated Clifford Weil groups

$$\mathcal{C}_1(p^2) := \langle \mathcal{C}(p^2), d_1 \rangle.$$

For $p = 2$ we could also restrict to doubly even self-dual codes which yield an additional invariance condition of the weight enumerator under the transformation $d_q : x_v \mapsto \zeta_8^{v^2} x_v$, where ζ_8 is a primitive 8th root of unity. We hence obtain

$$\mathcal{C}_q(4) := \langle \mathcal{C}(4), d_q \rangle \text{ and } \mathcal{C}_{1,q}(4) := \langle \mathcal{C}(4), d_q, d_1 \rangle.$$

The isomorphism type of these Clifford Weil groups may be read off from the description of the hyperbolic co-unitary groups in [12, Definition 5.2.4]. From this we obtain the following:

Remark 14. For odd primes p the group $\mathcal{C}(p^2)$ is isomorphic to $\mathrm{SL}_2(\mathbb{Z}/p^2\mathbb{Z}) \cong (\mathbb{Z}/p\mathbb{Z})^3 : \mathrm{SL}_2(\mathbb{Z}/p\mathbb{Z})$. The group $\mathcal{C}(4)$ is isomorphic to an extension of $(\mathbb{Z}/2\mathbb{Z})^2$ by a certain Sylow 2-subgroup $\mathcal{S}$ of $\mathrm{SL}_2(\mathbb{Z}/4\mathbb{Z})$, namely

$$\mathcal{S} = \left\{ \begin{pmatrix} a & b \\ c & d \end{pmatrix} \in \mathrm{SL}_2(\mathbb{Z}/4\mathbb{Z}) \mid ac \in 2\mathbb{Z}, bd \in 2\mathbb{Z} \right\}.$$

2.5 Equivalence of Codes

Definition 15. Let $\pi \in S_N$ be a permutation of $\{1, \ldots, N\}$ and let $a_1, \ldots, a_N \in R^*$. A *monomial map* is a map $((a_1, \ldots, a_N), \pi) : R^N \to R^N$ given by

$$((a_1, \ldots, a_N), \pi)(r_1, \ldots, r_N) := \left(a_1 r_{\pi(1)}, \ldots, a_N r_{\pi(N)}\right).$$

For the ring R, the set of all monomial maps is a group. We call this the *monomial group* and denote it by $\mathrm{Mon}_N(R)$.

Monomial maps are interesting in this context because of their connection to the Hamming weight. What follows is the MacWilliams extension theorem which was originally proved for codes over finite fields, but later extends to the more general setting of codes over finite rings by Wood in [18].

Theorem 16 (MacWilliams Extension Theorem). *Let $C \leq R^N$ and $\varphi : C \to R^N$ be a homomorphism which preserves the Hamming weight, i.e.,* $\mathrm{wt}(c) = \mathrm{wt}(\varphi(c))$ *for all $c \in C$, then there exists a monomial map $((a_1, \ldots, a_N), \pi)$ such that*

$$((a_1, \ldots, a_N), \pi)(c) = \varphi(c)$$

for all $c \in C$.

So if one only considers the Hamming weight, the natural notion of equivalence for codes is monomial equivalence, where two codes are called monomially equivalent if they are in the same orbit under the monomial group.

Monomial maps do not preserve the bilinear form b and hence they do not preserve self-duality. This leads to the notion of strong monomial equivalence: this is where we restrict the values $a_1, \ldots, a_N$ to the set $\{\pm 1\}$. In this case, $a_i^2 = 1$ and thus the bilinear form b is preserved.

In this paper we consider an even finer equivalence relation, the permutation equivalence:

Definition 17. Two codes $C, D \leq R^N$ are called (permutation) *equivalent*, $C \equiv D$, if there is a coordinate permutation $\pi \in S_N$ such that $\pi(C) = D$. Let

$$[C] := \{D \mid D \equiv C\}$$

denote the (permutation) equivalence class of the code C and

$$\mathrm{Aut}(C) := \{\pi \in S_N \mid \pi(C) = C\}$$

the *automorphism group* of C.

Note that permutation equivalent codes have the same complete weight enumerator.

No matter which type of equivalence we take, equivalent codes are always isomorphic as modules. However, two codes of the same module type need not be equivalent as codes. Thus

$$C \equiv D \Rightarrow C \cong D.$$

3 The Action of the Orthogonal Group

Definition 18. The orthogonal group $\mathcal{O}_N(R)$ is the group of all R-linear maps preserving the bilinear form b,

$$\mathcal{O}_N(R) := \{\varphi \in \mathrm{GL}_N(R) \mid b(\varphi(x), \varphi(y)) = b(x, y) \text{ for all } x, y \in R^N\}.$$

For $R = \mathbb{Z}/4\mathbb{Z}$ the theory becomes much smoother if we work with quadratic forms. Recall that $\mathcal{X} = \{x \in (\mathbb{Z}/4\mathbb{Z})^N \mid b(x, x) \in 2\mathbb{Z}\}$ with radical $\mathcal{X}^{\perp} = \langle 2 \cdot \mathbf{1} \rangle$. Then

$$\overline{q} : \mathcal{X} \to R, x \mapsto \frac{1}{2}|\{i \mid x_i \text{ is odd }\}| + 2|\{i \mid x_i = 2\}|$$

is a well-defined R-valued quadratic form, and $\mathcal{X}/\mathcal{X}^{\perp}$ is a non-degenerate quadratic space with respect to $\overline{q}$. Note that the module structure of $\mathcal{X}/\mathcal{X}^{\perp} \cong R^{N-2} \oplus S^2$ if N is even and it is R^{N-1} if N is odd.

For orthogonal groups of quadratic forms over fields, there is a famous theorem attributed to Ernst Witt, that orthogonal mappings from subspaces extend to orthogonal mappings of the full space.

Theorem 19 (Witt's Extension Theorem) (See, for instance, [7] or [17]). *Let (V, q) be a regular quadratic space over a field K and $U \leq V$. For any K-linear injective map $\varphi : U \to V$ with $q(\varphi(u)) = q(u)$ for all $u \in U$, there is $g \in \mathcal{O}(V, q)$ such that $\varphi = g_{|U}$.*

Martin Kneser [7] generalized this theorem to local rings, if U is a free module.

It is easy to see that Witt's extension theorem is wrong without this assumption that U be free: In our situation $R = \mathbb{Z}/p^2\mathbb{Z}$ we could choose $u := (p, 0, \ldots, 0)$ and $w := (\underbrace{p, \ldots, p}_{p}, 0, \ldots, 0)$. Then $b(u, u) = b(w, w) = 0$ but there is no $g \in \mathcal{O}_N(R)$ with $g(u) = w$, because any such g would map $\tilde{u} := \{x \in R^N \mid px = u\} = (1, 0, \ldots, 0) + pR^N$ to $\tilde{w} = (1, \ldots, 1, 0, \ldots, 0) + pR^N$ but the elements $x \in \tilde{u}$ satisfy $b(x, x) \in 1 + pR$ and the elements $y \in \tilde{w}$ satisfy $b(y, y) \in pR$.

Note that this is also true for odd primes p, so this is not caused by the problem of working with a bilinear form instead of a quadratic form.

Nevertheless, the following theorem is true:

Theorem 20. *Let $\mathcal{F}_{a,b} := \{C = C^{\perp} \leq R^N \mid C \cong_R R^a \oplus S^b\}$ so $a, b \in N_0$, $2a + b = N$. If $p \neq 2$, then $\mathcal{O}_N(R)$ acts transitively on $\mathcal{F}_{a,b}$.*

Proof. Let $C, D \in \mathcal{F}_{a,b}$ and choose subcodes $C' \leq C, D' \leq D$ so that $C' \cong R^a$, $D' \cong R^a$. Then C' is a free module, any R-isomorphism $\varphi : C' \to D'$ preserves the bilinear form (as this is 0 on C' and also on D'), and

$$\mathrm{Hom}_R(C', R) = \{c \mapsto b(c, x) \mid x \in R^N\}.$$

So by [7, Folgerung (4.4)] there is $g \in \mathcal{O}_N(R)$ such that $g_{|C'} = \varphi$. Then Corollary 5 implies that $g(C) = D$. q.e.d.

For $p = 2$ we want to proceed similarly as for odd primes but $(C' + \mathcal{X}^\perp)/\mathcal{X}^\perp \leq \mathcal{X}/\mathcal{X}^\perp$ does not satisfy the conditions from [7, Folgerung (4.4)], because $(C' + \mathcal{X}^\perp)/\mathcal{X}^\perp$ is usually not free.

Remark 21. If $p = 2$, then define

$$\mathcal{F}_{a,b}^{(q)} := \{C = C^\perp \leq R^N \mid q(C) = \{0\}, C \cong R^a \oplus S^b\}$$

$2a + b = N$ for $a, b \in N_0$. For $C \in \mathcal{F}_{a,b}^{(q)}$ there are two possibilities,

(1) $2 \cdot \mathbf{1} \in 2C$ and then $C/\mathcal{X}^\perp \cong R^{a-1} \oplus S^{b+1}$; or,
(2) $2 \cdot \mathbf{1} \notin 2C$ then $C/\mathcal{X}^\perp \cong R^a \oplus S^b$.

According to these two possibilities the orthogonal group $\mathcal{O}(\mathcal{X}/\mathcal{X}^\perp, \overline{q})$ has at least two orbits $\mathcal{F}_{a,b}^{(q)}(1)$ and $\mathcal{F}_{a,b}^{(q)}(2)$ on $\mathcal{F}_{a,b}^{(q)}$.

Remark 22. Keeping the notation of the previous remark, let $C' \leq C$ be a subcode of C that is free of rank a. Then $2C' = 2C$ and in the first case $\mathcal{X}^\perp \subseteq C'$ and $C'/\mathcal{X}^\perp = C_1 \oplus \langle v \rangle$ with $C_1 \cong R^{a-1}$ and $2v = 2 \cdot \mathbf{1}$. In the second case $\mathcal{X}^\perp \nsubseteq C'$ and $C' + \mathcal{X}^\perp/\mathcal{X}^\perp \cong C' \cong R^a$ is free. So here we may directly apply Witt's theorem for local rings [7, Folgerung (4.4)] to show that $\mathcal{O}(\mathcal{X}/\mathcal{X}^\perp, \overline{q})$ acts transitively on $\mathcal{F}_{a,b}^{(q)}(2)$.

For $\mathcal{F}_{a,b}^{(q)}(1)$ we can apply [7, Folgerung (4.4)] twice to obtain transitivity:

Lemma 23. *Let $C, D \in \mathcal{F}_{a,b}^{(q)}(1)$, C', D' free submodules of rank a as before, and $C' = C_1 \oplus \langle v \rangle$, $D' = D_1 \oplus \langle w \rangle$ as in Remark 22. Then there is $u \in \mathcal{O}(\mathcal{X}/\mathcal{X}^\perp, \overline{q})$ mapping $C'/\mathcal{X}^\perp$ onto $D'/\mathcal{X}^\perp$.*

Proof. We first want to find reflections in $\mathcal{O}(\mathcal{X}/\mathcal{X}^\perp, \overline{q})$ that map v to w. As both vectors v, w satisfy $2v = 2 \cdot \mathbf{1} = 2w$, we have $v, w \in \{1, -1\}^N$. The element $z_i := 2e_i = (0, \ldots, 0, 2, 0, \ldots, 0)$, with 2 at the i-th place lives in $\mathcal{X}$. We have $q(z_i) = 2$, $b_q(z_i, x) \in 2R$ for all $x \in \mathcal{X}$, so the map

$$s_{z_i} : \mathcal{X} \to \mathcal{X}, x \mapsto x - b_q(z_i, x)e_i$$

is a well-defined orthogonal mapping $s_{z_i} \in \mathcal{O}(\mathcal{X}/\mathcal{X}^\perp, q)$. If $x = (x_1, \ldots, x_N)$, then $b_q(z_i, x) = 2x_i$ and hence

$$q(s_{z_i}(x)) = q(x - x_i z_i) = q(x) + x_i^2 q(z_i) - x_i b_q(x, z_i) = q(x) + 2x_i^2 - 2x_i^2 = q(x).$$

Moreover s_{z_i} multiplies the i-th coordinate of v by -1, so a certain product of these reflections s_{z_i} will map v to w. So we may assume without loss of generality that $v = w$. We now replace $\mathcal{X}/\mathcal{X}^\perp$ by the subspace

$$E := \langle v \rangle^\perp = \{x + \mathcal{X}^\perp \mid x \in \mathcal{X}, b_q(x, v) = 0\} \leq \mathcal{X}/\mathcal{X}^\perp.$$

As $E^{\perp} = \langle v \rangle / \mathcal{X}^{\perp}$ any $g \in \mathcal{O}(E, \overline{q})$ preserves v. Moreover $C_1 \cong C_1 + \mathcal{X}^{\perp}/\mathcal{X}^{\perp} \leq E$ and $D_1 \cong D_1 + \mathcal{X}^{\perp}/\mathcal{X}^{\perp} \leq E$ are both free submodules of E. By Witt's theorem for free modules over local rings, there is such a mapping $g \in \mathcal{O}(E, \overline{q})$ with $g(C_1) = D_1$. q.e.d.

Corollary 24. *The orthogonal group $\mathcal{O}(\mathcal{X}/\mathcal{X}^{\perp}, \overline{q})$ has two orbits on the set $\mathcal{F}_{a,b}^{(q)}$:*

$$\mathcal{F}_{a,b}^{(q)}(1) := \{C \in \mathcal{F}_{a,b}^{(q)} \mid 2 \cdot \mathbf{1} \in 2C\}$$

and

$$\mathcal{F}_{a,b}^{(q)}(2) := \{C \in \mathcal{F}_{a,b}^{(q)} \mid 2 \cdot \mathbf{1} \notin 2C\}$$

One might replace $\mathcal{O}(\mathcal{X}/\mathcal{X}^{\perp}, \overline{q})$ by the group $\mathcal{O}_N(R, q)$ in this corollary, however we have not yet closed a necessary gap in the proof. It would also be interesting to have a similar result for $p = 2$ and $\mathcal{O}_N(\mathbb{Z}/4\mathbb{Z})$ for the set $\mathcal{F}_{a,b}$. This might be obtained along the lines of [15].

4 Hecke Operators

4.1 Survey of the Results Over Fields

The paper [10] defines and analyzes certain linear operators that are shown to be Hecke operators for the associated Clifford Weil groups in [9]. Let $\mathbb{F}$ be a finite field, $N \in 2\mathbb{N}$, and $\mathcal{F}$ be the set of all self-dual codes in $\mathbb{F}^N$. Assume that $C_1, \ldots, C_h$ represent the permutation equivalence classes of codes in $\mathcal{F}$. Let $\mathcal{V}$ denote the h-dimensional complex vector space

$$\mathcal{V} := \left\{ \sum_{i=1}^{h} a_i [C_i] \mid a_i \in \mathbb{C} \right\},$$

and define a Hermitian positive definite scalar product by

$$([C], [D]) := |\mathrm{Aut}(C)| \, \delta_{[C],[D]}$$

for all $C, D \in \mathcal{F}$.

Definition 25. (1) For $0 \leq k \leq N/2$, two codes $C, D \in \mathcal{F}$ are called *k-neighbors*, written as $C \sim_k D$, if $\dim(C \cap D) = \dim(C) - k$.

(2) Define a linear operator T_k on $\mathcal{V}$ by

$$T_k([C]) := \sum_{D \sim_k C} [D],$$

where the sum is over all k-neighbors $D \in \mathcal{F}$ of the code C. The operator T_k is called the *k-th Kneser–Hecke-operator* for $\mathcal{F}$.

(3) Let $T := T_1$ be the *Kneser–Hecke-operator* and call 1-neighbors simply *neighbors*.

The following result and its proof can be found in [10, Theorem 3], but we state it here in order to illustrate the approach for codes over fields as compared to codes over finite chain rings.

Theorem 26. *For $0 \leq k \leq N/2$, the operator T_k is a self-adjoint linear operator on the Hermitian vector space $\mathcal{V}$.*

The main result of [10] is an explicit computation of the T-eigenspace decomposition of $\mathcal{V}$ and the corresponding eigenvalues for all classical types of self-dual codes over finite fields.

In [9] the action of T on the space of genus m-weight enumerators of the codes in $\mathcal{F}$ coincides with the action of a certain linear combination of double cosets of the associated Clifford Weil groups.

There is also a nice representation theoretic interpretation of T (see [14, Chap. 5]): Let A denote the adjacency matrix of the neighboring graph whose vertices are the elements of $\mathcal{F}$ and two vertices C and D are connected by an edge, if and only if C and D are neighbors. The associated orthogonal group $\mathcal{O}(\mathbb{F}^N)$ acts on the set $\mathcal{F}$ and respects the neighboring relation. In particular, A is an element in the endomorphism ring of the corresponding permutation representation of $\mathcal{O}(\mathbb{F}^N)$. This endomorphism ring is well known and can be described in the framework of Bruhat–Tits theory using the Weil group of $\mathcal{O}(\mathbb{F}^N)$. In particular, it is shown in [14, Sect. 5.3.13] that the action of A coincides with the one of a certain (very natural) double coset of $\mathcal{O}(\mathbb{F}^N)$. This implies that A generates the endomorphism ring of this permutation representation as a $\mathbb{C}$-algebra, and in particular, this endomorphism ring is commutative and hence the permutation representation is multiplicity free.

The aim of the next section is to generalize some of these aspects to codes over finite chain rings, where we start with $R = \mathbb{Z}/p^2\mathbb{Z}$.

4.2 Kneser–Hecke-Operators for R

We now return to the situation where $R = \mathbb{Z}/p^2\mathbb{Z}$. As before, let $\mathcal{F} = \dot{\bigcup}_{2a+b=N} \mathcal{F}_{a,b}$ denote the set of all self-dual codes in R^N. Assume that $C_1, \ldots, C_h$ represent the permutation equivalence classes of codes in $\mathcal{F}$. Let $\mathcal{V}$ denote the h-dimensional complex vector space

$$\mathcal{V} := \left\{ \sum_{i=1}^{h} a_i[C_i] \mid a_i \in \mathbb{C} \right\},$$

and define a Hermitian positive definite scalar product by

$$([C],[D]) := |\mathrm{Aut}(C)|\, \delta_{[C],[D]}$$

for all $C, D \in \mathcal{F}$. Then, $\mathcal{V}$ is the orthogonal sum of all $\mathcal{V}_{a,b}$ with $2a + b = N$, where $\mathcal{V}_{a,b}$ is the subspace of $\mathcal{V}$ generated by all $[C]$ with $C \in \mathcal{F}_{a,b}$.

Definition 27. (1) Two codes $C, D \in \mathcal{F}$ are called *neighbors*, denoted $C \sim D$, if $C/(C \cap D) \cong D/(C \cap D) \cong pR$. We call them *free neighbors*, denoted $C \sim_R D$, if $C/(C \cap D) \cong D/(C \cap D) \cong R$.

(2) We define a graph Γ with vertex set $\mathcal{F}$. Two vertices C and D are joined by an edge in Γ, if $C \sim D$. Similarly we define the graph Γ_a as the restriction of Γ to $\mathcal{F}_{a,b}$.

In the following we will mainly be concerned with the neighboring relation $\sim$ and the graphs Γ_a. However $\sim_R$ might be the more suitable generalization of neighbors over fields, as this relation preserves the module isomorphism type.

Lemma 28. *If $C \sim_R D$, then $C \cong D$ as R-modules.*

Proof. By definition, we have the exact sequences

$$0 \to C \cap D \to C \to R \to 0, \quad 0 \to C \cap D \to D \to R \to 0.$$

As R is a free module, both sequences split and hence $C \cong C \cap D \oplus R \cong D$. □

Note that this lemma is not true if one replaces free neighbors by neighbors. Now we define a linear operator T on $\mathcal{V}$ by $T([C]) = \sum_{D \sim C} [D]$. By arranging the basis elements according to module isomorphism type, we have

$$T = \begin{bmatrix} T_0 & \cdots & & \\ \vdots & T_1 & & \\ & & \ddots & \vdots \\ & & \cdots & T_n \end{bmatrix},$$

where $n = \lfloor \frac{N}{2} \rfloor$ and

$$T_a : \mathcal{V}_{a,b} \to \mathcal{V}_{a,b}, T_a([C]) := \sum_{D \sim C, D \in \mathcal{F}_{a,b}} [D]$$

can be computed from the adjacency matrix of the neighboring graph Γ_a. We immediately observe that for any choice of $\mathcal{F}$, $T_0 = [0]$, since there is only one code in $\mathcal{F}$ of isomorphism type S^N. Note that our notation does not imply that T is a block diagonal matrix. On the contrary, as we will see below, for odd primes p the matrices T_a are all 0. So the T_a are only interesting for $p = 2$ and for odd primes one should probably consider free neighbors or the matrix T^2 to obtain interesting operators on $\mathcal{V}_{a,b}$.

Theorem 29. *The Kneser–Hecke-operators T (and hence all the T_a) and T_R are self-adjoint linear operators on the Hermitian space $\mathcal{V}$.*

Proof. Let us prove the self-adjointness of T, this implies that the T_a as the summands of $\mathcal{V}_{a,b}$ are orthogonal. By definition, T is linear. For basis vectors $[C]$, $[D]$ with $C, D \in \mathcal{F}$, one has

$$\frac{N!}{|\mathrm{Aut}(D)|}|\{C' \in \mathcal{F} \mid C' \sim D \text{ and } C' \equiv C\}|$$

$$= \sum_{\tilde{D}\equiv D} |\{C' \in \mathcal{F} \mid C' \sim \tilde{D} \text{ and } C' \equiv C\}|$$

$$= \sum_{\tilde{C}\equiv C} |\{D' \in \mathcal{F} \mid D' \sim \tilde{C} \text{ and } D' \equiv D\}|$$

$$= \frac{N!}{|\mathrm{Aut}(C)|}|\{D' \in \mathcal{F} \mid D' \sim C \text{ and } D' \equiv D\}|.$$

The middle equality follows since the neighboring relation is symmetric and invariant under equivalences. Therefore

$$(T([C]), [D]) = |\mathrm{Aut}(D)|\, |\{D' \in \mathcal{F} \mid D' \sim C \text{ and } D' \equiv D\}|$$

$$= |\mathrm{Aut}(C)|\, |\{C' \in \mathcal{F} \mid C' \sim D \text{ and } C' \equiv C\}| = ([C], T([D])).$$

Hence T is self-adjoint. The self-adjointness of T_R follows similarly. □

4.3 Connected Components of Γ_a

Although it is known from [8] that Γ is a connected graph, it does not follow that the Γ_a are connected. In fact, it is not difficult to compute explicit examples in which the Γ_a have multiple connected components; one such example will appear in Sect. 5. Therefore, it is of interest to understand the size and composition of the connected components of the Γ_a. To do this, we will begin by studying some of the natural lifts of neighboring codes over R to codes over $\mathbb{F}_p$, as described in Definition 3.

Lemma 30. *Let $C, D \in \mathcal{F}_{a,b}$ be neighbors, $C \sim D$. Then $Tor(C) = Tor(D)$ and $Res(C) = Res(D)$.*

Proof. Let $C \sim D$ and put $E := C \cap D$. Then C/E and D/E are two distinct minimal submodules of $E^{\perp}/E$, and hence $E^{\perp}/E \cong S \oplus S$ is not cyclic.

It suffices to show that $C \cap pR^N = D \cap pR^N$ as this implies $\mathrm{Tor}(C) = \mathrm{Tor}(D)$. The equality of the residue codes then follows from Lemma 30.

Seeking for a contradiction we suppose that $C \cap pR^N \neq D \cap pR^N$, we get

$$C = C \cap pR^N + E,$$

so we may choose $x = pv \in C \cap pR^N$ such that $C = E \oplus \langle x\rangle$.

Next, we will show that for any $w \in E^{\perp} \backslash C$, it follows that $pw \notin pE$. To show this, suppose that $w \in E^{\perp} \backslash C$. Then $\langle w, x\rangle \neq 0$, or else it would follow that $\langle w, e+x\rangle = 0$ for every $e \in E$, and hence $w \in C = C^{\perp}$. Now suppose there is some $y \in E$ so that $pw = py$, then $\langle y, x\rangle = 0$, since $y \in E \subseteq C = C^{\perp}$. But then,

$$0 = \langle y, x\rangle = \langle y, pv\rangle = \langle py, v\rangle = \langle pw, v\rangle = \langle w, pv\rangle = \langle w, x\rangle \neq 0,$$

a contradiction.

Since $C \cap pR^N \neq D \cap pR^N$, we may similarly say that $D = E \oplus \langle pw\rangle$ for some $pw \in D \cap pR^N$. But then clearly $pw \in E^{\perp} \setminus C$. However,

$$p \cdot pw = p^2 w = 0 \in pE$$

contradicting what was established in the preceding paragraph. Therefore, we may conclude that $C \cap pR^N = D \cap pR^N$. □

Now we define the following set,

$$N(C) := \{D \in \mathcal{F}_{a,b} \mid \mathrm{Tor}(D) = \mathrm{Tor}(C)\},$$

noting that in view of Lemma 30, $N(C)$ contains all neighbors of C which are of the same module isomorphism type. Recalling Definition 2, we have

$$\mathrm{Res}(C) := (C + pR^N)/pR^N \leq (R/pR)^N = \mathbb{F}_p^N,$$

and so from Lemma 30, we also have

$$N(C) = \{D \in \mathcal{F}_{a,b} \mid \mathrm{Res}(D) = \mathrm{Res}(C)\}.$$

For any $D \in N(C)$ it is clear that $N(C) = N(D)$. Furthermore, for $D \in N(C)$ we have $D + pR^N = C + pR^N$ from Lemma 30. Hence, any such a family $N(C)$ defines a unique non-degenerate bilinear $\mathbb{F}_p$-vector space

$$W := C + pR^N / C \cap pR^N \cong \mathbb{F}_p^{2a},$$

with the bilinear form $\langle .,. \rangle : W \times W \to \mathbb{F}_p$ given by

$$\langle c + px + (C \cap pR^N), d + py + (C \cap pR^N)\rangle = \frac{1}{p} b(c + px, d + py).$$

The $\mathbb{F}_p$-bilinearity is clear as b is bilinear over R, and note that this product is well defined because for all $c, d \in C$ and $x, y \in R^N$ we have

$$b(c + px, d + py) = b(c, d) + p(b(x, d) + b(c, y)) = p(b(x, d) + b(c, y)),$$

hence

$$\langle c + px + (C \cap pR^N), d + py + (C \cap pR^N)\rangle = b(x,d) + b(c,y) \pmod{p} \in \mathbb{F}_p,$$

and since $(C \cap pR^N) = (C + pR^N)^{\perp}$, this value is independent of choice of representative. Finally, the non-degeneracy follows from the fact that

$$\langle c + px + (C \cap pR^N), d + py + (C \cap pR^N)\rangle = 0$$

for all $d + py \in C + pR^N$, if and only if $c + px \in (C + pR^N)^{\perp} = C \cap pR^N$, and thus if and only if $c + px + (C \cap pR^N) = 0 \in W$.

Lemma 31. *$C/(C \cap pR^N)$ and $X := pR^N/(C \cap pR^N)$ are maximal isotropic subspaces of W with*

$$W = C/(C \cap pR^N) \oplus pR^N/(C \cap pR^N).$$

Proof. As $C = C^{\perp}$ and $pR^N = (pR^N)^{\perp}$ are maximal self-dual submodules of (R^N, b), their images are maximal self-dual subspaces of W. Note that both $\mathbb{F}_p$-vector spaces have dimension a and $2a = \dim(W)$. To see that the sum is direct, it is enough to show that their intersection is 0, but this is clear, as

$$C/(C \cap pR^N) \cap pR^N/(C \cap pR^N) = (C \cap pR^N)/(C \cap pR^N) = \{0\}.$$

□

Let $(e_1, \ldots, e_a)$ be any $\mathbb{F}_p$-basis of $C/(C \cap pR^N)$. By the non-degeneracy of $\langle .,. \rangle$ there are elements $(f_1, \ldots, f_a) \in X := pR^N/(C \cap pR^N)$ such that $\langle e_i, f_j \rangle = \delta_{ij}$. Then $(e_1, \ldots, e_a, f_1, \ldots, f_a)$ is a basis of W.

Let

$$M(C) := \{Y \leq W \mid \langle Y, Y\rangle = \{0\}, W = Y \oplus X\}$$

denote the set of all totally isotropic complements of X in W.

Lemma 32. *The mapping $\varphi : N(C) \to M(C)$, $D \mapsto D/(C \cap pR^N)$ is a bijection.*

Proof. For any code $D \in N(C)$, we have that $C \cap pR^N = D \cap pR^N$ and $N(D) = N(C)$. As we have already seen in the previous lemma that $C/(C \cap pR^N)$ is a totally isotropic complement of X in W, the same is true for $D/(C \cap pR^N)$. So φ is well defined.

That the map is a bijection follows from the homomorphism theorem: The map $R^N \to R^N/(C \cap pR^N)$, $x \mapsto x + (C \cap pR^N)$ is an R-module epimorphism with kernel $C \cap pR^N$, so it defines a bijection between the set of all submodules D of R^N that contain $C \cap pR^N$ and the submodules of $R^N/(C \cap pR^N)$. Such a submodule D lies in $N(C)$, if and only if $\dim(D/(C \cap pR^N)) = a$, $D/(C \cap pR^N)$ is isotropic and $D \cap pR^N = C \cap pR^N$, and thus if and only if $D/(C \cap pR^N)$ lies in $M(C)$. □

Lemma 33. *Any $Y \in M(C)$ has a unique basis*

$$\left(e_1 + \sum_{j=1}^{a} S_{1j}f_j, \ldots, e_a + \sum_{j=1}^{a} S_{aj}f_j\right)$$

for some matrix $S \in \mathbb{F}_p^{a \times a}$ such that $S + S^{tr} = 0$. Call this space $Y(S)$.

Proof. Let $Y \in M(C)$. Then Y is a complement of $X = \langle f_1, \ldots, f_a \rangle$, and, in particular, there are $b_1, \ldots, b_a \in Y$, $z_1, \ldots, z_a \in X$, such that $e_i = b_i - z_i$ for $i = 1, \ldots, a$. As $W = Y \oplus X$, these b_i and z_i are uniquely determined and $(b_1, \ldots, b_a)$ is a basis of Y. Moreover there are unique $S_{ij} \in \mathbb{F}_p$ such that $z_i = \sum_{j=1}^{a} S_{ij}f_j$. That S is skew symmetric follows from the fact that Y is isotropic:

$$\langle b_i, b_k \rangle = \left\langle e_i + \sum_{j=1}^{a} S_{ij}f_j, e_k + \sum_{j=1}^{a} S_{kj}f_j \right\rangle = S_{ik} + S_{ki} = 0.$$

□

Combining these two bijections we hence obtain a bijection

$$\text{skew} : N(C) \to \left\{S \in \mathbb{F}_p^{a \times a} \mid S + S^{tr} = 0\right\} = \text{Skew}_a.$$

We note that then $\text{skew}(C) = 0$ since C maps to $C/(C \cap pR^N)$ which already has $(e_1, \ldots, e_a)$ as a basis and consequently the S from Lemma 33 is the all zeros matrix.

Remark 34. Everything is completely analogous if we work with quadratic forms for $p = 2$, but then the image of skew lies in the space of alternating matrices, so that $S = S^{tr} \in \mathbb{F}_2^{a \times a}$, $S_{ii} = 0$ for all i.

Remark 35. Let $D \in N(C)$. Then $\dim(D/(D \cap C)) = \dim(C/(D \cap C)) = \dim(\varphi(D)/(\varphi(D) \cap \varphi(C))) = \dim(\varphi(C)/(\varphi(D) \cap \varphi(C))) = \text{rank}(\text{skew}(D))$.

Because the rank of an alternating matrix is always even, the code C has no neighbors in $N(C)$.

Corollary 36. *If p is odd or we deal with quadratic forms for $p = 2$ (i.e., doubly even quaternary codes), then $T_a = 0$ for all a.*

5 Some Results for $\mathbb{Z}/4\mathbb{Z}$

In this section we turn our attention to the special case where $p = 2$, and $R = \mathbb{Z}/4\mathbb{Z}$. In this case we are able to compute the adjacency matrix, T, associated with the linear operator on $\mathcal{V}$. With these computations we make observations about the eigenvalues associated with each orthogonal component, T_a, and eventually describe the associated eigenspace. Before doing so, we establish the following results.

Lemma 37. *For any doubly even binary code $H \leq \mathbb{F}_2^N$ of dimension a, there exists a code $C \in \mathcal{F}_{a,b}$ such that $Res(C) = H$. Hence, $N(C) = \{D \in \mathcal{F}_{a,b} \mid Res(D) = H\}$.*

Proof. Suppose that $H \leq \mathbb{F}_2^N$ is a doubly even binary code of dimension a with generator matrix $G \in \mathbb{F}_2^{a \times N}$. Then it is possible to lift G to an $a \times N$ matrix A over R. It is clear that $A \cdot A^{\mathrm{tr}} = 2Z$ for some $Z \in \mathbb{F}_2^{a \times a}$. Moreover, this matrix is symmetric with zeroes along the diagonal, since H is doubly even. Now we will show that for a suitable choice of A, we have $Z = 0$, and consequently we will have the generating matrix for a self-dual code in R^N. Replacing A with $A + 2B$ for some $B \in \mathbb{F}_2^{a \times N}$, we obtain $(A + 2B)(A + 2B)^{\mathrm{tr}} = 2Z + 2(A \cdot B^{\mathrm{tr}} + A^{\mathrm{tr}} \cdot B)$. But since A has rank a, its columns contain a basis for $\mathbb{F}_2^a$ and it is therefore possible to choose B so that $A \cdot B^{\mathrm{tr}} = [z_{ij}]$ where $z_{ij} = 0$ for $i \leq j$, and for $i > j$, z_{ij} is the entry in the ith row and j^{th} column of Z. From here it is clear that $(A+2B)(A+2B)^{\mathrm{tr}} = 0$, and therefore $A+2B$ is the generating matrix for a self-orthogonal code, say C_1, which is a free R-module of rank a. By Remark 6, there is a unique self-dual code $C = C_1 + 2C_1^{\perp} \in \mathcal{F}_{a,b}$ that contains C_1 as a maximal free submodule. □

Combining this lemma with the general discussion in the previous section we arrive at the following main results.

Theorem 38. *For $p = 2$, the connected components of the graph Γ_a are in bijection with the binary doubly even codes of length N and dimension a.*

Proof. To begin, suppose that $C, D \in \mathcal{F}_{a,b}$, with $C \sim D$, and let $H = \mathrm{Res}(C)$. From Lemma 30, it follows that $C + 2R^N = D + 2R^N$, and hence $\mathrm{Res}(D) = H$. Extending this argument, we can easily see that for any $C, D \in \mathcal{F}_{a,b}$ which are connected by a path of neighbors in $\mathcal{F}_{a,b}$, all codes in that path will lift to H. This proves one direction of the claim.

Now, suppose we have two arbitrary codes $C, D \in \mathcal{F}_{a,b}$ with $\mathrm{Res}(C) = \mathrm{Res}(D) = H$ where $H \leq \mathbb{F}_2^N$ is a doubly even binary code. From here we know that C and D are in $N(C)$, and from the general results in the previous section, including the bijectivity of the skew map, we can conclude that C and D are connected by a chain of neighbors. □

Clearly if H and H' are equivalent doubly even codes, then the equivalence between H and H' (which is just a permutation in S_N) gives rise to a simultaneous equivalence between any codes C and C' for which $\mathrm{Res}(C) = H$ and $\mathrm{Res}(C') = H'$. Recalling that the nodes of Γ_a are defined as the permutation equivalence classes of codes in $\mathcal{F}_{a,b}$, the entire discussion above holds up to permutation equivalence; that is to say, the equivalence classes of binary doubly even codes of length N and dimension a precisely described the connected components of Γ.

Let $H_1, \ldots, H_t$ be a system of representatives of equivalence classes of binary doubly even codes of length N and dimension a. When $a = N/2$ and N is not congruent to $0 \mod 8$, then it is an immediate consequence of Corollary 8 that $t = 0$, and therefore the associated eigenspace is trivial.

Corollary 39. *The maximal eigenvalue of T_a is $2^a - 1$. This occurs with multiplicity t and the eigenspace of T_a to the eigenvalue $2^a - 1$ has a basis $(\sigma_1, \ldots, \sigma_t)$ where*

$$\sigma_i = \sum_{C \in \pi^{-1}(H_i)} \frac{1}{|\operatorname{Aut}(C)|}[C].$$

Proof. In view of Theorem 38, we know that counting the neighbors of C in $\mathcal{F}_{a,b}$ is equivalent to counting the number of elements in $M(C)$, which will be equal to the number of $(a-1)$-dimensional subspaces of $\mathbb{F}_2^a$. Consequently, Γ_a is (2^a-1)-regular, and therefore has the all ones eigenvector and $2^a - 1$ is an eigenvalue. Moreover, from the Peron–Frobenius theorem we can be guaranteed that this is indeed the maximal eigenvalue.

Since the neighboring relation is symmetric, the connected components of Γ_a are strongly connected, therefore the multiplicity of the eigenvalue $2^a - 1$ is the same as the number of connected components, namely t. □

Now we will compute an explicit example when $N = 8$. Using Magma, we compute the permutation equivalence classes of the codes. There are 29 distinct permutation equivalence classes when $N = 8$, which we will enumerate by module isomorphism type, where $[4^a 2^b]_n$ will denote the nth permutation equivalence class of codes isomorphic to $R^a \oplus S^b$ as R-submodules. Then we have

$$[4^0 2^8]_1$$

$$[4^1 2^6]_n \text{ where } 1 \le n \le 4$$

$$[4^2 2^4]_n \text{ where } 1 \le n \le 8$$

$$[4^3 2^2]_n \text{ where } 1 \le n \le 9$$

$$[4^4 2^0]_n \text{ where } 1 \le n \le 7$$

Using Magma we compute the adjacency matrix associated with T and its component block matrices, the T_a, which we give below.

$$T_0 = \begin{bmatrix} 0 \end{bmatrix}, T_1 = \begin{bmatrix} 0&1&0&0 \\ 1&0&0&0 \\ 0&0&0&1 \\ 0&0&1&0 \end{bmatrix}, T_2 = \begin{bmatrix} 2&1&0&0&0&0&0&0 \\ 3&0&0&0&0&0&0&0 \\ 0&0&0&1&2&0&0&0 \\ 0&0&1&0&0&2&0&0 \\ 0&0&1&0&0&1&0&1 \\ 0&0&0&1&1&0&1&0 \\ 0&0&0&0&0&2&0&1 \\ 0&0&0&0&2&0&1&0 \end{bmatrix},$$

$$T_3 = \begin{bmatrix} 6&1&0&0&0&0&0&0&0 \\ 7&0&0&0&0&0&0&0&0 \\ 0&0&0&4&0&3&0&0&0 \\ 0&0&1&0&3&0&0&3&0 \\ 0&0&0&1&2&1&2&0&1 \\ 0&0&1&0&4&2&0&0&0 \\ 0&0&0&0&4&0&1&1&1 \\ 0&0&0&4&0&0&2&0&1 \\ 0&0&0&0&4&0&2&1&0 \end{bmatrix}, T_4 = \begin{bmatrix} 6&0&6&0&1&1&1 \\ 0&0&0&0&8&0&7 \\ 8&0&5&1&0&1&0 \\ 0&0&6&0&8&1&0 \\ 7&1&0&7&0&0&0 \\ 8&0&6&1&0&0&0 \\ 8&1&0&0&0&0&6 \end{bmatrix}.$$

Now we can view the graphs Γ_a associated with each T_a, in particular we take a closer look at T_2. Figure 1 is the graph associated with Γ_2, whose nodes are precisely the permutation equivalence classes of codes isomorphic to $R^2 \oplus S^4$.

Fig. 1 Graph of Γ_2

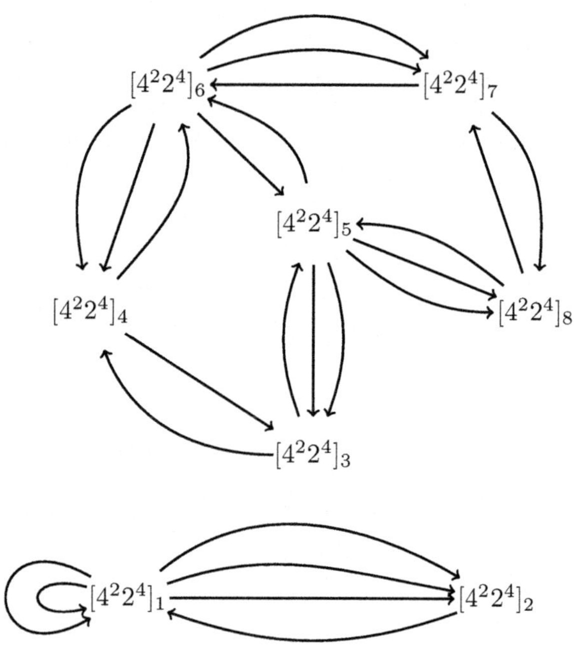

The two distinct connected components of Γ_2 are determined by the two distinct permutation equivalence classes of binary doubly even codes of length 8 and dimension 2, namely those with generator matrices

$$H = \begin{bmatrix} 1&1&1&1&0&0&0&0 \\ 0&0&0&0&1&1&1&1 \end{bmatrix} \text{ and } H' = \begin{bmatrix} 1&1&1&1&0&0&0&0 \\ 0&0&1&1&1&1&0&0 \end{bmatrix}.$$

The equivalence classes $[4^2 2^4]_1$ and $[4^2 2^4]_2$ contain codes with the following generator matrices,

$$G_1 = \begin{bmatrix} 2&0&0&0&0&0&0&0 \\ 0&2&0&0&0&0&0&0 \\ 0&0&1&0&1&1&0&1 \\ 0&0&0&1&1&1&1&2 \\ 0&0&0&0&2&0&2&2 \\ 0&0&0&0&0&2&2&2 \end{bmatrix} \text{ and } G_2 = \begin{bmatrix} 2&0&0&0&0&0&0&0 \\ 0&1&0&1&0&1&0&3 \\ 0&0&1&1&0&0&3&1 \\ 0&0&0&2&0&0&0&2 \\ 0&0&0&0&2&0&0&0 \\ 0&0&0&0&0&2&2&2 \end{bmatrix},$$

respectively. Let C_1 be the code with generator matrix G_1, and C_2 the code with generator matrix G_2. Then it's clear that both C_1 and C_2 lift to H'. Similarly, it can be shown each of the codes in the remaining permutation equivalence classes lifts to H.

Furthermore, for each a we compute the eigenvalues, λ, associated with each T_a, and their multiplicities t, which we give in the table below, noting that in any case the largest eigenvalue corresponds to $2^a - 1$.

a	$\langle \lambda, t \rangle$
0	$\langle 0, 1 \rangle$
1	$\langle 1, 2 \rangle$, $\langle -1, 2 \rangle$
2	$\langle 3, 2 \rangle$, $\langle 1, 2 \rangle$, $\langle -1, 3 \rangle$, $\langle -3, 1 \rangle$
3	$\langle 7, 2 \rangle$, $\langle 3, 2 \rangle$, $\langle -1, 4 \rangle$, $\langle -5, 1 \rangle$
4	$\langle 15, 1 \rangle$, $\langle 7, 2 \rangle$, $\langle -1, 3 \rangle$, $\langle -9, 1 \rangle$

References

1. Bachoc, C.: On harmonic weight enumerators of binary codes. Des. Codes Cryptogr. **18**, 11–28 (1999)
2. Bachoc, C.: Harmonic weight enumerators of non-binary codes and MacWilliams identities. In: Codes and Association Schemes (Piscataway, NJ, 1999). DIMACS Series in Discrete Mathematics and Theoretical Computer Science, vol. 56, pp. 1–23. American Mathematical Society, Providence, RI (2001)
3. Broué, M.: Codes correcteurs d'erreurs auto-orthogonaux sur le corps à deux éléments et formes quadratiques entières définies positives à discriminant +1. Discrete Math. **17**(3), 247–269 (1977)
4. Conway, J.H., Sloane, N.J.A.: Sphere Packings, Lattices and Groups, vol. 290. Springer, New York (1988)
5. Dougherty, S.T., Kim, J.-L., Liu, H.: Constructions of self-dual codes over finite commutative chain rings. Int. J. Inform. Coding Theory **1**(2), 170–190 (2010)
6. Ebeling, W.: Lattices and Codes. Springer Fachmedien, Wiesbaden (2013)
7. Kneser, M.: Quadratische Formen. Springer-Verlag, Berlin, Heidelberg (2002)
8. Meyer, A.: Automorphism groups of self-dual codes. PhD thesis, RWTH Aachen University (2009)
9. Nebe, G.: Finite Weil-representations and associated Hecke-algebras. J. Number Theory **129**, 588–603 (2009)

10. Nebe, G.: Kneser–Hecke-operators in coding theory. Abh. Math. Sem. Univ. Hamburg **76**, 79–90 (2006)
11. Nebe, G., Venkov, B.B.: On Siegel modular forms of weight 12. J. Reine Angew. Math. **531**, 49–60 (2001)
12. Nebe, G., Rains, E.M., Sloane, N.J.A.: Self-Dual Codes and Invariant Theory. Springer-Verlag, Berlin, Heidelberg (2006)
13. Norton, G.H., Salagean, A.: On the structure of linear and cyclic codes over finite chain rings. Appl. Algebra Eng. Commun. Comput. **10**, 489–506 (2000)
14. Nossek, E.: Kneser–Hecke-Operatoren in der Codierungstheorie, Diplomathesis, RWTH Aachen (2009). http://www.math.rwth-aachen.de/~Gabriele.Nebe/dipl.html
15. Pless, V.: On Witt's theorem for nonalternating symmetric bilinear forms over a field of characteristic 2 Proc. AMS **15**, 979–983 (1964)
16. Runge, B.: Codes and Siegel modular forms. Discrete Math. **148**, 175–204 (1996)
17. Wood, J.A.: Witt's extension theorem for mod four valued quadratic forms. Trans. AMS **336**, 445–461 (1993)
18. Wood, J.A.: Extension theorems for linear codes over finite rings. In: Applied Algebra, Algebraic Algorithms and Error-Correcting Codes (Toulouse, 1997). Lecture Notes in Computer Science, vol. 1255, pp. 329–340. Springer, Berlin (1997)

Ring-LWE Cryptography for the Number Theorist

Yara Elias, Kristin E. Lauter, Ekin Ozman, and Katherine E. Stange

Abstract In this paper, we survey the status of attacks on the *ring and polynomial learning with errors problems* (RLWE and PLWE). Recent work on the security of these problems (Eisentraeger et al., Weak Instances of PLWE. In: Proceedings of the selected areas of cryptography 2014. Lecture notes in computer science. Springer, New York, 2014; Elias Y., Lauter K., Ozman E., Stange K., Provably weak instances of ring-LWE. In: Advances in Cryptology – CRYPTO 2015. Springer, 2015 gives rise to interesting questions about number fields. We extend these attacks and survey related open problems in number theory, including spectral distortion of an algebraic number and its relationship to Mahler measure, the monogenic property for the ring of integers of a number field, and the size of elements of small order modulo q.

Keywords Ring and Polynomial learning with errors problems • Lattice based cryptography • Spectral distortion • Mahler measure • Monogenic Fields

Y. Elias (✉)
Department of Mathematics and Statistics, McGill University, Montreal, QC, Canada
e-mail: yara.elias@mail.mcgill.ca

K.E. Lauter
Microsoft Research, One Microsoft Way, Redmond, WA 98052, USA
e-mail: klauter@microsoft.com

E. Ozman
Department of Mathematics, Faculty of Arts and Science, Bogazici University, 34342 Bebek-Istanbul, Turkey
e-mail: ekin.ozman@boun.edu.tr

K.E. Stange
Department of Mathematics, University of Colorado, Campus Box 395, Boulder, CO 80309-0395, USA
e-mail: kstange@math.colorado.edu

© Springer International Publishing Switzerland 2016
E.E. Eischen et al. (eds.), *Directions in Number Theory*, Association for Women in Mathematics Series 3, DOI 10.1007/978-3-319-30976-7_9

1 Introduction

Public key cryptography relies on the existence of hard computational problems in mathematics: i.e., problems for which there are no known general polynomial-time algorithms. Hard mathematical problems related to lattices were first suggested as the basis for cryptography almost two decades ago [1, 2, 24]. While other better-known problems in public key cryptography such as factoring and the discrete logarithm problem are closely tied to computational number theory, lattice-based cryptography has seemed somewhat more distant. Recent developments, including the introduction of the ring-learning with errors problem instantiated in the ring of integers of a number field [30], have connected the area to new questions in computational number theory.

At the same time, lattice-based cryptography has seen a dramatic surge of activity. Since there are no known polynomial-time algorithms for attacking standard lattice problems on a quantum computer (in contrast to the case for widely deployed cryptographic systems such as RSA, discrete log, and elliptic curves), lattice-based cryptography is considered to be a promising cryptographic solution in a post-quantum world. One of the most exciting recent developments has been the construction of *fully homomorphic encryption* schemes [9–11, 18, 19] which allow meaningful operations to be performed on data without decrypting it: one can add and multiply encrypted numbers, returning the encrypted correct result, *without knowledge of the plaintext or private key*. The addition and multiplication of ciphertexts is possible due to the ring structure inherent in polynomial rings: these translate into AND and OR gates which can be used to build arbitrary circuits. Exciting applications include privacy problems in the health sector for electronic medical records, predictive analysis and learning from sensitive private data, and genomic computations [8, 20, 27, 28].

These new homomorphic encryption solutions are based on versions of hard "learning problems" with security reductions to and from standard lattice problems such as the shortest vector problem [36]. The idea behind the whole class of learning problems is that it is hard to "learn" a secret vector, given only sample inner products of that vector with other random vectors, provided these products are obscured by adding a small amount of Gaussian noise ("errors").

The ring version, which we call Ring-LWE or RLWE, was introduced in [30], presenting a fundamental hardness result which can be described informally as follows: for any ring of integers R, any algorithm that solves the (search version of the) Ring-LWE problem yields a comparably efficient *quantum* algorithm that finds approximately shortest vectors in *any* ideal lattice in R.

Soon after the introduction of Ring-LWE, an efficient cryptosystem allowing for homomorphic multiplication was proposed in [9] based on a variant of the RLWE problem, the Polynomial Learning With Errors problem (here denoted PLWE). Improvements to that cryptosystem (e.g. [11, 19]) have followed in the same vein, with the same hardness assumption. The reader should note that the terminology of "Ring-LWE" vs. "Poly-LWE" is not entirely standard, and some authors use "Ring-LWE" to refer to a larger class of problems including both.

We focus in this paper on PLWE, specified by the following choices:

(1) a polynomial ring $P_q = \mathbb{F}_q[x]/(f(x))$, with $f(x)$ a monic irreducible polynomial of degree n over $\mathbb{Z}$ which splits completely over $\mathbb{F}_q$,
(2) a basis for the polynomial ring, which will often be taken to be a power basis in the monogenic case (in particular, the choice of a basis can be used to endow the ring with the standard inner product on the ring),
(3) and a parameter specifying the size of the Gaussian noise to be added (the size of the "error"), spherical with respect to this inner product.

We also focus on RLWE obtained from the same setup, but with the inner product instead given by the Minkowski embedding of a ring of integers of the form $\mathbb{Z}[x]/(f(x))$. More general situations, including the case where the defining polynomial for the number ring does not split modulo q, or the case where q is composite, or the distribution is non-spherical or non-Gaussian, are considered in the cryptographic literature, but the setup above will suffice for our present purpose, which is to give a number theorist an entrée into the subject.

A key point is that for cryptographic applications, the errors must be chosen to be relatively small, to allow for correct decryption. For PLWE, "small" refers to the coefficient size (absolute value of the smallest residue), where the error is a polynomial, i.e., represented according to a polynomial basis for the ring. But to relate RLWE to other standard lattice problems, [30] considers the embedding of the ring $\mathbb{Z}[x]/(f(x))$ into the real vector space $\mathbb{R}^n$ under the Minkowski embedding (before reduction modulo q), and uses a Gaussian in $\mathbb{R}^n$; this induces an entirely different distribution on the error vectors for general number rings. It was shown in [12, 30] that in the case of 2-power cyclotomic rings, the distributions are the same. However, in [15] it was shown that in general rings the distortion of the distribution is governed by the largest singular value of the change-of-basis matrix between the Minkowski and the polynomial basis. Thus the RLWE and PLWE distributions are not equivalent in general rings.

Although RLWE and PLWE for cyclotomic rings, particularly two-power cyclotomic rings, are the current candidates for practical lattice-based homomorphic encryption with ideal lattices, it will be important for a full study of their security to consider the RLWE and PLWE problems for general rings. This includes studying the two problems independently, and analysing their relationships via the distortion of distributions just mentioned. One lesson from [15] is that deviating from these recommended candidates can lead to an insecure system.

The RLWE and PLWE problems are formulated as either 'search' or 'decision' problems (see Sect. 2 below). A security reduction was presented in [30] showing that, for any *cyclotomic* ring R, an algorithm for the decision version of the Ring-LWE problem yields a comparably efficient algorithm for the search version of the Ring-LWE problem. This search-to-decision reduction was subsequently extended to apply to any Galois field in [14].

In [14], an attack on PLWE was presented in rings $P_q = \mathbb{F}_q[x]/(f(x))$, where $f(1) \equiv 0 \pmod{q}$. In addition, [14] gives sufficient conditions on the ring so that the 'search-to-decision' reduction for RLWE holds, and also that RLWE instances can be translated into PLWE instances, so that the RLWE decision problem can be reduced to the PLWE decision problem. Thus, if a number field K satisfies the following six conditions simultaneously, then the results of [14] give an attack on the search version of RLWE:

(1) $K = \mathbb{Q}(\beta)$ is Galois of degree n.
(2) The ideal (q) splits completely in $R = \mathcal{O}_K$, the ring of integers of K, and $q \nmid [R : \mathbb{Z}[\beta]]$.
(3) K is monogenic, i.e., is generated by β over $\mathbb{Z}$.
(4) The transformation between the canonical embedding of K and the power basis representation of K is given by a scaled orthogonal matrix.
(5) If f is the minimal polynomial of β, then $f(1) \equiv 0 \pmod{q}$.
(6) The prime q can be chosen suitably large.

The first two conditions are sufficient for the RLWE search-to-decision reduction; the next two conditions are sufficient for the RLWE-to-PLWE reduction; and the last two conditions are sufficient for the attack on PLWE. Unfortunately, it is difficult to construct number fields satisfying all six conditions simultaneously. In [14] examples of number fields were given which are vulnerable to the attack on PLWE.

In [15], the attack on PLWE was extended by weakening the conditions on $f(x)$ and the reduction from RLWE to PLWE was extended by weakening condition (4). A large class of fields were constructed where the attack on PLWE holds and RLWE samples can be converted to PLWE samples, thus providing examples of weak instances for the RLWE problem.

Exciting number theory problems often arise from cryptographic applications. In this paper we survey and extend the attacks on the PLWE and RLWE problems and raise associated number-theoretic questions. In Sect. 2, we recall the PLWE and RLWE problems. In Sects. 3 and 4 we survey and extend the attacks on PLWE which were introduced in [14, 15]. In Sect. 5, we explain the reduction between the RLWE and PLWE problems. Finally in Sect. 6 we raise related questions in number theory; in particular, we investigate the spectral distortion of an algebraic number and its relationship to Mahler measure, the monogenic property for the ring of integers of a number field, and the size of elements of small order modulo q.

2 The Fundamental Hard Problems: PLWE and RLWE

2.1 PLWE

Take $f(x) \in \mathbb{Z}[x]$ to be monic and irreducible of degree n. Suppose q is a prime modulo which $f(x)$ factors completely (this is not necessary for the definition of the problem, but we will assume this throughout the paper). Write

$$P := \mathbb{Z}[x]/f(x), \quad P_q := P/qP \cong \mathbb{F}_q[x]/f(x).$$

Let $\sigma \in \mathbb{R}^{>0}$. By a *Gaussian distribution $\mathcal{G}_\sigma$ of parameter σ*, we mean a Gaussian of mean 0 and variance σ^2 on P which is spherical with respect to the power basis $1, x, x^2, \dots, x^{n-1}$ of P. The prime q is generally assumed to be polynomial in n, sometimes as large as 2^{50} but in some applications much smaller (even as small as 2^{12}), and σ is taken fairly small (perhaps $\sigma = 8$), so that in practice the tails of the Gaussian will decay to negligible size well before its variable reaches size q. Since P has integer coordinates, we must 'discretize' the Gaussian in an appropriate fashion; the result is simply referred to as a *discretized Gaussian*. We will not go into the technical details in this paper, but instead refer the reader to [30].

There are two standard PLWE problems, quoted here from [9]. The difficulty involves determining a secret obscured by a small *error* drawn from the discretized Gaussian.

Problem 2.1 (Search PLWE Problem). *Let $s(x) \in P_q$ be a secret. The* search PLWE problem, *is to discover $s(x)$ given access to arbitrarily many independent samples of the form $(a_i(x), b_i(x) := a_i(x)s(x) + e_i(x)) \in P_q \times P_q$, where for each i, $e_i(x)$ is drawn from a discretized Gaussian of parameter σ, and $a_i(x)$ is uniformly random.*

The polynomial $s(x)$ is the *secret* and the polynomials $e_i(x)$ are the *errors*. There is a decisional version of this problem:

Problem 2.2 (Decision PLWE Problem). *Let $s(x) \in P_q$ be a secret. The* decision PLWE problem *is to distinguish, with non-negligible advantage, between the same number of independent samples in two distributions on $P_q \times P_q$. The first consists of samples of the form $(a(x), b(x) := a(x)s(x) + e(x))$ where $e(x)$ is drawn from a discretized Gaussian distribution of parameter σ, and $a(x)$ is uniformly random. The second consists of uniformly random and independent samples from $P_q \times P_q$.*

Search-to-decision reductions were proved for cyclotomic number fields in [30] and extended to work for Galois number fields in [14]. Of course, the phrase 'to distinguish' must be interpreted to mean that the distinguisher's acceptance probabilities, given PLWE samples versus uniform samples, differ by a non-negligible amount.

2.2 RLWE

The original formulation of the hard learning problem for rings, RLWE, presented in [30], was based on the ring of integers, R, of a number field. The authors studied a general class of problems where the error distribution was allowed to vary.

Here we are concerned with only two choices of distributions. The first is to consider rings, R, which are isomorphic to a polynomial ring P, and study the PLWE problem (PLWE was stated as a "variant" of RLWE in [9, 30]). The distribution in this case is with respect to the polynomial basis of one of its polynomial representations.

The second is to choose the error according to a discretized Gaussian with respect to a special basis of the ambient space in which R was embedded via the Minkowski embedding. We will refer to this as RLWE. Therefore, in our language, when R is isomorphic to some polynomial ring P, RLWE differs from PLWE only in the error distribution.

We will state the fundamental RLWE problems and then discuss the relationship between the RLWE and PLWE problems. Let K be a number field of degree n with ring of integers R. Let $R^\vee$ denote the dual of R,

$$R^\vee = \{\alpha \in K : Tr(\alpha x) \in \mathbb{Z} \text{ for all } x \in R\}.$$

Let us write $R_q := R/qR$ and $R_q^\vee = R^\vee/qR^\vee$. We will embed K in $\mathbb{C}^n$ via the usual Minkowski embedding. The vector space $\mathbb{C}^n$ is endowed with a standard inner product, and we will use the spherical Gaussian with respect to this inner product, discretized to $R^\vee$, as the discretized Gaussian distribution. We will refer to this as the *canonical discretized Gaussian*. This will *not*, in general, coincide with the discretized Gaussian defined in PLWE for a $P \cong R$, and this is the fundamental difference between the two problems.

The standard RLWE problems for a canonical discretized Gaussian are as follows:

Problem 2.3 (Search RLWE Problem [30]). *Let $s \in R_q^\vee$ be a secret. The* search RLWE problem *is to discover s given access to arbitrarily many independent samples of the form $(a, b := as + e)$ where e is drawn from the canonical discretized Gaussian and a is uniformly random.*

Problem 2.4 (Decision RLWE Problem [30]). *Let $s \in R_q^\vee$ be a secret. The* decision RLWE problem *is to distinguish with non-negligible advantage between the same number of independent samples in two distributions on $R_q \times R_q^\vee$. The first consists of samples of the form $(a, b := as + e)$ where e is drawn from the canonical discretized Gaussian and a is uniformly random, and the second consists of uniformly random and independent samples from $R_q \times R_q^\vee$.*

An isomorphism between R and an appropriate polynomial ring P can be used to relate an instance of the RLWE problem to an instance of the PLWE problem. In particular, one requires R to be monogenic (having a power basis). Analysing the relationship between the two problems involves a close look at the change of basis under an isomorphism from R to the appropriate P. We will take up this issue in Sect. 5.

3 Summary of Attacks

In practice today, parameters for cryptosystems based on the RLWE and PLWE problems are set according to two known attacks, the *distinguishing attack* [31, 37] and the *decoding attack* [29]. These attacks work in general for learning-with-error problems and do not exploit the special structure of the ring versions of the problem.

In this paper, we will focus solely on the new attacks presented in [14, 15] that exploit the special number-theoretic structure of the PLWE and RLWE rings.

The attacks presented in [14, 15] can be described in terms of the ring homomorphisms from P_q to smaller rings. As $P_q \cong \mathbb{F}_q^n$, the only candidates are the projections to each factor:

$$\pi_\alpha : P_q \to \mathbb{F}_q, \quad p(x) \mapsto p(\alpha)$$

for each root α of $f(x)$. In P_q, the short vectors sampled by the Gaussian are easy to recognize since they have small coefficients. But they are hard to tease out of $b(x) = a(x)s(x) + e(x)$ without knowledge of $s(x)$, and the possibilities for $s(x)$ are too many to examine exhaustively. By contrast, in a small ring like $\mathbb{F}_q$, it is easy to examine the possibilities for $s(\alpha)$ exhaustively. And the ring homomorphism preserves the relationship of the important players: $b(\alpha) = a(\alpha)s(\alpha) + e(\alpha)$. Hence we can loop through the possibilities for $s(\alpha)$, obtaining for each the putative value

$$e(\alpha) = b(\alpha) - a(\alpha)s(\alpha).$$

The Decision Problem for PLWE, then, is solved as soon as we can recognize the set of $e(\alpha)$ that arise from the Gaussian.

Unfortunately (or fortunately), one does not expect to be able to do this in general. Heuristically, let $\mathcal{S} \subset P_q$ denote the subset of polynomials that are produced by the Gaussian with non-negligible probability. In P_q, parameters are such that this is a small set. But $\mathbb{F}_q$ is a much smaller ring and one expects that generically, the image of $\mathcal{S}$ will 'smear' across all of $\mathbb{F}_q$. Something quite special must happen if we expect the image of $\mathcal{S}$ to remain confined to a small subset of $\mathbb{F}_q$, and hence be recognizable.

That 'something special' is certainly possible, however: suppose that $\alpha = 1$. The polynomials $g(x) \in \mathcal{S}$ have small coefficients, and hence have small images $g(1)$ in $\mathbb{F}_q$. This is simply because n is much smaller than q, so that the sum of n small coefficients is still small modulo q. More generally, all of the attacks suggested in [14, 15] come down to considering α with certain advantageous properties, so that the image of $\mathcal{S}$ can be recognized.

The cyclotomic cases currently under consideration for PLWE and RLWE are uniquely protected against this occurrence: $\alpha = 1$ is never a root modulo q of a cyclotomic polynomial of degree > 1 when q is sufficiently large.

3.1 Attacking $\alpha = 1$

The approach described above and the $\alpha = 1$ attack was first presented in [14]. The details are as follows. Suppose

$$f(1) \equiv 0 \bmod q.$$

We are given access to a collection of samples $(a_i(x), b_i(x))$. We wish to determine if a sample is *valid*, of the form

$$b_i(x) = s(x)a_i(x) + e_i(x)$$

for $e_i(x)$ produced by a Gaussian, or *random* (uniformly random). The algorithm is as follows:

Algorithm 1:

(1) Let the set of valid guesses be $S = \mathbb{F}_q$.
(2) Loop through the available samples. For each sample:
    (a) Loop through guesses $s \in S$ for the value of $s(1)$. For each s:
        (i) Compute $e_i := b_i(1) - sa_i(1)$
        (ii) If e_i is *not* small in absolute value[1] modulo q, then conclude that the sample cannot be valid for s with non-negligible probability, and remove s from S.
(3) If $S = \emptyset$, conclude that the sample was random. If S is non-empty, conclude that the sample is valid.

If the guess s is correct, then $e_i = e_i(1) = \sum_{j=1}^{n} e_{ij}$ where e_{ij} are chosen from a Gaussian $\mathcal{G}_\sigma$ of parameter σ. It follows that $e_i(1)$ are approximately sampled from a Gaussian $\mathcal{G}_{\sqrt{n}\sigma}$ of parameter $\sqrt{n}\sigma$ where $n\sigma^2 \ll q$.

3.2 *Attacking α of Small Order*

The following attack described and developed in [14, 15] requires α to have small order mod q. The fundamental idea is the same as for the $\alpha = 1$ attack, except that to discern whether or not $e_i(\alpha)$ is a possible image of a Gaussian-sampled error is more complicated.

[1]Meaning residue of smallest absolute value.

Assume that $\alpha^r \equiv 1 \mod q$, then

$$e(\alpha) = \sum_{i=1}^{n} e_i \alpha^i = (e_r + e_{2r} + \cdots) + \cdots + \alpha^{r-1}(e_{r-1} + e_{2r-1} + \cdots).$$

If r is small enough, $e(\alpha)$ takes on only a small number of values modulo q. This set of values may not be easy to describe, but q is small enough that it can be enumerated and stored. The attack proceeds as for $\alpha = 1$ except that to determine if a sample is potentially valid for s in step (2)(a)(ii), we compare to the stored list of possible values.

4 Attacking α of Small Residue

A third attack described in [15] is based on the size of the residue $e_i(\alpha) \mod q$. This is more subtle. Here, the errors $e(\alpha)$ may potentially take on all values modulo $\mathbb{F}_q$ with non-negligible probability. But it may be possible to notice if the probability distribution across $\mathbb{F}_q$ is not uniform, given enough samples.

This method of attack differs from the previous ones, but is also applicable to $\alpha = 1$ and α of small order, so all cases will be treated together.

Assume that

$$f(\alpha) \equiv 0 \mod q \tag{1}$$

for some α. Let E_i be the event that

$$b_i(\alpha) - g a_i(\alpha) \mod q \text{ is in the interval } [-q/4, q/4)$$

for some sample i and guess g for $s(\alpha) \mod q$. We wish to compare the probabilities

$$P(E_i \mid \mathcal{D} = \mathcal{U}) \text{ and } P(E_i \mid \mathcal{D} = \mathcal{G}_\sigma).$$

Here, $\mathcal{D} = \mathcal{U}$ refers to the situation where b_i is uniformly random, while $\mathcal{D} = \mathcal{G}_\sigma$ refers to the situation where b_i is obtained as $a_i s + e_i$ for some secret s, where e_i follows a Gaussian $\mathcal{G}_\sigma$ truncated at 2σ (in practice, the Gaussian is truncated as the tails decay to negligible values). If $\mathcal{D} = \mathcal{U}$, then $b_i(\alpha) - g a_i(\alpha)$ is random modulo q for all guesses g, that is,

$$P(E_i \mid \mathcal{D} = \mathcal{U}) = \frac{1}{2}.$$

If $\mathcal{D} = \mathcal{G}_\sigma$, then $b_i(\alpha) - s(\alpha) a_i(\alpha) = e_i(\alpha) \mod q$. Indeed, the terms of $b_i(\alpha) - s(\alpha) a_i(\alpha)$ that are a multiple of f vanish at α modulo q by Assumption (1). We consider

$$e_i(\alpha) = \sum_{j=0}^{n-1} e_{ij}\alpha^j,$$

where e_{ij} is chosen according to the distribution $\mathcal{G}_\sigma$ and distinguish three cases corresponding to

(1) $\alpha = \pm 1$
(2) $\alpha \neq \pm 1$ and α has small order r modulo q
(3) $\alpha \neq \pm 1$ and α is not of small order r modulo q

We will now drop the subscript i for simplicity. In Case (1), the error $e(\alpha)$ is distributed according to $\mathcal{G}_{\bar{\sigma}}$ where

$$\bar{\sigma} = \sigma\sqrt{n}.$$

In Case (2), the error can be written as

$$e(\alpha) = \sum_{i=0}^{r-1} e_i\alpha^i = (e_0+e_r+\cdots)+\alpha(e_1+e_{r+1}+\cdots)+\cdots+\alpha^{r-1}(e_{r-1}+e_{2r-1}+\cdots)$$

where we assume that n is divisible by r for simplicity. For $j = 0, \cdots, r-1$, we have that

$$e_j + e_{j+r} + \cdots + e_{j+n-r}$$

is distributed according to $\mathcal{G}_{\tilde{\sigma}}$ where

$$\tilde{\sigma} = \sigma\sqrt{\frac{n}{r}}.$$

As a consequence $e(\alpha)$ is sampled from $\mathcal{G}_{\bar{\sigma}}$ where

$$\bar{\sigma}^2 = \sum_{i=0}^{r-1} \tilde{\sigma}^2\alpha^{2i} = \sum_{i=0}^{r-1} \frac{n}{r}\sigma^2\alpha^{2i} = \frac{n}{r}\sigma^2\frac{\alpha^{2r}-1}{\alpha^2-1}.$$

In Case (3), the error $e(\alpha)$ is distributed according to $\mathcal{G}_{\bar{\sigma}}$ where

$$\bar{\sigma}^2 = \sum_{i=0}^{n-1} \sigma^2\alpha^{2i} = \sigma^2\frac{\alpha^{2n}-1}{\alpha^2-1}.$$

If $\frac{q}{4} \geq 2\bar{\sigma}$, then errors always lie in $[-\frac{q}{4}, \frac{q}{4})$ and

$$P(E_i \mid \mathcal{D} = \mathcal{G}_\sigma) = 1.$$

Otherwise, assuming for simplicity that $N = \dfrac{2\bar{\sigma} - q/2}{q}$ is an integer, we have

$$P(E_i \mid \mathcal{D} = \mathcal{G}_\sigma) = \left(\int_0^{2\bar{\sigma}} \mathcal{G}_{\bar{\sigma}} \right)^{-1} \left(\int_0^{\frac{q}{4}} \mathcal{G}_{\bar{\sigma}} + \sum_{k=0}^{N-1} \int_{\frac{3q}{4}+kq}^{\frac{5q}{4}+kq} \mathcal{G}_{\bar{\sigma}} \right).$$

In the situation where this value exceeds $1/2$, i.e., $P(E_i \mid \mathcal{D} = \mathcal{G}_\sigma) = \dfrac{1}{2} + \epsilon$ with $\epsilon > 0$, the following algorithm attacks PLWE. Let

$$N = \left\lceil \frac{\ell q + \epsilon \ell}{2} \right\rceil$$

where ℓ is the number of samples observed. For each guess $g \mod q$, we compute $b_i - ga_i \mod q$ for $i = 1, \cdots, \ell$. We denote by C the number of elements obtained in the interval $[-q/4, q/4)$. If $C < N$, the algorithm outputs

$$\mathcal{D} = U,$$

otherwise, the algorithm outputs

$$\mathcal{D} = \mathcal{G}_\sigma.$$

In the analysis of the probability of success of the algorithm, we denote by B the binomial distribution and by F the cumulative Binomial distribution. If $\mathcal{D} = \mathcal{U}$, the algorithm is successful with probability

$$P(C < N | \mathcal{D} = U) = F(N-1; \ell q, \frac{1}{2}).$$

If $\mathcal{D} = \mathcal{G}_\sigma$, we denote by C_s the number of elements of the form $b_i - sa_i \mod q$ in the interval $[-q/4, q/4)$. In this case, the algorithm is successful with probability

$$P(C \geq N | \mathcal{D} = \mathcal{G}_\sigma) = \sum_{i=0}^{\ell} P(C - C_s \geq N - i) \times P(C_s = i)$$

$$= \sum_{i=0}^{\ell} (1 - F(N - i - 1, \ell q - \ell, 1/2)) \times B(i, \ell, 1/2 + \epsilon)$$

When $\epsilon > 0$, the algorithm is successful since

$$\frac{1}{2}(P(C < N | \mathcal{D} = U) + P(C \geq N | \mathcal{D} = \mathcal{G}_\sigma))$$

$$= \frac{1}{2}(P(C < N|\mathcal{D} = U) + 1 - P(C < N|\mathcal{D} = \mathcal{G}_\sigma))$$

$$= \frac{1}{2} + \frac{1}{2}(P(C < N|\mathcal{D} = U) - P(C < N|\mathcal{D} = \mathcal{G}_\sigma)) > \frac{1}{2}$$

Example 4.1. In Case (1), when $n = 2^{10}$, $q \approx 2^{50}$, and $\sigma = 8$, we can compute $\epsilon \approx 0.5$. Therefore, the attack is successful for any irreducible polynomial of degree 2^{10} and with a root 1 mod q.

In Case (2), when $n = 2^9$, $q \approx 2^{50}$, $\sigma = 8$, and $\alpha = q - 1$, α has order 2 and we can compute $\epsilon \approx 0.002$. This is particularly interesting since there is an irreducible polynomial with these properties that generates a power of 2 cyclotomic number field [15]; however, it is not the usual cyclotomic polynomial.

In Case (3), when $n = 2^6$, $q \approx 2^{60}$, $\sigma = 8$, and $\alpha = 2$, computations show that $\epsilon = 0.02$. Therefore, this attack is successful for any irreducible polynomial of degree 2^6 with a root $\alpha = 2$ modulo a prime $q \approx 2^{60}$.

5 RLWE-to-PLWE Reduction

Suppose that K is a number field, and R is its ring of integers. For technical reasons, we give a slight variant on the Minkowski embedding, which is as follows: $\theta : K \to \mathbb{R}^n$

$$\theta(r) := (\sigma_1(r), \ldots, \sigma_{s_1}(r), Re(\sigma_{s_1+1}(r)), \ldots\ldots Re(\sigma_{s_1+s_2}(r)),$$
$$Im(\sigma_{s_1+1}(r)), \ldots, Im(\sigma_{s_1+s_2}(r))).$$

where the σ_i are the $s_1 + s_2$ embeddings of K, ordered so that the s_1 real embeddings are first, and the s_2 complex embeddings are paired so that $\overline{\sigma_{s_1+k}} = \sigma_{s_1+s_2+k}$.

A spherical Gaussian of parameter σ with respect to the usual inner product on $\mathbb{R}^n$ can be discretized to the *canonical discretized Gaussian* on R or its dual $R^\vee$.

Suppose $R \cong P$ for some polynomial ring P under a map $\alpha \mapsto x$ for some root α of $f(x)$. Suppose further that R is monogenic. Then $R^\vee \cong P$ also as R-modules (as its different ideal is principal). For RLWE, $\mathbb{R} \otimes R^\vee$ is equipped with a basis $\mathbf{b}_i$, $i = 0, \ldots, n-1$ with respect to which the Gaussian is spherical (the standard basis of $\mathbb{R}^n$, pulled back by θ). For PLWE, $\mathbb{R} \otimes P$ is equipped with such a basis also, i.e., the standard power basis x^i, $i = 0, \ldots, n-1$. To relate the two problems, one must write down the change-of-basis matrix between them. It is the matrix

$$N_\alpha := \gamma M_\alpha^{-1} : \mathbb{R} \otimes R^\vee \to \mathbb{R} \otimes P$$

where γ is such that $R^\vee = \gamma R$, and where M_α is the matrix with columns $[\alpha^i]_{\mathbf{b}}$ (i.e., the i-th column is the element α^i represented with respect to the basis $\mathbf{b} = \{\mathbf{b}_i\}$).

The properties of N_α determine how much the Gaussian is distorted in moving from one problem to the other. If it is not very distorted, then solving one problem may solve the other.

Details are to be found in [15], but in short, the *normalized spectral norm* gives a good measure of 'distortion'. This is defined for an $n \times n$ matrix M by

$$||M||_2/\det(M)^{1/n}.$$

6 Number-Theoretical Open Problems

In this section we will describe a number of open problems in number theory that are motivated by attacks to PLWE and RLWE, some very speculative and some more precise.

6.1 Conditions for Smearing

As described in Sect. 3, we are concerned with the map

$$\pi : P_q \to \mathbb{F}_q, \quad g(x) \mapsto g(\alpha).$$

Question 6.1. For which subsets $\mathcal{S} \subset P_q$, is the image $\pi(\mathcal{S}) = \mathbb{F}_q$?

If $\pi(\mathcal{S}) = \mathbb{F}_q$, we will say that $\mathcal{S}$ *smears* under π.

Partial solutions to this problem may come in a wide variety of shapes. For example, can one prove that almost all $\mathcal{S}$ of a given size smear? Can one characterize the types of situations that lead to a negative answer (e.g. $\alpha = 1$ and $\mathcal{S}$ consisting of polynomials of small coefficients)? What if we restrict to the PLWE case, where $\mathcal{S}$ consists of polynomials with small coefficients? Or the RLWE case, where $\mathcal{S}$ is the image of a canonical discretized Gaussian?

6.2 The Spectral Distortion of Algebraic Numbers, and Mahler Measure

By Sect. 5, the normalized spectral norm of N_α is a property of any algebraic number α for which $\mathbb{Z}[\alpha]$ is a maximal order. We will therefore denote it ρ_α, and call it the *spectral distortion of* α. It measures the extent to which the power basis α^i is distorted from the canonical basis of the associated number field. Recall from Sect. 5 that for number rings with small spectral distortion we expect to have an equivalence between the RLWE and PLWE problems. For completeness, we state a slightly more general definition, separate from its cryptographic origins, as follows:

Definition 6.2. Let α be an algebraic number of degree n and $K = \mathbb{Q}(\alpha)$. Let M be the matrix whose columns are given by $\theta(\alpha^i)$, where $\theta : K \to \mathbb{R}^n$,

$$\theta(r) = (\sigma_1(r), \ldots, \sigma_{s_1}(r), Re(\sigma_{s_1+1}(r)), \ldots Re(\sigma_{s_1+s_2}(r)),$$
$$Im(\sigma_{s_1+1}(r)), \ldots, Im(\sigma_{s_1+s_2}(r)))$$

where the σ_i are the $s_1 + s_2$ complex embeddings of K, ordered so that the s_1 real embeddings are first, and the s_2 complex embeddings are paired so that $\overline{\sigma_{s_1+k}} = \sigma_{s_1+s_2+k}$. The *spectral distortion of* α is $||M||_2/(\det(M))^{\frac{1}{n}}$.

Question 6.3. What are possible spectral distortions of algebraic numbers?

It follows from the special properties of 2-power roots of unity that they have spectral distortion equal to 1. However, even other roots of unity do not have spectral distortion equal to 1 (and this is what necessitates the more elaborate RLWE-to-PLWE reduction argument given in [12] for cyclotomic rings which are not 2-power cyclotomics).

Is the spectral distortion a continuum, or is the collection of values discrete in regions of $\mathbb{R}$? Does this relate to Mahler measure?

The Mahler measure of a polynomial can be defined as the product of the absolute values of the roots which lie outside the unit circle in the complex plane, times the absolute value of the leading coefficient. For a polynomial

$$f(x) = a(x - \alpha_1)(x - \alpha_2)\cdots(x - \alpha_n)$$

the Mahler measure is

$$M(f) := |a| \prod_{|\alpha_i| \geq 1} |\alpha_i|.$$

The Mahler measure of an algebraic number α is defined as the Mahler measure of its minimal polynomial.

Interestingly, polynomials which have small Mahler measure (all roots very close to 1 in absolute value) seem to have small spectral distortion. For example, consider "Lehmer's polynomial", the polynomial with the smallest known Mahler measure greater than 1:

$$f(x) = x^{10} + x^9 - x^7 - x^6 - x^5 - x^4 - x^3 + x + 1.$$

The Mahler measure is approximately 1.176, and the spectral distortion for its roots is approximately 3.214. This spectral distortion is rather small, and compares favourably, for example, with the spectral distortion for 11th roots of unity, which is approximately 2.942. Other examples of polynomials with small Mahler measure also have small spectral distortion: $f(x) = x^3 - x + 1$ has Mahler measure approximately 1.324 and spectral distortion approximately 1.738.

To explain the phenomenon observed for polynomials with small Mahler measure and to relate the Mahler measure to the spectral norm, we need to have some estimate on the spectral norm in terms of the entries of the matrix. The entries of the matrix M are powers of the roots $\{\alpha_j\}$ of the minimal polynomial. When the Mahler measure is small, the entries of the matrix M have absolute value close to 1, since the absolute values of the roots are as close as possible to 1. To make the connection with the spectral norm more precise, [35] gives an improvement on Schur's bound and expresses the bound on the largest singular value in terms of the entries of the matrix. Thus we can use Schur's bound or this improvement to see that polynomials with small Mahler measure must have relatively small spectral norm.

It could also be interesting to look at other properties of M, such as the entire vector of singular values of M, its conditioning number, etc.

6.3 *Galois Versus Monogenic*

We say that K is *monogenic* if the ring of integers R of K is monogenic, i.e., a simple ring extension $\mathbb{Z}[\beta]$ of $\mathbb{Z}$. In this case, K will have an integral basis of the form $\{1, \beta, \beta^2, \ldots, \beta^{n-1}\}$ which is called a *power integral basis*. In this section we will focus on properties (1) and (4) from the introduction.

Example 6.4. The following are examples of number fields that are both Galois and monogenic:

- Cyclotomic number fields, $K = \mathbb{Q}(\zeta_n)$ where ζ_n is a primitive nth root of unity,
- Maximal real subfields of cyclotomic fields, $K = \mathbb{Q}(\zeta_n + \zeta_n^{-1})$,
- Quadratic number fields $K = \mathbb{Q}(\sqrt{d})$.

Question 6.5. Are there fields of cryptographic size which are Galois and monogenic, other than the cyclotomic number fields and their maximal real subfields? How can one construct such fields explicitly?

The problem of characterizing all number fields which are monogenic goes back to Hasse, however, a complete solution is not known to date. Here we will summarize some of the known related results.

Proposition 6.6 ([34]). *Let p be a prime and K a Galois extension of $\mathbb{Q}$ of degree n. Let e be the ramification index of p and f be the inertia degree of p. If one of the conditions below is satisfied, then K is not monogenic:*

- *If $f = 1$: $ep < n$*
- *If $f \geq 2$: $ep^f \leq n + e - 1$*

Let K be a Galois extension of prime degree ℓ. (Such extensions are called cyclic extensions.) The following result of Gras [23] states that cyclic extensions are often non-monogenic.

Theorem 6.7 ([23]). *Any cyclic extension K of prime degree $\ell \geq 5$ is non-monogenic except for the maximal real subfield of the $(2\ell + 1)$-th cyclotomic field with prime conductor $2\ell + 1$.*

Theorem 6.8 ([22]). *Let $n \geq 5$ be relatively prime to $2, 3$. There are only finitely many abelian number fields of degree n that are monogenic.*

For number fields of smaller degree it may be possible to give a complete characterization. For instance, for cyclic cubic extensions K, Gras [21] and Archinard [3] gave necessary and sufficient conditions for K to be monogenic.

Even though monogenic fields are rare in the abelian case for large degree, Dummit and Kisilevsky [13] have shown that there exist infinitely many cyclic cubic fields which are monogenic. A result of Kedlaya [25] implies that there are infinitely many monogenic number fields of any given signature. In fact, we expect monogenicity frequently: if f is an irreducible polynomial with squarefree discriminant, then the number field K obtained by adjoining a root of f to $\mathbb{Q}$ is monogenic. For polynomials of fixed degree ≥ 4 whose coefficients are chosen randomly, it is conjectured that with probability $\gtrsim 0.307$, the root will generate the ring of integers of the associated number field [25]. However, to require K also to be Galois is much more restrictive. Moreover, for fields of cryptographic size ($n \sim 2^{10}$), the discriminant of f is too large to test whether it is squarefree. Therefore testing whether an arbitrary number field of cryptographic size is monogenic is not known to be feasible in general.

6.4 *Finding Roots of Small Order Mod p*

We have seen that a root of small order of $f(x)$ modulo q provides a method of attack on the PLWE problem in the ring $\mathbb{Z}_q[x]/(f(x))$. The attack is even more effective if, in addition, this root is small as a minimal residue modulo q ('minimal' meaning the smallest in absolute value). See Example 4.1, Case(3) in Sect. 4. Cyclotomic fields are protected against this attack by the observation that the roots of a cyclotomic polynomial modulo q are of full order n. However for 'random' polynomials, there is a priori no particular reason to expect roots of any particular order modulo q, or to expect the roots to be small. Motivated by these two requirements, it is natural to ask the following question:

Question 6.9. For random polynomials $f(x)$ and random primes q for which $f(x)$ has a root α modulo q, what can one say about the order of α modulo q?

A special case of this question, for f monic of degree one, is to ask, for a fixed a, how often is a a primitive root modulo p? A famous conjecture of Artin states that this should happen for infinitely many p provided a is not a perfect square or -1, and describes the density of such primes. This has been the subject of much research, and the question above is a sort of number field analogue. Some investigations in the direction of a number field analogue of Artin's conjecture exist; for a gateway to the literature, see [33, 38].

Computationally, to locate polynomials having a small root of small order, it is easiest to start with the desired order, find a suitable q, and then build the polynomial. The algorithm is as follows:

Algorithm 2

Input: Integers r, n, q_0 such that $r > 2$ represents the desired order, $n \geq 1$ represents the desired degree, and $q_0 > \log_2(n)$ represents the desired bitsize of q.

(1) Let s be the degree of the cyclotomic polynomial $\Phi_r(x)$.
(2) Let $a = 1$ (our candidate for the element of order r mod q). Test $\Phi_r(a)$ for primality. If it is a prime of approximate bitsize q_0, let q be this prime. Otherwise, increment a and try again.
(3) Once a and q are fixed, choose a set S of n elements of $\mathbb{Z}/q\mathbb{Z}$ that includes a and the other $n-1$ smallest minimal residues (or choose any other subset of residues).
(4) Choose $i = 1$ and increment i until the polynomial

$$f(x) = \prod_{s \in S}(x - s) + qi$$

of degree n is irreducible.

Output: A monic irreducible polynomial $f(x) \in \mathbb{Z}[x]$, a prime q roughly of size q_0, such that f splits modulo q, and $a \in \mathbb{Z}/q\mathbb{Z}$ such that $f(a) \equiv 0 \pmod{q}$ and $a^r \equiv 1 \pmod{q}$.

Note that if one wishes to relax the condition that f splits modulo q, one could take $f(x) = (x-a)^n + q$, which is irreducible, to avoid Step 4.

Using this method, it is easy to find examples of (K, q) such that $f(x)$ has a root of small order modulo q. Among them, an example of cryptographic size is afforded by $n = 2^{10}, r = 3, a = 33554450, q = 1125901148356951$ and $i = 1$ (the polynomial is too unwieldy to print here). Using the last two parts of the method, one can, in fact, easily construct polynomials having as roots many elements of small order modulo q.

A simpler starting point is the following second question:

Question 6.10. What is the distribution of elements of small order among residues modulo q?

There is a significant body of research on the distribution of primitive roots (see Artin's conjecture) and quadratic residues. More recently there have been advances on the distribution of elements of small order. For example, the number of elements of bounded size and specified order is bounded above in [7]; see also [5, 6, 26]. More useful in our present context, for the purposes of finding elements of small order, would be a guarantee that such elements exist in some small interval.

A more precise question is as follows:

Question 6.11. What is the smallest residue modulo a prime q which has order exactly r ?

Let q be a prime and $r > 2$. Let $n_{r,q}$ represent the smallest residue modulo q which has order exactly r. A first observation is the following (which allows us to choose a more suitable starting point for a in the algorithm above).

Proposition 6.12. *Let $\varphi(r)$ represent the Euler function, giving the number of positive integers less than and coprime to r. Then, if r has at most two distinct prime factors, which are odd, then*

$$|n_{r,q}| \geq (q/\varphi(r))^{1/\varphi(r)}$$

Proof. The element $n_{r,q}$ is a root of the r-th cyclotomic polynomial, of degree $\varphi(r)$, modulo q. Since $\Phi_r(n_{r,q}) \neq 0$ as an integer relation, it must be that $|\Phi_r(n_{r,q})| \geq q$. It is known that under the given hypotheses on the factorization of r, the coefficients of Φ_r are chosen from $\{\pm 1, 0\}$ [32]. Therefore $|\Phi_r(n_{r,q})| \leq \varphi(r)|n_{r,q}^{\varphi(r)}|$ from which the result follows. □

In general, combining upper and lower bounds on $n_{r,q}$ would limit the search space for an element of small order.

Remark 6.13. (1) Other restrictions on the coefficients of Φ_r give rise to similar results. To derive an asymptotic statement, one could turn to asymptotic results such as [17].

(2) The case of $r = 3$, the study of $n_{3,q}$ gives the full story, as the cube roots of unity are of the form

$$1, n_{3,q}, -n_{3,q} - 1.$$

(3) In general, the primes q such that $n_{r,q} = a$ for a fixed a and r are among those dividing $\Phi_r(a)$, hence there are finitely many.

(4) Elliott has some results on k-th power residues [16].

We will call $n_{r,q}$ *minimal* if, in addition to being the smallest residue of order r modulo q, it also satisfies $\Phi_r(n_{r,q}) = \pm q$. For non-minimal $n_{r,q}$, the lower bound in Proposition 6.12 increases. A conjecture of Bouniakowski implies that minimality happens infinitely often.

Conjecture 6.14 (Bouniakowski, [4]). *Let $f(x) \in \mathbb{Z}[x]$ be a non-constant irreducible polynomial such that $f(x)$ is not identically zero modulo any prime p. Then $f(n)$ is prime for infinitely many $n \in \mathbb{Z}$.*

Proposition 6.15. *Let $r > 2$. If Bouniakowski's Conjecture holds, then there are infinitely many primes q for which $n_{r,q}$ is minimal.*

Proof of Proposition 6.15. The cyclotomic polynomials for $r > 1$ satisfy the Bouniakowski conditions, as they are irreducible and $\Phi_r(1) \not\equiv 0 \pmod{p}$ since 1 is not of exact order r modulo any p. Hence $\Phi_r(x)$ takes on infinitely many prime values; for such a prime q, the smallest such x in absolute value is $n_{r,q}$ and this is minimal. □

Acknowledgements The authors thank the organizers of the research conference Women in Numbers 3 (Rachel Pries, Ling Long and the fourth author) and the Banff International Research Station, for making this collaboration possible. The authors also thank the anonymous referee for detailed comments and suggestions to improve the paper, and Igor Shparlinski for useful feedback and references.

References

1. Ajtai, M.: Generating hard instances of lattice problems. In: Complexity of Computations and Proofs. Quadeni di Matematica, vol. 13, pp. 1–32 (2004). Preliminary version in STOC 1996
2. Ajtai, M., Dwork, C.: A public-key cryptosystem with worst-case/average-case equivalence. In: Proceedings of the Twenty-ninth Annual ACM Symposium on Theory of Computing, pp. 284–293 (1997)
3. Archinard, G.: Extensions cubiques cycliques de Q dont l'anneau des entiers est monogène. Enseignement Math. **20**(2), 179–203 (1974)
4. Bouniakowski, V.: Sur les diviseurs numériques invariables des fonctions rationelles entières. Mem. Acad. Sci. St. Petersburg **6**, 305–329 (1857)
5. Bourgain, J.: On the distribution of the residues of small multiplicative subgroups of $\mathbb{F}_p$. Israel J. Math. **172**, 61–74 (2009)
6. Bourgain, J., Konyagin, S.V., Shparlinski, I.E.: Product sets of rationals, multiplicative translates of subgroups in residue rings, and fixed points of the discrete logarithm. Int. Math. Res. Not. rnn 090 (2008)
7. Bourgain, J., Konyagin, S.V., Shparlinski, I.E.: Distribution of elements of cosets of small subgroups and applications. Int. Math. Res. Not. **9**, 1968–2009 (2012)
8. Bos, J.W., Lauter, K., Naehrig, M.: Private predictive analysis on encrypted medical data. J. Biomed. Inform. (2014). doi:10.1016/j.jbi.2014.04.003
9. Brakerski, Z., Vaikuntanathan, V.: Fully homomorphic encryption from RLWE and security for key dependent messages. In: Rogaway, P. (ed.) CRYPTO. Lecture Notes in Computer Science, vol. 6841, pp. 505–524. Springer, New York (2011)
10. Brakerski, Z., Vaikuntanathan, V.: Efficient fully homomorphic encryption from (standard) LWE. SIAM J. Comput. **43**(2), 831–871 (2014)
11. Brakerski, Z., Gentry, C., Vaikuntanathan, V.: Fully homomorphic encryption without bootstrapping. In: Proceedings of the 3rd Innovations in Theoretical Computer Science Conference, pp. 309–325. Association for Computing Machinery, New York (2011)
12. Ducas, L., Durmus, A.: RLWE in polynomial rings. In: Fischlin, M., Buchmann, J., Manulis, M. (eds.) 15th International Conference on Practice and Theory in Public Key Cryptography, PKC 2012. Lecture Notes in Computer Science, vol. 7293 (2012)
13. Dummit, D.S., Kisilevsky, H.: Indices in cyclic cubic fields. In: Number Theory and Algebra, pp. 29–42. Academic, London (1977)
14. Eisentraeger, K., Hallgren, S., Lauter, K.: Weak Instances of PLWE. In: Proceedings of Selected Areas of Cryptography 2014. Lecture Notes in Computer Science. Springer, New York (2014)

15. Elias, Y., Lauter, K., Ozman, E., Stange, K.: Provably weak instances of ring-LWE. In: Advances in Cryptology – CRYPTO 2015, pp. 63–92. Springer (2015). doi:10.1007/978-3-662-47989-6
16. Elliott, P.D.T.A.: A problem of Erdős concerning power residue sums. Acta Arith. **13**, 131–149 (1967/1968)
17. Erdos, P.: On the coefficients of the Erdos polynomial. Bull. Am. Math. Soc. **52**(2), 179–184 (1946)
18. Gentry, C.: Fully homomorphic encryption using ideal lattices. In: Proceedings of the 40th ACM Symposium on Theory of Computing, pp. 169–178 (2009)
19. Gentry, C., Halevi, S., Smart, N.P.: Fully homomorphic encryption with polylog overhead, Cryptology ePrint Archive, Report 2011/566, Eurocrypt 2012 (2011)
20. Graepel, T., Lauter, K., Naehrig, M.: ML Confidential: Machine Learning on Encrypted Data, International Conference on Information Security and Cryptology – ICISC 2012. Lecture Notes in Computer Science. Springer, Berlin (2012)
21. Gras, M.-N.: Sur les corps cubiques cycliques dont l'anneau des entiers est monogéne. Ann. Sci. Univ. Besan. Math. **3**(6), 26 (1973)
22. Gras, M.-N.: Condition necessaire de monogeneite de l'anneau des entiers d'une extension abelienne de $\mathbb{Q}$, Seminare de theorie des nombres(Paris, 1984/1985). Prog. in Math. Birkhauser, Basel
23. Gras, M.-N.: Non monogeneite de l'anneau des entiers des extensions cycliques de $\mathbb{Q}$ de degre premier $\ell \geq 5$. J. Number Theory **23**(3), 347–353 (1986)
24. Hoffstein, J., Pipher, J., Silverman, J.: NTRU: a ring based public key cryptosystem. In: Proceedings of ANTS-III. Lecture Notes in Computer Science, vol. 1423, pp. 267–288. Springer, Berlin (1998)
25. Kedlaya, K.: A construction of polynomials with squarefree discriminants. Proc. Amer. Math. Soc. **140**, 3025–3033 (2012)
26. Konyagin, S.V., Shparlinski, I.E.: Character sums with exponential functions and their applications. Cambridge Tracts in Mathematics, vol. 136. Cambridge University Press, Cambridge (1999)
27. Lauter, K., Naehrig, M., Vaikuntanathan, V.: Can homomorphic encryption be practical? In: CCSW 2011, ACM Cloud Computing Security Workshop (2011)
28. Lauter, K., Lopez-Alt, A., Naehrig, M.: Private Computation on Encrypted Genomic Data. LatinCrypt 2014 (GenoPri 2014)
29. Lindner, R., Peikert, C.: Better key sizes (and attacks) for LWE-based encryption. In: CT-RSA 2011 (2011)
30. Lyubashevsky, V., Peikert, C., Regev, O.: On ideal lattices and learning with errors over rings. In: Eurocrypt 2010. Lecture Notes in Computer Science (2010). See also journal version: J. ACM (JACM) **60** (6), 43
31. Micciancio, D., Regev, O.: Lattice-based cryptography. In: Post-Quantum Cryptography, pp. 147–191. Springer, Berlin (2009)
32. Migotti, A.: Zur theorie der kreisteilungsgleichung, Z. B. der Math.-Naturwiss. Classe der Kaiserlichen Akademie der Wissenschaften, Wein **87**, 7–14 (1883)
33. Murty, M.R., Petersen, K.L.: The Euclidean algorithm for number fields and primitive roots. Proc. Amer. Math. Soc. **141**(1), 181–190 (2013)
34. Nakahara, T., Shah, S.: Monogenesis of the rings of integers in certain imaginary abelian fields. Nagoya Math. J. **168**, 85–92 (2002)
35. Nikiforov, V.: Revisiting Schur's bound on the largest singular value (2007). http://arxiv.org/abs/math/0702722
36. Regev, O.: On lattices, learning with errors, random linear codes, and cryptography. J. ACM **56**(6), 40 (2009). Art. 34 (Preliminary version STOC 2005)
37. Rückert, M., Schneider, M.: Selecting secure parameters for lattice-based cryptography. Cryptology ePrint Archive, Report 2010/137 (2010)
38. Samuel, P.: About Euclidean rings. J. Algebra **19**, 282–301 (1971)

Asymptotics for Number Fields and Class Groups

Melanie Matchett Wood

Abstract This article is an exposition of some of the basic questions of arithmetic statistics (counting number fields and distribution of class groups) aimed at readers new to the area. Instead of a thorough treatment of the most general cases, it treats the simplest cases in detailed way, with an emphasis on connections and perspectives that are well known to experts but absent from the literature.

Keywords Number fields • Class groups • Density of discriminants • Malle's conjecture • Cohen–Lentra Heuristics • Arithmetic statistics

1 Counting Number Fields

We start by giving an introduction to some of the most basic questions of arithmetic statistics. A number field K is a finite extension of fields $K/\mathbb{Q}$. Associated with a number field is an integer $\operatorname{Disc} X$, its discriminant. (See your favorite algebraic number theory text, or, e.g., [51, p. 15].) We have the following classical result.

Theorem 1.1 (Hermite's theorem, see, e.g., III.2.16 in [51]). *Given a positive real number X, there are finitely many number fields K (up to isomorphism, or in a fixed algebraic closure $\bar{\mathbb{Q}}$ of $\mathbb{Q}$) with $|\operatorname{Disc} K| < X$.*

So the most basic question we can ask is how many are there? Let $D(X)$ be the set of isomorphism classes of number fields K such that $|\operatorname{Disc} K| < X$.

Question 1.2. *What are the asymptotics in X of*

$$N(X) := \#D(X)?$$

M.M. Wood (✉)
Department of Mathematics, University of Wisconsin-Madison,
480 Lincoln Drive, Madison, WI 53705, USA

American Institute of Mathematics, 360 Portage Ave, Palo Alto, CA 94306-2244, USA
e-mail: mmwood@math.wisc.edu

© Springer International Publishing Switzerland 2016

E.E. Eischen et al. (eds.), *Directions in Number Theory*, Association for Women in Mathematics Series 3, DOI 10.1007/978-3-319-30976-7_10

Remark 1.3. One can ask a version of this question in which $\mathbb{Q}$ is replaced by any global field, as well, and Disc K is replaced by the norm of the discriminant ideal.

It turns out that after seeing the heuristics in these talks and perhaps also looking at data, one might conjecture

$$N(X) = cX + o(X)$$

for some constant $c > 0$, but we are a long way from a proof of such a statement.

Notation. Given real-valued functions $f(X)$, $g(X)$, and $h(X)$ on some subset of $\mathbb{R}$ when we write

$$f(X) = g(X) + O(h(X)),$$

we mean that there exists a C such that for every $X \geq 1$ we have

$$|f(x) - g(x)| \leq CX.$$

Given such functions $f(X)$, $g(X)$, and $h(X)$ when we write

$$f(X) = g(X) + o(h(X)),$$

we mean that for every real number $\epsilon > 0$, there exists an N such that for $X > N$, we have

$$|f(X) - g(X)| < \epsilon h(X).$$

When we have $f(X) = g(X) + o(h(X))$ and also $\frac{h(X)}{g(X)} = O(1)$, i.e., $\lim_{X\to\infty} \frac{f(X)}{g(X)} = 1$, then we write

$$f(X) \sim g(X)$$

to denote that $f(X)$ and $g(X)$ are asymptotically equivalent.

One can approach Question 1.2 by filtering the number fields by other invariants.

1.1 Galois Group

Given a number field K of degree n—not necessarily Galois—with Galois closure $\tilde{K}$ over $\mathbb{Q}$, we define (by a standard abuse of language) the *Galois group of* K, or $\mathrm{Gal}(K)$, to be the permutation group given as the image of

$$\mathrm{Gal}(\tilde{K}/\mathbb{Q}) \to S_n$$

given by the action of the Galois group on the n homomorphisms of K into $\bar{\mathbb{Q}}$. For example, if $K = \mathbb{Q}(\theta)$, these n homomorphisms are given by mapping θ to each of its n Galois conjugates in $\bar{\mathbb{Q}}$. The Galois group is well defined as a permutation group, i.e., up to relabeling the n homomorphisms, or, equivalently, conjugation in S_n. Two permutation groups in S_n are isomorphic if they are S_n-conjugate.

Exercise 1.4. Show that the Galois group of a number field is a transitive permutation group, i.e., it has a single orbit on $\{1, 2, \ldots, n\}$.

Exercise 1.5. Show that if K is Galois, then its Galois group as defined above is isomorphic, as a group, to the usual notation of Galois group.

So given a transitive permutation group $\Gamma \subset S_n$, (i.e., a conjugacy class of subgroups of S_n, each with a single orbit), we can ask the following.

Question 1.6. *What are the asymptotics in X of*

$$N_\Gamma(X) := \#\{K \in D(X) \mid \mathrm{Gal}(K) \simeq \Gamma\}?$$

(where $\mathrm{Gal}(K) \simeq \Gamma$ *denotes an isomorphism of permutation groups).*

Note that since Γ determines n, any K with $\mathrm{Gal}(K) \simeq \Gamma$ is necessarily of degree n. Note also that we count fields up to isomorphism, not as subfields of $\bar{\mathbb{Q}}$, so, e.g., each isomorphism class of non-Galois cubic field is counted once, not three times.

Remark 1.7. One can of course *ask* these questions, but it should be clear that in general these questions are very hard (for example, they contain the inverse Galois problem). In this article we will discuss what one can do towards solving them in some cases.

Exercise 1.8. For each permutation group Γ, determine how $N_\Gamma(X)$ differs from the similar function that counts subfields of $\bar{\mathbb{Q}}$ with Galois group Γ.

Exercise 1.9. For each permutation group Γ, determine how $N_\Gamma(X)$ differs from the similar function that counts elements K of $D(X)$ *with* a choice of isomorphism of $\mathrm{Gal}(K)$ to Γ.

1.2 Local Behavior

Given a place p of $\mathbb{Q}$ and a number field K, we can form $K_p := K \otimes_{\mathbb{Q}} \mathbb{Q}_p$ (where $\mathbb{Q}_\infty = \mathbb{R}$).

Exercise 1.10. Prove that K_p is an étale $\mathbb{Q}_p$ algebra, equivalently a direct sum of field extensions of $\mathbb{Q}_p$.

In particular, if $\wp_i$ are the places of K dividing p, then $K_p = \bigoplus_i K_{\wp_i}$. So, in particular, we see that the algebra K_p contains the information of the splitting/ramification type of p. The n homomorphisms $K \to \bar{\mathbb{Q}}$ extend to n homomorphisms $K_p \to \bar{\mathbb{Q}}_p$, and we have a map then from the decomposition group $\mathrm{Gal}(\bar{\mathbb{Q}}_p/\mathbb{Q}_p) \subset \mathrm{Gal}(\bar{\mathbb{Q}}/\mathbb{Q})$ to Γ given by its actions on the n homomorphisms. Let the image of this map be H, as a permutation group, we can define it as $\mathrm{Gal}(K_p)$, the Galois group of the étale algebra (by yet another abuse of language). Note that H is the decomposition subgroup of $\mathrm{Gal}(\tilde{K}/\mathbb{Q}) \simeq \Gamma$.

We can also count number fields with a fixed local behavior.

Question 1.11. *Let Γ be a permutation group with sub-permutation group H and let M be an étale K_p algebra with* $\mathrm{Gal}(M) \simeq H$. *What are the asymptotics in X of*

$$N_{\Gamma,M}(X) := \#\{K \in D(X) \mid \mathrm{Gal}(K) \simeq \Gamma, K_p \simeq M, \mathrm{Gal}(K_p) = H\}?$$

Exercise 1.12. I wrote $\mathrm{Gal}(K_p) = H$ to denote that the isomorphism $\mathrm{Gal}(K) \simeq \Gamma$ should induce an isomorphism of $\mathrm{Gal}(K_p)$ onto the precise subgroup H of Γ, not just some other subgroup isomorphic to H. Find a case where this would make a difference.

Question 1.11 then contains, for example, the question of counting quadratic number fields that are split completely at 7.

Exercise 1.13. Answer Question 1.11 for the question of counting quadratic number fields that are split completely at 7. Can you extend your method to count non-Galois cubic ($\Gamma = S_3$) fields split completely at 7? What makes this problem harder?

We can further refine Question 1.11 to ask for local conditions at a set of primes (either a finite set of an infinite set).

Exercise 1.14. Answer this refinement for real quadratic fields split completely at 7. Answer this refinement for real quadratic fields split completely at 7 and ramified at 3.

1.3 Independence

By dividing the answers to the above questions, we can ask what proportion of number fields (with some Galois structure) have a certain local behavior. It is natural to phrase this proportion as a probability and define

$$\begin{aligned}&\mathbb{P}_{\mathrm{Disc}}(K \text{ with } \mathrm{Gal}(K) \simeq \Gamma \text{ has some local behavior})\\&= \lim_{X\to\infty} \frac{\#\{K \in D(X) \mid \mathrm{Gal}(K) \simeq \Gamma, K \text{ has that local behavior}\}}{\#\{K \in D(X) \mid \mathrm{Gal}(K) \simeq \Gamma\}}.\end{aligned}$$

This might now remind us of the Chebotarev density theorem, which is of similar flavor. That theorem tells us, for example, that if we fix a quadratic field K, then half of the primes of $\mathbb{Q}$ split completely in K and half are inert. We could phrase that as above as a question about a fixed field and a random prime. Note that the probability above is for a *fixed prime* and a *random field*. One thing that makes this version much harder is that it is much harder to enumerate the fields than the primes—with quadratic fields being an exception.

Imagine we make a big chart, listing all the quadratic fields by (absolute value of their) discriminant and all the primes, and marking which split (S), ramify (R), or are inert (I).

$$\begin{array}{r|ccccc}
\vdots & \vdots & \vdots & \vdots & \vdots & \iddots \\
\mathbb{Q}(\sqrt{-7}) & \text{S} & \text{I} & \text{I} & \text{R} & \cdots \\
\mathbb{Q}(\sqrt{5}) & \text{I} & \text{I} & \text{R} & \text{I} & \cdots \\
\mathbb{Q}(i) & \text{R} & \text{I} & \text{S} & \text{I} & \cdots \\
\mathbb{Q}(\sqrt{-3}) & \text{I} & \text{R} & \text{I} & \text{S} & \cdots \\
\hline
\text{quad. fields / primes} & 2 & 3 & 5 & 7 & \cdots
\end{array}$$

Above we defined a notion of probability for a random field. Now we do the same for a random prime. For a subset $\mathcal{S}$ of the primes, let

$$\mathbb{P}_p(p \in \mathcal{S}) := \lim_{P \to \infty} \frac{\#\{p \in \mathcal{S} \mid p < P\}}{\#\{p \text{ prime } \mid p < P\}}.$$

The Chebotarev density theorem (or simply Dirichlet's theorem on primes in arithmetic progressions plus quadratic reciprocity) tells us that for each quadratic field K

$$\mathbb{P}_p(p \text{ prime, splits in } K) = 1/2$$

$$\mathbb{P}_p(p \text{ prime, inert in } K) = 1/2$$

$$\mathbb{P}_p(p \text{ prime, ramifies in } K) = 0.$$

This says, in each row of the chart, if we go out far enough to the right, about half the entries are S and have are I. If we ask for $\mathbb{P}_{\text{Disc}}(7 \text{ splits})$, that is asking in the 7th column of the chart, as we look far enough up, what proportion of S's do we get. We can do a simple calculation to see that in $\mathbb{P}_{\text{Disc}}$ there is a *positive* probability of ramification, in contrast to the Chebotarev Density Theorem. So, for example, we have (as the result of a slightly more complicated but classical calculation, like the exercise above)

$$\mathbb{P}_{\text{Disc}}(K \text{ with } \operatorname{Gal}(K) \simeq S_2 \text{ splits completely at } 7) = 7/16$$

$$\mathbb{P}_{\text{Disc}}(K \text{ with } \operatorname{Gal}(K) \simeq S_2 \text{ inert at } 7) = 7/16$$

$$\mathbb{P}_{\text{Disc}}(K \text{ with } \operatorname{Gal}(K) \simeq S_2 \text{ ramified at } 7) = 1/8.$$

If we condition on K being unramified, in this case we do obtain the "same probabilities" as in the Chebotarev Density Theorem. We will see this in other cases below, and in generality for abelian extensions in Theorem 9.1.

Note that $\mathbb{P}_{\text{Disc}}$ and $\mathbb{P}_p$ are not probability measures in the usual sense (perhaps more precisely they are "asymptotic probabilities") because they are not always countably additive. However, it is a useful piece of language here, because it lets us phrase the following important question.

Question 1.15. *For a finite set of distinct places, are probabilities of local behaviors independent at those places?*

It is interesting to compare this to the Chebotarev version. Suppose we look in two rows of the above chart. How often do we see

S	vs. S	vs. I	vs. I
S	I	S	I?

If the two rows were independent, then we would see each of these possibilities $1/4$ of the time, asymptotically. Indeed, we can see that is the case by applying the Chebotarev density theorem to the composite of the two quadratic fields.

Exercise 1.16. Do that application of the Chebotarev density theorem.

More generally, the Chebotarev density theorem tells us that even if we included all Galois number fields in our list on the left of our chart, two rows are independent if they correspond to number fields K, L that do not contain a common subfield larger than $\mathbb{Q}$. For example, we have the following.

Proposition 1.17. *Let $K, L \subset \bar{\mathbb{Q}}$ be Galois number fields. Then for a random rational prime p the events "p splits completely in K" and "p splits completely in L" are independent, i.e.,*

$$\mathbb{P}_p(p \text{ splits completely in } K)\mathbb{P}_p(p \text{ splits completely in } L) = \mathbb{P}_p(p \text{ splits completely in } K \text{ and } L),$$

if and only if $K \cap L = \mathbb{Q}$.

Proof. By the Chebotarev Density Theorem, for a Galois number field F we have

$$\mathbb{P}_p(p \text{ splits completely in } F) = [F : \mathbb{Q}]^{-1}.$$

We have, since K is Galois, $[K : \mathbb{Q}][L : \mathbb{Q}] = [KL : \mathbb{Q}]$ if and only if $K \cap L = \mathbb{Q}$. □

Exercise 1.18. Prove the analog of Proposition 1.17 for other local behaviors besides "splits completely."

If K and L are not Galois, the question of independence is more subtle (see, e.g., the notion of Kronecker equivalence in [39, Chap. 2]). If K, L do contain a common subfield F larger than $\mathbb{Q}$, then it is easy to understand heuristically that the rows should not be independent because they both have a dependence on local behaviors in F. One can moreover see that the common subfield exactly accounts for the dependence as in the following.

Proposition 1.19. *Let $K, L \subset \bar{\mathbb{Q}}$ be Galois number fields. Then for a random rational prime p, conditional on p's splitting type in $K \cap L$, the events "p splits completely in K" and "p splits completely in L" are independent, i.e.,*

$$\begin{aligned} &\mathbb{P}_p(p \text{ splits completely in } K | p \text{ splits completely in } K \cap L) \\ &\cdot \mathbb{P}_p(p \text{ splits completely in } L | p \text{ splits completely in } K \cap L) \\ = &\mathbb{P}_p(p \text{ splits completely in } K \text{ and } L | p \text{ splits completely in } K \cap L). \end{aligned}$$

Proof. We have

$$\mathbb{P}_p(p \text{ splits completely in } K | p \text{ splits completely in } K \cap L) = [K : \mathbb{Q}]^{-1}[K \cap L : \mathbb{Q}],$$

and similarly for the other two conditional probabilities. Since K is Galois, $[K : \mathbb{Q}]^{-1}[K \cap L : \mathbb{Q}][L : \mathbb{Q}]^{-1}[K \cap L : \mathbb{Q}] = [KL : \mathbb{Q}]^{-1}[K \cap L : \mathbb{Q}]$. □

We can prove the analogous fact for any local behaviors. So the dependence of two rows for K and L is determined by whether K and L have a subfield in common. You can try to imagine what the analog should be for primes. What should be analogous for primes, to two fields containing a common subfield?

Exercise 1.20. What is the probability that a quadratic field is split completely at 3? at 5? at 3 and 5? is there independence? What more general statement along these lines can you prove for quadratic fields?

2 Counting Class Groups

Similarly, we can ask what proportion of number fields have a certain class group behavior. Here, even phrasing the right question in the most general case is complicated, so we will start with the simplest case.

Question 2.1. *Given an odd prime p and a finite abelian p-group G, what proportion of imaginary quadratic number fields K, ordered by discriminant, have $Cl(K)_p$ (denoting the Sylow p-subgroup of the class group) isomorphic to G, i.e., what is the limit*

$$\lim_{X \to \infty} \frac{\#\{K \in D(X) \mid K \text{ imag quad}, Cl(K)_p \simeq G\}}{\#\{K \in D(X) \mid K \text{ imag quad}\}}$$

if it exists?

The answer to this question is not known for any G.

Exercise 2.2. Let G be a finite abelian group. What are the asymptotics of the number of imaginary quadratic number fields up to discriminant X with class group isomorphic to G? Can you see why we restrict to the Sylow p-subgroup in Question 2.1?

We can also ask for various averages of the class groups of number fields.

Question 2.3. *Given an odd prime p, what are the limits, if they exist:*

$$\lim_{X\to\infty} \frac{\sum_{K\in D(X) K \text{ imag quad}} |Cl(K)/pCl(K)|}{\#\{K\in D(X) \mid K \text{ imag quad}\}}? \qquad \textit{(average size of } p\textit{-torsion)} \tag{1}$$

$$\lim_{X\to\infty} \frac{\sum_{K\in D(X) K \text{ imag quad}} |Cl(K)/pCl(K)|^k}{\#\{K\in D(X) \mid K \text{ imag quad}\}}? \qquad \textit{(k-th moment of } p\textit{-torsion)} \tag{2}$$

$$\lim_{X\to\infty} \frac{\sum_{K\in D(X) K \text{ imag quad}} |\operatorname{Sur}(Cl(K),A)|}{\#\{K\in D(X) \mid K \text{ imag quad}\}}? \qquad \textit{(A-th moment)}, \tag{3}$$

where A is a finite odd abelian group and Sur *denotes surjections.*

Note that the denominators here are examples of the kind of counting number field functions we considered above (but in an example we can answer!).

Exercise 2.4. If $p = 2$, what is the answer to Eq. (1)? (Hint: learn about genus theory if you haven't already.)

Exercise 2.5. Can you relate the case $A = (\mathbb{Z}/p\mathbb{Z})^m$ of Eq. (3) to Eq. (2)? (Hint: first try the analog of Eq. (3) for homomorphisms instead of surjections. Can you say how many homomorphisms a finite abelian group G has to $(\mathbb{Z}/p\mathbb{Z})^m$ in terms of the size of the p-torsion of G? Then use the fact that homomorphisms are all surjections on to their image.)

One might hope that if you knew the answer to Question 2.1 for every G, then you could average over G to obtain the answers to Question 2.3. However, one cannot switch the sum over G with the limit in X without an argument, of which none has been suggested (if you know one, let me know!).

Questions of the form (3) are the most natural to approach from a certain angle, and in fact the $A = \mathbb{Z}/3\mathbb{Z}$ case (which is also the $p = 3$ case of (1)) is a theorem of Davenport and Heilbronn that we will discuss.

3 Relation Between Counting Number Fields and Class Groups Statistics

The questions discussed above, of counting number fields and averages for class groups, are in fact deeply intertwined. We give the first example of their relationship. Let K be an imaginary quadratic field. Then surjections from $Cl(K)$ to $\mathbb{Z}/3\mathbb{Z}$, by class field theory, correspond to unramified $\mathbb{Z}/3\mathbb{Z}$-extensions L of K with an isomorphism $\mathrm{Gal}(L/K) \simeq \mathbb{Z}/3\mathbb{Z}$. Also by class field theory, we can show that such an L is necessarily Galois over $\mathbb{Q}$, with Galois group S_3. So for each imaginary quadratic K, we have a bijection

$$\mathrm{Sur}(Cl(K), \mathbb{Z}/3\mathbb{Z}) \leftrightarrow \{L/K \text{ unramified}, \phi : \mathrm{Gal}(L/K) \simeq \mathbb{Z}/3\mathbb{Z}\}$$

and all L appearing on the right are Galois over $\mathbb{Q}$ with Galois group S_3.

Exercise 3.1. Show it.

Conversely, let $L/\mathbb{Q}$ be an S_3 Galois sextic field (so with Galois group $S_3 \subset S_6$ via the regular representation) with quadratic subfield K, such that K is imaginary, and L/K is unramified. Then we have two corresponding surjections from $Cl(K)$ to $\mathbb{Z}/3\mathbb{Z}$. So, counting surjections from $Cl(K)$ to $\mathbb{Z}/3\mathbb{Z}$ for imaginary quadratic fields K is the same (up to a multiple of 2) as counting S_3 cubic fields that are unramified over their quadratic subfields (that are also imaginary). And since L/K is unramified $|\operatorname{Disc} L| = |\operatorname{Disc} K|^3$. The condition that the quadratic subfield of L is imaginary is a condition on L_∞, and the condition that L is unramified over its quadratic subfield is a condition on L_p for every p.

So we see here how the class group average of (3) is related to a question of counting S_3 Galois sextic number fields with local conditions (at all primes). This was first explained by Hasse [36]. (In particular, we now see the numerator is a question of this flavor. We already noted the denominator was a question of counting number fields.) We can further translate the class group average to an even simpler question of counting number fields. A Galois sextic S_3-field corresponds to exactly one non-Galois cubic field, up to isomorphism (of which it is the Galois closure). So the question of counting $S_3 \subset S_6$ sextic extensions is equivalent to counting S_3 cubic extensions. In general, it is delicate to relate the discriminant of the non-Galois cubics and the discriminant of their Galois closure (we will discuss this further below), but in this case, the local conditions we are imposing make the relationship simple.

Exercise 3.2. What are the different possible ramification types of a prime p in a non-Galois cubic field K_3? Can you tell from the ramification type whether the Galois closure $\tilde{K}_3$ is ramified over its quadratic subfield? How? If $\tilde{K}_3$ is unramified over its quadratic subfield, how are $\operatorname{Disc} \tilde{K}_3$ and $\operatorname{Disc} K_3$ related?

Exercise 3.3. Can you tell from $K_3 \otimes_{\mathbb{Q}} \mathbb{R}$ if the quadratic subfield of $\tilde{K}_3$ is imaginary? How?

Exercise 3.4. Do the same translation of the class group average question (3) to a question of counting number fields with local conditions for an arbitrary finite abelian group A.

This relationship to counting number fields is one reason that the A-moments of (3) are a particularly nice class group statistic. (We will see another reason later.) See [41] for a more in-depth discussion and results explaining the connections between asymptotics of number fields and averages of class groups.

4 Different Counting Invariants

So far in this entire discussion we have *ordered by discriminant*, i.e., counted number fields of discriminant up to X asymptotically in X. We could replace discriminant by other real-valued invariants and ask the same questions. We will see below how this can change the answers, both quantitatively and qualitatively.

In fact, some of the questions we have already asked about counting number fields ordered by discriminant can be seen as a question of counting other number fields ordered by a different counting invariant. For example, the question of counting all $S_3 \subset S_6$ sextic extensions by discriminant is equivalent to counting S_3 cubic extensions by the discriminant of their Galois closure, which without the local conditions imposed in the last chapter is truly a different counting invariant.

5 Cohen–Lenstra Heuristics

We will discuss heuristics (conjectural answers) for the questions we have considered so far, starting with the class group question. The Cohen–Lenstra heuristics start from the observation that structures often occur in nature with frequency inversely proportional to their number of automorphisms.

Exercise 5.1. Consider a graph G on n vertices. Of the 2^n graphs on n labeled vertices ("nature"—you might imagine the vertices are n computers actually out there in the world, and we are considering all possible network structures between them) how many are isomorphic to G?

Exercise 5.2. Consider all multiplication tables on n elements $a_1, \ldots, a_n$ that satisfy the group axioms. How many are isomorphic to a given group G of order n?

Exercise 5.3. Fix a degree n number field K. Consider all subfields of $\bar{\mathbb{Q}}$ with degree n over $\mathbb{Q}$—how many of these are isomorphic to K?

Now that you are convinced of this principle, the idea of the Cohen–Lenstra heuristics is that for odd p, the group $Cl(K)_p$ is a finite abelian p-group occurring in nature, that we know nothing else about, so we will guess it is distributed in this same way. (For $p = 2$, when K is imaginary quadratic, genus theory tells us something about the form of $Cl(K)_2$, or more precisely about $Cl(K)_2/2Cl(K)_2$, but nothing about $2Cl(K)_2$, so as suggested by Gerth we can make the same guess for $2Cl(K)_2$.) Let $G(n)$ be the set of all finite abelian groups of order n.

Conjecture 5.4 (Cohen and Lenstra [18], $p = 2$ part from [33, 34]). *For any "reasonable" function f on finite abelian groups we have*

$$\lim_{X\to\infty} \frac{\sum_{K\in D(X) K imag\ quad} |f(2Cl(K))|}{\#\{K \in D(X) \mid K imag\ quad\}} = \lim_{n\to\infty} \frac{\sum_{i=1}^{n} \sum_{G\in G(i)} \frac{f(G)}{\#\mathrm{Aut}(G)}}{\sum_{i=1}^{n} \sum_{G\in G(i)} \frac{1}{\#\mathrm{Aut}(G)}}.$$

I expect that we should interpret this as saying not that both limits exist, but that either both limits do not exist or they both exist and are equal. Of course, everything hangs on what "reasonable" means. Cohen and Lenstra [18] said that "reasonable" should probably include all non-negative functions, but Poonen pointed out to me that it may not make sense to include all non-negative functions as "reasonable." See page 22 of [14] for one idea of what one might include as "reasonable." The main examples that Cohen and Lenstra in [18] apply their conjecture to are f that only depend on the Sylow p-subgroups of G for finitely many p (which most people think should be "reasonable," [14] agrees), and f the characteristic function of cyclic groups (which is not in the class of "reasonable" functions considered in [14]).

Notably, Conjecture 5.4 is known for almost no non-trivial f. One exception is when $f = \#\operatorname{Sur}(-, \mathbb{Z}/3\mathbb{Z})$, in which case Conjecture 5.4 is known by Davenport and Heilbronn's work [25] (discussed in depth below) on counting cubic fields (see Sect. 3 for the connection to counting cubic fields). This result, of course, predates the conjecture. The other exception is Fouvry and Klüners results [29, 30] that show Conjecture 5.4 when $f = \#\operatorname{Sur}(-, (\mathbb{Z}/2\mathbb{Z})^k)$ or f is the indicator function of having a particular 2-rank, confirming conjectures of Gerth [33, 34].

In order words, the class group average equals the Aut^{-1} weighted group average. We define, if it exists,

$$M(f) := \lim_{n\to\infty} \frac{\sum_{i=1}^{n} \sum_{G\in G(n)} \frac{f(G)}{\#\operatorname{Aut}(G)}}{\sum_{i=1}^{n} \sum_{G\in G(i)} \frac{1}{\#\operatorname{Aut}(G)}}$$

(the right-hand side of Conjecture 5.4).

It is useful to know (see exercises above or Sect. 8) that the denominator on the left-hand side of Conjecture 5.4 is $\sim \frac{6}{\pi^2}X$.

Exercise 5.5. Show that

$$\lim_{n\to\infty} \sum_{i=1}^{n} \sum_{G\in G(n)} \frac{1}{\#\operatorname{Aut}(G)} = \infty.$$

In particular, then, Conjecture 5.4 implies that each group G appears as a class group of imaginary quadratic fields asymptotically 0 % of the time. In fact, it was first shown by Heilbronn [37] that each group G appears as a class group of an imaginary quadratic field only finitely many times.

Cohen and Lenstra in [18] compute the right-hand side of Conjecture 5.4 for many interesting functions f (which is purely a problem in "finite group theory statistics" instead of "arithmetic statistics"). If f is the indicator function of cyclic groups, they show

$$M(f) \approx .977575.$$

Given an odd prime p, if f is the indicator function for when $p \mid \#G$, then

$$M(f) = 1 - \prod_{i\geq 1}(1 - p^{-i}),$$

and for $p = 3$ this is $\approx .43987$. Perhaps most striking is the following, which follows from combining lemmas of [18] but is not highlighted by them.

Proposition 5.6. *If A is a finite abelian group and $f(G) = \#\operatorname{Sur}(G, A)$, then*

$$M(f) = 1.$$

So the Cohen–Lenstra heuristics suggest that the expected number of surjections from an imaginary quadratic class group to A is 1, regardless of the group A. We saw above, these A-moments were related to questions of counting number fields, and now we see they have particularly nice predicted values. (See also Ellenberg's 2014 Arizona Winter School lectures for more on the interpretation of these moments in the function field analog.)

Note that $f(G) = \#\operatorname{Sur}(G, A)$ only depends on finitely many Sylow p-subgroups of G. Let P be a finite set of primes, and let G_P be the sum of the Sylow p-subgroups of G for $p \in P$. Let $G_P(n)$ be the set of finite abelian groups G of order n such that $G = G_P$.

Proposition 5.7 (Cohen and Lenstra [18]). *Let P be a finite set of primes. Let f be a function depending on only the Sylow p-subgroups of G for $p \in P$. Then*

$$\lim_{n\to\infty} \sum_{i=1}^{n} \sum_{G\in G_P(i)} \frac{1}{\#\operatorname{Aut}(G)} = \prod_{p\in P}\prod_{i\geq 1}(1 - p^{-i})^{-1} =: c_P,$$

and

$$M(f) = \frac{\sum_{i=1}^{\infty} \sum_{G\in G_P(i)} \frac{f(G)}{\#\operatorname{Aut}(G)}}{c_P}.$$

Exercise 5.8. If F is the indicator function for groups that have cyclic Sylow 3-subgroup and cyclic Sylow 5-subgroup, compute $M(f)$.

5.1 Further Class Group Heuristics

We have only discussed conjectures for class groups of imaginary quadratic fields. There are conjectures for much more general situations. Cohen and Lenstra [18] also formulate precise conjectures for real quadratic class groups, and more generally for class groups of totally real number fields with some fixed abelian Galois group.

Cohen and Martinet [19] further extended these conjectures to arbitrary extensions of an arbitrary global field. It is in general a subtle question for which primes p the general form of the conjecture can be made for $Cl(K)_p$ (e.g., all odd primes in the imaginary quadratic case). See [1, 20, 31, 48, 49] for work on this question, and further modifications of the conjecture for p where the base field contains pth roots of unity. In all of these cases, the conjectures are based on the same sort of principles as those above, but are modified to take into account further information about the fields.

6 Galois Permutation Representations

In order to formulate the conjectures for counting number fields, it is useful to translate from the language of number fields to Galois (permutation) representations. For a field F, let $G_F := \mathrm{Gal}(\bar{F}/F)$, the Galois group of a separable closure of F (if $charF = 0$, then $\bar{F}$ is equivalently an algebraic closure of F). An étale F-algebra is a direct sum of finitely many finite separable field extensions of F. Two étale F-algebras are isomorphic if they are isomorphic as algebras. The degree of an étale algebra is its dimension as an F-vector space, or equivalently the sum of the degrees of the field extensions. Then given a permutation representation, i.e., a continuous homomorphism

$$G_F \to S_n,$$

we can pick representatives $a_i \in \{1, \ldots, n\}$ of the orbits, and let $H_i = \mathrm{Stab}(a_i) \subset G_F$. If K_i is the fixed field of H_i, then we can form an étale F-algebra $\bigoplus_i K_i$. Conversely, given an étale F-algebra $M = \bigoplus_i K_i$, where K_i/F are finite, separable field extensions whose degrees sum to n, we have an action of G_F on the n homomorphisms $M \to \bar{F}$, which gives a permutation representation of G_F. We say two permutation representations are isomorphic if they differ by relabeling the n elements, i.e., by conjugacy in S_n.

Proposition 6.1. *The above constructions gives a bijection between isomorphism classes of permutation representations $G_F \to S_n$ and isomorphism classes of degree n étale F-algebras. In this bijection, transitive permutation representations correspond to field extensions of F.*

Exercise 6.2. Prove the above proposition.

Exercise 6.3. If we restrict $G_{\mathbb{Q}} \to S_n$ (corresponding to a field extension $K/\mathbb{Q}$) to a decomposition group $G_{\mathbb{Q}_p} \to S_n$, show that K_p is the étale algebra corresponding to $G_{\mathbb{Q}_p} \to S_n$.

We will now consider the case when $F = \mathbb{Q}$. Fix a transitive permutation group $\Gamma \subset S_n$. We will be interested in counting $\rho : G_{\mathbb{Q}} \to \Gamma$, whose corresponding field extension K has $|\mathrm{Disc}\, K| < X$. We define $\mathrm{Disc}\, \rho := |\mathrm{Disc}\, K|$.

If we factor $|\operatorname{Disc} K| = \prod_i p_i^{e_i}$ with p_i distinct primes, then recall that the ideal $(\operatorname{Disc} K_{p_i}/\mathbb{Q}_{p_i}) = (p_i)^{e_i}$. Further, if we write K_{p_i} as a direct sum of field extensions K_j then $(\prod_j \operatorname{Disc} K_j/\mathbb{Q}_{p_i}) = (\operatorname{Disc} K_{p_i}/\mathbb{Q}_{p_i})$. For an étale $\mathbb{Q}_p$-algebra M, we define $d(M)$ to be the discriminant exponent so that $(\operatorname{Disc} M) = (p)^{d(M)}$. If M corresponds to $\rho_p : \mathbb{G}_{\bar{\mathbb{Q}}_p} \to S_n$, we define $\operatorname{Disc} \rho_p = p^{d(M)}$.

For an étale $\mathbb{Q}_p$-algebra M associated with $\rho : G_{\bar{\mathbb{Q}}_p} \to S_n$, recall that $d(M)$ is the Artin conductor of the composition of ρ with the permutation representation $S_n \to \mathrm{GL}_n(\mathbb{C})$. This allows us to compute $d(M)$. For example, we have the following.

Lemma 6.4. *If $M/\mathbb{Q}_p$ is a tame étale extension corresponding to $\rho : G_{\bar{\mathbb{Q}}_p} \to S_n$, and y is a generator of tame inertia (i.e., a generator of the quotient of the inertia subgroup by its unique pro-p-Sylow subgroup) in $G_{\bar{\mathbb{Q}}_p}$, and c is the number of cycles in $\rho(y)$, then*

$$d(M) = n - c.$$

Exercise 6.5. Prove this lemma.

7 Tauberian Theorem

Before we get to the conjectures about counting number fields, we will review an important tool in asymptotic counting. We give an example of a Tauberian theorem, which can be found as Corollary p. 121 of [50].

Theorem 7.1. *Let $f(s) = \sum_{n\geq 1} a_n n^{-s}$ with $a_n \geq 0$ be convergent for $\Re s > a > 0$. Assume that in the domain of convergence $f(s) = g(s)(s-a)^{-w} + h(s)$ holds, where $g(s), h(s)$ are holomorphic functions in the closed half-plane $\Re s \geq a$, and $g(a) \neq 0$, and $w > 0$. Then*

$$\sum_{1\leq n\leq X} a_n = \frac{g(a)}{a\Gamma(w)} x^a (\log x)^{w-1} + o(x^a (\log x)^{w-1}).$$

For example, if $f(s)$ converges for $\Re(s) > 1$ and has a meromorphic continuation to $\Re(s) \geq 1$ with a simple pole at $s = 1$ with residue r, then

$$\sum_{1\leq n\leq X} a_n = rX + o(X).$$

Upon seeing Theorem 7.1, you might think whenever you are counting something asymptotically, you should just make it into a Dirichlet series and study the pole behavior of the function. However, it should be emphasized that in order to gain any traction with this method one must *produce an analytic continuation* of $f(s)$ beyond where the Dirichlet series converges. In general, producing such an analytic continuation can be as hard as any question in mathematics (e.g., one defines the L function of an elliptic curve as a Dirichlet series, and the analytic continuation is a consequence of the modularity theorem, used, for example, in the proof of Fermat's Last Theorem—and that's a case where you have the benefit of the Langlands program telling you where the analytic continuation should come from!). Even if you produce an analytic continuation it may be very non-trivial to understand its rightmost poles (e.g., this is the case when counting quintic number fields by discriminant). Nonetheless, this is a tool that can help us understand some asymptotic counting questions, and it gives us a framework for thinking about the questions that we can't answer.

8 Counting Abelian Number Fields

We will next apply our Tauberian theorem to count some abelian number fields. The results in this section will be special cases of more general results which we will cite more properly in the next section.

Let $J_{\mathbb{Q}}$ be the idèle class group of $\mathbb{Q}$, so

$$J_{\mathbb{Q}} = \left(\prod_p{}' \mathbb{Q}_p^*\right) / \mathbb{Q}^*$$

where the product is over places p of $\mathbb{Q}$, and the product is restricted so that an element of $J_{\mathbb{Q}}$ must have all but finitely many terms as units $\mathbb{Z}_p^*$ in the local ring of integers.

We will first consider quadratic extensions. Though these can be approached more directly, the advantage of our method is that it, with enough work, will generalize to any abelian group. Class field theory tells us that the abelianization $G_{\mathbb{Q}}^{ab}$ is isomorphic, as a topological group to the profinite completion $\widehat{J_{\mathbb{Q}}}$, and in particular that continuous homomorphisms

$$G_{\mathbb{Q}} \to \mathbb{Z}/2\mathbb{Z}$$

correspond exactly to continuous homomorphisms

$$J_{\mathbb{Q}} \to \mathbb{Z}/2\mathbb{Z}.$$

So, we will focus on maps $\phi : J_{\mathbb{Q}} \to \mathbb{Z}/2\mathbb{Z}$. Note that ϕ restricts to a map

$$\phi_0 : \prod_p \mathbb{Z}_p^* \to \mathbb{Z}/2\mathbb{Z},$$

where we use $\mathbb{Z}_\infty$ to denote the positive real numbers. Moreover, we will see that any such ϕ_0 extends to a unique $\phi : J_\mathbb{Q} \to \mathbb{Z}/2\mathbb{Z}$. Given ϕ_0, to define an extension we must define $\phi(1,\ldots,1,p,1,\ldots)$ (where the p is in the p place), and we see that the quotient by $\mathbb{Q}^*$ forces

$$\phi(1,\ldots,1,p,1,\ldots)=\phi(p^{-1},\ldots,p^{-1},1,p^{-1},\ldots)=\phi_0(p^{-1},\ldots,p^{-1},1,p^{-1},\ldots).$$

Similarly, for $\phi(1,\ldots,-1)$ (with a 1 in every finite place),

$$\phi(1,\ldots,-1) = \phi(-1,\ldots,1) = \phi_0(-1,\ldots,1).$$

So we conclude that ϕ is determined by ϕ_0. Moreover, for any ϕ_0, we can check that the above construction gives a well-defined ϕ.

Moreover, we can compute $\operatorname{Disc}\phi$ (defined to be the discriminant of the corresponding Galois representation to S_2) in terms of ϕ_0. In our map $\phi : J_\mathbb{Q} \to \mathbb{Z}/2\mathbb{Z}$, the image of the decomposition group at p is the image of $\mathbb{Q}_p^*$ and the inertia group is the image of $\mathbb{Z}_p^*$. In particular, since the discriminant (viewed as an Artin conductor) only depends on the inertia group, we see that we can recover $\operatorname{Disc}\phi$ from ϕ_0.

What are the maps $\mathbb{Z}_p^* \to \mathbb{Z}/2\mathbb{Z}$? Since the kernel of the map $\mathbb{Z}_p^* \to (\mathbb{Z}/p\mathbb{Z})^*$ is pro-p, for p odd, a map $\mathbb{Z}_p^* \to \mathbb{Z}/2\mathbb{Z}$ must factor through $(\mathbb{Z}/p\mathbb{Z})^* \to \mathbb{Z}/2\mathbb{Z}$. There are of course 2 such maps, depending on whether a generator is sent to 1 or 0. When $p = 2$, a map $\mathbb{Z}_p^* \to \mathbb{Z}/2\mathbb{Z}$ factors through

$$\mathbb{Z}_2^* \to (\mathbb{Z}/8\mathbb{Z})^* \simeq \mathbb{Z}/2 \times \mathbb{Z}/2$$

(to conclude this, one must understand the structure of $\mathbb{Z}_2^*$ as a profinite group). In particular, there are four maps $\mathbb{Z}_2^* \to \mathbb{Z}/2\mathbb{Z}$.

Given a map $\mathbb{Z}_p^* \to \mathbb{Z}/2\mathbb{Z}$, how do we compute the discriminant? One approach is to use the conductor-discriminant formula. This requires knowing the conductor of the map $\mathbb{Z}_p^* \to \mathbb{Z}/2\mathbb{Z}$, which is k where p^k is minimal such that the map factors through $(\mathbb{Z}/p^k\mathbb{Z})^*$.

Exercise 8.1. Show that $\operatorname{Disc}\rho$ for the $\rho : \mathbb{Q}_p^* \to \mathbb{Z}/2\mathbb{Z}$ restricting to the $\mathbb{Z}_p^* \to \mathbb{Z}/2\mathbb{Z}$ discussed above are $1, p$ for odd p, and $1, 2^2, 2^3, 2^3$ for $p = 2$.

So, if a_n is the number of continuous homomorphisms $\rho : G_\mathbb{Q} \to \mathbb{Z}/2\mathbb{Z}$ with $\operatorname{Disc}\rho = n$, we have

$$f(s) = \sum_{n\geq 1} a_n n^{-s} = (1 + 2^{-2s} + 2 \cdot 2^{-3s}) \prod_{p \text{ odd prime}} (1 + p^{-s}).$$

We note that

$$f(s) = h(s)\frac{\zeta(s)}{\zeta(2s)},$$

where $h(s) = (1 + 2^{-2s} + 2 \cdot 2^{-3s})/(1 + 2^{-s})$ and $\zeta(s) = \sum_{n \geq 1} n^{-s}$. In particular, we can apply Theorem 7.1. Since the residue of $f(s)$ at $s = 1$ is $\zeta(2)^{-1} = 6/\pi^2$, we have that

$$N_{S_2}(X) \sim \frac{6}{\pi^2} X.$$

(There is a single non-surjective $\rho : G_{\mathbb{Q}} \to \mathbb{Z}/2\mathbb{Z}$ so this doesn't affect the asymptotics.) So we have counted, you might say the long way around, quadratic extensions by discriminant. (Such a result is of course classical.)

An important feature of our approach is that it works more generally. Suppose we want to count cyclic cubic fields (as was done by Cohn [16] in the same way as we will do). Each field corresponds to 2 surjective maps $\mathbb{G}_{\mathbb{Q}} \to \mathbb{Z}/3\mathbb{Z} \subset S_3$. Let a_n be the number of continuous homomorphisms $\rho : G_{\mathbb{Q}} \to \mathbb{Z}/3\mathbb{Z}$ with $\operatorname{Disc} \rho = n$, then by a similar analysis to the above we have

$$f(s) = \sum_{n \geq 1} a_n n^{-s} = (1 + 2 \cdot 3^{-4s}) \prod_{p \equiv 1 \pmod 3 \text{ prime}} (1 + 2p^{-2s}).$$

We can show that the function $f(s)$ has rightmost pole at $s = 1/2$, where it has a simple pole. Let χ be a Dirichlet character modulo 3 such that $\chi(1) = 1$ and $\chi(2) = -1$. Then note that

$$\prod_{p \text{ prime, not } 3} (1 + p^{-2s}) \prod_{p \text{ prime, not } 3} (1 + \chi(p) p^{-2s})$$
$$= \prod_{p \equiv 1 \pmod 3 \text{ prime}} (1 + 2p^{-2s} + p^{-4s}) \prod_{p \equiv 2 \pmod 3 \text{ prime}} (1 - p^{-4s}).$$

By comparison to $\zeta(2s)$ and $L(2s, \chi)$, the top has rightmost pole a simple pole at $s = 1/2$. We can see that the bottom has the same pole behavior as $f(s)$ to $s = 1/2$. We conclude the result of Cohn [16]

$$N_{\mathbb{Z}/3 \subset S_3}(X) \sim \frac{1}{2} \lim_{s \mapsto +1/2} (1 + 2 \cdot 3^{-4s}) \prod_{p \equiv 1 \pmod 3 \text{ prime}} (1 + 2p^{-2s}) \zeta(2s)^{-1} X^{1/2}, \quad (4)$$

where the limit in s is a constant.

Note here, the main input, after we have applied class field theory in the set-up above, is to find appropriate L functions to compare our Dirichlet series with so that we can analyze the pole behavior and get analytic continuation so as to apply Theorem 7.1. However, so far we are counting all maps $\mathbb{G}_{\mathbb{Q}} \to G$, and when G is not $\mathbb{Z}/\ell$ for some prime ℓ, there will be more than 1 non-surjective map to subtract. We will come back to this issue.

8.1 Local Conditions

Let's go back to considering $\rho : \mathbb{G}_{\mathbb{Q}} \to \mathbb{Z}/2\mathbb{Z}$. What if we wanted to impose local conditions Σ such that 3 was split completely? Taylor [54] attributes the question of the distribution of splitting types of a given prime in random G-extensions to Fröhlich, who was motivated by the work of Davenport and Heilbronn [25]. We discuss work of Taylor [54] and the author [60] on this question below.

Since in the isomorphism $\mathbb{G}_{\mathbb{Q}_p}^{ab} \simeq \widehat{\mathbb{Q}_p^*}$ of class field theory, we have that $\mathrm{Frob}_p \mapsto p$, this is equivalent to counting $\rho : J_{\mathbb{Q}} \to \mathbb{Z}/2$ such that $\rho(1, 3, 1, \dots) = 0$ (where the 3 is in the $\mathbb{Q}_3$ place). Since we have that $\rho(1, 3, 1, \dots) = \rho(3, 1, 3, \dots)$, we can check this just from the maps from $\mathbb{Z}_p^*$. Where does a local map $\chi : \mathbb{Q}_p^* \to \mathbb{Z}/2$ send 3 when p is not 3? If χ is trivial, then $\chi(3) = 1$. If p is odd and χ is not trivial, then $\chi(p) = 0$ if 3 is a square mod p and $\chi(p) = 1$ if 3 is not a square mod p. Let Ψ be the Dirichlet character that is 1 on odd primes p such that 3 is a square mod p and -1 on odd primes p such that 3 is not a square (which exists and is of modulus 12 by quadratic reciprocity).

Let b_n be the number of continuous homomorphisms $\rho : G_{\mathbb{Q}} \to \mathbb{Z}/2\mathbb{Z}$ with $\mathrm{Disc}\,\rho = n$ and sending $(1, 3, 1, \dots)$ to 0. Then

$$\begin{aligned} g(s) &= \sum_{n\geq 1} b_n n^{-s} \\ &= \frac{1}{2}\left((1+2^{-2s}+2\cdot 2^{-3s}) \prod_{p \text{ odd prime}} (1+p^{-s}) + (1-2^{-2s})(1+3^{-s}) \prod_{p \text{ prime}>5} (1+\Psi(p)p^{-s}) \right). \end{aligned}$$

The first term counts all $\rho : G_{\mathbb{Q}} \to \mathbb{Z}/2\mathbb{Z}$ with coefficient 1. The second counts $\rho : G_{\mathbb{Q}} \to \mathbb{Z}/2\mathbb{Z}$ that send $(3, 1, 3, \dots) \mapsto 0$ with coefficient 1 and those that send $(3, 1, 3, \dots) \mapsto 1$ with coefficient -1. (We have checked at $p = 2$ that the discriminant 2^2 map sends $3 \mapsto 1$ and of the two discriminant 2^3 maps, one sends $3 \mapsto 1$ and one $3 \mapsto 0$.) We will compare g to the L function

$$L(s, \Psi) = \prod_{p \text{ prime}} (1 - \Psi(p)p^{-s})^{-1}.$$

We have that

$$g(s) = \frac{1}{2}h(s)\frac{\zeta(s)}{\zeta(2s)} + \frac{1}{2}k(s)\frac{L(s, \Psi)}{L(2s, \Psi)},$$

where $k(s)$ is analytic on $\Re s \geq 1$. Since $\frac{L(s,\Psi)}{L(2s,\Psi)}$ is holomorphic on $\Re s \geq 1$, we have

$$N_{S_2,\Sigma}(X) \sim \frac{1}{2}\frac{6}{\pi^2}X.$$

We can go about this more systematically. Above, we essentially argued that

$$\Big(\prod_{p \text{ finite}} \mathbb{Z}_p^*\Big) \times \mathbb{R}_+ \simeq J_{\mathbb{Q}}.$$

However, let S be a finite set of places, then we also have

$$\left(\prod_{p\in S} \mathbb{Q}_p^* \times \prod_{p\notin S} \mathbb{Z}_p^*\right) / \mathbb{Z}_S^* \simeq J_{\mathbb{Q}}$$

where $\mathbb{Z}_S^*$ denotes S-units, i.e., integers in $\mathbb{Z}_p^*$ at all primes $p \notin S$. (We let $\mathbb{Z}_\infty^* = 1$.) So we will let S be a finite set of places on which we want to make local specifications. We will make a Dirichlet series counting maps $\prod_{p\in S} \mathbb{Q}_p^* \times \prod_{p\notin S} \mathbb{Z}_p^* \to G$ for abelian G, and then we will use multisection (i.e., use the above trick with 1 and -1) to pull out the maps that are trivial on $\mathbb{Z}_S^*$. First we have

$$F_G(s) = \prod_{p\in S} \left(\sum_{\rho_p : \mathbb{Q}_p^* \to G} p^{-d(\rho_p)s} \right) \prod_{p\notin S} \left(\sum_{\rho_p : \mathbb{Z}_p^* \to G} p^{-d(\rho_p)s} \right),$$

which counts all maps $\prod_{p\in S} \mathbb{Q}_p^* \times \prod_{p\notin S} \mathbb{Z}_p^* \to G$. Now, for simplicity, we will let $G = \mathbb{Z}/n\mathbb{Z}$. Let A be a set of representatives for $\mathbb{Z}_S^*/(\mathbb{Z}_S^*)^n$. Let ζ be a primitive nth root of unity. Note that if $\rho : \prod_{p\in S} \mathbb{Q}_p^* \times \prod_{p\notin S} \mathbb{Z}_p^* \to G$, then

$$\frac{1}{\#A} \sum_{a\in A} \zeta^{\rho(a)}$$

is 1 if $\rho(\mathbb{Z}_S^*) = 0$ and is 0 otherwise. So, we have

$$H_G(s) = \frac{1}{\#A} \sum_{a\in A} \left(\prod_{p\in S} \left(\sum_{\rho_p : \mathbb{Q}_p^* \to G} \zeta^{\rho_p(a)} p^{-d(\rho_p)s} \right) \prod_{p\notin S} \left(\sum_{\rho_p : \mathbb{Z}_p^* \to G} \zeta^{\rho_p(a)} p^{-d(\rho_p)s} \right) \right),$$

which counts all maps $\prod_{p\in S} \mathbb{Q}_p^* \times \prod_{p\notin S} \mathbb{Z}_p^* \to G = \mathbb{Z}/n\mathbb{Z}$ that are trivial on $\mathbb{Z}_S^*$, or equivalently, maps $J_{\mathbb{Q}} \to G = \mathbb{Z}/n\mathbb{Z}$ with this property. Now, if we have a local specific Σ at places $p \in S$, we can form

$$H_{G,\Sigma}(s) = \frac{1}{\#A} \sum_{a\in A} \left(\prod_{p\in S} \left(\sum_{\substack{\rho_p : \mathbb{Q}_p^* \to G \\ \rho_p \in \Sigma_p}} \zeta^{\rho_p(a)} p^{-d(\rho_p)s} \right) \prod_{p\notin S} \left(\sum_{\rho_p : \mathbb{Z}_p^* \to G} \zeta^{\rho_p(a)} p^{-d(\rho_p)s} \right) \right), \tag{5}$$

which counts all maps $J_{\mathbb{Q}} \to G = \mathbb{Z}/n\mathbb{Z}$ satisfying the local conditions Σ. We see now the advantage of taking the full map from $\mathbb{Q}_p^*$ at the places where we want to specify. *Now imposing local conditions is just a matter of taking the terms we want from those factors.*

Exercise 8.2. See that the above gives the same analytic functions for the question of counting quadratic extensions split completely at 3.

Above we see that $H_{G,\Sigma}(s)$ is a sum of #*A* Euler products. In an ideal world, (*) the rightmost pole would occur in exactly 1 of those Euler products (presumably the $a = 1$ term, since we could hope the roots of unity help the other terms be smaller), and we would have an analytic continuation beyond the line of that rightmost pole so we could apply Theorem 7.1. If this were true, then the discriminant probability *among all maps, not necessarily surjective,* $J_{\mathbb{Q}} \to G$ of any Σ with specifications on a finite set of primes s would be

$$\frac{\prod_{p\in S}\sum_{\substack{\rho_p:\mathbb{Q}_p^*\to G\\ \rho_p\in\Sigma_p}} p^{-d(\rho_p)s}}{\prod_{p\in S}\sum_{\rho_p:\mathbb{Q}_p^*\to G} p^{-d(\rho_p)s}},$$

and, in particular, because only one Euler product contributed to the asymptotic count we would certainly have independence of probabilities of local behaviors at a finite set of places.

However, the world is not always ideal. It turns out (*) is true when G is abelian of prime exponent, but is not true in general. Also, there is a distinction between the question above and the question of counting number fields, which would correspond to surjective ρ. One can use inclusion–exclusion to subtract out ρ with smaller image. In an ideal world, (**) the Dirichlet series coming from the maps with smaller images would be holomorphic past the rightmost pole of H_G. However, (**) is only true when G is abelian of prime exponent as well. (See [60] for the proofs of all of these claims.)

We always consider abelian groups in their regular permutation representation. When $G = (\mathbb{Z}/\ell)^k$ for some prime ℓ, then as we have said (but certainly not proven!) above, one obtains an asymptotic for N_G and for $N_{G,\Sigma}$ for each set Σ of local specifications with restrictions only at finitely many primes. In particular, the probabilities of local conditions at distinct primes are independent. The probabilities are simple values we can read off from above.

Now we consider again an abelian G. If we impose local conditions that only depend on the restriction $\rho_p : \mathbb{Z}_p^* \to G$ (for example, whether the map is ramified or unramified at p), then we can pick out the appropriate terms in the Dirichlet series

$$\prod_p \left(\sum_{\rho_p:\mathbb{Z}_p^*\to G} p^{-d(\rho_p)s} \right),$$

and again in this case show independence of such local conditions among all (not necessarily surjective) $\rho_p : \mathbb{Z}_p^* \to G$. (Note we have not shown this here, one still has to do the analysis of the Dirichlet series by comparison to appropriate L functions.)

Exercise 8.3. Let A be an event with positive probability not equal to 1. If E and F are independent, independent given A and we have that $\mathbb{P}(E|A) \neq \mathbb{P}(E)$ and $\mathbb{P}(F|A) \neq \mathbb{P}(F)$, then, given not-$A$, the events E and F are not independent.

Using the exercise above, it is shown in [60] that since ramification is independent, e.g., in $\mathbb{Z}/\ell$ extensions (for ℓ prime), and for maps $\rho : G_{\mathbb{Q}} \to \mathbb{Z}_\ell^2$, then it cannot be independent in $\mathbb{Z}/\ell^2$ extensions. (This requires, among other things, knowing that non-surjective $\rho : G_{\mathbb{Q}} \to \mathbb{Z}/\ell^2\mathbb{Z}$ occur with positive probability.) For example, we have the following.

Proposition 8.4 (Proposition 1.4 of [60]). *Let p, q_1, and q_2 be primes with $q_i \equiv 1 \pmod{p}$ for $i = 1, 2$. Then q_1 ramifying and q_2 ramifying in a random $\mathbb{Z}/p^2\mathbb{Z}$-extension are not (discriminant) independent.*

One might hope to then simplify things by not considering G-extensions but rather all maps $\mathbb{G}_{\mathbb{Q}} \to G$. Then, we wouldn't have to subtract out non-surjective maps. In this situation, for local conditions only depending on the map of the inertia subgroups, we do indeed get independence of local behaviors and easy to read off probabilities. However, if we consider all local conditions, then the other terms in the Eq. (5) sum over $a \in A$ with the same rightmost pole as the $a = 1$ term necessarily have an effect on the answer.

8.2 Grunwald–Wang

It follows from Wang's [57] counterexample to Grunwald's "Theorem" [35] (this is a great story to read about if you don't know it already), that there is no $\mathbb{Z}/8$ extension K of $\mathbb{Q}$ for which K_2 is an unramified extension of $\mathbb{Q}_2$ of degree 8. However, there is certainly a local Galois representation $\rho_2 : \mathbb{Q}_2^* \to \mathbb{Z}/8$ that is unramified and sends $2 \mapsto 1$. This local Galois representation does not come from a global $\mathbb{G}_{\mathbb{Q}} \to \mathbb{Z}/8$, and considering all $\mathbb{G}_{\mathbb{Q}} \to \mathbb{Z}/8$ or just surjective $\mathbb{G}_{\mathbb{Q}} \to \mathbb{Z}/8$ does not change a thing about this fact. So, we realize that some term in the Eq. (5) sum over $a \in A$ with the same pole as the $a = 1$ term must in fact cancel that pole entirely.

Exercise 8.5. Show there is no $\mathbb{Z}/8$ extension K of $\mathbb{Q}$ for which K_2 is an unramified extension of $\mathbb{Q}_2$ of degree 8.

Exercise 8.6. When $G = \mathbb{Z}/8\mathbb{Z}$ and Σ is the restriction at 2 that $\mathbb{Q}_2^* \to G$ be unramified and surjective, write out the sum of Euler products for $H_{G,\Sigma}$ and see the cancellation explicitly.

It can be completely classified when local Galois representations to abelian groups do not occur as restrictions of global Galois representations (a convenient reference is [2, Chap. 10]). Over $\mathbb{Q}$, problems only occur at 2. However, the Wang counterexamples are in fact the nicest possible behavior that occurs when the Eq. (5) sum over $a \in A$ has multiple terms with the same pole. In general, say for restrictions only at odd places, there are still multiple terms that contribute to the sum, and the result is terrible looking probabilities, and lack of independence, whether one considers all $\mathbb{G}_{\mathbb{Q}} \to G$ or just surjective ones.

8.3 Counting by Conductor

One can ask all the same questions for counting abelian extensions, but instead of counting by discriminant, can count by conductor. The answers to the questions above are much nicer in this situation. (In fact, in [60] a notion of *fair* counting function is introduced which includes conductor, and, e.g., the product of ramified primes, for which all the same qualitative results hold as for conductor.) When counting by conductor, as in [60], the terms from subtracting non-surjective maps do not have any poles in the relevant region, so (**) is not a problem.

However, we know that some of the other terms in the Eq. (5) sum $a \in A$ must have the same rightmost pole as the $a = 1$ term, because counting by a different invariant is never going to eliminate Wang's counterexample. Amazingly, when counting by conductor, while there are indeed multiple terms with the same rightmost pole, they all differ by simple rational constants and can be combined simply. In a precise sense (described in [60, Sect. 1]), there are no obstructions to simple probabilities and independence other than the completely classified Wang counterexamples. All of the local behaviors that do occur from global extensions still occur with the same relative probabilities that one would expect from their contribution to the $a = 1$ term of the Dirichlet series—just the ones that are impossible occur with probability 0. When counting abelian extensions of $\mathbb{Q}$, since 2 is the only place that has Wang counterexamples, there is independence of local behaviors when counting by conductor.

Over other fields with multiple places dividing 2, there are sometimes local behaviors at two different places that are both possible, but not possible together. (Again, these belong to the completely classified Wang counterexamples.) So when $\mathbb{Q}$ is replaced one of these fields, using any counting function for that field, independence has to fail. However, as described explicitly below, when counting by conductor the simplest possible thing happens given this. We can build a model from the Dirichlet series, and then restrict to what is possible, and that gives the answer.

9 Abelian Results

These asymptotics of $N_G(X)$ for G abelian ($G \subset S_{|G|}$ in its regular representation) were determined completely for abelian Galois groups by Mäki [44]. Mäki [45] also determined the asymptotics of the number of extensions of $\mathbb{Q}$ with fixed abelian Galois group and bounded conductor. Wright [67] proved the analogous asymptotics for counting G-extensions for a fixed abelian G by discriminant over any number field or function field (in tame characteristic). Wright also showed that any local restrictions that occur at all (i.e., are not Wang counter examples) occur with positive asymptotic probability. Wright, however, noted that the probabilities seem quite complicated. Before the work of Mäki and Wright, there were many papers that worked on these questions for specific abelian groups. See [67] for an overview of this literature.

In [60], we give the asymptotics of the number of G-extensions with bounded conductor (or any fair counting function) for a finite abelian group G over any number field. We also give the constant in the asymptotic more explicitly than it appears in [45]. In [60], we also completely determine the probabilities of local conditions when counting G-extensions by conductor for some fixed abelian G. We also more carefully analyze the probabilities when counting by discriminant to prove that indeed they are as bad as Wright suspected.

9.1 Some Explicit Results

If we count abelian number fields by their conductor (in the sense of class field theory [51, Chap. VI, 6.4]), we can define $\mathbb{P}_{cond}$ analogously to our definition of $\mathbb{P}_{\mathrm{Disc}}$ in Sect. 1 above.

Theorem 9.1 (Wood [60]). *Let G be a finite abelian group in its regular permutation representation, and q be a fixed rational prime (not 2 if $|G|$ is even). Then for a random L with Galois group G, a fixed K with Galois group G, and a random rational prime p*

$$\mathbb{P}_{cond}(q \text{ splits into } r \text{ primes in } L \mid q \text{ unramified in } L) =$$
$$\mathbb{P}_p(p \text{ splits into } r \text{ primes in } K).$$

Taylor [54] first proved the result of Theorem 9.1 in the special case that $G = \mathbb{Z}/n\mathbb{Z}$, and assuming that $4 \nmid n$. Wright [67] proves an analog of Theorem 9.1 for discriminant probability in the case that $G = (\mathbb{Z}/p\mathbb{Z})^b$ for p prime, and for these G the discriminant is a fixed power of the conductor, and thus discriminant probability is the same as conductor probability. Theorem 9.1 relates the row probabilities to the column probabilities of the sort of big chart we made in Sect. 1. In fact in [60], the probabilities of all (ramified or unramified) local behaviors are determined.

Further it is shown $|G|$ is even and $p = 2$ the probabilities of splitting types that ever occur in a random G-extension that occur in the same proportions as they occur in the Chebotarev density theorem for a fixed extension and random prime. Of course, there will always be the contrast, seen already for quadratic extensions in Sect. 1, that for a fixed p and a random G-extension L, the prime p will be ramified with positive probability, while for a fixed number field K a random prime q is ramified with probability 0.

Further, in [60] it is shown that these local probabilities for splitting in a random G-extension are independent for different primes.

Theorem 9.2 (Wood [60]). *Let G be a finite abelian group in its regular permutation representation. For any finite set S of places of $\mathbb{Q}$ and any choice of local $\mathbb{Q}_v$-algebras T_v for $v \in S$, for a random L with Galois group G (counted by conductor), the events $L \otimes_{\mathbb{Q}} \mathbb{Q}_v \simeq T_v$ are independent.*

Wright [67] showed that all $\mathbb{Q}_v$-algebras that ever occur as $L \otimes_{\mathbb{Q}} \mathbb{Q}_v$ for a number field L with Galois group G occur with positive discriminant probability, but noted that the probabilities are apparently very complicated. The discriminant probability analog of Theorem 9.1 does not hold. If K is a fixed number field with Galois group $\mathbb{Z}/9\mathbb{Z}$ and p a random rational prime, then

$$\mathbb{P}_p(p \text{ splits completely in } K) = \frac{1}{9}.$$

However, we have the following.

Proposition 9.3 (Wood [60]). *Let $q = 2, 3, 5, 7, 11,$ or 13. Then for a random L with Galois group $\mathbb{Z}/9\mathbb{Z}$,*

$$\mathbb{P}_{\mathrm{Disc}}(q \text{ splits completely in } L \mid q \text{ unramified in } L) < \frac{1}{9}.$$

The situation for abelian extensions and their local behaviors is quite interesting over a base number field K other than $\mathbb{Q}$. Wang counterexamples occur only at primes dividing 2, but now there can be more than one such prime, and so these examples affect even independence. Given an abelian G, it is possible that the K_v-algebra T_v and the $K_{v'}$-algebra $T'_{v'}$ both occur from global G-extensions, but never occur simultaneously. This suggests that we should not expect T_v and $T'_{v'}$ to be independent events. However, given obstructions of this sort, which were completely determined in [57] (or see [2, Chap. 10]), we have the best possible behavior of the local probabilities. This is described in detail in [60, Sect. 1], but roughly if you build a model where local behaviors are predicted to have their heuristic probabilities (see Sect. 10), and then restrict to the combinations of local behaviors that ever occur, the model gives the correct predictions.

10 The Malle–Bhargava Principle

Malle [46, 47] has given a conjecture for Question 1.6 and Bhargava [12] has given some heuristics (stated as a question—when do these heuristics apply?) for the more refined Questions 1.11 that extend Malle's original conjecture. However, there are counterexamples to both the original conjecture (see Klüners [40]), and to Bhargava's more refined heuristics (e.g., some of the abelian counting results of [60] above), so we will refer to these conjectures/heuristics as a principle. (Though notably, there are not any known counterexamples to the weaker form of Malle's conjecture given in [46] that says $K_\Gamma X^{1/a(\Gamma)} \leq N_\Gamma(X) \leq K_{\Gamma,\epsilon} X^{1/a(\Gamma)+\epsilon}$ for a specified constant $a(\Gamma)$ and unspecified constants $K_\Gamma, K_{\Gamma,\epsilon}$.) An important open question is to even make a good conjecture about when exactly the principle should apply.

We now explain the principle, following [12] (similar, but not identical, heuristic reasoning is given by Malle [47]). Let $\Gamma \subset S_n$ be a permutation group. For each place p of $\mathbb{Q}$, let Σ_p be a set of continuous homomorphisms $G_{\mathbb{Q}_p} \to \Gamma$. Let Σ be the collection of these Σ_p. If Σ_p is all the maps $G_{\mathbb{Q}_p} \to \Gamma$ for all but finitely many p, we say Σ is nice. (We may want to call other Σ nice as well.) We define a Dirichlet series as an Euler product over places p

$$D_{\Gamma,\Sigma}(s) := C_\Gamma \prod_p \left(\frac{1}{\#\Gamma} \sum_{\rho_p \in \Sigma_p} (\operatorname{Disc} \rho_p)^{-s} \right)$$

for some as yet unspecified constant C_Γ, where the product is over all places p of $\mathbb{Q}$. Let $D_{\Gamma,\Sigma}(s) = \sum_{n \geq 1} d_n n^{-s}$.

For nice Σ, the principle predicts that the asymptotics of

$$\#\{\rho : G_{\mathbb{Q}} \to \Gamma \mid \Gamma \text{ surjective, } \operatorname{Disc} \rho < X, \rho_p \in \Sigma_p \text{ for all } p\}$$

in X are the same as the asymptotics of

$$\sum_{n=1}^{X} d_n$$

in X. (Equivalently, it predicts that the limit of their ratios is 1.) Perhaps stated another way, the principle would suggest that $D_{\Gamma,\Sigma}(s)$ and a Dirichlet series counting surjective $G_{\mathbb{Q}} \to \Gamma$ by discriminant would have the same rightmost pole and residue at that pole and analytic continuation just beyond the pole, so that the previous version of the principle would follow from Theorem 7.1.

10.1 Local Factors

To begin to digest this principle, we will consider a single local factor

$$\frac{1}{\#\Gamma} \sum_{\rho_p \in \Sigma_p} (\text{Disc}\,\rho_p)^{-s}.$$

Finitely many factors have $p|\#\Gamma$ (these are the terms where the local extension might be wild), but we see that these finitely many factors cannot introduce any poles to $D_{\Gamma,\Sigma}(s)$. So we expect the important analytic behavior of $D_{\Gamma,\Sigma}(s)$ to be present in just the tame factors, and for the rest of this section we assume $p \nmid \#\Gamma$.

Let $\mathbb{Q}_p^t$ be the maximal tame extension of $\mathbb{Q}_p$, and $G_{\mathbb{Q}_p}^t := \text{Gal}(\mathbb{Q}_p^t/\mathbb{Q}_p)$ be the tame quotient of the absolute Galois group of $\mathbb{Q}_p$. Let F be the free group on x and y with the relation

$$xyx^{-1} = y^p.$$

Let $\hat{F}$ be the profinite completion of F. Then we have an isomorphism

$$\hat{F} \simeq G_{\mathbb{Q}_p}^t,$$

where y goes to a topological generator of the inertia subgroup, and x goes to Frob (see, e.g., [52, Theorem 7.5.2]). Given a $y \in \Gamma$, let $c(y)$ be the number of orbits of y on $\{1, \ldots, n\}$, and let $d(y) = n - c(y)$. So

$$\sum_{\rho_p \in \Sigma_p} (\text{Disc}\,\rho_p)^{-s} = \sum_{\substack{x,y \in \Gamma \\ xyx^{-1} = y^p}} p^{-d(y)s}.$$

Given $y \in \Gamma$, how many x are there in Γ such that $xyx^{-1} = y^p$? If y and y^p are conjugate, then there are $\#\Gamma/\#\{\text{conj. class of } y\}$. If y and y^p are not conjugate, then there are no such x. Let $\sim$ denote "is conjugate to." Then we have

$$\sum_{\rho_p \in \Sigma_p} (\text{Disc}\,\rho_p)^{-s} = \sum_{\substack{y \in \Gamma \\ y \sim y^p}} \frac{\#\Gamma}{\#\{\text{conj. class of } y\}} p^{-d(y)s}.$$

We can see why we must include the factor $1/\#\Gamma$ to get a reasonable principle. From the $y = 1$ term, the above sum has constant term $\#\Gamma$, and so the $1/\#\Gamma$ factor is necessary so that we could even have a chance of the product over p converging in some right half-plane. Let Γ_p be the set of conjugacy classes of Γ of the form $[y]$ such that $y \sim y^p$. So

$$\frac{1}{\#\Gamma} \sum_{\rho_p \in \Sigma_p} (\text{Disc}\,\rho_p)^{-s} = \sum_{[y] \in \Gamma_p} p^{-d(y)s}.$$

Note that we always have plenty of primes $p \equiv 1 \pmod{\#\Gamma}$, so that every $y \sim y^p$, and Γ_p is simply the set of conjugacy classes of Γ. A nice special case is when $\Gamma = S_n$. For $p > n$, we have $y \sim y^p$ for every $y \in S_n$, so for every tame prime p in this case Γ_p is simply the set of conjugacy classes of Γ.

Computing these local factors in the wild cases is much more complicated, and some interesting phenomena arise (see [12, 38, 59, 66]).

10.2 Malle's Conjecture

This computation of the local factors leads to Malle's conjecture [46, 47] that

$$N_\Gamma(X) \sim K_\Gamma X^{1/a(\Gamma)}(\log X)^{b(\Gamma)}$$

for some constant K_Γ and $a = \min_{y \in \Gamma \setminus \{1\}} d(y)$ and where $b(\Gamma)$ can also be given explicitly. Klüners [40] has given a counterexample when $\Gamma = C_3 \wr C_2$, where the value of $b(\Gamma)$ is too small, or said another way, there are more extensions than the conjecture predicts by a $\log X$ factor. Turkelli, by analogy with heuristics on the function field side, has given a "corrected" version of Malle's conjecture that takes into account Klüners counterexample [56] to give a different prediction for the log factor.

The weak conjecture of Malle [46] says $K_\Gamma X^{1/a(\Gamma)} \leq N_\Gamma(X) \leq K_{\Gamma,\epsilon} X^{1/a(\Gamma)+\epsilon}$ for some constants $K_\Gamma, K_{\Gamma,\epsilon}$. There are no known counterexamples to this conjecture, and it is known in a wide variety of cases (see below).

10.3 Independence and Local Behaviors

The principle suggests that when counting extensions with Galois group Γ, for any finite set of places, local conditions at those places should be independent. We can see this from the mechanics of how the Tauberian theorem applies. Changing a finite number of the factors in the Euler product for $D_{\Gamma,\Sigma}$ just changes the constant in the asymptotic by the product of the ratios of the new factors over the old factors, all evaluated at the rightmost pole.

More precisely, let $a(\Gamma)$ be as above. Recall, the definition of $D_{\Gamma,\Sigma}(s)$ above as an Euler product built as a heuristic model for the Dirichlet series counting extensions with Galois group Γ and local behaviors Σ.

Exercise 10.1. Show that $D_{\Gamma,\Sigma}(s)$ converges in $\Re(s) > 1/a(\Gamma)$.

Exercise 10.2. Show that $D_{\Gamma,\Sigma}(1/a(\Gamma))$ does not converge.

So, if there is a meromorphic continuation past $\Re(s) > 1/a(\Gamma)$, we expect a pole at $1/a(\Gamma)$. (To show more precisely that such a pole had to exist, we could find

$\lim_{s\to+1/a(\Gamma)} D_{\Gamma,\Sigma}(s)$.) So, supposing that $D_{\Gamma,\Sigma}$ has a meromorphic continuation to $\Re(s) \geq 1/a(\Gamma)$ (which one can actually show for every Γ), if D_Γ is the Dirichlet series where all local behaviors are allowed with coefficients d_n, and Σ makes specifications at a finite set S of places, and $D_{\Gamma,\Sigma}$ has coefficients $d(\Sigma)_n$, then we have

$$\lim_{X\to\infty} \frac{\sum_{n=1}^X d(\Sigma)_n}{\sum_{n=1}^X d_n} = \prod_{p\in S} \frac{\frac{1}{\#\Gamma}\sum_{\rho_p\in\Sigma_p}(\operatorname{Disc}\rho_p)^{-1/a(\Gamma)}}{\frac{1}{\#\Gamma}\sum_{\rho_p:G_{\mathbb{Q}_p}\to\Gamma}(\operatorname{Disc}\rho_p)^{-1/a(\Gamma)}}.$$

Exercise 10.3. Show this (assuming a continuation to $\Re(s) \geq 1/a(\Gamma)$ with no other poles except at $1/a(\Gamma)$).

So we see the probabilities of local behaviors at different places would be independent and given by relatively simple fractions (with all the principles/assumptions above).

10.4 When the Principle Holds

It is an important open question to even make a good guess as to when the above principle holds. The work of Mäki [44] (and also Wright [67]) shows that Malle's conjecture holds when Γ is abelian in its regular representation, i.e., the order of magnitude is right when counting all Γ extensions. Moreover, Wright shows that the order of magnitude is right for any local condition that is not a Wang counterexample (in which case the principle's prediction must be wrong since the actual count is 0). However, we saw above that [60] shows that the general principle fails when Γ is not abelian with prime exponent because the independence of local conditions fails. (But all can be recovered if one counts by conductor instead of discriminant!) Somewhat orthogonal to the focus of the principle above, but also interesting, is work of Cohen, Diaz y Diaz, and Oliver [22] that finds the constant K_Γ very explicitly when Γ is cyclic of prime degree.

For $\Gamma = S_3$, it is a theorem of Davenport and Heilbronn [25], that we will discuss below, that the entire principle holds. Moreover, when $\Gamma = S_3 \subset S_6$ via the regular representation (i.e., counting Galois sextic S_3 fields) Bhargava and the author [13] have shown that the entire principle holds.

For $\Gamma = S_4, S_5$, theorems of Bhargava [9, 11] show that the entire principle holds. For $\Gamma = D_4 \subset S_4$, Cohen, Diaz y Diaz, and Oliver [21] have shown that Malle's conjecture holds, but the evidence suggests that the entire principle does not hold. (In fact, Cohen [17] counts D_4 extensions with local conditions at infinity, from which it could probably be seen explicitly that the entire principle does not hold.) Given the above story with abelian Γ, perhaps the right question is how can we count extensions so that the principle holds, and in [59] a different way of counting D_4 extensions is suggested for which the principle might hold. Klüners [42] has shown that Malle's conjecture holds for groups $C_2 \wr H$, under mild assumptions on the count of H-extensions.

As mentioned above, for some groups even Malle's conjecture fails. (See [40] and [56].) However, the weak conjecture of Malle is proven in many cases and has no known counterexamples. Besides those mentioned above it is also known for nilpotent groups in their regular representation by Klüners and Malle [43].

10.5 Other Base Fields

Of course, in these questions we could replace $\mathbb{Q}$ with any number field or function field. As suggested by Bhargava [12], the principle above could be formulated over any global field, and in particular it refines Malle's conjecture [46, 47] which is stated over an arbitrary number field. Of the work above, [67] is over any number field or function field of tame characteristic and [22, 42, 43, 60] are over any number field. Also when $\Gamma = \mathbb{Z}/\ell\mathbb{Z}$ and the base field is a rational function field, the counting has been done in [4], and in particular cases relevant to the application at hand there it is shown that the full principle with local conditions also holds. The work [25] has been generalized to any number field or function field of tame characteristic by Datskovsky and Wright in [23], and the result of [13] is given over any number field. As far as the current author is aware, the results [9, 11, 21] are known only over $\mathbb{Q}$.

11 Davenport–Heilbronn

In this section, we will discuss how to answer Question 1.6 for $\Gamma = S_3$, i.e., to count non-Galois cubic number fields. This is originally a result of Davenport and Heilbronn [25]. Since we know from Cohn [16] (see Eq. (4)) that

$$N_{\mathbb{Z}/3}(X) \sim cX^{1/2}$$

and the results of Davenport and Heilbronn [25] will say that

$$N_{S_3}(X) \sim c'X,$$

we may equally well consider counting all cubic number fields. We will outline a modern proof (largely following [15]) of this statement, which has the same main ideas as the original proof, but with improvements and simplifications. There is an excellent exposition of such a modern proof in Sects. 2–5 and 8 of the paper of Bhargava, Shankar, and Tsimerman [15] (which, in the other sections, proves a *secondary* term for this asymptotic count—see also [55] where the secondary term is proven with very different methods). So in these notes we will mostly emphasize aspects that are complementary to what is given in that exposition, and might be best read along with [15]. There is also a do-it-yourself exposition of the proof in the Arizona Winter School 2014 problem set, as a series of problems that should complement this article nicely.

11.1 The Parametrization

We begin by considering *cubic rings*, which are commutative rings whose underlying additive groups structure is isomorphic to $\mathbb{Z}^3$ (equivalently, locally free rank 3 $\mathbb{Z}$-algebras, or a finite, flat degree 3 cover of $\operatorname{Spec}\mathbb{Z}$). We will be able to count cubic rings, and then we will specialize to cubic rings that are domains and maximal. The maximal cubic domains correspond exactly to the rings of integers in cubic number fields. We give a parametrization of cubic rings originally due to Delone and Faddeev [27] (and refined by Gan, Gross, and Savin [32]).

Let R be a cubic ring.

Exercise 11.1. Show that 1 generates a direct summand of R.

Let $1, W, T$ be a $\mathbb{Z}$ basis of R. Since

$$WT = q + rW + sT,$$

for some $q, r, s \in \mathbb{Z}$, we can take $\omega = W - s$ and $\theta = T - r$ and have $1, \omega, \theta$ a $\mathbb{Z}$ basis of R with $\omega\theta \in \mathbb{Z}$. We call such a basis a *normalized* basis. Next, we write down a multiplication table for a normalized basis:

$$\begin{aligned} \omega\theta &= n \\ \omega^2 &= m - b\omega + a\theta \\ \theta^2 &= \ell - d\omega + c\theta, \end{aligned} \tag{6}$$

where $n, m, \ell, a, b, c, d \in \mathbb{Z}$. However, not all values of n, m, ℓ, a, b, c, d are possible.

Exercise 11.2. Show that the associative law exactly corresponds to the equations

$$n = -ad \quad m = ac \quad \ell = -bd. \tag{7}$$

That means that not only is Eq. (7) necessarily true for any cubic ring R with normalized basis $1, \omega, \theta$, but also if we have a free rank 3 $\mathbb{Z}$-module with generators $1, \omega, \theta$, we can put a commutative, associative multiplication structure on the module using Eqs. (6) and (7). So rank 3 rings with a choice of normalized basis are parametrized by $\mathbb{Z}^4$ using (a, b, c, d). We can package an element $(a, b, c, d) \in \mathbb{Z}^4$ as a binary cubic form $ax^3 + bx^2y + cxy^2 + dy^3$, and thereby such forms parametrize rank 3 rings with a choice of normalized basis. Given a form f, let R_f be the associated cubic ring with a choice of normalized basis. One important fact is that the construction above preserves discriminants (see [32] or any of the other references on this parametrization).

Exercise 11.3. Find a binary cubic form associated with $\mathbb{Z}[\sqrt[3]{2}]$. Find another one (using a different choice of basis). Find a binary cubic form associated with $\mathbb{Z}[i] \oplus \mathbb{Z}$ (with component-wise multiplication). Find a binary cubic form associated with $\mathbb{Z} \oplus \mathbb{Z} \oplus \mathbb{Z}$. What cubic ring is associated with the zero form?

A choice of normalized basis of R is equivalent to a choice of $\mathbb{Z}$ basis of $R/\mathbb{Z}$. The action of $\mathrm{GL}_2(\mathbb{Z})$ on bases of $R/\mathbb{Z}$ gives a $\mathrm{GL}_2(\mathbb{Z})$ action on $\mathbb{Z}^4$, such that the orbits are in bijection with cubic rings as given above. Because of the normalization, this action is slightly annoying to work out by hand, though perhaps not so bad to at least see it is linear. The action turns out to be (almost) the 4-dimensional representation of $\mathrm{GL}_2(\mathbb{Z})$ on binary cubic forms. Let $f(x,y) = ax^3 + bx^2y + cxy^2 + dy^3$. Let $g \in \mathrm{GL}_2(\mathbb{Z})$. Then we can let $\mathrm{GL}_2(\mathbb{Z})$ act on binary cubic forms via

$$(gf)(x,y) = \frac{1}{\det(g)} f((x,y)g), \tag{8}$$

where $(x,y)g$ is the multiplication of a row vector by a matrix on the right. This action exactly translates into the action on the parameters (a,b,c,d) of cubic rings given by action on the choice of basis of $R/\mathbb{Z}$. There are several ways to see this more intuitively, we will see one later. So, we will face the problem of counting $\mathrm{GL}_2(\mathbb{Z})$ classes of binary cubic forms, with the action given in Eq. (8).

An important feature of the parametrization is that in fact many important properties of the cubic ring can be read off from the associated binary cubic form. One way to see this is to understand that the parametrization is more general than what is written above over $\mathbb{Z}$. We will in fact see that it is far more general.

Let B be a commutative ring (which will replace $\mathbb{Z}$). A cubic B-algebra is a commutative B-algebra which is locally free rank 3 as a B-module. (There are two common notions of locally free, an algebraic one where "locally" means at localizations at prime ideals, and a geometric one where "locally" means in Zariski opens of $\operatorname{Spec} B$, but the rank 3 condition ensures that these two notions are the same in this case.) If R is a cubic B-algebra, then $\operatorname{Spec} R$ is a finite, flat, degree 3 cover of $\operatorname{Spec} B$, and conversely, every finite, flat degree 3 cover is $\operatorname{Spec} R$ for a cubic algebra R. Similarly, for any scheme S, we can define a *cubic cover* of S to be a finite, flat, degree 3 cover of S, or equivalently a locally free rank 3 $\mathcal{O}_S$-algebra (which we call a *cubic* $\mathcal{O}_S$*-algebra*).

Over a scheme S, we define a *binary cubic form* to be a locally free rank 2 $\mathcal{O}_S$-module V (i.e., a rank 2 vector bundle on S) and a global section $f \in H^0(S, \operatorname{Sym}^3 V \otimes \wedge^2 V^*)$, where all tensors and exterior powers are over $\mathcal{O}_S$ and for an $\mathcal{O}_S$-module W, we write W^* for the dual $\mathcal{O}_S$ module $\mathcal{H}om(W, \mathcal{O}_S)$. (These might be more properly called "twisted" binary cubic forms because of the twist by the locally free rank 1 $\wedge^2 V^*$, but they are the only ones we will talk about, so we will leave out the "twisted.") Over a ring B then (applying the above definition in the case $S = \operatorname{Spec} B$), we see that a binary cubic form is a locally free rank 2 B-module V, and an element $f \in \operatorname{Sym}^3 V \otimes \wedge^2 V^*$. An isomorphism of binary cubic forms is given by an isomorphism $V \to V'$ that takes $f \to f'$.

Given a binary cubic form (V,f) over S, we can construct a cubic cover of S as follows. This construction is due to Deligne [26] (see [63, Sects. 2.3, 2.4, 3] for a generalization of this construction from binary cubic forms to binary n-ic forms for any n). Let $\mathbb{P}(V) = \operatorname{Proj} \operatorname{Sym} V$, so $\pi : \mathbb{P}(V) \to S$ is a $\mathbb{P}^1$ bundle over S.

If f is not a zero-divisor, then f cuts out a codimension 1 subscheme $S_f \subset \mathbb{P}(V)$. Then $\pi_*(\mathcal{O}_{S_f})$ is a locally free rank 3 $\mathcal{O}_s$-algebra. This construction does not work when f is identically the 0 form.

Let $\mathcal{O}(k)$ denote the usual sheaf of $\mathbb{P}(V)$. We then have a complex of sheaves

$$C_f : \mathcal{O}(-3) \otimes \pi^* \wedge^2 V \xrightarrow{f} \mathcal{O},$$

(in degrees -1 and 0). If f is not a zero-divisor, then the map above is injective, so this complex is quasi-isomorphic to

$$0 \to \mathcal{O}/f(\mathcal{O}(-3) \otimes \pi^* \wedge^2 V),$$

and the term on the right is just $\mathcal{O}_{S_f}$. The hypercohomology

$$H^0 R\pi_*(C_f)$$

has a product given by the product on the complex (which is a Koszul complex so has that natural product) and inherits the structure of a cubic $\mathcal{O}_s$-algebra (see [63, Sects. 2.4, 3] for more details on this structure, and this entire construction). When f is not a zero-divisor, by the quasi-isomorphism above, we see that this agrees with $\pi_*(\mathcal{O}_{S_f})$.

Theorem 11.4 (Theorem 2.4 of [63]). *When V is a free $\mathcal{O}_S$-module (e.g., this is always the case when B is a P.I.D. and $S = \operatorname{Spec} B$), this geometric/hypercohomological construction of a cubic $\mathcal{O}_S$-algebra from a binary cubic form $(\mathcal{O}_S^{\oplus 2}, f)$ agrees with R_f constructed from Eqs.* (6) *and* (7) *above.*

We restrict to V a free $\mathcal{O}_S$-module only because that is the case for which the first construction is defined.

Theorem 11.5 (Corollary 4.7 of [63]). *Given a scheme S, the geometric/hypercohomological construction above gives an isomorphism of categories between the category of binary cubic forms over S (where the morphisms are isomorphisms) and the category of cubic $\mathcal{O}_S$-algebras (where the morphisms are isomorphisms). Thus, over a scheme S, the construction gives a bijection between isomorphism classes of cubic algebras and isomorphism classes of binary cubic forms. If a cubic algebra R corresponds to a binary cubic form (V,f), then as $\mathcal{O}_S$-modules, we have $R/\mathcal{O}_S \simeq V^*$. The functor described above from binary cubic forms to cubic $\mathcal{O}_s$-algebras commutes with base change in S.*

(See also [53] which proves an isomorphism of categories using a rigidification by a choice of basis and the construction from Eqs. (6) and (7) above.)

We will now unpack some of the features of this result. Let B be a principal ideal domain. Then the only choice of a locally free rank 2 B-module V is $V = B^2$, and we will take generators x and y. Let x^*, y^* be the associated dual basis of $V^* = \operatorname{Hom}(V, B)$. Then $\operatorname{Sym}^3 V$ is a free rank 4 B-module generated by

x^3, x^2y, xy^2, y^3. The element $x^* \wedge y^*$ gives an isomorphism $B \simeq \wedge^2 V^*$. So then a binary cubic form over B is a choice $a, b, c, d \in B$ for

$$(ax^3 + bx^2y + cxy^2 + dy^3) \otimes (x^* \wedge y^*).$$

However, since we had to pick a basis x, y of B, the choice of a, b, c, d is only well-defined up to the $\mathrm{GL}_2(B)$ action on x, y.

Exercise 11.6. When B is a PID, show that this notion of isomorphism classes of binary cubic forms over B agrees with the notion of $\mathrm{GL}_2(B)$ classes of binary cubic forms that was given above over $\mathbb{Z}$ (but could be just as well interpreted over B).

Exercise 11.7. Suppose B is the ring of integers of a number field but is not a PID. Show that our notion of isomorphism classes of binary cubic forms over B does not agree with the notion of $\mathrm{GL}_2(B)$ classes of binary cubic forms given that was given above over $\mathbb{Z}$ (but could be just as well interpreted over B).

The twist by $\wedge^2 V^*$ may look somewhat irrelevant, especially if one is working only over $\mathbb{Z}$, where the determinant of an element in $\mathrm{GL}_2(\mathbb{Z})$ can only be ± 1. Even over B, for $\lambda \in B^*$ the matrix λI multiplies each of a, b, c, d by λ. (In particular, this remark shows that the $\mathrm{GL}_2(B)$ classes of binary cubic forms over B would be the same even if we didn't take the twisted action.) However, it is actually quite important. Without the twist, the GL_2 action on the form does not agree with the GL_2 action of the choice of basis of R/B. This agreement is important because we will need to use the fact that the automorphism groups of corresponding objects agree, which comes from the equivalence of categories statement above or can be seen more concretely from the fact that the GL_2 actions agree.

For any integer ℓ, over a general base scheme S, for $f \in H^0(S, \mathrm{Sym}^3 V \otimes (\wedge^2 V)^\ell)$ one can make a cubic algebra via an analog of the above construction [63, p. 219]. (Note that since $(\wedge^2 V)^{-1} = (\wedge^2 V)^*$ the above construction is the case $\ell = -1$.) This always gives a functor from (ℓ-twisted) binary cubic forms to cubic algebras over S that commutes with base change [63, p. 219]. However only in the cases $\ell = -1$ and $\ell = -2$ is this functor an isomorphism of categories. For other ℓ, even over $S = \mathbb{P}^1$ one can see that not all cubic algebras arise (see the last paragraph of page 227 of [63]).

Over a field K, we see that a non-zero binary cubic form f just cuts out a degree 3 subscheme of $\mathbb{P}^1_K$. If $a \neq 0$ (and if the form is non-zero, we can always change basis so that this is the case unless $K = \mathbb{F}_2$), then the cubic K-algebra is $K[\alpha]/(a\alpha^3 + b\alpha^2 + c\alpha + d)$.

Exercise 11.8. Show that this agrees with each of the above constructions (the construction originally given over $\mathbb{Z}$ in terms of the multiplication table, and the geometric description involving the global functions on the scheme cut out by the form).

From this, using the base change from $\mathbb{Z}$ to $\mathbb{Q}$, it follows, over $\mathbb{Z}$, that the cubic ring R_f (associated with f) is a domain if and only if f is irreducible (over $\mathbb{Q}$).

In this case, we call f *irreducible*, and otherwise we call it *reducible*. We have that f is irreducible if and only if R_f is an order in a cubic number field. The number field then is given by the dehomogenized version of the binary cubic form.

We can use the base change from $\mathbb{Z}$ to $\mathbb{Z}_p$ to understand which forms associated with orders in cubic fields are associated with orders maximal at p (meaning they have index relatively prime to p in the maximal order). In fact, maximality at p is determined by the class of the form modulo p^2. See [15, Lemma 13] for more details on this.

One can, in fact, see more explicitly what the geometric construction gives over $\mathbb{Z}$ (base change from $\mathbb{Z}$ to $\mathbb{Z}_p$ also simplifies matters in understanding this, see also [63, Sect. 2]). If f is non-zero, then it cuts out a 1-dimensional subscheme S_f of $\mathbb{P}^1_{\mathbb{Z}}$. If for some prime p, we have $p|a, b, c, d$, then S_f has a vertical fiber over p, and, in particular, S_f is not finite, flat degree 3 over $\mathbb{Z}$. However, the global functions $H^0(S_f, \mathcal{O}_{S_f})$ are still a cubic ring, and this is the associated cubic ring R_f. So $\operatorname{Spec} R_f$ is a finite, flat, degree 3 cover (in particular, we have gotten rid of the vertical fibers). We have a map $S_f \to \operatorname{Spec} R_f$, in which $\operatorname{Spec} R_f$ is the universal affine variety that S_f maps to. Over p, the scheme $\operatorname{Spec} R_f$ has a singularity that cannot be embedded in $\mathbb{P}^1_{\mathbb{Z}}$. If $(a, b, c, d) = 1$, we call the form *primitive*, and then in fact S_f is affine and is the cubic cover of $\operatorname{Spec} \mathbb{Z}$ associated with f.

We can use base change from $\mathbb{Z}$ to $\mathbb{Z}/p\mathbb{Z}$ to see that R_f/p is the cubic ring given by reducing f modulo p. Over $\mathbb{Z}/p\mathbb{Z}$, there are not so many options for f up to isomorphism. It either is 0, has three distinct roots over $\mathbb{Z}/p\mathbb{Z}$, has 1 single root over $\mathbb{Z}/p\mathbb{Z}$, has 1 double roots, and 1 single root over $\mathbb{Z}/p\mathbb{Z}$, or has 1 triple root over $\mathbb{Z}/\mathbb{Z}$, and in each of these cases we can read off the cubic algebra from the analysis for a general field K above. So in the case when R_f is a maximal order, this lets us read off R_f/p and thus the splitting type of p in R_f from the splitting of f modulo p.

The parametrization of cubic rings given above seems so simple, that at first glance one might think that one could parametrize all rank n rings this way. However, things get much more complicated quickly. As mentioned above, the construction of a rank n ring from a binary n-ic form works for all n [63], but for $n > 3$ not all rings are obtained this way. The paper [63] proves that the rings obtained this way are the ones with certain special kinds of ideal classes. For more reading on parametrizations of rings and ideal classes in rings, see [6–8, 10, 28, 61–65].

11.2 Fundamental Domain

Let $V_{\mathbb{Z}}$ be the space of binary cubic forms over $\mathbb{Z}$ (in the first sense, so $V_{\mathbb{Z}} = \mathbb{Z}^4$), and $V_{\mathbb{R}}$ be the space of binary cubic forms over $\mathbb{R}$. Let F be a fundamental domain for the action of $\mathrm{GL}_2(\mathbb{Z})$ on binary cubic forms. Let F_X be the intersection of F with $\{v || \operatorname{Disc}(v)| < X\}$. Given the above parametrization, we would like to count $\mathrm{GL}_2(\mathbb{Z})$ orbits of $V_{\mathbb{Z}}$ of discriminant up to X, with additional conditions we will come back to later. This is the same as counting lattice points of $V_{\mathbb{Z}}$

in F_X, asymptotically in X. To solve this problem, we will use geometry of numbers techniques. From the beginning, we should note that F_X is *not* expanding homogeneously in X. The $|\operatorname{Disc}(v)| < X$ boundaries are homogeneously expanding, but the boundaries of F are not expanding as X grows. So the geometry of numbers is more complicated than the first examples ones sees in number theory.

Davenport [24] writes down explicit equations for a fundamental domain F using Hermite's reduction theory of binary cubic forms. Let

$$A = b^2 - 3ac, \quad B = bc - 9ad, \quad \text{and } C = c^2 - 3bd. \tag{9}$$

Then the points in F are those satisfying

$$-A < B \leq A < C \text{ or } 0 \leq B \leq A = C.$$

These are fairly simple quadratic inequalities defining F. Davenport uses this fundamental domain to (successfully) count binary cubic forms, and, eventually, with Heilbronn [25], to count cubic fields. However, we are going to use the approach of Bhargava [9] (in his work counting quartic fields) to find a different fundamental domain.

Let $\mathcal{F}$ be Gauss's fundamental domain for $\mathrm{GL}_2(\mathbb{Z})\backslash \mathrm{GL}_2(\mathbb{R})$. We can write

$$\mathcal{F} = \{na'k\lambda \mid n \in N'(a'), a' \in A', k \in K, \lambda \in \Lambda\},$$

where

$$N'(a') = \left\{\begin{pmatrix} 1 & \\ n & 1 \end{pmatrix} : n \in \nu(a')\right\}, \quad A' = \left\{\begin{pmatrix} t^{-1} & \\ & t \end{pmatrix} : t \geq \sqrt[4]{3}/\sqrt{2}\right\},$$

$$\Lambda = \left\{\begin{pmatrix} \lambda & \\ & \lambda \end{pmatrix} : \lambda > 0\right\},$$

and K is as usual the (compact) real special orthogonal group $\mathrm{SO}_2(\mathbb{R})$; here $\nu(a')$ is the union of either one or two subintervals of $[-\frac{1}{2}, \frac{1}{2}]$ depending only on the value of $a' \in A'$. Furthermore, if a' is such that $t \geq 1$, then $\nu(a') = [-\frac{1}{2}, \frac{1}{2}]$. (There is an unfortunate coincidence that a is the first coefficient of our binary cubic form and a' a diagonal matrix. This however is consistent with the literature, except for when the literature calls them both a.)

To recognize this fundamental domain as something we all know and love (the fundamental domain for $\mathrm{SL}_2(\mathbb{Z})$ on the upper half-plane), consider the action on shapes of lattices in $\mathbb{C}$, realized as elements of the upper half-plane $\mathbb{H}$. (So τ in the upper half-plane corresponds to the shape of the lattice $\mathbb{Z} \oplus \tau\mathbb{Z}$.) Let $\mathrm{GL}_2^+(\mathbb{R})$ be the elements of positive discriminant.

Exercise 11.9. Show that a fundamental domain for $\mathrm{SL}_2(\mathbb{Z})\backslash \mathrm{GL}_2^+(\mathbb{R})$ gives a fundamental domain for $\mathrm{GL}_2(\mathbb{Z})\backslash \mathrm{GL}_2(\mathbb{R})$. Conversely, show that a fundamental domain for $\mathrm{GL}_2(\mathbb{Z})\backslash \mathrm{GL}_2(\mathbb{R})$ that is contained in $\mathrm{GL}_2^+(\mathbb{R})$ gives a fundamental domain for $\mathrm{SL}_2(\mathbb{Z})\backslash \mathrm{GL}_2^+(\mathbb{R})$.

So we consider the action of $\mathrm{GL}_2^+(\mathbb{R})$ on the upper half-plane $\mathbb{H}$ given by

$$\left(\begin{smallmatrix} A & B \\ C & D \end{smallmatrix}\right) \circ \tau = \frac{D\tau + C}{B + A\tau}$$

for $\tau \in \mathbb{H}$. (This is the conjugate of the usual action by $\left(\begin{smallmatrix} 0 & 1 \\ 1 & 0 \end{smallmatrix}\right)$.) We see that Λ is in the kernel of this action. In the quotient $\mathrm{GL}_2^+(\mathbb{R})/\Lambda$, the group K is the stabilizer of $i \in \mathbb{H}$. (If we consider not just shapes of lattices up to homothety, but actual lattices, then the lattices are equivalent to binary quadratic forms, and the $\mathrm{GL}_2^+(\mathbb{R})$-action can be lifted in the obvious way. The binary quadratic form x^2+y^2 corresponds to the square lattice shape, which is represented by the point $i \in \mathbb{H}$. Then $\mathrm{Stab}_{\mathrm{GL}_2^+(\mathbb{R})}(x^2 + y^2) = \mathrm{SO}_2(\mathbb{R}) = K$.) So, we have that $\mathcal{F} \circ i = \{na \circ i \mid n \in N'(a), a' \in A'\}$. This is the familiar fundamental domain for the action of $\mathrm{SL}_2(\mathbb{Z})$ on $\mathbb{H}$.

The matrix $a' \in A'$ scales points in the upper half-plane by t^2. The group $N'(a')$ translates points in the upper half-plane to the left and right by at most $1/2$ in each direction. So $na \circ i = t^2 i + n$. Now we see that the inequalities defining $N'(a')$ and A' look familiar from our usual fundamental domain for $\mathrm{SL}_2(\mathbb{Z})$ on the upper half-plane.

The action of $\mathrm{GL}_2(\mathbb{R})$ on $V_{\mathbb{Z}}$ has 2 4-dimensional orbits, corresponding to the cubic rings $\mathbb{R}^3$ and $\mathbb{R} \oplus \mathbb{C}$. These cover all the points of $V_{\mathbb{Z}}$ with non-zero discriminant. Let V^+ be the orbit where the points have positive discriminant (corresponding to $\mathbb{R}^3$) and V^- be the orbit where the points have negative discriminant (corresponding to $\mathbb{R} \oplus \mathbb{C}$). Let $v \in V^-$. Then, roughly, the idea is that $\mathcal{F}v$ is a fundamental domain for the action of $\mathrm{GL}_2(\mathbb{Z})$ on V^-. We can do everything similarly for V^+, so for now we will restrict our attention to V^-.

The presence of stabilizers means that $\mathcal{F}v$ is not a fundamental domain, but rather a "multi-fundamental domain." More precisely, we consider $\mathcal{F}v$ as the multiset $\{fv \mid f \in \mathcal{F}\}$.

Exercise 11.10. For $x \in V_{\mathbb{Z}}^-$, show that the number of times an element of $\mathrm{GL}_2(\mathbb{Z})x$ appears in $\mathcal{F}v$ is

$$\frac{\# \mathrm{Stab}_{\mathrm{GL}_2(\mathbb{R})}\, x}{\# \mathrm{Stab}_{\mathrm{GL}_2(\mathbb{Z})}\, x}.$$

We note that for $x \in V_{\mathbb{Z}}^-$, we have $\mathrm{Stab}_{\mathrm{GL}_2(\mathbb{R})}\, x = \mathrm{Aut}_{\mathbb{R}}(\mathbb{R} \oplus \mathbb{C}) = \mathbb{Z}/2\mathbb{Z}$. Also, $\mathrm{Stab}_{\mathrm{GL}_2(\mathbb{Z})}\, x$ are the automorphisms of the corresponding cubic ring, which will be trivial if the ring is a domain (in V^- there are no cyclic cubic fields), and will be $\mathbb{Z}/2\mathbb{Z}$ if the cubic ring is a subring of $K \oplus \mathbb{Q}$ for an imaginary quadratic field $\mathbb{Q}$. So $\mathcal{F}v$ covers all the points we are interested in twice, which is just as good as a fundamental domain (as long as we divide by 2 in the end).

Let $\mathcal{R}_X(v) = \mathcal{F}v \cap \{w \mid 1 \leq |\mathrm{Disc}(w)| < X\}$.

11.3 Geometry of Numbers

To count points in this very problem, Davenport [24] proved a general result on counting points in regions defined by polynomial inequalities. Let R be a region in $\mathbb{R}^n$ and let $N(R)$ be the number of lattice points in R. We would like to say

$$N(R) \approx \text{Vol}(R),$$

and the real content of such a statement is an estimate for the error

$$|N(R) - \text{Vol}(R)|.$$

We will now imagine what can contribute to such an error. In 2 dimensions, clearly, the perimeter of R is closely tied to this error. Consider a square that is $N+\epsilon$ by $M+\epsilon$. It may have as many as $(N+1)(M+1)$ or as few as NM lattice points, compared to its area which is near NM. Considering a $1/N \times 1/N$ box, which may have 1 or 0 points, we see we need a 1 in the error as well. In fact in two dimensions

$$|N(R) - \text{Vol}(R)| \leq 4(L+1),$$

where L is the length of the boundary of R.

In $\mathbb{R}^n$, a similar construction with a box shows that the volumes of the projections onto $\mathbb{R}^{n-1}$ by dropping a coordinate are related to the error $|N(R) - \text{Vol}(R)|$. A box which is approximately $N_1 \times \cdots \times N_n$ will have volume $N_1 \ldots N_n$, but could have $N_1 \cdots N_n$ points or could have

$$N_1 \cdots N_n + N_2 \cdots N_n$$

points. However, these $n-1$ dimensional projections are not enough. Consider a $1/N \times 1/N \times N$ box. This may have as few as 0 points or as many as N. However the 2-dimensional projections have size $1, 1, 1/N^2$, so are not large enough to account for this discrepancy, even allowing for constant factors. However, there is a 1-dimensional projection of size N. Davenport shows that these volumes of projections are indeed enough to bound the error.

Lemma 11.11 (Davenport [24]). *Let R be a closed, bounded region in $\mathbb{R}^n$ defined by k polynomials of degree at most d, and let $h = kd/2$. Then*

$$|N(R) - \text{Vol}(R)| \leq \sum_{m=0}^{n-1} h^{n-m} R_m,$$

where R_m is the sum of the m-dimensional volumes of the projections of R onto any of the coordinate spaces obtained by forgetting some $n-m$ coordinates, and $R_0 = 1$.

We also write this as

$$|N(R) - \text{Vol}(R)| = O(\text{Vol}(\text{Proj}(R)), 1),$$

where the constant in the big O notation depends on the dimension n of R and the number and maximum degree of the defining inequalities. The notation $\text{Proj}(R)$ is supposed to denote all the projections, and we write the 1 as not to forget the 0-dimensional projection.

11.4 Averaging

One advantage of taking a fundamental domain $\mathcal{F}v$ as we have done above (instead of Davenport's explicit F) is that this generalizes more easily to finding fundamental domains in more complicated situations, such as those Bhargava encounters when counting quartic and quintic fields [9, 11]. Another advantage is we can use an averaging technique due to Bhargava (in [9], where he counts quartic fields) which significantly improves the geometry of numbers results. (Yet another advantage is that it will be simpler to compute the volume of our fundamental domain.)

The idea of averaging is that $\mathcal{F}v$ is a fundamental domain for any choice of v, so in fact we have many fundamental domains. Now since they are all fundamental domains, the number of lattice points in $\mathcal{R}_X(v)$ does not depend on v (for $v \in V^-$). So we will average over many such v, and obtain the same count as in one $\mathcal{F}v$.

At first, it might seem counterintuitive that this can help. The key is that the answer is the same, but we have more tools to estimate the averaged number of points. It sometimes helps to first think about a discrete averaging scenario. Image a fundamental domain for the action of $\mathbb{Z}$ on $\mathbb{R}^2$ by addition in the y coordinate, and we would like to count lattice points with $|x| \leq X$, asymptotically in X (and we are ignoring constant factors). This is like counting lattice points in a $1 \times X$ box, so the volume is order X, but there is a projection of size X. We cannot get a useful result from Davenport's geometry of numbers Lemma 11.11 in this case, so we have to use more information. Given the volume of the box and the size of its projections, it could have as many as 0 points or as many as X (up to constants). Suppose instead we take X fundamental domains, all stacked on top of each other to form an $X \times X$ box. Now, the fact that they make a nice box exactly (e.g., each was, say, closed on the bottom and open on the top, and they perfectly fit together) essentially comes from the fact that they are fundamental domains. So we get $X^2 + O(X)$ points in our X fundamental domains, and so $X + O(1)$ in a single fundamental domain. We could not have gotten an error term this small without using the fact that our region was a fundamental domain.

Now continuous averaging can be even more powerful. We will average over v in some bounded compact region B. You should think of B as a ball. So for each $v \in B$, we get some number of lattice points in $\mathcal{R}_X(v) = \mathcal{F}v \cap \{w \mid 1 \leq |\text{Disc}(w)| < X\}$. We could average this over v in B according to any measure we choose on B, and the answer would be the same (the same as the number of lattice points in a single fundamental domain), but the measure we choose will affect whether we can find said answer.

For each $v \in B$, we are counting lattice points in $\mathcal{F}v \cap \{w \mid 1 \leq |\operatorname{Disc}(w)| < X\}$ and then we are adding up (integrating) over v. Intuitively, it is not hard to imagine this is related to counting lattice points in $gB \cap \{w \mid 1 \leq |\operatorname{Disc}(w)| < X\}$ for each $g \in \mathcal{F}$. To actually change variables from an integral over $v \in B$ to an integral over $g \in \mathcal{F}$ is somewhat delicate (see problem 62 in the 2014 Arizona Winter School problem set), and in particular in order to end up with a reasonable measure on $g \in \mathcal{F}$ for integrating with respect to, we need to have carefully chosen our measure on $v \in B$. We will take the measure on $v \in B$ to be the pushforward of the Haar measure on $\mathrm{GL}_2(\mathbb{R})$, and the result is that we need to average the number of points in gB over $g \in \mathcal{F}$ with respect to Haar measure on $\mathcal{F}$.

The question that remains is what the gB look like and if they are easy to count points in. Write $g \in \mathcal{F}$ as $g = na'k\lambda$. We are interested in $na'k\lambda B$. First it is useful to understand what g does to the discriminant, and how X is involved. The matrix λ multiplies each of a, b, c, d by λ, and thus scales the discriminant by λ^4.

Exercise 11.12. Show that the actions of n, a', k do not change the discriminant of a binary cubic form.

We can, for example, pick our B so that all the discriminants of points $v \in B$ are between 1 and 2. Since we would only like to count lattice points with absolute discriminant between 1 and X, that means λ is ranging between 2–4 and $X^{1/4}$, or up to multiplicative constants between 1 and $X^{1/4}$. Once $\lambda \geq X^{1/4}$, there are no points in gB of discriminant $< X$. This is the main place that X comes in to play.

Now λ just scales our ball B, so in particular does not change its shape, so λB is a good shape for geometry of numbers. We have $\operatorname{Vol}(\lambda B)$ is of order λ^4 (up to multiplicative constants, that depend on B) and the largest projection is $O(\lambda^3)$.

Now for k, we can just pick a B that is fixed by K so that $kB = B$ for each $k \in K$. (For this it is important that K is compact. If we had chosen a B that was not fixed by K, we could just average it over K to obtain a choice that was fixed by K.)

The element $a' \in A'$ will stretch our ball B by different amounts in different directions. We have that

$$a' \circ (a, b, c, d) = (t^{-3}a, t^{-1}b, tc, t^3 d).$$

This is exactly the kind of thing that can make a terrible region for geometry of numbers. Suppose t is some large number Y. Then we have that $a'B$ is a region of volume $\operatorname{Vol}(a'B) = \operatorname{Vol}(B)$. However, the projection onto the d coordinate is order of magnitude Y^3, which is much bigger than the volume (as, say, we take Y larger and larger). It turns out that this reflects a reality about the shape of our fundamental domains $\mathcal{F}v$ that we can't get around by averaging. They have a long, skinny cusp around $a = 0$.

What saves us here is that the forms with $a = 0$ correspond to cubic rings that are not domains (as we saw above). So we don't want to count points with $a = 0$ in the end. So, instead of counting points in $\mathcal{R}_X(v) = \mathcal{F}v \cap \{w \mid 1 \leq |\operatorname{Disc}(w)| < X\}$, we will intersect that region with $|a| \geq 1$ and count points there.

Before we keep going with the geometry of numbers, we want to point out an issue. All of our fundamental domains contained the same number of points, because they all correspond to orbits of something (counted with $1/\#\operatorname{Stab}$ weight). However, when we add the condition $|a| \geq 1$ this is no longer true. The fundamental domains intersected with $|a| \geq 1$ may now include different numbers of reducible orbits (we only got rid of reducible forms when we eliminated the case $a = 0$). In the end we will be able to deal with this—we were going to have to get rid of reducible forms anyway at some point.

Now, let C be the maximum $|a|$ for any form in B. So we have that in $a'k\lambda B$, the maximum x^3 coefficient is $t^{-3}\lambda C$.

Exercise 11.13. Show that the action of n does not change the a coordinate.

Thus, we only need to consider g with $\lambda \leq X^{1/4}$, and t so that $C\lambda/t^3 \geq 1$. This means that t cannot get too large, and so a' does not make our region too stretched out.

11.5 Counting Lattice Points After a Lower Triangular Transformation

Now we come to n. A method of the author [58] allows one to completely generally ignore lower triangular transformations when applying Davenport's geometry of numbers Lemma 11.11 (even if the lower triangular transformations are not bounded). Let's first consider a simple example. Suppose you have a roughly X by X region H (so it has around X^2 lattice points, with $O(X)$ error). Now we apply a shear mapping s that sends a point with coordinate $(x, y) \mapsto (x, y + X^2 x)$. So the sheared region $s(H)$ has the same volume X^2 as H, but now the projection onto the second coordinate is size X^3, so we can't use Davenport's geometry of numbers lemma to count points in this region.

However, since a slice for each x coordinate was just shifted up in the plane, we can see that each slice of $s(H)$ has almost the same number of lattice points as H (differing by at most 1). In fact, if we merely translate a region R in $\mathbb{R}^n$ to $t(R)$ since we don't change the volume or the size of the projections, we can only change the number of lattice points from R to $t(R)$ by $O(\operatorname{Proj}(R) + 1)$.

More generally we have the following. Recall for a region E in $\mathbb{R}^n$, we let $N(E)$ be the number of points in $\mathbb{Z}^n \cap E$.

Theorem 11.14 (Wood [58]). *If E is a semi-algebraic region of $\mathbb{R}^n$ and T is an upper (or lower) triangular linear transformation of $\mathbb{R}^n$ with ones along the diagonal, then*

$$|N(E) - N(T(E))| = O(\operatorname{Vol}\operatorname{Proj} E + 1),$$

where Vol Proj E *denotes the volume of the largest (in volume) projection of E onto any proper coordinate hyperplane (of any dimension* $< n$*). The constant in the O notation depends only on n and the degree of the semi-algebraic region.*

Proof. Write $T = t_k \circ \cdots \circ t_2$, where t_i is a linear transformation that fixes the value of x_j for $j \neq i$, and for a point P we have $x_i(T(P)) = x_i + \lambda_{i-1}x_{i-1} + \lambda_{i-2}x_{i-2} + \cdots + \lambda_1 x_1$.

Lemma 11.15. *For any semi-algebraic region E of* $\mathbb{R}^n$

$$|N(E) - N\,(T(E))\,| = O\,(N\,\mathrm{Proj}\,E)\,,$$

where NProj(E) is the number of $\mathbb{Z}^n$ *points in the projection of E onto a proper coordinate hyperplane that has the most lattice points. The constant in the O notation depends on n and the degree of E.*

Proof. We induct on the number of terms t_i in T. First, we assume $T = t_i$. For a given set of $c_j \in \mathbb{Z}$, we consider the line $L = \{x_j = c_j : j \neq i\}$. The region $R \cap L$ is a union of intervals on the line. Since T only affects the coordinate x_i, we have that T preserves L and $T(E) \cap L = T(E \cap L)$. The transformation T just translates the intervals of $R \cap L$ along L. Since an interval of length ℓ has between $\ell - 1$ and $\ell + 1$ integral points, the difference $|N(E \cap L) - N(T(E) \cap L)|$ is at most twice the number of intervals of $E \cap L$, which is bounded by the degree of E. It remains to estimate the number of choices of c_j (for all $j \neq i$) such that $E \cap L$ is non-empty. This is exactly the number of lattice points in the projection of E onto $x_i = 0$, which is $O\,(N\,\mathrm{Proj}\,E)$.

Assuming the inductive hypothesis, we write $T = t_k \circ \cdots \circ t_2$ and $T' = t_{k-1} \circ \cdots \circ t_2$. We have

$$N(E) - N(T(E)) = N(E) - N(T'(E)) + N(T'(E)) - N(t_k \circ T'(E)),$$

and by the inductive hypothesis, we have

$$|N(E) - N(T'(E))| \leq O\,(N\,\mathrm{Proj}\,E)\,. \tag{10}$$

By our work above, for the case $k = 1$ we have

$$|N(T'(E)) - N(t_k \circ T'(E))| \leq O\left(N\left(T'(E)_k\right)\right),$$

where $T'(E)_k$ is the projection of $T'(E)$ onto the hyperplane cut out by $x_k = 0$. For a point P, the values of $x_i(T'(P))$ for $i \neq k$ do not depend on $x_k(P)$, we have that $T'(E)_k$ is just T' applied to the projection E_k of E onto the hyperplane $x_k = 0$. Thus,

$$|N(T'(E)) - N(t_k \circ T'(E))| \leq O\left(N\left(T'(E_k)\right)\right). \tag{11}$$

By the inductive hypothesis, we have

$$N(T'(E_k)) - N(E_k) \leq O\left(N \operatorname{Proj} E_k\right). \tag{12}$$

Thus combining Equations (10), (11), and (12), we have

$$|N(E) - N\left(T(E)\right)| \leq O\left(N \operatorname{Proj} E + N(E_k) + N \operatorname{Proj} E_k\right),$$

proving the lemma. (Though at each step of the induction, the constants in the O notation might grow, they are bounded in terms of n). Note that if E has infinitely many lattice points, then so does some projection of E onto a coordinate hyperplane. □

The number of points in the projection E' of E onto some proper coordinate plane is $\operatorname{Vol} E' + O(\operatorname{Proj} E' + 1)$. Since every projection of E' onto a proper coordinate hyperplane is also a projection of E, we have that $|N(E) - N\left(T(E)\right)| = O(\operatorname{Proj} E + 1)$, as desired. □

Though this general result is often quoted, in some cases, including our problem at hand (counting binary cubic forms), one can use a simpler argument that involves the specific lower triangular transformation and the shape of the region it is being applied to (see Exercise 46 in the 2014 Arizona Winter School problem set). In any case, we conclude that we only need to estimate the size of the projections of $a'k\lambda B$ in order to bound the error between the number of lattice points in gB and the volume of gB.

11.6 Estimating the Projections

We have that $a'k\lambda B = a'\lambda k B = a'\lambda B$. We have

$$a'\lambda \circ (a, b, c, d) = (\lambda t^{-3} a, \lambda t^{-1} b, \lambda t c, \lambda t^3 d),$$

and λ, t are bounded below by constants. Also, recall from the end of Sect. 11.4, we had $\lambda = O(X^{1/4})$ and $t = O(\lambda^{1/3})$. So we can bound the volume of each of the projections of $a'k\lambda B$, using the fact that B is fixed. For example, the projection onto the last three coordinates has size $O(\lambda t^{-1} \cdot \lambda t \cdot \lambda t^3) = O(\lambda^3 t^3)$. The projection onto the last coordinate is size $O(\lambda t^3)$, but this is $O(\lambda^3 t^3)$ since λ is bounded below by a constant. Similarly, we can bound each of the lower dimensional projections and conclude

$$\operatorname{Vol} \operatorname{Proj}(a'k\lambda B) = O(\lambda^3 t^3).$$

We then integrate over the Haar measure dg (see Problem 6.4 on the problem set for the determination of the Haar measure). We have (where the below ignores constant factors in all inequalities)

$$\int_{g\in\mathcal{F},gB\cap\{1\leq|\operatorname{Disc}(-)|<X\}\cap\{|a|\geq 1\}\neq\emptyset}$$

$$\begin{aligned}
\operatorname{Vol}\operatorname{Proj}(a'k\lambda B)dg &\leq \int_{g\in\mathcal{F},1\leq\lambda\leq X^{1/4},1\leq t\leq\lambda^{1/3}} O(\lambda^3t^3)\lambda^{-1}t^{-3}d\lambda dtdndk\\
&= \int_{g\in\mathcal{F},1\leq\lambda\leq X^{1/4},1\leq t\leq\lambda^{1/3}} O(\lambda^2)d\lambda dtdndk\\
&= \int_{(n,a',k,\lambda)\in\mathcal{F},1\leq\lambda\leq X^{1/4}} O(\lambda^{7/3})d\lambda dndk\\
&= \int_{n\in N',k\in K} O(X^{5/6})dndk.
\end{aligned}$$

The final integral is $O(X^{5/6})$ because we have that N' and K are compact and fixed (where N' is the union of all the $N'(a')$, i.e., $N' = N'(a')$ for a value of a' with $t \geq 1$).

11.7 Putting the Count Together

To put all the pieces together, we need to also compute the (average over g) volumes of our gB's, which we can do by reversing the change of variables we did above and instead computing volumes of fundamental domains $\mathcal{F}v$. For this, we can use a well-known calculation of the volume of $\mathcal{F}$. The average volume is order of X, and so the error term $O(X^{5/6})$ above is indeed small enough.

Also, our count now includes some reducible binary cubic forms (but not all of them since we cut out the $a = 0$ forms in each fundamental domain). First of all, notice that if we had included all of them, it would have contributed to our main term. For each quadratic field K with $|\operatorname{Disc}(K)| < X$, we have a cubic ring R which is the maximal order in $K \oplus \mathbb{Q}$, and has $|\operatorname{Disc}(R)| < X$. So if we counted all classes of reducible binary cubic forms, we would be seeing the reducible ones contribute to our main term (as $N_{S_2}(X) \sim cX$). (This perhaps would not be so bad, because we have an exact main term for $N_{S_2}(X)$.) However, it turns out that by discarding the $a = 0$ forms, we have thrown out 100 % (asymptotically in X) of the reducible forms (though the precise number we have thrown out in $\mathcal{F}v$ can depend a bit on v). See Problems 69 and 70 in the problem set or Lemma 21 of [15].

11.8 Sieving for Maximal Rings

So far, we have outlined the proof of counting classes of binary cubic forms that correspond to orders in cubic fields. The final step to count cubic fields is to sieve for maximal orders. We discussed above that maximality at p of the cubic ring can be given by a condition on the form modulo p^2. One can compute [15, Lemma 19] the proportion $\mu(\mathcal{U}_p)$ of binary cubic forms modulo p that correspond to cubic rings that are maximal at p. One then can do all the above counting, but instead of using integral binary cubic forms, only using ones with "maximal" reductions modulo p^2. One would get the main term multiplied by $\mu(\mathcal{U}_p)$, and the same error term *with the constant in the error term now depending on p*. For finitely many primes, one can sieve like this, but for infinitely many primes, the constant in the error term could in theory go off to infinity. One needs a uniform error bound to sieve at infinitely many primes.

This is a sieve in the spirit of the sieve for squarefree integers that is discussed in Poonen's Arizona Winter School 2014 lectures and in Problems 53–55 of the problem set, but the error bounds one needs as input are more sophisticated. (The sieve itself is carried out in Problem 78 of the problem set, but assuming the required input bound—see [15, Sect. 8.2] for the required bound.) One can finally obtain the following result of Davenport and Heilbronn.

Theorem 11.16. *We have*

$$N_{S_3}(X) \sim \frac{1}{3\zeta(3)}X.$$

If instead, we sieve for cubic rings that are not only maximal at p but also not totally ramified at p, we are counting 2-torsion in class groups of cubic fields (see Sect. 3). One final result of Davenport and Heilbronn [25] is

$$\lim_{X\to\infty} \frac{\sum_{K\in D(X) K \text{ imag quad}} |Cl(K)/3Cl(K)|}{\#\{K \in D(X) \mid K \text{ imag quad}\}} = 2.$$

We will see a sieve like that is necessary for either of these cases in the next section.

12 Counting Galois S_3 Sextics

In this section, we outline a result from [13] giving the asymptotics of $N_{S_3\subset S_6}(X)$, where S_3 is acting via its regular representation. As discussed above, this gives another example of counting fields by an invariant other than their discriminant. It will also show the kind of sieve that is necessary to sieve for even maximal cubic rings in Davenport and Heilbronn's result above. We are counting non-Galois cubic fields by the discriminant of their Galois closure. Degree 6 number fields with Galois group $S_3 \subset S_6$ are called S_3-*sextic fields*, and they are Galois over $\mathbb{Q}$. We are able to obtain an exact asymptotic in this case.

Theorem 12.1 (Bhargava and Wood [13]). *We have* $N_{S_3 \subset S_6}(X) \sim \left(\frac{1}{3}\prod_p c_p\right) X^{1/3}$, *where the product is over primes,* $c_p = (1 + p^{-1} + p^{-4/3})(1 - p^{-1})$ *for* $p \neq 3$, *and* $c_3 = (\frac{4}{3} + \frac{1}{3^{5/3}} + \frac{2}{3^{7/3}})(1 - \frac{1}{3})$.

Theorem 12.1 was also obtained (independently) by Belabas–Fouvry [3] as a result of a deeper study of S_3-sextic fields. To compare with Davenport and Heilbronn's theorem

$$N_{S_3}(X) \sim \frac{1}{3\zeta(3)} X,$$

we may write $\frac{1}{\zeta(3)} = \prod_p (1 + p^{-1} + p^{-2})(1 - p^{-1})$.

Any S_3-sextic field K_6 contains a unique (up to conjugation) non-Galois cubic subfield K_3. Conversely, any non-Galois cubic field K_3 has a unique Galois closure K_6, which is an S_3-sextic field. Let K_3 and K_6 be such for the rest of the section. Let $v_p(n)$ denote the exponent of the largest power of p that divides n. If K_3 is nowhere totally or wildly ramified, then $d(K_6) = d(K_3)^3$. If this were always the case, then the asymptotics of $N(X, S_3(6))$ would follow immediately from Theorem 11.16. However, at a tame rational prime p, the possible ramification types in K_3 are $\mathfrak{p}_1^2\mathfrak{p}$ *and* $\mathfrak{p}^3$, and $v_p(d(K_3))$ is 1 and 2 in these cases, respectively. These give, respectively, splitting types $\wp_1^2\wp_2^2\wp_3^2$ and $\wp_1^3\wp_2^3$ in K_6, and hence $v_p(d(K_6))$ is 3 and 4 in these cases, respectively. We call primes p of the second type *overramified* in K_3 (or K_6).

If tamely overramified primes exist, then evidently $d(K_3)^3 \neq d(K_6)$. To prove Theorem 12.1, we define transitional discriminants $d_Y(K_6)$ for K_6 that "correct" $d(K_3)^3$ to $d(K_6)$ at the primes less than Y:

Definition. Let $d_Y(K_6)$ be the positive integer such that

$$v_p(d_Y(K_6)) = \begin{cases} v_p(d(K_6)) & \text{if } p \text{ is a prime } < Y \\ v_p(d(K_3)^3) & \text{if } p \text{ is a prime } \geq Y \end{cases}.$$

In fact, a stronger version of Theorem 11.16 is true, and is proven with essentially the same methods. For an integer m, let ϕ_m be the map that takes binary cubic forms with integer coefficients to binary cubic forms with coefficients in $\mathbb{Z}/m\mathbb{Z}$, via reduction of the coefficients modulo m. For each of *finitely many* rational primes p_i, let us specify a set Σ_i of étale cubic $\mathbb{Q}_{p_i}$-algebras and let Σ_i' be the corresponding set of maximal $\mathbb{Z}_{p_i}$-orders. For some $m = \prod_i p_i^{n_i}$, there are sets U and S of binary cubic forms with coefficients in $\mathbb{Z}/m\mathbb{Z}$ such that $\phi_m^{-1}(U)$ is the set of forms which correspond to rings which are maximal at all p_i and $\phi_m^{-1}(S)$ is the set of forms which correspond to rings which are maximal at all p_i and whose p_i-adic completions are in Σ_i'. We define the *relative density* of $\{\Sigma_i\}_i$ to be $\#S/\#U$. By the Chinese remainder theorem, this relative density is simply the product of the *relative* p_i*-adic densities* of the individual Σ_i, defined as above in the case that $\{p_i\}_i$ has one element. The strengthened version of Theorem 11.16 we require is that

$$\#\{K \in \mathcal{F}(S_3) \mid d(K) < X \text{ and } K \otimes \mathbb{Q}_{p_i} \in \Sigma_i \text{ for all } i\} \sim (\text{relative density of } \{\Sigma_i\}_i) \cdot \frac{X}{3\zeta(3)}. \tag{13}$$

(For further details on this strengthened Theorem 11.16, see [25, Sect. 5] or [15, Theorem 31].) Since $d_Y(K_6)$ only differs from $d(K_3)^3$ at finitely many primes, we can compute the asymptotics of $N_Y(X) := \#\{K_6 \mid d_Y(K_6) < X\}$ directly from such a strengthened version of Theorem 11.16.

In [13] we compute these relative densities and obtain

$$\lim_{X\to\infty} \frac{N_Y(X)}{X^{1/3}} = \frac{c_2c_3}{(1-2^{-3})(1-3^{-3})} \prod_{3<p<Y} \frac{1+p^{-1}+p^{-4/3}}{1+p^{-1}+p^{-2}} \frac{1}{3\zeta(3)}.$$

Taking the limit in Y, we obtain

$$\lim_{Y\to\infty} \lim_{X\to\infty} \frac{N_Y(X)}{X^{1/3}} = \frac{c_2c_3}{3\zeta(3)} \prod_{p>3} \frac{1+p^{-1}+p^{-4/3}}{1+p^{-1}+p^{-2}}. \tag{14}$$

12.1 Proof of Theorem 12.1

We now compute the asymptotics of $N(X) := N_{S_3\subset S_6}(X)$. Note that for $Y > 3$, we have $N_Y(X) \le N(X)$ and thus

$$\lim_{Y\to\infty} \lim_{X\to\infty} \frac{N_Y(X)}{X^{1/3}} \le \liminf_{X\to\infty} \frac{N(X)}{X^{1/3}}. \tag{15}$$

To obtain an upper bound, let $M(n,X) = \#\{K_3 \text{ overramified at all primes } p|n,\ d(K_3) < X\}$. From [5, Lemma 3.3], we know $M(n,X) = O(n^{-2+\epsilon}X)$. If K_6 is an S_3-sextic field with $d(K_6) < X$, and n is the product of the primes where K_6 is overramified, then $d(K_3) < cn^{2/3}X^{1/3}$ for some finite absolute constant c given by the behavior of the finitely many 2-adic and 3-adic cubic algebras (in fact, we may take $c = 36$). Thus,

$$N(X) \le N_Y(X) + \sum_{n\ge Y} M(n, cn^{2/3}X^{1/3}) \le N_Y(X) + d\sum_{n\ge Y} \frac{X^{1/3}}{n^{4/3+\epsilon}}$$

for some constant d. Taking limits in X, we obtain

$$\limsup_{X\to\infty} \frac{N(X)}{X^{1/3}} \le \lim_{X\to\infty} \frac{N_Y(X)}{X^{1/3}} + d\sum_{n\ge Y} \frac{1}{n^{4/3+\epsilon}},$$

and then taking limits in Y, we conclude

$$\limsup_{X\to\infty}\frac{N(X)}{X^{1/3}} \le \lim_{Y\to\infty}\lim_{X\to\infty}\frac{N_Y(X)}{X^{1/3}}. \qquad (16)$$

Equations (15) and (16) combine to prove $\lim_{X\to\infty}\frac{N(X)}{X^{1/3}} = \lim_{Y\to\infty}\lim_{X\to\infty}\frac{N_Y(X)}{X^{1/3}}$, which together with Eq. (14) proves Theorem 12.1.

In [13, Sect. 5], we give an interpretation of the constants c_p above, which are exactly as predicted by the Malle–Bhargava principle discussed above. Further the work in [13] shows that the Malle–Bhargava principle holds in full generality, with local behaviors, in the case $G = S_3 \subset S_6$.

Acknowledgements This article was developed from lecture notes for a series of five lectures at the Arizona Winter School on "Arithmetic statistics" held March 15–19, 2014 in Tucson, Arizona. Thank you to Bjorn Poonen, Michiel Kosters, Jonah Leshin, and especially the anonymous referee for helpful comments. This work was done with the support of an American Institute of Mathematics 5-Year Fellowship and National Science Foundation grants DMS-1147782 and DMS-1301690.

References

1. Adam, M., Malle, G.: A class group heuristic based on the distribution of 1-eigenspaces in matrix groups. J. Number Theory **149**, 225–235 (2015)
2. Artin, E., Tate, J.: Class Field Theory. W. A. Benjamin, Inc., New York/Amsterdam (1968)
3. Belabas, K., Bhargava, M., Pomerance, C.: Error estimates for the Davenport-Heilbronn theorems. Duke Math. J. **153**(1), 173–210 (2010)
4. Bucur, A., David, C., Feigon, B., Kaplan, N., Lalín, M., Ozman, E., Wood, M.M.: The distribution of $\mathbb{F}_q$-points on cyclic ℓ-covers of genus g (2015). arXiv:1505.07136 [math]
5. Belabas, K., Fouvry, É.: Discriminants cubiques et progressions arithmétiques. Int. J. Number Theory **6**(7), 1491–1529 (2010)
6. Bhargava, M.: Higher composition laws. I. A new view on Gauss composition, and quadratic generalizations. Ann. Math. Second Ser. **159**(1), 217–250 (2004)
7. Bhargava, M.: Higher composition laws. II: on cubic analogues of Gauss composition. Ann. Math. **159**(2), 865–886 (2004)
8. Bhargava, M.: Higher composition laws III: the parametrization of quartic rings. Ann. Math. **159**(3), 1329–1360 (2004)
9. Bhargava, M.: The density of discriminants of quartic rings and fields. Ann. Math. **162**(2), 1031–1063 (2005)
10. Bhargava, M.: Higher composition laws. IV. The parametrization of quintic rings. Ann. Math. Second Ser. **167**(1), 53–94 (2008)
11. Bhargava, M.: The density of discriminants of quintic rings and fields. Ann. Math. **172**(3), 1559–1591 (2010)
12. Bhargava, M.: Mass formulae for extensions of local fields, and conjectures on the density of number field discriminants. Int. Math. Res. Not. MRN 2007, no. 17, Art. ID rnm052, 20 pp (2010)
13. Bhargava, M., Wood, M.M.: The density of discriminants of S_3-sextic number fields. Proc. Am. Math. Soc. **136**(5), 1581–1587 (2008)

14. Bhargava, M., Kane, D.M., Lenstra Jr. H.W., Poonen, B., Rains, E.: Modeling the distribution of ranks, Selmer groups, and Shafarevich-Tate groups of elliptic curves (2013). arXiv:1304.3971 [math]
15. Bhargava, M., Shankar, A., Tsimerman, J.: On the Davenport–Heilbronn theorems and second order terms. Invent. Math. **193**(2), 439–499 (2013)
16. Cohen, H.: The density of abelian cubic fields. Proc. Am. Math. Soc. **5**, 476–477 (1954)
17. Cohen, H.: Enumerating quartic dihedral extensions of $\mathbb{Q}$ with signatures. Université de Grenoble. Annales de l'Institut Fourier **53**(2), 339–377 (2003)
18. Cohen, H., Lenstra Jr. H.W.: Heuristics on class groups of number fields. In: Number Theory, Noordwijkerhout 1983 (Noordwijkerhout, 1983). Lecture Notes in Mathematics, vol. 1068, pp. 33–62. Springer, Berlin (1984)
19. Cohen, H., Martinet, J.: Étude heuristique des groupes de classes des corps de nombres. J. die Reine Angew. Math. **404**, 39–76 (1990)
20. Cohen, H., Martinet, J.: Heuristics on class groups: some good primes are not too good. Math. Comput. **63**(207), 329–334 (1994)
21. Cohen, H., Diaz y Diaz, F., Olivier, M.: Enumerating quartic dihedral extensions of $\mathbb{Q}$. Compos. Math. **133**(1), 65–93 (2002)
22. Cohen, H., Diaz y Diaz, F., Olivier, M.: On the density of discriminants of cyclic extensions of prime degree. J. die Reine Angew. Math. **550**, 169–209 (2002)
23. Datskovsky, B., Wright, D.J.: The adelic zeta function associated to the space of binary cubic forms. II. Local theory. J. die Reine Angew. Math. **367**, 27–75 (1986)
24. Davenport, H.: On a principle of Lipschitz. J. Lond. Math. Soc. Second Ser. **26**, 179–183 (1951)
25. Davenport, H., Heilbronn, H.: On the density of discriminants of cubic fields. II. Proc. R. Soc. London, Ser. A Math. Phys. Eng. Sci. **322**(1551), 405–420 (1971)
26. Deligne, P.: letter to W. T. Gan, B. Gross and G. Savin. November 13, 2000
27. Delone, B.N., Faddeev, D.K.: The Theory of Irrationalities of the Third Degree. Translations of Mathematical Monographs, vol. 10. American Mathematical Society, Providence (1964)
28. Erman, D., Wood, M.M.: Gauss Composition for P^1, and the universal Jacobian of the Hurwitz space of double covers (2011). arXiv:1111.0498 [math]
29. Fouvry, É., Klüners, J.: Cohen–Lenstra heuristics of quadratic number fields. In: Hess, F., Pauli, S., Pohst, M. (eds.) Algorithmic Number Theory. Lecture Notes in Computer Science, vol. 4076, pp. 40–55. Springer, Berlin/Heidelberg (2006)
30. Fouvry, É., Klüners, J.: On the 4-rank of class groups of quadratic number fields. Invent. Math. **167**(3), 455–513 (2006)
31. Garton, D.: Random matrices, the Cohen-Lenstra heuristics, and roots of unity (2014). arXiv:1405.6083 [math]
32. Gan, W.T., Gross, B., Savin, G.: Fourier coefficients of modular forms on G_2. Duke Math. J. **115**(1), 105–169 (2002)
33. Gerth, F. III.: Densities for ranks of certain parts of p-class groups. Proc. Am. Math. Soc. **99**(1), 1–8 (1987)
34. Gerth, F. III.: Extension of conjectures of Cohen and Lenstra. Expo. Math. Int. J. Pure Appl. Math. **5**(2), 181–184 (1987)
35. Grunwald, W.: Ein allgemeines Existenztheorem für algebraische Zahlkörper (German). J. Reine Angew. Math. 169, 103–107 (1933)
36. Hasse, H.: Arithmetische Theorie der kubischen Zahlkörper auf klassenkörpertheoretischer Grundlage. Math. Z. **31**(1), 565–582 (1930)
37. Heilbronn, H.: On the class-number in imaginary quadratic fields. Q. J. Math. **os-5**(1), 150–160 (1934)
38. Kedlaya, K.S.: Mass formulas for local Galois representations. Int. Math. Res. Not. IMRN 2007, no. 17, Art. ID rnm021, 26 pp (2007); With an appendix by Daniel Gulotta
39. Klingen, N.: Arithmetical Similarities. Oxford Mathematical Monographs. The Clarendon Press/Oxford University Press, New York (1998); Prime decomposition and finite group theory, Oxford Science Publications
40. Klüners, J.: A counterexample to Malle's conjecture on the asymptotics of discriminants. C.R. Math. Acad. Sci. Paris **340**(6), 411–414 (2005)

41. Klüners, J.: Asymptotics of number fields and the Cohen–Lenstra heuristics. J. Théorie Nombres Bordeaux **18**(3), 607–615 (2006)
42. Klüners, J.: The distribution of number fields with wreath products as Galois groups. Int. J. Number Theory **8**(3), 845–858 (2012)
43. Klüners, J., Malle, G.: Counting nilpotent Galois extensions. J. die Reine Angew. Math. **572**, 1–26 (2004)
44. Mäki, S.: On the density of abelian number fields. Ann. Acad. Sci. Fenn. Ser. A I. Math. Dissertationes no. 54, 104 pp (1985)
45. Mäki, S.: The conductor density of abelian number fields. J. Lond. Math. Soc. Second Ser. **47**(1), 18–30 (1993)
46. Malle, G: On the distribution of Galois groups. J. Number Theory **92**(2), 315–329 (2002)
47. Malle, G: On the distribution of Galois groups. II. Exp. Math. **13**(2), 129–135 (2004)
48. Malle, G.: Cohen–Lenstra heuristic and roots of unity. J. Number Theory **128**(10), 2823–2835 (2008)
49. Malle, G.: On the Distribution of Class Groups of Number Fields. Exp. Math. **19**(4), 465–474 (2010)
50. Narkiewicz, W.: Number Theory. World Scientific Publishing Co., Singapore (1983); Translated from the Polish by S. Kanemitsu
51. Neukirch, J.: Algebraic Number Theory. Grundlehren der Mathematischen Wissenschaften [Fundamental Principles of Mathematical Sciences], vol. 322. Springer, Berlin (1999); Translated from the 1992 German original and with a note by Norbert Schappacher, With a foreword by G. Harder
52. Neukirch, J., Schmidt, A., Wingberg, K.: Cohomology of Number Fields. Grundlehren der Mathematischen Wissenschaften [Fundamental Principles of Mathematical Sciences], vol. 323. Springer, Berlin (2000)
53. Poonen, B.: The moduli space of commutative algebras of finite rank. J. Eur. Math. Soc. (JEMS) 10(3), 817–836 (2008)
54. Taylor, M.J.: On the equidistribution of Frobenius in cyclic extensions of a number field. J. Lond. Math. Soc. Second Ser. **29**(2), 211–223 (1984)
55. Taniguchi, T., Thorne, F: Secondary terms in counting functions for cubic fields. Duke Math. J. **162**(13), 2451–2508 (2013)
56. Turkelli, S.: Connected components of Hurwitz schemes and Malle's conjecture (2008). arXiv:0809.0951 [math]
57. Wang, S.: On Grunwald's theorem. Ann. Math. Second Ser. **51**, 471–484 (1950)
58. Wood, M.M.: Lemma on the number of lattice points after triangular transformation (2006), preprint
59. Wood, M.M.: Mass formulas for local Galois representations to wreath products and cross products. Algebra Number Theory **2**(4), 391–405 (2008)
60. Wood, M.M.: On the probabilities of local behaviors in abelian field extensions. Compos. Math. **146**(1), 102–128 (2010)
61. Wood, M.M.: Gauss composition over an arbitrary base. Adv. Math. **226**(2), 1756–1771 (2011)
62. Wood, M.M.: Parametrizing quartic algebras over an arbitrary base. Algebra Number Theory **5**(8), 1069–1094 (2011)
63. Wood, M.M.: Rings and ideals parameterized by binary n-ic forms. J. Lond. Math. Soc. **83**(1), 208–231 (2011)
64. Wood, M.M.: Quartic rings associated to binary quartic forms. Int. Math. Res. Not. **2012**(6), 1300–1320 (2012)
65. Wood, M.M.: Parametrization of ideal classes in rings associated to binary forms. J. die Reine Angew. Math. (Crelles J.) **2014**(689), 169–199 (2014)
66. Wood, M.M., Yasuda, T.: Mass formulas for local Galois representations and quotient singularities. I: a comparison of counting functions. Int. Math. Res. Not. IMRN 2015, (23), 12590–12619 (2015)
67. Wright, D.J.: Distribution of discriminants of abelian extensions. Proc. Lond. Math. Soc. Third Ser. **58**(1), 17–50 (1989)

Printed by Printforce, the Netherlands